Contemporary Debates in Applied Ethics

Contemporary Debates in Philosophy

In teaching and research, philosophy makes progress through argumentation and debate. Contemporary Debates in Philosophy provides a forum for students and their teachers to follow and participate in the debates that animate philosophy today in the western world. Each volume presents pairs of opposing viewpoints on contested themes and topics in the central subfields of philosophy. Each volume is edited and introduced by an expert in the field, and also includes an index, bibliography, and suggestions for further reading. The opposing chapters, commissioned especially for the volumes in the series, are thorough but accessible presentations of opposing points of view.

1. Contemporary Debates in Philosophy of Religion
   edited by Michael L. Peterson and Raymond J. Vanarragon
2. Contemporary Debates in Philosophy of Science
   edited by Christopher Hitchcock
3. Contemporary Debates in Epistemology
   edited by Matthias Steup and Ernest Sosa
4. Contemporary Debates in Applied Ethics
   edited by Andrew I. Cohen and Christopher Heath Wellman
5. Contemporary Debates in Aesthetics and the Philosophy of Art
   edited by Matthew Kieran
6. Contemporary Debates in Moral Theory
   edited by James Dreier
7. Contemporary Debates in Cognitive Science
   edited by Robert Stainton
8. Contemporary Debates in Philosophy of Mind
   edited by Brian McLaughlin and Jonathan Cohen
9. Contemporary Debates in Social Philosophy
   edited by Laurence Thomas
10. Contemporary Debates in Metaphysics
    edited by Theodore Sider, John Hawthorne, and Dean W. Zimmerman
11. Contemporary Debates in Political Philosophy
    edited by Thomas Christiano and John Christman
12. Contemporary Debates in Philosophy of Biology
    edited by Francisco J. Ayala and Robert Arp
13. Contemporary Debates in Bioethics
    edited by Arthur L. Caplan and Robert Arp
14. Contemporary Debates in Epistemology, Second Edition
    edited by Matthias Steup, John Turri, and Ernest Sosa
15. Contemporary Debates in Applied Ethics, Second Edition
    edited by Andrew I. Cohen and Christopher Heath Wellman

Contemporary Debates in Applied Ethics

SECOND EDITION

Edited by

Andrew I. Cohen
Christopher Heath Wellman

WILEY Blackwell

This edition first published 2014
© 2014 John Wiley & Sons, Inc., except for Chapter 18 © 2014 John Corvino

Registered Office
John Wiley & Sons Ltd, The Atrium, Southern Gate, Chichester, West Sussex, PO19 8SQ, UK

Editorial Offices
350 Main Street, Malden, MA 02148-5020, USA
9600 Garsington Road, Oxford, OX4 2DQ, UK
The Atrium, Southern Gate, Chichester, West Sussex, PO19 8SQ, UK

For details of our global editorial offices, for customer services, and for information about how to apply for permission to reuse the copyright material in this book please see our website at www.wiley.com/wiley-blackwell.

The right of Andrew I. Cohen and Christopher Heath Wellman to be identified as the authors of the editorial material in this work has been asserted in accordance with the UK Copyright, Designs and Patents Act 1988.

Wiley also publishes its books in a variety of electronic formats. Some content that appears in print may not be available in electronic books.

Designations used by companies to distinguish their products are often claimed as trademarks. All brand names and product names used in this book are trade names, service marks, trademarks or registered trademarks of their respective owners. The publisher is not associated with any product or vendor mentioned in this book.

Limit of Liability/Disclaimer of Warranty: While the publisher and author(s) have used their best efforts in preparing this book, they make no representations or warranties with respect to the accuracy or completeness of the contents of this book and specifically disclaim any implied warranties of merchantability or fitness for a particular purpose. It is sold on the understanding that the publisher is not engaged in rendering professional services and neither the publisher nor the author shall be liable for damages arising herefrom. If professional advice or other expert assistance is required, the services of a competent professional should be sought.

Library of Congress Cataloging-in-Publication Data applied for

Paperback ISBN: 9781118479391

A catalogue record for this book is available from the British Library.

Cover design by Cyan Design.

Set in 10/12.5 pt Photina by Toppan Best-set Premedia Limited

1   2014

Contents

Notes on Contributors

Fritz Allhoff is an Associate Professor in the Department of Philosophy at Western Michigan University and a Senior Research Fellow in the Centre for Applied Philosophy and Public Ethics (Canberra, Australia). He has held visiting posts at the American Medical Association, University of Michigan, University of Oxford, and the University of Pittsburgh. His primary fields of research are applied ethics, ethical theory, and philosophy of biology/science. He has published work in the *American Journal of Bioethics*, *Cambridge Quarterly of Healthcare Ethics*, *International Journal of Applied Philosophy*, and *Kennedy Institute of Ethics Journal*, among other places. His latest books include *What Is Nanotechnology and Why Does It Matter?: From Science to Ethics* (Wiley-Blackwell, 2010; with Patrick Lin and Daniel Moore) and *Terrorism, Ticking Time-Bombs, and Torture* (University of Chicago, 2012).

Andrew Altman is Distinguished University Professor and Professor of Philosophy at Georgia State University. He specializes in legal and political philosophy and applied ethics. His publications include *Critical Legal Studies* (Princeton University Press, 1989) and *A Liberal Theory of International Justice* (with Christopher Heath Wellman; Oxford University Press, 2009). His articles on such topics as hate speech, sexual harassment, and international criminal law have appeared in *Philosophy & Public Affairs* and *Ethics*, among other leading journals. He is currently working on a book on pornography (with Susan J. Brison).

Arthur Isak Applbaum is Adams Professor of Political Leadership and Democratic Values at Harvard University. Applbaum's work on legitimate political authority, civil and official disobedience, and role morality has appeared in journals such as *Philosophy & Public Affairs, Journal of the American Medical Association, Harvard Law Review, Ethics,* and *Legal Theory.* He is the author of *Ethics for Adversaries* (Princeton University Press, 2000), a book about the morality of roles in public and professional life.

Bernard Boxill is Pardue Distinguished Professor of Philosophy at the University of North Carolina, Chapel Hill. He is author of *Blacks and Social Justice* (Rowman & Allanheld, 1984), editor of *Race and Racism* (Oxford University Press, 2001), and has published numerous articles on themes in ethics, the history of political thought, and social and political philosophy.

Bob Brecher is Professor of Moral Philosophy at the University of Brighton, and Director of its Centre for Applied Philosophy, Politics & Ethics. He has published over sixty articles in moral theory, applied ethics and politics, healthcare and medical ethics, sexual politics, terrorism and the politics of higher education. His latest book, *Torture and the Ticking Bomb* (Wiley-Blackwell, 2007) is the first book-length rebuttal of calls to legalize interrogational torture. Currently he is working on a theory of morality as practical reason, building on his earlier *Getting What You Want? A Critique of Liberal Morality* (Routledge, 1997). A past president of the Association for Legal & Social Philosophy, he is also on the Board of a number of academic journals as well as being a member of several Research Ethics Committees.

Susan J. Brison is Associate Professor of Philosophy at Dartmouth College and has held visiting appointments at Tufts, New York University, and Princeton. She is author of *Aftermath: Violence and the Remaking of the Self* (Princeton University Press, 2002) and *Speech, Harm, and Conflicts of Rights* (Princeton University Press, forthcoming) and co-editor of *Contemporary Perspectives on Constitutional Interpretation* (Westview Press, 1993).

Daniel Callahan is Senior Research Scholar and President Emeritus of the Hastings Center. He was its co-founder in 1969 and served as its president between 1969 and 1996. He is also co-director of the Yale-Hastings Program in Ethics and Health Policy. Over the years his research and writing have covered a wide range of issues, from the beginning until the end of life. His recent books include *What Price Better Health: Hazards of the Research Imperative* (University of California Press, 2003) and *Taming the Beloved Beast: How Medical Technology Costs Are Destroying Our Health Care System* (Princeton University Press, 2009).

Andrew I. Cohen is Director of the Jean Beer Blumenfeld Center for Ethics and Associate Professor of Philosophy at Georgia State University. His research focuses on contractarian social and political theory, themes in global justice, and reparations and apologies for historic injustice. His work has appeared in journals such as *Philosophy and Public Affairs*, *The Journal of Social Philosophy*, and *The Journal of Moral Philosophy*.

John Corvino is Associate Professor and Chair of Philosophy at Wayne State University in Detroit. His is the co-author (with Maggie Gallagher) of *Debating Same-Sex Marriage* (2012) and the author of *What's Wrong with Homosexuality?* (2013), both from Oxford University Press. A frequent speaker on sexuality, ethics, and marriage, Corvino posts articles and video clips at www.johncorvino.com.

Stephen L. Darwall is Andrew Downey Orrick Professor of Philosophy at Yale University. He is the author of *Impartial Reason* (Cornell University Press, 1983), *The British*

Moralists and the Internal "Ought": 1640–1740 (Cambridge University Press, 1995), *Philosophical Ethics* (Westview Press, 1998), *Welfare and Rational Care* (Princeton University Press, 2002), *The Second-Person Standpoint: Respect, Morality and Accountability* (Harvard University Press, 2006).

R.G. Frey was Professor of Philosophy at Bowling Green State University. He specialized in ethical and political philosophy and was the author of numerous books and articles on applied ethics, normative theory, and the history of eighteenth-century British moral philosophy, including *Interests and Rights* (Oxford University Press, 1980) and *Euthanasia and Physician-Assisted Suicide* (with Gerald Dworkin and Sissela Bok; Cambridge University Press, 1998).

Robert P. George holds the McCormick Chair in Jurisprudence and is the founding director of the James Madison Program at Princeton University. He has served on the President's Council on Bioethics and as a presidential appointee to the United States Commission on Civil Rights. He has also served on UNESCO's World Commission on the Ethics of Science and Technology, of which he continues to be a corresponding member. He is a former Judicial Fellow at the Supreme Court of the United States, where he received the Justice Tom C. Clark Award. Among many other publications, he is the author of *In Defense of Natural Law* (Oxford University Press, 2001), *Making Men Moral: Civil Liberties and Public Morality* (Oxford University Press, 1995), and *The Clash of Orthodoxies: Law, Religion and Morality in Crisis* (Intercollegiate Studies Institute, 2002).

Sherif Girgis is a PhD candidate in Philosophy at Princeton and a JD candidate at Yale Law School, where he is an editor of the *Yale Law Journal*. He has lectured and debated on social issues, and has published in academic and popular venues including the *New York Times*, the *Yale Law Journal*, the *Harvard Journal of Law and Public Policy*, the *Wall Street Journal*, *Public Discourse*, *National Review*, and *Commonweal*. He is the lead author of the book *What Is Marriage? Man and Woman: A Defense*, published by Encounter Books in 2012.

Deborah Hellman is Professor of Law at the University of Virginia School of Law. Prior to joining the UVA faculty in 2012, she was the Jacob France Research Professor at the University of Maryland School of Law. She writes about discrimination and equality, campaign finance and obligations of professional role. Hellman is the author of *When is Discrimination Wrong?* (Harvard University Press, 2008), which lays out a theory of discrimination, and a co-editor of *The Philosophical Foundations of Discrimination Law* (Oxford University Press, forthcoming).

Douglas Husak (PhD, JD) is Professor of Philosophy and Law at Rutgers University where he is co-Director of the Institute for Law and Philosophy. He is the author of *The Philosophy of Criminal Law: Selected Essays* (Oxford University Press, 2010), *Overcriminalization: The Limits of the Criminal Law* (Oxford University Press, 2008), *Legalize This! The Case for Decriminalizing Drugs* (Verso, 2002), and *Drugs and Rights* (Cambridge University Press, 1992). He is the Editor-in-Chief of *Criminal Law and Philosophy*.

Chandran Kukathas is Chair of Political Theory at the London School of Economics. He is the author of numerous articles and books, including *Hayek and Modern Liberalism* (Oxford University Press, 1989) and *The Liberal Archipelago: A Theory of Diversity and Freedom* (Oxford University Press, 2003).

Patrick Lee is the John N. and Jamie D. McAleer Professor of Bioethics and Director of the Institute of Bioethics at Franciscan University of Steubenville. In addition to numerous articles, he is the author of *Abortion and Human Life* (Catholic University Press, 1996) and *Body-Self Dualism in Contemporary Ethics and Politics* (with Robert P. George; Cambridge University Press, 2008).

Margaret Olivia Little is Director of the Kennedy Institute of Ethics, and Associate Professor in the Philosophy Department, at Georgetown University. She has written widely on issues of meta-ethics, normative ethics, and applied ethics. She has recently co-edited (with David Bakhurst and Brad Hooker) *Thinking About Reasons* (Oxford University Press, 2013), a compilation of essays honoring Jonathan Dancy.

Peter de Marneffe is Professor of Philosophy at Arizona State University. He is the author of *Liberalism and Prostitution* (Oxford, 2010) and, with Doug Husak, *The Legalization of Drugs* (Cambridge, 2005).

David Miller is Professor of Political Theory at the University of Oxford and an Official Fellow of Nuffield College. His books include *On Nationality* (Clarendon Press, 1995), *Principles of Social Justice* (Harvard University Press, 1999), *Citizenship and National Identity* (Polity Press, 2000), and *National Responsibility and Global Justice* (Oxford University Press, 2007).

Albert Mosley is Professor of Philosophy at Smith College. In addition to writing numerous articles and book chapters, he is the author of *Affirmative Action: Social Justice or Unfair Preference?* (with Nicholas Capaldi; Rowman & Littlefield, 1996) and *An Introduction to Logic: From Everyday Life to Formal Systems* (with Eulalio Baltazar; Ginn Press, 1984). He is also the editor of *African Philosophy: Selected Readings* (Prentice Hall, 1995).

Stephen Nathanson is Professor of Philosophy at Northeastern University. His is the author of *An Eye for an Eye? The Immorality of Punishing by Death* (Rowman & Littlefield, 1987), *The Ideal of Rationality* (Open Court, 1994), *Should We Consent to be Governed? A Short Introduction to Political Philosophy* (Wadsworth, 1992), *Economic Justice* (Prentice Hall, 1998), and *Terrorism and the Ethics of War* (Cambridge University Press, 2010).

Nahshon Perez, is an assistant professor at the Department of Political Studies, Bar Ilan University, and a holder of a European Union Marie Curie Re-integration grant. A political theorist, his first book, *Freedom from Past Injustices*, was published by Edinburgh University Press in 2012. Aside from his research on past wrongs, rectification and property rights, his fields of research include pluralism, toleration and religion and state. His articles have appeared in peer-reviewed journals such as *Social Theory and*

Practice, Canadian Journal of Political Science, the *Journal of Church and State* and the *Journal of Applied Philosophy* among others.

Louis P. Pojman was Professor of Philosophy at the United States Military Academy. He authored numerous works on a wide variety of subjects in social and political philosophy.

Tom Regan is Emeritus Professor of Philosophy at North Carolina State University. He is the author of hundreds of articles and more than twenty books, including *The Case for Animal Rights* (Routledge & Kegan Paul, 1983), *The Struggle for Animal Rights* (International Society for Animal Rights, 1987), *Defending Animal Rights* (University of Illinois Press, 2001), *Animal Rights, Human Wrongs: An Introduction to Moral Philosophy* (Rowman & Littlefield, 2003), and *Empty Cages: Facing the Challenge of Animal Rights* (Rowman & Littlefield, 2003). Upon his retirement in 2001, he received the Alexander Quarles Holladay Medal, the highest honor North Carolina State University can bestow on one of its faculty.

Fernando Tesón, a native of Buenos Aires, is the Tobias Simon Eminent Scholar at Florida State University College of Law. He is known for his scholarship relating political philosophy to international law (in particular his defense of humanitarian intervention), and his work on political rhetoric. He has authored *Humanitarian Intervention: An Inquiry into Law and Morality* (3rd edition fully revised and updated, Transnational Publishers, 2005); *Rational Choice and Democratic Deliberation* (with Guido Pincione; Cambridge University Press, 2006), *A Philosophy of International Law* (Westview Press, 1998); and many articles in law, philosophy, and international relations journals and collections of essays. At the moment he is completing a book on global justice entitled *Justice at a Distance* (with Loren Lomasky). Before joining FSU in 2003 he taught for 17 years at Arizona State University. He has served as visiting professor at Cornell Law School, Indiana University School of Law, University of California Hastings College of Law, the Oxford-George Washington International Human Rights Program, and is Permanent Visiting Professor, Universidad Torcuato Di Tella, Buenos Aires, Argentina. He has dual US and Argentine citizenship.

Michael Tooley is Distinguished College Professor of Philosophy at the University of Colorado. He is co-editor (with Ernest Sosa) of *Causation* (Oxford University Press, 1993) and editor of the five-volume anthology *Analytic Metaphysics* (Garland, 1999), and the author of *Abortion and Infanticide* (Clarendon Press, 1983), *Causation: A Realist Approach* (Clarendon Press, 1987), *Time, Tense and Causation* (Clarendon Press, 1996), and *Knowledge of God* (with Alvin Plantinga; Blackwell Publishing, 2008).

Bas van der Vossen is Assistant Professor in Philosophy at University of North Carolina, Greensboro. He research focuses on political philosophy and the philosophy of law. His work has appeared in journals such as *Law and Philosophy*, *Politics, Philosophy, and Economics*, and the *Oxford Journal of Legal Studies*.

Christopher Heath Wellman is Professor of Philosophy at Washington University in St Louis. He works in ethics, specializing in political and legal philosophy. His most recent books include *A Liberal Theory of International Justice* (with Andrew Altman; Oxford University Press, 2009), *Debating the Ethics of Immigration: Is There a Right to Exclude?* (with Phillip Cole; Oxford University Press, 2011), and *Liberal Rights and Responsibilities: Essays on Citizenship and Sovereignty* (Oxford University Press, 2013).

Celia Wolf-Devine is Emerita Associate Professor at Stonehill College. She is the author of *Descartes on Seeing: Epistemology and Visual Perception* (Southern Illinois University Press, 1993) and *Diversity and Community in the Academy: Affirmative Action in Faculty Appointments* (Rowman & Littlefield, 1997), and is co-editor (with Philip Devine) of *Sex and Gender: A Spectrum of Views* (Wadsworth, 2003).

Acknowledgments

We are grateful to Bernard R. Boxill, Dorothy Denning, R.G. Frey, Deborah G. Jonson, Hugh LaFollette, and Jeffrey Rosen for advice in the early stages. Jeff Dean, Nirit Simon, and Jennifer Bray at Wiley-Blackwell have been immensely supportive and patient. Most importantly, we would like to thank Adam Adler, Brad Champion, Chetan Cetty, Ryan McWhorter and Carson Young for providing crucial research and editorial assistance.

Introduction

Andrew I. Cohen and Christopher Heath Wellman

Contemporary Debates in Applied Ethics presents fourteen pairs of chapters by some of the leading theorists working in the field today. Philosophers, social theorists, and legal scholars take opposing sides on issues of enduring and special contemporary importance. This second edition includes several new topics (including drugs, humanitarian intervention, profiling, reparations, same-sex marriage, and torture), as well as an additional introductory chapter on ethical theory by Stephen Darwall. The authors draw on recent developments in moral and political theory, economics, science, and public policy. Their chapters are written in plain, jargon-free language so as to be accessible to introductory students, but they also feature cutting-edge, rigorous arguments that will demand the attention of scholars currently working on these important issues.

Issues of Life and Death

Patrick Lee and **Robert P. George** argue that abortion often wrongly kills a human being. A fetus, they claim in "The Wrong of Abortion," is a morally significant and distinct entity who is internally programmed to become an independent and mature human being – unless stopped by some disease or act of violence. A fetus is the same *kind* of thing as you are, but only at an earlier developmental stage. The authors discuss how we are living bodily entities, and as such we come to be long before birth. We become morally significant at the moment of conception; at that point each of us becomes the sort of entity who has the potential to develop and exercise higher mental functions. We do not find such capacities among mere parts of human beings or among nonhuman animals. Human beings enjoy rights in virtue of being a certain *kind* of entity, but their moral status is not a function of the extent to which they exhibit certain qualities. Lee and George consider a "bodily rights argument," which holds that women

Contemporary Debates in Applied Ethics, Second Edition. Edited by Andrew I. Cohen and Christopher Heath Wellman.
© 2014 John Wiley & Sons, Inc. Published 2014 by John Wiley & Sons, Inc.

are not required to give the use of their bodies to gestating fetuses. But the authors reject this view, holding that nonconsensual relationships sometimes generate moral responsibilities. Except in cases where a mother's life is threatened, the sacrifice a mother must perform when carrying a fetus to term is far less serious than the harm involved in killing a fetus.

Margaret Olivia Little defends abortion as often morally permissible, but not because developing embryos are morally inert bundles of cells. In her chapter, "The Moral Permissibility of Abortion," Little discusses how morality – and the reasons it furnishes – should not be forced into metaphysical views that regard steady states as the only possible explanatory categories. Rather, a more nuanced metaphysics acknowledges scalar qualities and ongoing development as key to an adequate picture of the world. Little notes that arguments investing moral significance in fetal potential are often importantly misleading. We must acknowledge that any such potential is only meaningful (and crucially depends upon) some woman's choices. Little argues that fetuses are not morally inert; their developing status does confer a developing moral significance. Nevertheless, in her view, sometimes aborting a fetus is a permissible withdrawal of sustenance and support for a developing life that would not have existed but for a woman's active support in the first place. This is part of the reason why some abortions do not violate any rights. Little then proposes reframing the abortion discussion into one of the *ethics of gestation*. Gestating a child and becoming a mother are momentous projects with profound moral implications. Besides entailing considerable medical risks and physical burdens, these projects involve significant reformulations of one's practical identity. In order to protect the intimacy crucial to personal identity and meaningfulness, Little argues, gestation and motherhood are and must be an individual's significant moral prerogative. Acknowledging such a prerogative no more diminishes the value of motherhood and babies than acknowledging sexual prerogatives diminishes the value of marriage and family. Even with a moral prerogative to terminate a pregnancy, however, there may still be important moral reasons not to abort in certain circumstances. Little then considers issues regarding the ethics of *creation* and how they relate to an ethics of *gestation*.

In "A Defense of Voluntary Active Euthanasia and Assisted Suicide," **Michael Tooley** writes that euthanasia refers to "any action where a person is intentionally killed or allowed to die because it is believed that the individual would be better off dead than alive – or else, as when one is in an irreversible coma, at least no worse off." Tooley surveys several key distinctions relevant to discussions of euthanasia and proceeds to defend as morally permissible voluntary active euthanasia. Under certain circumstances, a person may be justified in committing suicide. In such cases, others would be justified in assisting that person to commit suicide. Where assisted suicide would be permissible, Tooley argues, so too would voluntary active euthanasia. Tooley considers various possible objections and finds that appeals to God or religious authority are unhelpful, and if suicide is in one's interests and violates no one's rights, then assisting someone in taking her life is morally permissible. Voluntary active euthanasia should also be legally permitted; slippery slope arguments against legalization often rest on poorly drawn distinctions and clash with empirical evidence. Allowing such euthanasia would also provide more skilled aid and comfort for those with the greatest need for it.

In "A Case against Euthanasia," **Daniel Callahan** notes that euthanasia does not fit into traditional categories for the justified taking of a life. Suicide is doubtless an option for many who suffer, but few make the choice. Callahan suggests that this is commendable; pain is a necessary part of human life, and "human life is better, even nobler, when we human beings put up with the pain and travail that come our way." He rejects the idea that principles of freedom and self-determination justify protecting a choice to end one's life or seek assistance in doing so. Callahan objects to physician-assisted suicide; by enlisting a doctor's aid, euthanasia is not merely a private act. It becomes a social act by enlarging the field of permissible killings. This would have dangerous consequences: it would violate a long established norm that those with the power to save lives should not have the power to end them. Callahan also questions defenses of euthanasia that attempt to collapse the distinction between active and passive euthanasia. Removing legal obstacles to euthanasia, Callahan further argues, would "teach the wrong kind of lesson" by changing the role of physician and generating vast enforcement problems. More sharply, legalizing euthanasia would entrench as public policy the idiosyncratic preferences of a small minority who mistakenly believe human dignity is incompatible with suffering.

In his "Empty Cages," **Tom Regan** argues against research on animals – whether for education, medical studies, or product testing. Many uses of animals, he notes, are unnecessary or otherwise gratuitous. Even in cases where the use of animals *seems* crucial, Regan argues that a key moral principle – moral rights – typically blocks us from using animals for our benefit. He then presents a series of arguments to show that animals possess moral rights. He discusses and rejects views that morally privilege human beings over other animals. Logical consistency demands that nonhuman animals enjoy certain fundamental rights in just the way that human beings do. **R.G. Frey**, on the other hand, believes using animals for some research purposes is permissible. In "Animals and Their Medical Use," he stakes out a position between an animal rights view that would forbid any experimentation on animals and an "anything goes" view that permits all but gratuitously cruel uses of animals. Against the former, Frey argues that regarding animals as bearers of rights protects them from experimental use with claims so strong that it would no longer be possible to state a defense of animal research. Their rights would cut off from the start any appeal to prospective human benefit – no matter how large the benefit. Against the "anything goes" position, Frey argues that we go to great lengths to justify inflicting suffering on animals, precisely because we rightly think that they count morally. But, not all living creatures have the same moral value. *If* experiments on living creatures are crucial for advancing human welfare, then we should "use the life of lower quality in preference to the life of higher quality." Just as there can be better or worse lives among creatures of one species, so too we can say that the life of a typical adult human is more valuable than that of an animal because a human being has more capacities for self-development, and so can have a richer life. Frey focuses particularly on how human *agency* adds moral value to a life. He then develops two accounts of moral community and applies his model to determining how and to what extent animals are morally considerable.

Issues in Justice

In "A Defense of Affirmative Action," **Albert Mosley** defends policies that take race into account as a means of increasing the ability of minorities to take advantage of employment, educational, and investment opportunities. Mosley considers and rejects in turn several arguments by critics of affirmative action policies: racial minorities are owed nothing by the innocent beneficiaries of racial injustices, aptitude and IQ tests prove that racial minorities are less competent on average, race is a bogus concept, and race-conscious policies are a form of reverse discrimination prohibited by the constitution. He argues that measures to increase racial diversity are morally justified as steps to undo entrenched unjust norms and to promote a better justified distribution of goods and services to underserved communities. **Celia Wolf-Devine**'s main disagreement with Mosley concerns the merits of *preferential* affirmative action policies. Such policies privilege some applicants simply in virtue of their being members of certain historically under-represented groups. In "Preferential Policies Have Become Toxic," Wolf-Devine considers some key contemporary arguments for such policies and finds them all inadequate. Preferential affirmative action policies might be cast as *compensation* for past injustices, but, she argues, such policies are often misguided and poorly targeted. She also cautions against devising policies to bring about proportionate representation of all groups in all professions, since, as she points out, there might be important cultural differences (independently of the effects of past oppression) that explain why particular racial and ethnic groups gravitate toward certain careers. *Corrective* defenses of preferential affirmative action hope to fix current bias, but, Wolf-Devine claims, we should be wary of generalizing findings of bias from one situation to others. Wolf-Devine then considers various *forward-looking* defenses of preferential affirmative action, but she worries about their unintended consequences, such as: fostering perverse pressures toward group conformity on some beneficiaries of the policies, further confusing race and class in remedial social policies, increasing the drop-out rate for black college students, and perpetuating negative racial stereotypes. Wolf-Devine ultimately argues that preferential policies are "divisive because they are zero-sum." She applauds recent evidence of the withering of racial categories, and defends social policies that target poverty instead of race.

In "A Defense of the Death Penalty," **Louis Pojman** employs both *forward-looking* and *backward-looking* arguments. Forward-looking arguments maintain that capital punishment deters commission of murder and so helps ensure the best consequences overall in the long run. Pojman also discusses how several forms of evidence support the deterrent effects of the death penalty. Backward-looking arguments see punishment as a form of proportionate retribution. Such arguments do not appeal to the consequences in any straightforward sense but hold that murderers violate the dignity of their victims and so *deserve* to die. When responding to several objections to capital punishment, Pojman distinguishes retribution from vengeance and explains how the state has authority to inflict the death penalty. Pojman also responds to worries about mistaken death sentences and over-representation of certain groups among those sentenced to die.

By arguing that the "factual and moral beliefs on which death penalty support depends are mistaken," **Stephen Nathanson** maintains that neither deterrence nor retribution justify capital punishment. Deterrence alone is an inadequate justification, Nathanson writes in "Why We Should Put the Death Penalty to Rest," because it can license barbarically draconian punishments and the use of force on innocent persons. Standard "eye for an eye" arguments also fail because they are committed to reciprocating barbarity with barbarity. Such arguments, Nathanson worries, are also inconsistent with many of our considered moral judgments, and they give little guidance in determining appropriate punishments. We also find substantial evidence that capital punishment is unfairly applied in practice. Whether one receives the punishment is often a function of morally irrelevant factors such as one's race, class, and the quality of one's legal counsel. Maintaining the death penalty, Nathanson then argues, fosters a lack of concern about the loss of human life.

Much human history is marked by depredations, oppression, and often, wholesale slaughter. As people reckon with the past, some writers defend attempting to undo historic injustice. The remedy is no small task. As time goes on, identifying the perpetrators and victims, and identifying how much is owed, are often extremely difficult if not impossible. In "Compensation and Past Injustice," **Bernard Boxill** argues that justice sometimes requires that victims must be compensated for their losses. This often involves making the world into what it would have been had the wrong not occurred. Such corrective justice, however, must be constrained by the rights of all affected parties. With close attention to several examples, Boxill considers whether, how, and why justice might require some form of compensation for past losses. He defends compensation both for past injustice and for misfortune. Since both disrupt our life plans, and since justice aims to protect our interests in having, revising, and pursuing a conception of the good, justice demands compensation for our losses provided doing so does not itself entail an injustice. This qualification means that sometimes justice will forbid doing what is required to compensate victims of injustice.

In "Must We Provide Material Redress for Past Wrongs?", **Nahshon Perez** argues that there are often daunting obstacles to justifying any compensation for past wrongs. Among them is the "non-identity problem," which undermines the compensation claims of persons born subsequent to a wrong when that wrong was a condition of their existence. Perez considers two likely replies to this challenge. One appeals to transgenerational cultural identity. The other appeals to the existence of persons who are supposedly victimized by a past wrong but whose existence does not depend on that wrong. Both approaches, Perez argues, raise far more serious problems than they might solve. Many claims to redress for historic wrongs also seem to fade over time. Ultimately, Perez suggests, it is often best not to dwell in the past.

Critics of recent responses to the threat of terrorism have worried that security and policy forces unfairly target persons because they fit a certain profile. In "Bayesian Inference and Contractualist Justification on Interstate 95," **Arthur Applbaum** explores whether race-based generalizations are unjust. Some such generalizations, Applbaum argues, are instrumentally rational if they promote a specific goal. To show some such profiling may be morally acceptable, he draws on contractualist moral theory to ask what burdens are reasonable to impose. Inconveniencing someone to promote public safety might be a small inconvenience that any reasonable person would accept. Applbaum discusses how such profiling might work given a background of historical racism. Some statistical generalizations, he argues, do not give reasons for action. Some race-based profiles, then, need not entail disrespectful behaviors or attitudes. Since social meanings evolve, uses of race need not repeat the mistakes of the past.

Deborah Hellman acknowledges that some uses of profiling are innocuous, but claims that others are morally problematic. In her chapter, "Racial Profiling and the Meaning of Racial Categories," Hellman argues that many generalizations are worrisome not because of generalization itself but because of what it expresses. The category "black" is fraught with history and meaning. Racial generalizations then threaten to reinforce racist views and may convey a demeaning message to persons in the group. She identifies how profiling by government officials is often even more problematic than that by private actors.

Recent disclosures show western governments have used "harsh interrogation" tactics on some prisoners. In some cases, prisoners were "rendered" to countries with fewer restrictions on the treatment of inmates. These developments have revived debates about the merits of torture. Discussions of torture often revolve around what are sometimes called "ticking-time bomb" cases. In "Ticking Time-bombs and Torture," **Fritz Allhoff** closely examines such cases and the questions they raise about the justifiability of torture. He first traces the history of the debate and then examines the utilitarian account of torture. He argues that utilitarianism offers a straightforward justification for torture in ticking-time bomb cases. He then critiques deontological views on torture. Next, Allhoff distinguishes between claims that torture is morally impermissible in principle from claims that it is impermissible in practice. He believes that the former sorts of claims are extremely difficult to justify. He concludes by making some remarks about problems facing claims that torture is always impermissible in practice. Allhoff notes that arguments against torture ever being acceptable in practice rely on several important and controversial empirical assumptions.

In "Torture and its Apologists," **Bob Brecher** rejects the possibility of justifying torture. Though not a proponent of consequentialist moral theorizing, Brecher argues that such reasoning will not support torture. He discusses how appeals to ticking-time bomb cases typically make many unwarranted assumptions about suspects, the reliability of what we believe about suspects, and the effectiveness of torture. Interrogators cannot know in advance that torture is necessary. Furthermore, permitting interrogational torture generates insurmountable moral and institutional problems. Brecher warns that by endorsing interrogational torture, philosophers, lawyers and other academics open the door for public officials to implement the practice – and not merely in a few exceptional cases. Against ticking-time bomb scenarios, Brecher notes that sometimes we are simply "too late" to avoid catastrophe morally. Because of what torture does to us as *persons*, Brecher argues, it is the worst thing human beings can do to one another, and so it is morally impermissible.

Issues of Privacy and the Good

Many writers object to same-sex marriage on grounds that altering the institution would undermine an important social institution and jeopardize the well-being of children. In "Same-Sex Marriage and the Definitional Objection," **John Corvino** does not specifically respond to such worries (though he does in other published work); instead, he explores a conceptual objection. On this objection, same-sex marriage does not count as marriage, given what it means for something to be a marriage. On Corvino's view, this "definitional objection" is unsuccessful. Corvino argues that it is not part of the definition of marriage, for instance, that successful reproduction is possible to the partners. After setting out likely responses from natural law theorists, Corvino argues that marriage is a social institution embracing various sorts of practices; coitus is not a necessary condition of all of them. He defends an "inclusivist" view of marriage, where some same-sex relationships can count as marriages. He concludes that the debate about same-sex marriage is not about the meaning of marriage so much as about substantive questions regarding what gay and lesbian persons deserve as members of our society.

In "Making Sense of Marriage," **Sherif Girgis** objects to views such as Corvino's, which are examples of what he calls the "revisionist view" of marriage. Girgis's alternative is a "conjugal view" in which marriage is a comprehensive union that must include the possibility of coitus. The revisionists are unable to explain why marriage must involve sex, why it must be restricted to pairs of persons, and why marriages must involve commitments to exclusivity. Instead, on Girgis's conjugal view, marriage involves a bodily union of a man and a woman in a deep and exclusive commitment to achieve certain shared ends that encompass both partners. Marriage, in particular, is ordered toward the rearing of children. But this does not preclude the possibility that infertile couples cannot marry.

In "The Right to Get Turned On: Pornography, Autonomy, Equality," **Andrew Altman** defends rights to produce, sell, and view pornography – including pornography depicting sexual violence. He hinges these rights not on free speech considerations but more on what he calls *sexual autonomy*. A suitably constrained right to such autonomy confers neither a right to coerce anyone into sexual acts nor a right to entice minors into sexual encounters. But this right does protect people who choose to produce pornography or consume the final product. They have this right even if (as might often be true) they are deficient in some human virtues. Altman argues that violent pornography *might* be a candidate for prohibition *if* there were conclusive evidence connecting it to violent imitative acts. But the evidence is far too weak to exclude violent pornography from the protection of a right to sexual autonomy. In the meantime, Altman calls for improved education regarding sexual violence as well as more vigorous prosecution and serious punishment for criminals guilty of such crimes. Some critics may still worry that pornography nevertheless fosters attitudes contributing to the degradation and subordination of women. Altman questions the connection there, noting that the liberal democracies that protect a freedom to produce and consume pornography tend to be societies with the best opportunities for the social and economic advancement of women.

Susan J. Brison rejects the notion of a right to produce and consume degrading or violent pornography. In her chapter, " 'The Price We Pay'?: Pornography and Harm," Brison offers detailed accounts of the exploitation and suffering of women involved in the pornography industry. She argues that many participants in pornography cannot be understood to have offered genuine consent. But even if they have given free consent, their participation in the industry has morally significant effects on social norms regarding sex roles. The industry harms nonparticipants, both male and female, by teaching and perpetuating discriminatory attitudes and by further injuring those previously victimized by sexual violence. Indeed, pornography's connection to subordination and degradation is not incidental; as Brison notes, pornography arouses precisely because of images of subordination. Brison then considers whether there can be a moral right to pornography – especially if, as Altman concedes, exercising such a right may be a sign of some moral vice. Brison argues, against Altman, that the harms pornography causes are sufficient to deny a right to produce and consume it.

In his chapter, **Douglas Husak** argues "In Favor of Drug Decriminalization." He specifically supports ending state punishments for persons who use drugs. He offers utilitarian and rights-based arguments for decriminalization, but then shifts the burden back to proponents of criminalization. He further disputes many common arguments in favor of punishing drug users, especially regarding many ill-effects of use. States must take great care when administering punishment since it is among the harshest things states may do to us. The consumption of known drugs, Husak argues, does not seem to warrant any state punishment. In "Against the Legalization of Drugs," **Peter de Marneffe**, however, opposes drug legalization in the sense of removing all criminal penalties on drug manufacture, sale, and use. He considers several leading objections to criminalizing drugs. He replies to concerns about effectiveness, paternalism, the social/economic costs of the drug war, the racial dimension of drug laws, the impact on violence in society, the potentially corrupting impact on foreign states, and other objections about altering the norms and institutions regarding drugs. In each case, de Marneffe argues that the criticisms misfire. Legalizing drugs, he argues, may have deeply worrisome consequences. Pointing to problems with current institutions does not show that drugs should be legalized; it may instead suggest that we should reform our institutions and laws.

Issues of Cosmospolitanism and Community

Freedom of movement is clearly a basic human right, **David Miller** admits in his chapter, "Immigration: The Case for Limits," but whether that translates into a right to move to any physical space of one's choosing is another matter. There are often important reasons for restricting a freedom of movement, and many times such restrictions do not impede any *right* to move freely. So long as individuals have access to an *adequate* range of choices for satisfying significant interests, their interest in migrating elsewhere is not protected by right. A "right of exit" may give political societies reasons not to abuse members, but given the diversity of contemporary political societies, such a right does not translate into an unlimited right to go to *any* state. Much then hinges on what Miller calls the *scope* of distributive justice. Do principles of distributive justice apply *within* or *across* societies? If global justice furnishes any moral reasons, these reasons likely fall short of requiring equal distribution of any particular good. We should note that immigration also invariably changes the public culture of a political community, but native people have legitimate interests in controlling such a culture. Immigration also raises significant issues in population growth, and given economic and ecological considerations, nations have good reason to restrict the influx of immigrants. Miller defends refugees' rights to move elsewhere for greater security, but he also argues for states' autonomy in deciding how to handle asylum requests. He upholds the prerogative of political communities to admit a non-refugee on the basis of the prospective benefit for granting entry as well as the migrant's interest in moving.

Chandran Kukathas defends free immigration and open borders in "The Case for Open Immigration." He discusses why states of various sorts may have interests in limiting immigration. While some authoritarian states wish to curtail the dissent that may challenge government authority, even liberal democratic states may have complex economic and political reasons to restrict the influx of immigrants. Though he admits that an open borders policy is unlikely without reconsidering the notion of the modern state, Kukathas defends the policy by appealing to principles of freedom and humanity. Immigration is often a crucial avenue for fulfilling moral duties, pursuing economic opportunities, fleeing injustice, or striving for the improvement of oneself or one's family. Kukathas discusses possible consequences to open borders and argues that, in the end, there is no compelling *economic* argument for restricting immigration. An appeal to *nationality* may suggest arguments against open borders, either because immigration will undermine a society's distinct cultural character, undermine natives' abilities to prosper through a distinct way of life, or jeopardize a political community's ability to implement shared principles of social justice. But Kukathas argues against all such considerations. Cultural transformations are often entirely beneficial, and, he argues, it is unclear that the nation state should be the locus for social justice or that implementing principles of social justice should take precedence over humanitarian concerns for helping the poor and the oppressed. While security concerns may give us pause, immigration restrictions are often poorly targeted and represent significant threats to personal liberty.

We live in a world of nation states. Each state claims sovereignty over a territory, including a right to govern its own internal affairs. Many writers, though, think sovereignty rights do not protect states that violate the rights of their residents. In "The Moral Structure of Humanitarian Intervention," **Fernando Tesón** argues that military intervention for humanitarian aims may be justified as a defensive use of force on behalf of innocent victims. Such force must not impose costs that are disproportionate to the moral significance of the rights violations it hopes to remedy. Sovereignty, Tesón argues, does not protect oppressive regimes from intervention. He further considers when and whether liberal states are permitted to use their own forces to rescue victims of injustices abroad. Tesón denies that authorization by current international institutions is necessary for morally permissible humanitarian intervention. As long as the principle of proportionality is respected, it is sometimes permissible to save people from oppression.

In "The Morality of Humanitarian Intervention," **Bas van der Vossen** explores whether states have rights against intrusion by outside forces. Many writers argue that just as individuals are permitted to undo injustices inflicted on residents in foreign lands (provided doing so does not harm innocent others), then so, too, may nations rescue unjustly harmed residents of foreign lands through humanitarian intervention. Van der Vossen rejects this inference. His main concern is that states are imprecise (and often bloody) agents of change abroad. He argues that moral rules both guide and constrain conduct, so allowing for a right of humanitarian intervention may invite grievous harms and escalations of violence. Indeed, such a right threatens even the justified actions of legitimate states. Judgments of legitimacy, van der Vossen adds, are less prone to dispute than are judgments of justice. Sometimes legitimate states have rights against intervention even when they are guilty of some important wrongs against their residents. Van der Vossen closes by considering what sorts of conditions license humanitarian interventions in illegitimate states.

In his chapter, "Famine Relief: The Duties We Have to Others," **Christopher Heath Wellman** argues that one has a moral duty to rescue persons in dire need when one can do so without incurring unreasonable costs. Wellman is careful to note that one's duties of rescue need not take precedence over responsibilities to those near and dear.

Still, when a person must choose between devoting resources to frivolous pursuits and providing easy rescue, she has a duty to lend a hand. The *proximity* of emergency is morally irrelevant if one can provide easy rescue or solicit others to do the same. Just as we often have a duty to save babies drowning within our *sight*, so too we often have a duty to direct modest amounts of our resources toward famine relief. Indeed, needy people sometimes have "samaritan rights" of rescue against persons who can provide assistance without unreasonable sacrifice. Modern communications have so expanded our knowledge of distant conditions that persons unwilling to donate a modest amount toward famine relief are often morally no different than persons who refuse easy rescue of infants drowning at their feet. Beyond our responsibilities to offer modest aid, we should also take steps to disassociate ourselves from unjust institutions – especially those that benefit us. Part of the reason we have our wealth, Wellman argues, is that we profit from an economic system that uses natural resources bought from oppressive governments abroad – and those governments "create the political conditions that play a causal role in the world's worst famines."

Appeals to babies drowning at our feet, **Andrew I. Cohen** argues in "Famine Relief and Human Virtue," tell us little about our responsibilities to alleviate world hunger. Hunger is a chronic problem calling for reflection on causes and alternative solutions. Such reflection is usually inappropriate for dire emergencies immediately in front of us. Cohen then explores the place for charity in a good life. He writes that persons need a protected opportunity *not* to be charitable in order for them to have the best chance properly to develop and cultivate the virtue of charity. Charity cannot be coerced. Cohen defends a limited "right to do wrong" with respect to withholding resources that good persons would otherwise have provided in similar circumstances. Such a right is important for giving persons the space to become virtuous, and it is crucial for maximizing the chance that there will be fewer needy people in the long term. He discusses several problems with enforcing positive duties to give to the needy. Such enforcement clashes with other moral values, jeopardizes satisfying other relevant moral demands, hinders personal virtue, and it is often dangerously ineffective at alleviating hunger. Cohen notes that reasonable persons disagree not only about how best to satisfy need but about what the good life is and how one ought best to strive for it. The liberty to live our own lives should take precedence over the aims of busybodies and autocrats who believe they know better how we should allocate our precious resources. Cohen further argues that a good human life is marked by moral demands from many sources, and distant human need is but one possible claim on one's resources and time. We in the West best help needy persons by curtailing misguided relief policies, eradicating government price supports that unfairly privilege the wealthy, and trading with people overseas.

ETHICAL THEORY

CHAPTER ONE

Theories of Ethics

Stephen L. Darwall

Ethics is customarily divided into two parts: meta-ethics and normative ethics, with the latter being divided further into normative theory and "applied ethics," the area with which we are concerned in this volume. This last term may not be especially apt, however, since it suggests a relation to normative theory like that applied to pure mathematics, where theories are derived independently and, only then, applied to cases. When it comes to normative ethics, theories are often formulated and evaluated by reflecting on the ethically relevant features of cases. Thus some philosophers maintain that we can appreciate the general moral relevance of a distinction between killing and letting die (or, more generally yet, between causing evils and letting them happen) by reflecting on a specific case like Judith Thomson's famous "trolley problem," in which a driver can choose between letting his runaway train kill a certain number of people and diverting it on to a track where it would kill a smaller number (Thomson, 1976). In thinking about this case, it seems to be relevant that by diverting the train the driver would be killing people or causing their deaths himself, whereas, if he let the train continue undiverted, he would only be allowing deaths to occur. By seeing this in a specific case, it is argued, we can appreciate a distinction of general theoretical relevance.

Another term for our area, "practical ethics," avoids these associations but is misleading in a different way, since it suggests that the only cases of interest concern practical questions of what to do. Frequently, what we want to know is not what someone should do (or should have done), but what to think or feel about someone's character or about his having done something out of certain motives. And there are many other ethical questions that are not primarily practical either, even if they have practical implications: do all living species have intrinsic worth? Is aesthetic appreciation a more valuable form of human experience than the relief of a scratched itch? And so on.

Contemporary Debates in Applied Ethics, Second Edition. Edited by Andrew I. Cohen and Christopher Heath Wellman.
© 2014 John Wiley & Sons, Inc. Published 2014 by John Wiley & Sons, Inc.

Case Ethics

A better term for our area might be "case ethics." Just as there is "case law," the findings of judges about the issues brought before them, including, crucially, the reasoning or *ratio* that led to their conclusions, so also is there case ethics: our considered judgments about specific ethical issues or cases along with the reasons or principled reflections that underlie our judgments. When it comes to the law, of course, only properly vested judges can render authoritative judgments. With moral and ethical discussion and debate, however, we all have standing to take part, and no individual has the authority to make final judgments, although we may freely accord a kind of authority to those we think especially thoughtful and judicious.

When judges render legal judgments, it is not enough for them simply to find for one side or the other. They must also support their judgments with applicable law and legal principles, aiming to show how anyone reflecting on these laws and principles might reasonably have come to the same conclusion. This rarely involves anything like a deductive proof. As Aristotle remarked, we cannot sensibly demand more logical rigor than "the subject-matter admits of" (Aristotle, 1998). It is enough if judges point to laws, principles, and ideals under which the case might reasonably be subsumed and which, when applied, might be seen to support their finding in the case.

No doubt it is a mistake to view all of ethics on the model of law, as a number of philosophers have noted (Anscombe, 1958; McDowell, 1979; Dancy, 1993). Some ethical questions require a sensitivity and insight that seems more akin to aesthetic appreciation than to anything that can be supported by theoretical principles. None the less, even art critics feel bound to give reasons for their judgments, citing considerations that can help others enter into their reflections and see the work as they see it. Moreover, many of the cases of applied or case ethics with which we are most often occupied are questions of public *morality*, where we implicitly understand our discussions and inquiries to take place in a democratic society in which everyone has standing to participate. Moreover, when these issues concern moral *obligations* – the area where, as John Stuart Mill pointed out, we hold one another accountable through formal and informal sanctions – the same considerations that lead us to demand a publicly formulated justification for legal judgments would seem to apply to moral judgments also, if not, perhaps, quite so urgently (Mill, 1957). A restriction of liberty of some kind will seem to be involved, calling for a principled justification that could be seen to be acceptable to our fellow citizens who would, in our view, be bound by them.

Normative Ethical Theory

Philosophers use the term "normative ethical theory" to refer broadly to principles, concepts, and ideals that can be cited in support of ethical judgments about cases. In this broad sense, we commit ourselves implicitly to some theory (or range of theories) whenever we give reasons to support our judgments. Furthermore, there is a sense in which we commit ourselves to the existence of some justifying background theory whenever we even *make* an ethical judgment. This is because of an important feature of ethical concepts and properties that we might call their reason- or warrant-dependence. When, for example, I judge that something is good, I say, not just that I value it, but that there is reason to value it – that valuing it is warranted, an attitude one ought to have. As a logical matter, however, this can be true only if something has *other* properties: the reasons for valuing it. And such reasons cannot simply consist in the property that it is good, since that is itself the property of there being such reasons. Unlike, say, the property of yellowness, which might attach to something all by itself, as it were, ethical properties require, by their very nature, completion by further properties that are their reasons or grounds. If I judge a certain experience to be valuable, I must think it has aspects that make it good, features that are the grounds of its value. Or if I think that a certain action is morally required, I must think there are certain characteristics of the action and the situation that make it morally obligatory, features that are the grounds of its obligatoriness. And these thoughts commit me to the existence of background normative theories. I am committed to thinking there are truths that relate an experience's having certain properties to its value, such that any experience that had exactly those (and no other ethically relevant) properties would be valuable also, other things being equal. Or, similarly, I am committed to thinking there exists some valid moral principle that relates an action's having certain features to its being morally required.

In this broad sense, then, the investigation of normative ethical theories is unavoidable if we are to think about ethical issues with any care. Any judgment we make about an individual case will be no better than the background theories we commit ourselves to in making it. Moreover, there are special considerations that commit us to normative theories of a distinctively robust sort when the judgments we make concern questions of moral right and wrong. This is because, as I noted briefly above, the part of ethics we call morality is modeled on law, even if other parts are not. What is wrong is what we are appropriately held accountable for doing, what warrants blame unless we have some adequate excuse. Practices of accountability are in their nature directive, and frequently even coercive. As with a judge's legal findings, therefore, we properly feel some pressure to articulate principles (or theories) that are capable both of justifying our judgments and of being publicly addressed to and accepted by other members of the moral community.

Someone we hold accountable for wrongdoing, we think, should be capable, in some sense, of accepting our judgment, of being brought to see that it is a reasonable judgment to have made. This is very different from other ethical assessments, as when, for example, we feel disdain for someone as a coward, or hold some human pursuits to be less worthy than others. Disdain for cowardice does not attempt to direct the coward or to hold him accountable, and because there is no prescriptive or liberty-limiting element, there is no thought that disdain is appropriate only if its object should be able to accept it and see things the same way. To the contrary, disdain may only increase if its object cannot "get it." If, however, we judge someone to be incapable of assessing his own conduct morally, this can lead us to think that he is not a fit object of moral evaluation, since he is incapable of entering into a mutually accountable moral community. It is therefore reasonable to accept a burden of being able to formulate public justifications for judgments of moral right and wrong that is similar to one we impose on judges' legal findings. Normative moral theory is in this way part and parcel of public moral discourse.

Meta-ethics

Our main interest in this chapter will be with the major normative ethical theories: contractualism, consequentialism, deontology, and virtue theory. Before discussing these, however, we need first to introduce briefly the other main area of ethical theory: meta-ethics. Unlike normative theories, which concern themselves with substantive normative questions – such as "What is valuable?" and "What is morally obligatory?" – meta-ethical theories are concerned with more abstract philosophical issues that underlie these. We can distinguish four different kinds: (a) questions in the philosophy of language concerning the meaning and content of ethical judgments; (b) related issues in the philosophy of mind concerning what mental states ethical judgments express or what it is to hold an ethical view; (c) metaphysical issues concerning the possibility and nature of ethical truth; and (d) epistemological questions concerning the possibility and nature of ethical knowledge and how we can justify our ethical views.

Why, however, should *we* care about meta-ethics? Some people think that case ethics can be divorced entirely from meta-ethics. They may grant that analyzing ethical cases ultimately calls on *normative* theory, but hold that this is entirely independent of meta-ethics. I think this view is mistaken and that the sharp separation sometimes made between meta-ethical and normative thought is a distortion both of how the great systematic ethical thinkers (like Aristotle or Kant) proceeded as well as of how we do and should proceed in contemporary moral debate.

Consider, for example, issues that arise in environmental ethics concerning the moral claims that other living species make on us. How much should we weigh harm to other species – either to individuals or to the species themselves – in our moral deliberations? It is impossible to think carefully about such questions without engaging meta-ethical issues. One concerns the nature of harm. To be able to be harmed, something must have a good or welfare. But what is it for something to be good or bad *for* some being?

On one common view about the good of human beings, a person's good consists in the satisfaction of desires (alternatively, the desires she would have if fully informed, or would, if fully informed, have for herself as she actually is) (Railton, 1986). This view is sometimes put forward, not just as a normative claim, but as a meta-ethical position concerning what personal welfare or benefit *is*. But such a meta-ethics of welfare rules out the possibility that any species lacking desire can be benefited or harmed. Roughly, nothing will be good or bad *for* some being unless it too can be good or bad *to* it (as it might appear through desire). I believe this meta-ethical theory to be the mistaken result of a line of thought that takes it for granted that a person's good is what he aims at in so far as he is rational. Once we see that the concept of welfare or benefit has no such privileged status from the perspective of an agent deliberating about what to do, but is one we require, rather, when we *care* for some being or thing for its sake (ourselves included), we can appreciate why harm and benefit are not restricted to beings with desires (Darwall, 1997, 2002). To have concern or care for members of another species is just to desire for *their* sake that they do well or flourish. We can sensibly regard a species as capable of being benefited or harmed, therefore, if we can care for them for their sake.

But, again, what sort of claim does harm make when we understand it in this way? Do we have a *moral obligation* not to harm members of any species? Or is the fact that an action would harm another person relevant to its being wrong in a way that harm to other species is not? Here again, we cannot answer this question without taking a stand on meta-ethical issues concerning what morality and moral obligation *are*, if only implicitly. If we understand moral questions broadly enough, then it may seem that harm is harm and is no less morally relevant whether the being harmed is a person or a snail darter. If, however, we think of morality as a system of reciprocity or mutual accountability, where norms of right and wrong mediate a moral community of free and equal moral persons, then harm to other persons will seem to have an intrinsic *moral* relevance that harm to other species does not. For then what is morally wrong will be what one can be held accountable by others for doing, in accordance with norms that must, in some sense, be acceptable to all from a perspective of equality. So viewed, harm to persons is not simply harm to members of a certain species, but harm to a member of the moral community to whom norms of right and wrong must be justifiable.

This is only one example of how questions of meta-ethics are implicitly involved in issues of normative ethical theory and, therefore, in case ethics. Ultimately, we have no alternative but to pursue *philosophical ethics*, that is, to attempt to work out a comprehensive outlook that integrates normative ethics *and* meta-ethics (Darwall, 1998).

Contractarianism/Contractualism

We can turn now to a review of different normative theories and begin with one that can be grounded in the meta-ethical theory of morality as reciprocity or mutual accountability just mentioned. This is the idea that whether an action is right or wrong depends on whether it accords with or violates principles that would be the object of an agreement, contract, or choice made under certain conditions by members of the moral community. The general idea can be developed in a variety of ways, depending on how the choice or agreement, the parties who make it, and the conditions under which it is made are characterized. One broad distinction is between *contractarianism*, under which the choice of moral principles is self-interested, and *contractualism*, which grounds it in a moral ideal of reciprocity, reasonableness, or fairness.

It may seem strange to think that moral principles can in any respect be agreed upon or chosen. How can a moral proposition be made true by any choice or agreement? Only rarely, however, do contractarians or contractualists claim that right and wrong are determined by *actual* choices or agreements (Harman, 1975). More frequently, what they hold is that moral principles are those that *would* be rationally or reasonably chosen or agreed to under certain (frequently, counterfactual) conditions.

Contractarianism

Contractarianism was initially formulated by Thomas Hobbes (see Hobbes, 1994). Hobbes begins by considering the situation of an agent deliberating independently of others from the perspective of his own desires or interests. Each person, he thinks, sees what he desires as good, as giving him reason to seek it. But what results if all of us, together, pursue our respective desires and interests? Although each person's doing so may actually result in his interests being best promoted, *given* the conduct of others, it does not follow that *everyone's* pursuing his respective interests, rather than everyone's pursuing some other aim, or acting on some principle other than self-interest, will actually result in everyone's (or even anyone's) interests or desires being best promoted. In situations where this is not the case, where the collective pursuit of self-interest leads to an outcome that is worse for each, we have what is known as a *collective action problem*.

This is illustrated by the game-theoretic example known as the Prisoner's Dilemma, in which two individuals are jailed on suspicion of robbery. The district attorney tells each that he lacks enough evidence to convict either him or his partner of robbery, but can easily convict each of breaking and entering, giving each a sentence of one year. He offers each a deal: if one confesses and his partner does not, the confessor will go free and the partner will get twenty years. If both confess, both get five years.

Suppose that each cares only about doing the least time. The structure of the situation then is as follows. If *A* confesses but *B* does not, then *A* gets his first-ranked outcome and *B* his fourth-ranked (worst). And vice versa, if *B* confesses but *A* does not: fourth-ranked for *A*, first-ranked for *B*. If both confess, both get their third-ranked outcome. And if both do not confess, both get their second-ranked.

What should each do? Reason first from *A's* perspective. *B* will act independently of *A* and either confess or not. It seems, therefore, that *A* should confess, since whatever *B* does, *A* will do better if he confesses. If *B* confesses, then *A* will get his third- as opposed to his fourth-best outcome by confessing. And if *B* does not confess, then *A* will get his first- as opposed to his second-best outcome by confessing. So *A* should confess. *A* will do better by confessing, whatever *B* does.

But *B's* situation is exactly analogous to *A's*, so any reasons for *A* to confess apply equally to *B*. So if *A* would do best to confess, then so would *B*. It may now be evident why this is called a collective action problem. *A's* and *B's* actions, although likeliest to achieve the best outcome for each when taken individually, taken together yield an outcome that is worse for each. If both prisoners do what would be best for him, given the actions of the other, both will confess. But that yields each one's third-ranked outcome, whereas they could have both achieved their second-ranked outcomes by not confessing. Although the jailhouse context makes this sound strange, not confessing is actually the *cooperative* strategy for *A* and *B*. If *A* and *B* could cooperate to their mutual advantage, they would both not confess and end up with their second-ranked outcome rather than the third-ranked outcome that independently promoting their interests will achieve.

People cooperate when they forgo the pursuit of their own independent interests and follow rules or roles, the collective following of which promotes everyone's interests better than would have been done by everyone pursuing their own interests *independently*. Obviously, cooperation is required for many, many things that are valuable in life, perhaps especially in complex modem societies, in which we cannot assume that genuinely common interests, shaped by common cultural or religious traditions will stretch across all areas of significant interaction. *Morality* can be thought of as an especially broad and pervasive form of cooperation. Principles of moral right and wrong would then be whatever rules, specifying requirements, permissions, and so on feature in the broadest possible form of cooperation, namely, cooperation involving not just this or that group, community, or political unit, but all competent human agents. (Actually, Hobbes's view was that cooperation among large groups was impossible without political authority, since otherwise uncertainty of others' participation would undermine the assurance necessary for it to make sense for one to do one's part.)

According to contractarianism, then, whether an action is right or wrong is determined by rules of cooperation of the broadest sort, that is, between all human moral agents. Take, for example, the rule that it is wrong not to come to the aid of others in need, so long as the sacrifice involved is not too great (say, as long as it is not above some level α and/or the ratio of sacrifice to need is not above some level β). Arguably, there exist some α and β, such that it would promote everyone's interests more for everyone to follow the resulting rule, than it would for everyone to pursue their own interests independently. If that is so, then, for starters, contractarianism will hold that it would be wrong not to follow this rule.

To a first approximation, contractarianism holds that what it is right to do depends on what rules it would be in everyone's interest for everyone to accept and be guided by in their deliberations and moral practice. However, what if various *different* possible rules for a given kind of situation have the property that everyone's interests would be promoted better by everyone's following that rule than they would be if everyone attempted to promote their own interests independently? This is where the idea of an agreement or contract enters the contractarian picture. Taking as a benchmark the "no agreement" point in which all regard themselves as bound by nothing but their own interests and values, contractarians treat the question of which principles we actually are morally bound by as the solution to a rational bargaining problem from this benchmark, in which we all have a greater interest in agreeing to some mutually advantageous principles, thereby avoiding the "no agreement" point, but have differing interests in exactly which principles are actually agreed (Gauthier, 1986). How favorably the resultant principles treat the different negotiating agents will depend on who has the most to lose if there is no agreement. Consider, for example, what principle of mutual aid would be agreed to. If those with fewer resources and greater vulnerabilities have more to lose from the lack of agreement than those with greater resources and fewer vulnerabilities, then rational bargaining may lead to a less onerous principle of mutual aid than would result if everyone were as vulnerable as those with less.

This, then, is contractarianism's basic framework for assessing moral issues. To work out our moral obligations in a specific case, say, the obligations of rich and poor countries in reducing global warming, we have to think about what agreement on principles for dealing with the issue would result from a negotiation from the "no agreement" point in which each party attempts to advance its own interests and values. In acting on the principles that would be agreed, however, the parties are not simply promoting their interests; they are cooperating. Cooperation promotes everyone's advantage, but, as in the Prisoner's Dilemma, it does so by requiring individuals to forgo promoting their own interests. Each would prefer schemes in which the necessary sacrifices are borne in greater measure by others, but each party is prepared to do its part as required by principles of cooperation that everyone could rationally agree to.

Contractualism

Contractualism has a similar structure. It too understands principles of right conduct as the object of a rational agreement. But whereas contractarianism takes moral principles to result from rational bargaining, contractualism sees the agreement on principles as governed by a moral ideal of equal respect, one that would be inconsistent with bargaining over fundamental terms of association. From contractualism's point of view, the problem with contractarianism is that it must assume that individuals have, in effect, a moral claim to the resources they would have if there were no agreed rules of cooperation. Otherwise, the rules that result by bargaining from that position will have no moral force. But why assume that people have such a moral claim? From a moral point of view this seems entirely arbitrary, unless some background theory of natural rights is assumed. And contractarianism cannot justify *that* assumption, since its own moral force would have already to depend upon it.

This problem emerges from another direction if we consider how individuals might get from self-interested practical reasoning to contractarian moral reasoning. If each agent reasons in terms of her own independent interests, then, from a situation in which there are no established rules of cooperation, she will think it rational to bargain to an agreement with others to be bound by certain rules. But how can this give her a reason actually to follow the rules? The reasons of interest she has for agreeing to follow the rules cannot give her a reason actually to follow them since the whole point and function of rules of cooperation requires that they *constrain* her pursuit of her interests. For her to be able to reason as these rules require, she must already accept moral reasons of cooperation. It may even be that it is in her interest to be someone who does accept contractarian moral reasons of cooperation, but while this would give her reasons to *want* to accept the moral reasons, she could not *accept* the moral reasons *for these reasons*.

The animating idea of contractualism is implicit in Kant's "kingdom of ends" formulation of his Categorical Imperative. Kant maintains that anyone subject to the moral law must be able to be regarded also as "making the law" (Kant, 1998). Only thus can the moral law be thought of as a common law for a community of free moral agents, subject only to laws they legislate themselves. This is a version of Rousseau's idea of legitimate political community as an association in which each, "while uniting with all, nevertheless obeys only himself" (Rousseau, 1987). According to Rousseau, this is only possible if laws express what he calls the "general will," the will of each *as* a free and equal member. Similarly, Kant conceives of moral laws as "made" by each moral agent when each would "legislate" it *as* a free and equal member of the "kingdom of ends." Here we have the central difference with contractarianism. Moral principles of right are not rules that individuals would prescribe, and attempt to gain acceptance for, from their *different* individual perspectives, bargaining out of self-interest. They are, rather, rules individuals would prescribe (and agree to) from a common perspective as one free and equal person among others.

But how, more specifically, is this perspective to be understood? Kant gives some hints, saying that we can conceive a "systematic union of rational beings under common objective laws," only if we "abstract from the personal differences between rational beings, and also from all the content of their private ends" (Kant, 1998). This suggests the contemporary contractualist (John Rawls), idea that principles of justice are those it would be rational to choose in an "original position" behind a "veil of ignorance" regarding any features that individuate different persons or their societies (Rawls, 1971, pp. 136–142). In particular, the choosing parties are ignorant of their individual resources, abilities, talents, gender, race, socioeconomic position, *and* their own interests or individual values. Rawls assumes that the parties have an interest in autonomously choosing and pursuing their interests (whatever these turn out to be) and, therefore, that they value the "primary goods" that are necessary for these: freedom, opportunities, wealth, and the "social bases" of self-respect.

Rawls's idea then is that justice is determined by whichever principles the parties would choose from behind the veil of ignorance, that is, as one free and equal person among others. Rawls does assume that this choice is self-interested within the constraints placed by the veil, but this does not reduce moral reasoning to self-interest in any way. To see this, suppose the parties were motivated not by self-interest but by concern for a single other individual. Since the veil of ignorance deprives them of any information that would let them tailor principles to *any* particular person's interest, there is no functional difference between assuming the parties to be self-interested and assuming them to be trustees for another individual. The original position is, in effect, the perspective of *a*, that is, an arbitrary, free and equal individual.

Rawls argues that the rational choice in the original position would be two principles of justice, ranked in order of priority: a first that requires the existence of certain basic civil and political rights and freedoms and a second, the "difference principle," which assures fair equality of opportunity and that remaining primary goods, such as wealth, are distributed by the basic institutions of society in ways that work to the greatest advantage of those who are least advantaged. In effect, this says that inequalities are justifiable only to the extent that they work as a social resource (for example, by providing incentives) from which everyone benefits, including the least advantaged. If we take seriously the possibility that we could be anyone, and have no way of estimating probabilities of ending up in any particular position, the rational thing would be to protect against the worst possibilities by choosing the two principles, including the difference principle.

Rawls put these ideas forward as a theory of justice: "justice as fairness," he called it. More recently, he has stressed that it is to be understood as a political, rather than a more general moral, theory (Rawls, 1993). In his earlier work, however, he suggested that it might also be conceived as a moral theory: "rightness as fairness" (Rawls, 1971, p. 111). To do this, we would have to ask: what principles of individual conduct (or of the conduct of groups, nation states, and so on) would it be rational to choose to govern everyone's conduct (as well as moral criticism and practices of accountability concerning it), from the original position. This would then give us a framework within which to consider which principles ought to apply to any specific case we might have under discussion.

A second contractualist approach can be motivated by thinking about what it is to make a claim on someone as an equal. When one person claims something in this way, she attempts to give another person a reason based on her needs as an equal. It is as if she says, "This is a reasonable claim for you to grant to me, as you can see were you to put yourself in my shoes and consider that it would be reasonable for you to make it of me." Such a claim implicitly invokes the idea of principles of conduct that reciprocally recognizing equals can accept or, at least, not reasonably reject. In developing such a contractualist approach, T.M. Scanlon assumes a community whose members wish to be able to justify their conduct to each other by principles that others could not reasonably reject, in so far as they also have this aim (Scanlon, 1998, pp. 147–257). Principles of moral right and wrong can then be thought of as norms that structure a mutually accountable community of equals.

To apply this criterion, we must make judgments about what is reasonable. How can we make these? There seems no alternative to putting ourselves into others' shoes and seeing whether we would regard a certain claim, or objection against a proposed principle, as one we would reasonably make if we were in their situation. This is a complex judgment. It is not simply the prediction that we would make the same claim or objection. We might think we would, but that it would be unreasonable. To make the requisite judgment, it seems, we must attempt to enter into the other's perspective impartially, as anyone, to see whether we would endorse the claim or objection as a reasonable one to make to another, reciprocally recognizing equal. Suppose, for example, we are trying to determine what principles should govern reduction in nations' use of fossil fuels to combat global warming. Contractualism will hold that developed and developing countries alike should govern their conduct by principles that none of them could reasonably reject. If a proposed standard is in fact rejected by developing countries, for instance, it is necessary then to judge whether this would be a reasonable objection to make were one in their shoes. And vice versa for standards that might actually be rejected by developed countries.

Consequentialism

Whereas contractarianism and contractualism begin from within conceptions of morality, as mutually advantageous cooperation or as reciprocity between equals, respectively, consequentialism begins with values it holds to be *prior* to morality. Even if there were no moral right and wrong, some things would still be good and others bad. When we judge the pain and suffering caused by a cataclysmic earthquake to be bad, for example, we are not making a moral evaluation, even if our judgment has implications for morality. Neither the pain nor the earthquake need have involved agency or character in any way. Rather, we are judging that it is a bad thing that the suffering *happened*, that such suffering is a bad state of affairs, a bad thing *to happen*. The idea is not just that suffering is bad for the sufferer, but that it is a bad thing to occur period. As these values and disvalues are independent of morality, they are called *non-moral*.

Consequentialist moral theories start with a *non-moral value theory*: a normative theory of which states of the world (things that can happen) have intrinsic value, which have disvalue, and some account of how these values compare, either with an ordinal ranking or with some cardinal metric. What makes these values non-moral, again, is that they are not evaluations of moral agency or character, but of outcomes or states – ways the world might be. Of course, such states might include agency and character. But even here the evaluation of the state (as something that *happens*) can be distinguished from the evaluation of the act or character trait that is a constituent of that state. Thus one might consistently think it would have been good if Hitler had been assassinated, that that would have been a good thing to have happened, say because of the lives it would have saved, even if such an assassination would, perhaps, have been morally wrong. As we will see, consequentialists might deny that such a killing would be wrong, but the point is that there would be no incoherence in holding it to be wrong and, at the same time, thinking that the state of the world of Hitler's being assassinated would, on balance, have been a good thing to have occurred.

Consequentialist moral theories all agree that the moral rightness and wrongness of acts are determined by the non-moral goodness of *relevant consequences*. There are, however, two kinds of issues on which consequentialist theories divide. First, and most obviously, they can disagree by being based on different theories of non-moral value. A consequentialist with a hedonist value theory, according to which pleasure is the only intrinsic good, will disagree, for example, with one who holds that preserving species or, perhaps, historical cultural treasures can be good things in themselves. Second, consequentialist theories can also disagree by holding that consequences of different sorts are relevant to determining moral right and wrong. *Act-consequentialism* holds that whether a given act is right depends on the value of the consequences of *that act*, compared with the value of the consequences of any other act the agent could do in the circumstances. According to *rule-consequentialism*, on the other hand, the rightness of acts depends on the consequences, not of the act, but of the social acceptance of a rule requiring, forbidding, or permitting the act, compared with the consequences of accepting other possible rules for that kind of case. If accepting a rule requiring an act of that kind would have the best consequences, then the act is morally required. And consequentialism can take other forms too.

All forms of consequentialism, however, understand moral evaluation to be an assessment of *instrumental* or *extrinsic* value at the most fundamental level. All are based on theories of the intrinsic, non-moral value of outcomes, and all assess the moral status of acts and character by determining which acts, social rules, or traits of character are the best instruments for promoting the most valuable states. For act-consequentialism, a morally right act is the agent's best available instrument for producing non-moral value. And rule-consequentialism judges the rightness of acts by the verdicts of socially realizable rules that are, via their participation in social practices of moral reasoning and criticism, the best instruments of that kind for producing non-moral value.

In principle, virtually any theory of outcome value can be harnessed to a consequentialist moral theory. Historically, however, consequentialism has been advanced most frequently by philosophers who have thought that valuable outcomes must somehow involve the lives of conscious beings. We might call *benefit consequentialism* the view that valuable states all concern the good or welfare of some being or other, and that moral assessment must ultimately be based on this. It is possible to believe, however, that something can benefit or harm a being by affecting something other than the quality of its experience or conscious mental states. For example, perfectionists sometimes assert that a being's approximating an ideal for its kind is intrinsically beneficial to it. This is what leads to the conclusion of Aristotle's famous "function" argument that human good or flourishing consists in excellent, distinctively human activity (Aristotle, 1998). By and large, however, benefit consequentialists have tended to hold that people are benefited or harmed, respectively, by what positively or negatively affects their mental lives, that is, to hold either *hedonistic* or *desire-based* forms of consequentialism.

Theories of Ethics   23

The most popular form historically has been *utilitarianism*, which is distinguished by three features. First, utilitarians are benefit consequentialists who, because they hold either hedonistic or desire-based conceptions of benefit, maintain either hedonistic or desire-based consequentialism. Second, utilitarians hold that the non-moral value of outcomes is determined by summing the benefits and costs to all affected parties. And, third, utilitarians believe that the moral rightness of action or the moral goodness of character traits depends on what would produce the greatest overall value, determined by such a sum. The classical *hedonistic* utilitarian formulation, in Bentham for example, holds that happiness is an experienced state and that people can be benefited only by the intrinsic qualities of their conscious lives; that is, by the degree of pleasure that they experience compared to their pain or suffering (Bentham, 1970). A different kind of utilitarianism, based as much, perhaps, on a free-standing value of autonomy as on a conception of happiness, holds that an individual's welfare is determined by his own desires and preferences. Since people can have preferences for things other than the intrinsic qualities of their own conscious states, this *desire-satisfaction* form of utilitarianism has rather different implications from a hedonist version. For example, someone might strongly desire the survival of a certain wilderness area. A desire satisfaction form of utilitarianism would weigh this fact in favor of saving the area even if saving it made no contribution to the quality of any being's experience (say, because the individual in question did not know anything about the area's survival).

Although consequentialists have usually been utilitarians or benefit consequentialists of some sort, there is nothing in the logic of consequentialism that restricts it to these versions. Philosophers have frequently argued that such things as knowledge, understanding, friendship, love, beauty, and artistic and other cultural activity and creation have intrinsic values that cannot be reduced to the benefits they bring to human (or other sentient) life. After all, some of our deepest satisfactions seem themselves to involve the *appreciation* of these values, so the values cannot wholly consist in these satisfactions. When it comes to controversial issues of case ethics, for example in environmental or medical ethics, it is open to consequentialists to argue that relevant values include such things as the existence of a species or of a relationship of a certain kind between doctors and patients. The consequentialist's test for moral relevance will simply be whether a given state of affairs' existing makes a positive contribution to the value instantiated in the world, whether it is a good or bad thing that it exist or happen.

This structure enables consequentialism to take account, in principle, of a wider range of considerations than contractarianism or contractualism. Most notably, there is nothing in the consequentialist conception of morality that ties it specially to the condition of other members of the moral community, or that restricts its consideration to human beings in any way. If pain or suffering is a bad thing, then, it seems, it would be bad whether the being that suffers is capable of moral agency or not. For this reason, advocates for the interests of animals frequently cite Bentham's dictum: "The question is not, can they reason? . . . but, can they suffer?" (Bentham, 1970, p. 282). And consequentialists who believe that intrinsically valuable states are not restricted to those in which beings benefit can take advantage of a wider array of considerations yet. Of course, moral reasoning is not simply a matter of deploying rhetorical resources. Any moral conception will face the burden of defending the relevance of the considerations it advances and, in the end, anyone who adopts that conception will need to think through how the moral relevance of such considerations can be situated in a philosophically adequate conception of moral obligation.

An important feature of any consequentialist view is what philosophers call the *agent-neutrality* of its fundamental values. Since the values are held to derive simply from the existence of the relevant states, they provide a justification for *any* actions, policies, or practices that might promote them, irrespective of *the agent's* relation to these states. An example will clarify this idea. Suppose you think that among the intrinsically bad things that can happen is someone's being betrayed by a friend. The thought here is not that it is wrong to betray friends, or even that this is a bad thing to do, but that someone's being betrayed by a friend is a bad thing to happen. And again, it is not just that this is bad for the person being betrayed, but that it is a bad thing period for someone to be hurt in this way. If you think this is a bad thing to happen, you should also think that it would be good, other things being equal, for actions, policies, and practices to be taken that would prevent this. Suppose that there are two people, Jones and Smith, who are contemplating betraying their friends. Suppose also that circumstances are such that if you betray your friend, Jones and Smith will be so horrified that they will not betray their friends, although they would have otherwise. We can now put the idea of agent-neutrality this way. From the point of view of the intrinsic badness of friends being betrayed, you would seem to have reason to betray your friend since it would go further towards minimizing the intrinsically disvaluable states of betrayals than would your not betraying your friend. Of course, this apparently runs against moral common sense. That is because we commonly believe that friends have duties to each other that are *agent-relative* rather than *agent-neutral*, that a moral agent has a duty not to betray *his* friends that is not reducible to preventing the (agent-neutral) evil of friends being betrayed.

The idea that there are moral obligations that are agent-relative in this way is a hallmark of *deontological* moral theories. According to deontologists, agency and action are not simply instruments for producing valuable states. Rather, actions are based on reasons and principles, and some important moral principles crucially involve the agent's relation to various persons (or other beings) in the outcomes she affects. It is commonly thought to be a wrong-making feature of an action, for example, that it will involve *one's* (that is, *the agent's*) harming others, betraying a friend, breaking a promise, divulging a confidence, and so on.

Some consequentialist theories, although not all, can agree with these aspects of moral common sense. Rule-consequentialism will agree if, and only if, it produces the greatest overall value (assessed agent-neutrally) for there to exist social practices of moral criticism and psychological patterns of moral reasoning that are themselves guided by agent-relative rules according to which it is a wrong-making feature of an action that it involves one's (the agent's) harming others, betraying friends, and so on. Moreover, it is widely agreed among consequentialists that this is so. Consequentialists generally agree that the most effective way to produce the greatest overall value is *indirectly*. Were everyone to be guided by act-consequentialism in their deliberations and moral criticism, the results would be much worse for many different reasons. Shared rules are necessary to coordinate complex cooperation, establish reliable expectations, diminish self-serving rationalizing and special pleading when the long-run effects of particular actions are unclear, and so on. In the end, however, even rule-consequentialists will agree that the fundamental reason for accepting such agent-relative rules and principles is that this is instrumentally useful in promoting states whose value is agent-neutral.

Deontology

Deontological theories depart from consequentialism on this fundamental point. They hold that what is morally right and wrong is not determined at any level of analysis by what would promote the best outcomes or states, assessed agent-neutrally. They may even be skeptical of the very possibility of pre- or non-moral evaluations of states that are both agent-neutral and morally relevant. Deontology disagrees with act-consequentialism in holding that producing good or bad outcomes is not the only thing that tends to make an act right or wrong. Deontology also disagrees with rule-consequentialism in holding that the reason why this is so is not that believing it to be so itself produces the best outcomes. Deontologists hold that at least some fundamental moral principles or ideas are agent-relative "all the way down."

Contractualism is one example of a deontological theory, since it holds that moral principles are grounded in the fundamental, agent-relative idea of living with others on terms of mutual respect. As this shows, our categories are overlapping in various ways. The situation is roughly as follows. Deontology and consequentialism, as we are defining them, are mutually exclusive and exhaustive of normative theories of moral right and wrong. Generally, contractualism and contractarianism are deontological theories, for reasons explained in this paragraph. Virtue theories are a mixed bag. Some are advanced not as *moral* theories at all, but as supplements or, in some cases, as replacements when put forward as part of a critique of morality (see section on "Virtue Theory"). Some are deontological theories. And some are consequentialist in at least some important respects. Thus Francis Hutcheson (see below) advanced a virtue theory according to which universal benevolence was the highest virtue and argued for act-utilitarianism on the grounds that this theory comports with the deliberative reasoning involved in this highest virtue. But deontological theories and principles are often defended directly, without attempting to ground them in some theory or idea that is held to be somehow more basic. Historically, these versions of deontology have been called *intuitionist* or species of *intuitionism*. What characterizes intuitionism in general is the view that there is an irreducible plurality of different right- or wrong-making features whose moral relevance cannot be derived from some more fundamental principle or reasoning but can only be confirmed by moral reflection or "intuition." This might be done directly as when, for example, it can seem obvious on reflection that the fact that an action would amount to a betrayal or a broken promise must count against it morally. Or it can occur in thinking about or analyzing a specific case as, for instance, when we reflect on the "trolley problem," it can seem evident that causing harm and allowing it to happen are morally different.

Another example, defended by some deontologists, is the "doctrine of double effect," according to which there is a moral difference between causing harm or evil as an unintended side-effect of an intended action or policy and intending the harm or evil directly, either as an end or as a means to an end. Thus, although it is a terrible thing whenever innocent civilians die during wartime, for example, when bombing military targets causes even a small number of casualties, it would seem to be worse to try to kill the same number of civilians directly, even if doing so would produce the same valuable end of victory over a repressive, aggressive regime. One issue of case ethics in which this principle has played an important role is the controversial issue of abortion. Since abortion aims directly at the death of the fetus, it is sometimes argued that it is morally worse than another action would be which caused the fetus's death only as an unintended side-effect. While it might be permissible to perform a medical procedure that is necessary to save a pregnant woman's life even at the risk of killing the fetus, it is argued that aborting a fetus to save the woman's life is morally wrong none the less because it is an impermissible intentional killing.

Deontological intuitionists have defended a wide variety of independent principles or doctrines of right- or wrong-making features of conduct. In addition to the doctrine of double effect and the distinction between "doing and allowing," there have been claimed to be: duties of beneficence or mutual aid, duties of non-maleficence ("do no harm" – along with the idea that these are weightier, other things being equal, than duties of beneficence), duties of gratitude for benevolence shown, duties of restitution for wrongs and injuries done, duties of fidelity relating to promise and contract, duties of personal relationship (including those of friends, parents, children, family members, and caretakers more generally), professional duties, duties owing to desert (what people deserve), duties of reciprocity and fair play, further duties of justice, duties to other animals (to the extent that these have not been included already), duties to ourselves, and various others. Of course, intuitionists do not agree about every doctrine or principle, not even about all we have mentioned. But they are none the less agreed that *some* such list of independent principles or doctrines is correct and that the principles on the list cannot be derived from some more fundamental principle or theory, such as contractualism or, even more so, consequentialism.

Both intuitionist and contractualist deontologists hold that the right is, in Rawls's phrase, "prior to the good" (Rawls, 1971, pp. 30–33). They believe that any attempt to derive the right from agent-neutral outcome value is bound to fail, since the question of what it is right or wrong to do is one that faces agents from a place *within* the world, defined by a complex set of relations to others who make widely varying claims on us owing to these different relations. What states of affairs would be good to exist, considered as from some agent-neutral observer's standpoint, may, if there are such values, be among the considerations that are relevant to what a moral agent should do. But our duties depend on our place *within* the states an observer might contemplate, specifically, on the myriad relations we stand in, and that our actions bring us to stand in, to other agents and patients, to our own past acts, to the histories of those with whom we interact, and so on.

Many of these relations were listed in passing above, but we should have them before us more explicitly. We cannot begin to exhaust them, but it will be helpful to give some idea of their range.

1   *Duties of beneficence and non-maleficence.* Like consequentialists, many deontologists believe that how our actions affect the good of others (other persons, at least, and perhaps any other being who can have a welfare or good) always has some relevance to what we should do morally. But the relevance, again, is not just that these are valuable or disvaluable outcomes we can promote or prevent. It also matters what antecedent relations we have to the affected parties (and

what relations our actions bring us to have). Harming another is worse (an injury done to the other), other things being equal, than forbearing to benefit. It is not just the causing of a disvaluable state; it is doing harm *to* some being. Doing harm is worse, also, than failing to prevent it. And directly intending harm is worse than causing it as an unintended side-effect.

2   *Duties of special care.* Various special relations of caretaking give rise to special obligations of beneficence. Thus parents have obligations to promote the welfare of their children that are much greater than the duties of beneficence we have to others in general. And similarly for trustees and other relations of more specialized concern, such as doctors, teachers, and so on, who are responsible for their patients' or students' medical or educational welfare.

3   *Duties of honesty and fidelity.* Obligations not to lie or intentionally mislead, to keep promises, not to violate contracts, and, more generally, not to encourage expectations we intend not to meet all fit under the general category of keeping faith and not violating trust. Various personal relationships, like those of friends, lovers, and spouses, can be placed under this rubric as well.

4   *Duties deriving from agents' and patients' histories of conduct.* When we wrong and injure others, we incur duties to them to acknowledge fault and offer restitution (*agent-fault*). When others benefit us, we acquire duties of gratitude toward them (*patient-benevolence*). A person's past conduct may call for some appropriate response, especially from those who have special responsibilities to respond appropriately to merit and desert, such as judges of various kinds (*patient desert*).

5   *Duties of reciprocity and fair play.* There is a duty to do one's part in mutually advantageous cooperation, especially when one voluntarily accepts cooperative benefits. Contractarians/contractualists see this duty as fundamental. For intuitionists, it is simply one independently important duty among others.

6   *Further duties of justice.* Various further duties of justice derive from political relations; for example, from that of equal citizenship. Here we have duties to support a just political order that establishes and protects basic rights and achieves distributive justice. Where actual political relations are lacking, as in the international context, justice may require that we do our part to help establish justice more widely through more extensive political forms.

7   *Duties to other species.* Here again, our duties depend on complex relations. In addition to duties of beneficence and non-maleficence, we can acquire special obligations to members of certain species owing to our history of interaction with them. Even if other animals cannot be full partners in cooperative schemes, we can acquire duties to them owing to the ways in which we have involved them in our lives and ourselves in theirs. Pets are an obvious example, but no less significant may be cases where species are themselves shaped and cultivated for human purposes in ways that give them special needs and vulnerabilities.

This list, again, is hardly exhaustive. It should be obvious even at this point, however, that the decisions we face in actual cases will inevitably involve, not just a single principle or right- or wrong-making consideration, but complex combinations of principles or considerations. Since intuitionistic deontology rejects the idea that some overarching principle, idea, or process of reasoning exists in terms of which these principles or considerations might be integrated or prioritized, how do intuitionists believe that the messy business of moral reflection is to proceed in thinking about concrete cases?

W.D. Ross distinguished between the claim that a given duty or right- or wrong-making consideration holds prima facie, that is, other things being equal, and the proposition, that, in some actual circumstance, something or other is our moral duty, all things considered, or, as he put it, *sans phrase* (Ross, 1963). (Since "prima facie" suggests something epistemological, philosophers nowadays are as likely to use the term *"pro tanto"* ["as far as it goes"]. The central idea is that a right- or wrong-making consideration is one that makes an act right or wrong, *other things being equal*, such that were that the only morally relevant feature then the action would be right or wrong, all things considered or "sans phrase.") It was claims of the former sort that Ross held to be self-evident to intuition. To render a moral judgment in any actual case, however, it is necessary to reflect on all of the morally relevant features and, moreover, on how they interact. To take a familiar kind of case, one may have promised to do something of relatively mundane importance only to find oneself placed in a position to give another aid without which he or she may die. Here both promise and need continue to have weight, but one is weightier and so overrides. It would be wrong not to render life-saving assistance, but the moral force of the promise continues, giving rise to a residual obligation to compensate the promisee in some way. But this is not the only way in which moral considerations can interact. Sometimes one consideration can wholly defeat another. When, for example, a benefit one is in a position to provide is tainted by injustice, this may cancel the positive reason to provide the benefit and not just outweigh it.

Ultimately, on an intuitionist picture, there is simply no substitute to carefully considering ethical cases in all of their complexity. Analyzing or "factoring" a case into various right- or wrong-making features or prima facie duties is an important part of the process. But even here, because these can interact in ways that intuitionists believe defy general formulation, one can do no better than to come to grips with these complex interactions in a way that leads to a reflective sense of what moral verdict they will ultimately support.

Virtue Theory

Contractarianism/contractualism, consequentialism, and deontology are all moral theories. Moreover, since case ethics is predominantly concerned with practical questions, we have been considering these theories as accounts of morally right conduct. An approach called *virtue ethics*, frequently associated with Aristotle, is orthogonal to these theories in both respects. First, virtue is concerned primarily with character rather than conduct – with how we should *be* rather than what we should do. And, second, virtue ethics is frequently advanced, not as moral theory, but as accounts of other, ethically deep aspects of human life that are, it is sometimes argued, potential rivals to and perhaps replacements for morality and its distinctive forms. The conception of morality as a set of universal and finally authoritative norms or laws by which all moral agents are categorically *obligated* is far from the only form that ethical reflection can take (Anscombe, 1958; MacIntyre, 1981; Williams, 1985; Slote, 1992). The modern idea of morality derives from a distinctive historical tradition, the

Judeo-Christian-Islamic idea of divinely ordained law, to which it is a secular successor. Some philosophers have argued that this conception of morality is seriously defective in various ways and that our ethical reflections would more profitably take other forms. Several who have made this argument have looked to Aristotle's *Nicomachean Ethics* for a more promising ethical conception. For Aristotle, the fundamental question is not, as for Mill, Hobbes, or Kant, "What is the fundamental principle of moral right or duty and how might this be defended philosophically?" Aristotle asks, rather, "What is the goal of human life?" "What kind of life is best for human beings?"

Aristotle's is a distinctive kind (a paradigm, perhaps) of *non-moral virtue ethics* – "non-moral," again, because, although Aristotle's translators frequently use "moral virtue" to signal that he is talking about excellences of character that are concerned with choice, Aristotle does not relate these virtues to any conception of a moral law under which all are accountable as equals. Virtues, for Aristotle, are dispositions to choose what is fine or noble *(kalon)* for its own sake, and to avoid what is base. The operative notion is what Nietzsche called a "rank-ordering" *ideal* with respect to which one can be better or worse, not a norm or law that one complies with or violates. For Aristotle, the operative ethical emotions are shame, esteem, pride, and disdain or contempt, not guilt, respect, self-respect, and moral indignation.

Virtues are excellences *(areté)*, traits, that is, that make something an excellent instance of its kind. It is a virtue in a knife, for example, that it have a sharp edge so that it can cut well. In general, we reckon which traits are excellences (excellent-making) in relation to a thing's function *(ergon)* or characteristic activity. As Aristotle believes that the characteristic activity distinctive of human beings is action *(praxis)* that expresses a distinctively human form of choice (of actions, valued in themselves as noble or fine *(kalon)*), he concludes that the virtues are traits of character, that is, settled dispositions to choose certain actions and avoid others as intrinsically noble or base. We might put his point by saying that human excellences are states of character concerned with choices that are themselves guided by an ideal of human excellence.

In general, a non-moral virtue ethics is any such (non-moral) human ideal. Although it may be tied, as is Aristotle's, to a teleological or perfectionist view of human nature, according to which there is something that human beings are inherently *for* or *to be*, it need not be so based. A non-moral virtue ethics may be put forward simply as a normative view about which traits in human beings are worthy of esteem (or disdain). Analogously, a *moral virtue ethics* is a theory of what is worthy of distinctively *moral* esteem, that is, traits that are worthy of esteem *in a moral agent*. Exemplars of such a view can be found in Leibniz and the eighteenth-century Scottish philosopher, Francis Hutcheson. Thus Hutcheson argued that the basic moral phenomenon is esteem for benevolence, the desire to benefit others and make them happy. Moral esteem, he held, is not primarily for any outcome, but for a motive or trait of character, namely, the desire to produce good outcomes for human beings and other sentient beings.

There are various ways in which a theory of virtue ethics might bear on questions of case or practical ethics. First, non-moral virtue ethics reminds us that questions of right and wrong are far from the only, or perhaps even the most important, ethical questions we can ask in specific cases. Thus, it might be that failing to provide significantly more aid to relieve world hunger and suffering, although not seriously morally wrong, none the less manifests vices of complacency and self-satisfaction. Or even if, suppose, the environment is not something that can be wronged or unjustly treated,

30   **Stephen L. Darwall**

clear-cutting may still manifest an inappropriate attitude toward the environment or unlovely traits that are at odds with living a fully satisfying human life. Second, conceptions of the virtues can give us an independent purchase on what action it is appropriate to take in specific cases. In considering what to do, it may be helpful to ask what a virtuous person, or someone with a specific virtue (say, generosity), would do in that case. This may simply be a useful heuristic, but it may also reflect the Aristotelian view that there is no way of formulating ethical insight in terms that can be grasped and more generally applied by someone who lacks the wisdom or "sense" of the virtuous person. As Louis Armstrong is reputed to have said about jazz, "If you have to ask, you'll never know."

Third, virtue ethicists may put forward conceptions of virtue, not simply as guides to appropriate (or morally right) action, but as accounts of what *makes* an action appropriate or morally right. Thus, it can be held that an action is the right or appropriate thing to do in some case or circumstance just in case it is what the virtuous person would (characteristically) do in that circumstance (Hursthouse, 1999). Such a view might depart from the letter of Aristotle's position, since he identified virtues as settled dispositions to choose specific actions for their own sake (as noble). This would seem to make which traits are virtuous depend on which actions are noble, not vice versa. Nevertheless, since it would hold that no access to the appropriateness of action is possible save through the wisdom or conduct of a virtuous person, such a view would remain quite close to Aristotle's in fundamental spirit. What is common to any virtue ethics is the idea that guidance on controversial questions of case ethics can be gained only by looking to the virtues or the virtuous person as a model.

Writers on case ethics therefore look to virtue ethics less frequently than they do to other ethical theories, especially when they are concerned with issues of moral right and wrong. If judgments of moral obligation are implicitly directive, holding others accountable for compliance, it will be reasonable to demand that justifications for these judgments be formulated in terms that those subject to the judgments can, in principle, accept. It is this demand that has led to the kinds of normative moral theories that have been advanced by contractarian/contractualists, consequentialists, and deontologists. In each case, the goal has been to articulate action-guiding principles of right conduct that can be grasped and applied without any special virtue other than the judgment of normally competent moral agents.

References

Anscombe, G.E.M. (1958) Modern moral philosophy. *Philosophy* 33: 1–19.
Aristotle (1998) *Nicomachean Ethics*, trans. W.D. Ross. New York: Oxford University Press.
Bentham, J. (1970) *Introduction to the Principles of Morals and Legislation*, ed. J.H. Burns and H.L.A. Hart. London: Athlone.
Dancy, J. (1993) *Moral Reasons*. Oxford: Blackwell.
Darwall, S. (1997) Self-interest and self-concern. *Social Philosophy and Policy* 14: 158–178. Also in Ellen F. Paul, ed. (1997) *Self-interest*. Cambridge: Cambridge University Press.
Darwall, S. (1998) *Philosophical Ethics*. Boulder, CO: Westview Press.
Darwall, S. (2002) *Welfare and Rational Care*. Princeton, NJ: Princeton University Press.
Gauthier, D. (1986) *Morals by Agreement*. Oxford: Clarendon Press.
Harman, G. (1975) Moral relativism defended. *The Philosophical Review* 84: 3–22.

Hobbes, T. (1994) *Leviathan*, ed. E.M. Curley. Indianapolis, IN: Hackett.

Hursthouse, R. (1999) *On Virtue Ethics*. Oxford: Oxford University Press.

Kant, I. (1998) *Groundwork of the Metaphysics of Morals*, ed. and trans. M. Gregor. New York: Cambridge University Press.

MacIntyre, A. (1981) *After Virtue*. Notre Dame, IN: University of Notre Dame.

McDowell, J. (1979) Virtue and reason. *The Monist* 62: 331–350.

Mill, J.S. (1957) *Utilitarianism*, ed. O. Piest. Indianapolis, IN: Hobbs-Merrill.

Railton, P. (1986) Moral realism. *The Philosophical Review* 95: 163–207.

Rawls, J. (1971) *A Theory of Justice*. Cambridge, MA: Belknap Press of Harvard University.

Rawls, J. (1993) *Political Liberalism*. New York: Columbia University Press.

Ross, W.D. (1963) *The Right and the Good*. Oxford: Clarendon Press.

Rousseau, J.-J. (1987) The social contract. In *The Basic Political Writings*, trans. D.A. Cress, Indianapolis, IN: Hackett.

Scanlon, T.M. (1998) *What We Owe to Each Other*. Cambridge, MA: Belknap Press of Harvard University.

Slote, M. (1992) *From Morality to Virtue*. New York: Oxford University Press.

Thomson, J. (1976) Killing, letting die, and the trolley problem. *The Monist* 59: 204–217.

Williams, B. (1985) *Ethics and the Limits of Philosophy*. Cambridge, MA: Harvard University Press.

Further Reading

Darwall, S., ed. (2002) *Contractarianism/Contractualism*. Oxford: Blackwell.

Darwall, S. (2002) *Deontology*. Oxford: Blackwell.

Darwall, S., ed. (2002) *Virtue Ethics*. Oxford: Blackwell.

ISSUES IN LIFE AND DEATH

Abortion

CHAPTER TWO

The Wrong of Abortion

Patrick Lee and Robert P. George

Much of the public debate about abortion concerns the question whether deliberate feticide ought to be unlawful, at least in most circumstances. We will lay that question aside here in order to focus first on the question: is the choice to have, to perform, or to help procure an abortion morally wrong?

We shall argue that the choice of abortion is objectively immoral. By "objectively" we indicate that we are discussing the choice itself, not the (subjective) guilt or innocence of someone who carries out the choice: someone may act from an erroneous conscience, and if he is not at fault for his error, then he remains subjectively innocent, even if his choice is objectively wrongful.

The first important question to consider is: what is killed in an abortion? It is obvious that some living entity is killed in an abortion. And no one doubts that the moral status of the entity killed is a central (though not the only) question in the abortion debate. We shall approach the issue step by step, first setting forth some (though not all) of the evidence that demonstrates that what is killed in abortion – a human embryo – is indeed a human being, then examining the ethical significance of that point.

Human Embryos and Fetuses are Complete (though Immature) Human Beings

It will be useful to begin by considering some of the facts of sexual reproduction. The standard embryology texts indicate that in the case of ordinary sexual reproduction the life of an individual human being begins with complete fertilization, which yields a genetically and functionally distinct organism, possessing the resources and active disposition for internally directed development toward human maturity.[1] In normal conception, a sex cell of the father, a sperm, unites with a sex cell of the mother, an

Contemporary Debates in Applied Ethics, Second Edition. Edited by Andrew I. Cohen and Christopher Heath Wellman.
© 2014 John Wiley & Sons, Inc. Published 2014 by John Wiley & Sons, Inc.

ovum. Within the chromosomes of these sex cells are the DNA molecules which constitute the information that guides the development of the new individual brought into being when the sperm and ovum fuse. When fertilization occurs, the 23 chromosomes of the sperm unite with the 23 chromosomes of the ovum. At the end of this process there is produced an entirely new and distinct organism, originally a single cell. This organism, the human embryo, begins to grow by the normal process of cell division – it divides into 2 cells, then 4, 8, 16, and so on (the divisions are not simultaneous, so there is a 3-cell stage, and so on). This embryo gradually develops all of the organs and organ systems necessary for the full functioning of a mature human being. His or her development (sex is determined from the beginning) is very rapid in the first few weeks. For example, as early as eight or ten weeks of gestation, the fetus has a fully formed, beating heart, a complete brain (although not all of its synaptic connections are complete – nor will they be until sometime *after* the child is born), a recognizably human form, and the fetus feels pain, cries, and even sucks his or her thumb.

There are three important points we wish to make about this human embryo. First, it is from the start *distinct* from any cell of the mother or of the father. This is clear because it is growing in its own distinct direction. Its growth is internally directed to its own survival and maturation. Second, the embryo is *human:* it has the genetic makeup characteristic of human beings. Third, and most importantly, the embryo is a *complete* or *whole* organism, though immature. The human embryo, from conception onward, is fully programmed actively to develop himself or herself to the mature stage of a human being, and, *unless prevented by disease or violence, will actually do so, despite possibly significant variation in environment* (in the mother's womb). None of the changes that occur to the embryo after fertilization, for as long as he or she survives, generates a new direction of growth. Rather, *all* of the changes (for example, those involving nutrition and environment) either facilitate or retard the internally directed growth of this persisting individual.

Sometimes it is objected that if we say human embryos are human beings, on the grounds that they have the potential to become mature humans, the same will have to be said of sperm and ova. This objection is untenable. The human embryo is radically unlike the sperm and ova, the sex cells. The sex cells are manifestly not *whole* or *complete* organisms. They are not only genetically but also functionally identifiable as parts of the male or female potential parents. They clearly are destined either to combine with an ovum or sperm or die. Even when they succeed in causing fertilization, they do not survive; rather, their genetic material enters into the composition of a distinct, new organism.

Nor are human embryos comparable to somatic cells (such as skin cells or muscle cells), though some have tried to argue that they are. Like sex cells, a somatic cell is functionally only a part of a larger organism. The human embryo, by contrast, possesses from the beginning the internal resources and active disposition to develop himself or herself to full maturity; all he or she needs is a suitable environment and nutrition. The direction of his or her growth *is not extrinsically determined*, but the embryo is internally directing his or her growth toward full maturity.

So, a human embryo (or fetus) is not something distinct from a human being; he or she is not an individual of any nonhuman or intermediate species. Rather, an embryo (and fetus) is a human being at a certain (early) stage of development – the embryonic (or fetal) stage. In abortion, what is killed is a human being, a whole living member of the species *homo sapiens*, the same *kind* of entity as you or I, only at an earlier stage of development.

No-Person Arguments: The Dualist Version

Defenders of abortion may adopt different strategies to respond to these points. Most will grant that human embryos or fetuses are human beings. However, they then distinguish "human being" from "person" and claim that embryonic human beings are not (yet) *persons*. They hold that while it is wrong to kill persons, it is not always wrong to kill human beings who are not persons.

Sometimes it is argued that human beings in the embryonic stage are not persons because embryonic human beings do not exercise higher mental capacities or functions. Certain defenders of abortion (and infanticide) have argued that in order to be a person, an entity must be self-aware (Tooley, 1983; Warren, 1984; Singer, 1993). They then claim that, because human embryos and fetuses (and infants) have not yet developed self-awareness, they are not persons.

These defenders of abortion raise the question: Where does one draw the line between those who are subjects of rights and those that are not? A long tradition says that the line should be drawn at *persons*. But what is a person, if not an entity that has self-awareness, rationality, etc.?

This argument is based on a false premise. It implicitly identifies the human person with a consciousness which inhabits (or is somehow associated with) and uses a body; the truth, however, is that we human persons are particular kinds of physical organisms. The argument here under review grants that the human organism comes to be at conception, but claims nevertheless that you or I, the human person, comes to be only much later, say, when self-awareness develops. But if this human organism came to be at one time, but *I* came to be at a later time, it follows that I am one thing and this human organism with which *I* am associated is another thing.

But this is false. We are not consciousnesses that *possess or inhabit* bodies. Rather, we are living bodily entities. We can see this by examining the kinds of action that we perform. If a living thing performs bodily actions, then it is a physical organism. Now, those who wish to deny that we are physical organisms think of *themselves*, what each of them refers to as "*I*," as the subject of self-conscious acts of conceptual thought and willing (what many philosophers, ourselves included, would say are non-physical acts). But one can show that this "I" is identical to the subject of physical, bodily actions, and so is a living, bodily being (an organism). Sensation is a bodily action. The act of seeing, for example, is an act that an animal performs with his eyeballs and his optic nerve, just as the act of walking is an act that he performs with his legs. But it is clear in the case of human individuals that it must be the same entity, the same single subject of actions, that performs the act of sensing and that performs the act of understanding. When I know, for example, that "That is a tree," it is by my understanding, or a self-conscious intellectual act, that I apprehend what is meant by "tree," apprehending what it is (at least in a general way). But the subject of that proposition, what I refer to by the word "That," is apprehended by sensation or perception. Clearly, it must be the same thing – the same I – which apprehends the predicate and the subject of a unitary judgment.

So, it is the same substantial entity, the same agent, which understands and which senses or perceives. And so what all agree is referred to by the word "I" (namely, the subject of conscious, intellectual acts) is identical with the physical organism which is the subject of bodily actions such as sensing or perceiving. Hence the entity that I am, and the entity that you are – what you and I refer to by the personal pronouns "you" and "I" – is in each case a human, physical organism (but also with non-physical capacities). Therefore, since you and I are *essentially* physical organisms, *we* came to be when these physical organisms came to be. But, as shown above, the human organism comes to be at conception.[2] Thus you and I came to be at conception; we once were embryos, then fetuses, then infants, just as we were once toddlers, pre-adolescent children, adolescents, and young adults.

So, how should we use the word "person"? Are human embryos persons or not? People may stipulate different meanings for the word "person," but we think it is clear that what we normally mean by the word "person" is that substantial entity that is referred to by personal pronouns – "I," "you," "she," and so on. It follows, we submit, that a person is a distinct subject with the natural capacity to reason and make free choices. That subject, in the case of human beings, is identical with the human organism, and therefore that subject comes to be when the human organism comes to be, even though it will take him or her months and even years to actualize the natural capacities to reason and make free choices, natural capacities which are already present (albeit in radical, i.e. root, form) from the beginning. So it makes no sense to say that the human organism came to be at one point but the person – you or I – came to be at some later point, To have destroyed the human organism that you are or I am even at an early stage of our lives would have been to have killed you or me.

No-Person Arguments: The Evaluative Version

Let us now consider a different argument by which some defenders of abortion seek to deny that human beings in the embryonic and fetal stages are "persons" and, as such, ought not to be killed. Unlike the argument criticized in the previous section, this argument grants that the being who is you or I came to be at conception, but contends that you and I became valuable and bearers of rights only much later, when, for example, we developed the proximate, or immediately exercisable, capacity for self-consciousness. Inasmuch as those who advance this argument concede that you and I once were human embryos, they do not identify the self or the person with a non-physical phenomenon, such as consciousness. They claim, however, that being a person is an accidental attribute. It is an accidental attribute in the way that someone's being a musician or basketball player is an accidental attribute. Just as you come to be at one time, but become a musician or basketball player only much later, so, they say, you and I came to be when the physical organisms we are came to be, but we became persons (beings with a certain type of special value and bearers of basic rights) only at some time later (Dworkin, 1993; Thomson, 1995). Those defenders of abortion whose view we discussed in the previous section disagree with the pro-life position on an ontological issue, that is, on what *kind of entity* the human embryo or fetus is. Those who advance the argument now under review, by contrast, disagree with the pro-life position on an evaluative question.

Judith Thomson argued for this position by comparing the right to life with the right to vote: "If children are allowed to develop normally they will have a right to vote; that does not show that they now have a right to vote" (1995). According to this position, it is true that we once were embryos and fetuses, but in the embryonic and fetal stages of our lives we were not yet valuable in the special way that would qualify us as having a right to life. We acquired that special kind of value and the right to life that comes with it at some point after we came into existence.

We can begin to see the error in this view by considering Thomson's comparison of the right to life with the right to vote. Thomson fails to advert to the fact that some rights vary with respect to place, circumstances, maturity, ability, and other factors, while other rights do not. We recognize that one's right to life does not vary with place, as does one's right to vote. One may have the right to vote in Switzerland, but not in Mexico. Moreover, some rights and entitlements accrue to individuals only at certain times, or in certain places or situations, and others do not. But to have the right to life is to have *moral status at all*; to have the right to life, in other words, is to be the sort of entity that can have rights or entitlements to begin with. And so it is to be expected that *this* right would differ in some fundamental ways from other rights, such as a right to vote.

In particular, it is reasonable to suppose (and we give reasons for this in the next few paragraphs) that having moral status at all, as opposed to having a right to perform a specific action in a specific situation, follows from an entity's being the *type of thing* (or substantial entity) it is. And so, just as one's right to life does not come and go with one's location or situation, so it does not accrue to someone in virtue of an acquired (i.e., accidental) property, capacity, skill, or disposition. Rather, this right belongs to a human being at all times that he or she exists, not just during certain stages of his or her existence, or in certain circumstances, or in virtue of additional, accidental attributes.

Our position is that we human beings have the special kind of value that makes us subjects of rights in virtue of *what* we are, not in virtue of some attribute that we acquire some time after we have come to be. Obviously, defenders of abortion cannot maintain that the accidental attribute required to have the special kind of value we ascribe to "persons" (additional to being a human individual) is an *actual* behavior. They, of course, do not wish to exclude from personhood people who are asleep or in reversible comas. So, the additional attribute will have to be a capacity or potentiality of some sort.[3] Thus, they will have to concede that sleeping or reversibly comatose human beings will be persons because they have the potentiality or capacity for higher mental functions.

But human embryos and fetuses also possess, albeit in radical form, a capacity or potentiality for such mental functions; human beings possess this radical capacity in virtue of the kind of entity they are, and possess it by coming into being as that kind of entity (viz., a being with a rational nature). Human embryos and fetuses cannot of course *immediately* exercise these capacities. Still, they are related to these capacities differently from, say, how a canine or feline embryo is. They are the kind of being – a natural kind, members of a biological species – which, if not prevented by extrinsic causes, in due course develops by active self-development to the point at which capacities initially possessed in root form become immediately exercisable. (Of course, the capacities in question become immediately exercisable only some months or years after the child's birth.) Each human being comes into existence possessing the internal resources and active disposition to develop the immediately exercisable capacity for higher mental functions. Only the adverse effects on them of other causes will prevent this development.

So, we must distinguish two sorts of capacity or potentiality for higher mental functions that a substantial entity might possess: first, an immediately (or nearly immediately) exercisable capacity to engage in higher mental functions; second, a basic, natural capacity to develop oneself to the point where one does perform such actions. But on what basis can one require the first sort of potentiality – as do proponents of the position under review in this section – which is an accidental attribute, and not just the second? There are three decisive reasons against supposing that the first sort of potentiality is required to qualify an entity as a bearer of the right to life.

First, the developing human being does not reach a level of maturity at which he or she performs a type of mental act that other animals do not perform – even animals such as dogs and cats – until at least several months after birth. A six-week-old baby lacks the immediately (or nearly immediately) exercisable capacity to perform characteristically human mental functions. So, if full moral respect were due only to those who possess a nearly immediately exercisable capacity for characteristically human mental functions, it would follow that six-week-old infants do not deserve full moral respect. If abortion were morally acceptable on the grounds that the human embryo or fetus lacks such a capacity for characteristically human mental functions, then one would be logically committed to the view that, subject to parental approval, human infants could be disposed of as well.

Second, the difference between these two types of capacity is merely a difference between stages along a continuum. The proximate or nearly immediately exercisable capacity for mental functions is only the development of an underlying potentiality that the human being possesses simply by virtue of the kind of entity it is. The capacities for reasoning, deliberating, and making choices are gradually developed, or brought towards maturation, through gestation, childhood, adolescence, and so on. But the difference between a being that deserves full moral respect and a being that does not (and can therefore legitimately be disposed of as a means of benefiting others) cannot consist only in the fact that, while both have some feature, one has more of it than the other. A mere *quantitative* difference (having more or less of the same feature, such as *the development* of a basic natural capacity) cannot by itself be a justificatory basis for treating different entities in *radically* different ways. Between the ovum and the approaching thousands of sperm, on the one hand, and the embryonic human being, on the other hand, there *is* a clear difference in kind. But between the embryonic human being and that same human being at any later stage of its maturation, there is only a difference in degree.

Note that there *is* a fundamental difference (as shown) between the gametes (the sperm and the ovum), on the one hand, and the human embryo and fetus, on the other. When a human being comes to be, a substantial entity that is identical with the entity that will later reason, make free choices, and so on, begins to exist. So, those who propose an accidental characteristic as qualifying an entity as a bearer of the right to life (or as a "person" or being with "moral worth") are *ignoring* a radical difference among groups of beings, and instead fastening on to a mere quantitative difference as the basis for treating different groups in radically different ways. In other words, there are beings a, b, c, d, e, and so on. And between a and b groups on the one hand and c, d and e groups on the other hand, there is a fundamental difference, a difference in kind not just in degree. But proponents of the position that being a person is an accidental characteristic ignore that difference and pick out a mere difference in degree between, say, d and e, and make that the basis for radically different types of treatment. That violates the most basic canons of justice.

Third, being a whole human being (whether immature or not) is an either/or matter – a thing either is or is not a whole human being. But the acquired qualities that could be proposed as criteria for personhood come in varying and continuous degrees: there is an infinite number of degrees of the *development of* the basic natural capacities for self-consciousness, intelligence, or rationality. So, if human beings were worthy of full moral respect (as subjects of rights) only because of such qualities, and not in virtue of the kind of being they are, then, since such qualities come in varying degrees, no account could be given of why basic rights are not possessed by human beings in varying degrees. The proposition that all human beings are created equal would be relegated to the status of a superstition. For example, if developed self-consciousness bestowed rights, then, since some people are more self-conscious than others (that is, have developed that capacity to a greater extent than others), some people would be greater in dignity than others, and the rights of the superiors would trump those of the inferiors where the interests of the superiors could be advanced at the cost of the inferiors. This conclusion would follow no matter which of the acquired qualities generally proposed as qualifying some human beings (or human beings at some stages) for full respect were selected. Clearly, developed self-consciousness, or desires, or so on, are arbitrarily selected degrees of development of capacities that all human beings possess in (at least) radical form from the coming into existence of the human being until his or her death. So, it cannot be the case that some human beings and not others possess the special kind of value that qualifies an entity as having a basic right to life, by virtue of a certain degree of development. Rather, human beings possess that kind of value, and therefore that right, in virtue of what (i.e., the kind of being) they are; and *all* human beings – not just some, and certainly not just those who have advanced sufficiently along the developmental path as to be able immediately (or almost immediately) to exercise their capacities for characteristically human mental functions – possess that kind of value and that right.[4]

Since human beings are valuable in the way that qualifies them as having a right to life in virtue of what they are, it follows that they have that right, whatever it entails, from the point at which they come into being – and that point (as shown in our first section) is at conception.

In sum, human beings are valuable (as subjects of rights) in virtue of what they are. But what they are are human physical organisms. Human physical organisms come to be at conception. Therefore, what is intrinsically valuable (as a subject of rights) comes to be at conception.

The Argument that Abortion is Justified as Non-intentional Killing

Some "pro-choice" philosophers have attempted to justify abortion by denying that all abortions are intentional killing. They have granted (at least for the sake of argument) that an unborn human being has a right to life but have then argued that this right does not entail that the child *in utero* is morally entitled to the use of the mother's body for life support. In effect, their argument is that, at least in many cases, abortion is not a case of intentionally killing the child, but a choice not to provide the child with assistance, that is, a choice to expel (or "evict") the child from the womb, despite the likelihood or certainty that expulsion (or "eviction") will result in his or her death (Thomson, 1971; McDonagh, 1996; Little, 1999).

Various analogies have been proposed by people making this argument. The mother's gestating a child has been compared to allowing someone the use of one's kidneys or even to donating an organ. We are not *required* (morally or as a matter of law) to allow someone to use our kidneys, or to donate organs to others, even when they would die without this assistance (and we could survive in good health despite rendering it). Analogously, the argument continues, a woman is not morally required to allow the fetus the use of her body. We shall call this "the bodily rights argument."

It may be objected that a woman has a special responsibility to the child she is carrying, whereas in the cases of withholding assistance to which abortion is compared there is no such special responsibility. Proponents of the bodily rights argument have replied, however, that the mother has not voluntarily assumed responsibility for the child, or a personal relationship with the child, and we have strong responsibilities to others only if we have voluntarily assumed such responsibilities (Thomson, 1971) or have consented to a personal relationship which generates such responsibilities (Little, 1999). True, the mother may have voluntarily performed an act which she knew may result in a child's conception, but that is distinct from consenting to gestate the child if a child is conceived. And so (according to this position) it is not until the woman consents to pregnancy, or perhaps not until the parents consent to care for the child by taking the baby home from the hospital or birthing center, that the full duties of parenthood accrue to the mother (and perhaps the father).

In reply to this argument we wish to make several points. We grant that in some few cases abortion is not intentional killing, but a choice to expel the child, the child's death being an unintended, albeit foreseen and (rightly or wrongly) accepted, side effect. However, these constitute a small minority of abortions. In the vast majority of cases, the death of the child *in utero* is precisely the object of the abortion. In most cases the end sought is to avoid being a parent; but abortion brings that about only by bringing it about that the child dies. Indeed, the attempted abortion would be considered by the woman requesting it and the abortionist performing it to have been *unsuccessful* if the child survives. In most cases abortion *is* intentional killing. Thus, even if the bodily rights argument succeeded, it would justify only a small percentage of abortions.

Still, in some few cases abortion is chosen as a means precisely toward ending the condition of pregnancy, and the woman requesting the termination of her pregnancy would not object if somehow the child survived. A pregnant woman may have less or more serious reasons for seeking the termination of this condition, but if that is her objective, then the child's death resulting from his or her expulsion will be a side effect, rather than the means chosen. For example, an actress may wish not to be pregnant because the pregnancy will change her figure during a time in which she is filming scenes in which having a slender appearance is important; or a woman may dread the discomforts, pains, and difficulties involved in pregnancy. (Of course, in many abortions there may be mixed motives: the parties making the choice may intend both ending the condition of pregnancy and the death of the child.)

Nevertheless, while it is true that in some cases abortion is not intentional killing, it remains misleading to describe it simply as choosing not to provide bodily life support. Rather, it is actively expelling the human embryo or fetus from the womb. There is a significant moral difference between *not doing* something that would assist someone, and *doing* something that causes someone harm, even if that harm is an unintended (but foreseen) side effect. It is more difficult morally to justify the latter than it is the former. Abortion is the *act* of extracting the unborn human being from the womb – an extraction that usually rips him or her to pieces or does him or her violence in some other way.

It is true that in some cases causing death as a side effect is morally permissible. For example, in some cases it is morally right to use force to stop a potentially lethal attack on one's family or country, even if one foresees that the force used will also result in the assailant's death. Similarly, there are instances in which it is permissible to perform an act that one knows or believes will, as a side effect, cause the death of a child *in utero*. For example, if a pregnant woman is discovered to have a cancerous uterus, and this is a proximate danger to the mother's life, it can be morally right to remove the cancerous uterus with the baby in it, even if the child will die as a result. A similar situation can occur in ectopic pregnancies. But in such cases, not only is the child's death a side effect, but the mother's life is in proximate danger. It is worth noting also that in these cases *what is done* (the means) is the correction of a pathology (such as a cancerous uterus, or a ruptured uterine tube). Thus, in such cases, not only the child's death, but also the ending of the pregnancy, are side effects. So, such acts are what traditional casuistry referred to as *indirect* or *non-intentional*, abortions.

But it is also clear that not every case of causing death as a side effect is morally right. For example, if a man's daughter has a serious respiratory disease and the father is told that his continued smoking in her presence will cause her death, it would obviously be immoral for him to continue the smoking. Similarly, if a man works for a steel company in a city with significant levels of air pollution, and his child has a serious respiratory problem making the air pollution a danger to her life, certainly he should move to another city. He should move, we would say, even if that meant he had to resign a prestigious position or make a significant career change.

In both examples (a) the parent has a special responsibility to his child, but (b) the act that would cause the child's death would avoid a harm to the parent but cause a significantly worse harm to his child. And so, although the harm done would be a side effect, in both cases the act that caused the death would be an *unjust* act, and morally wrongful *as such*. The special responsibility of parents to their children requires that they *at least* refrain from performing acts that cause terrible harms to their children in order to avoid significantly lesser harms to themselves.

But (a) and (b) also obtain in intentional abortions (that is, those in which the removal of the child is directly sought, rather than the correction of a life-threatening pathology) even though they are not, strictly speaking, intentional killing. First, the mother has a special responsibility to her child, in virtue of being her biological mother (as does the father in virtue of his paternal relationship). The parental relationship itself – not just the voluntary acceptance of that relationship – gives rise to a special responsibility to a child.

Proponents of the bodily rights argument deny this point. Many claim that one has full parental responsibilities only if one has voluntarily assumed them. And so the child, on this view, has a right to care from his or her mother (including gestation) only if the mother has accepted her pregnancy, or perhaps only if the mother (and/or the father?) has in some way voluntarily begun a deep personal relationship with the child (Little, 1999).

But suppose a mother takes her baby home after giving birth, but the only reason she did not get an abortion was that she could not afford one. Or suppose she lives in a society where abortion is not available (perhaps very few physicians are willing to do the grisly deed). She and her husband take the child home only because they had no alternative. Moreover, suppose that in their society people are not waiting in line to adopt a newborn baby. And so the baby is several days old before anything can be done. If they abandon the baby and the baby is found, she will simply be returned to them. In such a case the parents have not voluntarily assumed responsibility; nor have they consented to a personal relationship with the child. But it would surely be wrong for these parents to abandon their baby in the woods (perhaps the only feasible way of ensuring she is not returned), even though the baby's death would be only a side effect. Clearly, we recognize that parents do have a responsibility to make sacrifices for their children, even if they have not voluntary assumed such responsibilities, or given their consent to the personal relationship with the child.

The bodily rights argument implicitly supposes that we have a primordial right to construct a life simply as we please, and that others have claims on us only very minimally or through our (at least tacit) consent to a certain sort of relationship with them. On the contrary, we are by nature members of communities. Our moral goodness or character consists to a large extent (though not solely) in contributing to the communities of which we are members. We ought to act for our genuine good or flourishing (we take that as a basic ethical principle), but our flourishing involves being in communion with others. And communion with others of itself – even if we find ourselves united with others because of a physical or social relationship which precedes our consent – entails duties or responsibilities. Moreover, the contribution we are morally required to make to others will likely bring each of us some discomfort and pain. This is not to say that we should simply ignore our own good, for the sake of others. Rather, since what (and who) I am is in part constituted by various relationships with others, not all of which are initiated by my will, my genuine good includes the contributions I make to the relationships in which I participate. Thus, the life we constitute by our free choices should be in large part a life of mutual reciprocity with others.

For example, I may wish to cultivate my talent to write and so I may want to spend hours each day reading and writing. Or I may wish to develop my athletic abilities and so I may want to spend hours every day on the baseball field. But if I am a father of minor children, and have an adequate paying job working (say) in a coal mine, then my clear duty is to keep that job. Similarly, if one's girlfriend finds she is pregnant and one is the father, then one might also be morally required to continue one's work in the mine (or mill, factory, warehouse, etc.).

In other words, I have a duty to do something with my life that contributes to the good of the human community, but that general duty becomes specified by my particular situation. It becomes specified by the connection or closeness to me of those who are in need. We acquire special responsibilities toward people, not only by *consenting* to contracts or relationships with them, but also by having various types of union with them. So, we have special responsibilities to those people with whom we are closely united. For example, we have special responsibilities to our parents, and brothers and sisters, even though we did not choose them.

The physical unity or continuity of children to their parents is unique. The child is brought into being out of the bodily unity and bodies of the mother and the father. The mother and the father are in a certain sense prolonged or continued in their offspring. So, there is a natural unity of the mother with her child, and a natural unity of the father with his child. Since we have special responsibilities to those with whom we are closely united, it follows that we in fact do have a special responsibility to our children anterior to our having voluntarily assumed such responsibility or consented to the relationship.[5]

The second point is this: in the types of case we are considering, the harm caused (death) is much worse than the harms avoided (the difficulties in pregnancy). Pregnancy can involve severe impositions, but it is not nearly as bad as death – which is total and irreversible. One need not make light of the burdens of pregnancy to acknowledge that the harm that is death is in a different category altogether.

The burdens of pregnancy include physical difficulties and the pain of labor, and can include significant financial costs, psychological burdens, and interference with autonomy and the pursuit of other important goals (McDonagh, 1996, ch. 5). These costs are not inconsiderable. Partly for that reason, we owe our mothers gratitude for carrying and giving birth to us. However, where pregnancy does not place a woman's life in jeopardy or threaten grave and lasting damage to her physical health, the harm done to other goods is not total. Moreover, most of the harms involved in pregnancy are not irreversible: pregnancy is a nine-month task – if the woman and man are not in a good position to raise the child, adoption is a possibility. So the difficulties of pregnancy, considered together, are in a different and lesser category than death. Death is not just worse in degree than the difficulties involved in pregnancy; it is worse in kind.

It has been argued, however, that pregnancy can involve a unique type of burden. It has been argued that the *intimacy* involved in pregnancy is such that if the woman must remain pregnant without her consent then there is inflicted on her a unique and serious harm. Just as sex with consent can be a desired experience but sex without consent is a violation of bodily integrity, so (the argument continues) pregnancy involves such a close physical intertwinement with the fetus that not to allow abortion is analogous to rape – it involves an enforced intimacy (Little, 1999, pp. 300–303; Boonin, 2003, p. 84).

However, this argument is based on a false analogy. Where the pregnancy is unwanted, the baby's "occupying" the mother's womb may involve a harm; but the child is committing no injustice against her. The baby is not forcing himself or herself on the woman, but is simply growing and developing in a way quite natural to him or her. The baby is not performing any action that could in any way be construed as aimed at violating the mother.[6]

It is true that the fulfillment of the duty of a mother to her child (during gestation) is unique and in many cases does involve a great sacrifice. The argument we have presented, however, is that being a mother *does* generate a special responsibility, and that the sacrifice morally required of the mother is less burdensome than the harm that would be done to the child by expelling the child, causing his or her death, to escape that responsibility. Our argument equally entails responsibilities for the father of the child. His duty does not involve as direct a bodily relationship with the child as the mother's, but it may be equally or even more burdensome. In certain circumstances, his obligation to care for the child (and the child's mother), and especially his obligation to provide financial support, may severely limit his freedom and even require months or, indeed, years, of extremely burdensome physical labor. Historically, many men have rightly seen that their basic responsibility to their family (and country) has entailed risking, and in many cases, losing, their lives. Different people in different circumstances, with different talents, will have different responsibilities. It is no argument against any of these responsibilities to point out their distinctness.

So, the burden of carrying the baby, for all its distinctness, is significantly less than the harm the baby would suffer by being killed; the mother and father have a special responsibility to the child; it follows that intentional abortion (even in the few cases where the baby's death is an unintended but foreseen side effect) is unjust and therefore objectively immoral.

Notes

1   See, for example: Carlson (1994, chs 2–4); Gilbert (2003, pp. 183–220, 363–390); Larson (2001, chs 1–2); Moore and Persaud (2003, chs 1–6); Muller (1997, chs 1–2); O'Rahilly and Mueller (2000, chs 3–4).

2   For a discussion of the issues raised by twinning and cloning, see George and Gomez-Lobo (2002).

3   Some defenders of abortion have seen the damaging implications of this point for their position (Stretton, 2004), and have struggled to find a way around it. There are two leading proposals. The first is to suggest a mean between a capacity and an actual behavior, such as a disposition. But a disposition is just the development or specification of a capacity and so raises the unanswerable question of why just that much development, and not more or less, should be required. The second proposal is to assert that the historical fact of someone having exercised a capacity (say, for conceptual thought) confers on her a right to life even if she does not now have the immediately exercisable capacity. But suppose we have baby Susan who has developed a brain and gained sufficient experience to the point that just now she has the immediately exercisable capacity for conceptual thought, but she has not yet exercised it. Why should she be in a wholly different category than say, baby Mary, who is just like Susan except she did actually have a conceptual thought? Neither proposal can bear the moral weight assigned to it. Both offer criteria that are wholly arbitrary.

4   In arguing against an article by Lee, Dean Stretton claims that the basic natural capacity of rationality also comes in degrees, and that therefore the argument we are presenting against the position that moral worth is based on having some accidental characteristic would apply to our position also (Stretton, 2004). But this is to miss the important distinction between having a basic natural capacity (of which there are no degrees, since one either has it or one does not), and the *development of that capacity* (of which there are infinite degrees).

5   David Boonin claims, in reply to this argument – in an earlier and less developed form, presented by Lee (1996, p. 122) – that it is not clear that it is impermissible for a woman to destroy what is a part of, or a continuation of, herself. He then says that to the extent the unborn human being is united to her in that way, "it would if anything seem that her act is *easier* to justify than if this claim were not true" (Boonin, 2003, p. 230). But Boonin fails to grasp the point of the argument (perhaps understandably since it was not expressed very

clearly in the earlier work he is discussing). The unity of the child to the mother is the basis for this child being related to the woman in a different way from how other children are. We ought to pursue our own good *and the good of others with whom we are united in various ways*. If that is so, then the closer someone is united to us, the deeper and more extensive our responsibility to the person will be.

6   In some sense being bodily "occupied" when one does not wish to be *is* a harm; however, just as the child does not (as explained in the text), neither does the state inflict this harm on the woman, in circumstances in which the state prohibits abortion. By prohibiting abortion the state would only prevent the woman from performing an act (forcibly detaching the child from her) that would unjustly kill this developing child, who is an innocent party.

References

Boonin, D. (2003) *A Defense of Abortion*. New York: Cambridge University Press.

Carlson, B. (1994) *Human Embryology and Developmental Biology*. St Louis, MO: Mosby.

Dworkin, R. (1993) *Life's Dominion: An Argument about Abortion, Euthanasia, and Individual Freedom*. New York: Random House.

George, R. and Gomez-Lobo, G. (2002) Statement of Professor George (joined by Dr Gomez-Lobo). In *Human Cloning and Human Dignity: An Ethical Inquiry. Report by the President's Council on Bioethics*. New York: Public Affairs, pp. 258–266.

Gilbert, S. (2003) *Developmental Biology*, 7th edn. Sunderland, MA: Sinnauer Associates.

Larson, W.J. (2001) *Human Embryology*, 3rd edn. New York: Churchill Livingstone.

Lee, P. (1996) *Abortion and Unborn Human Life*. Washington, DC: Catholic University of America Press.

Little, M.O. (1999) Abortion, intimacy, and the duty to gestate. *Ethical Theory and Moral Practice* 2: 295–312.

McDonagh, E. (1996) *Breaking the Abortion Deadlock: From Choice to Consent*. New York: Oxford University Press.

Moore, K. and Persaud, T.V.N. (2003) *The Developing Human, Clinically Oriented Embryology*, 7th edn. New York: W.B. Saunders.

Muller, W.A. (1997) *Developmental Biology*. New York: Springer Verlag.

O'Rahilly, R. and Mueller, F. (2000) *Human Embryology and Teratology*, 3rd edn. New York: John Wiley & Sons.

Singer, P. (1993) *Practical Ethics*, 2nd edn. Cambridge: Cambridge University Press.

Stretton, D. (2004) Essential properties and the right to life: a response to Lee. *Bioethics* 18 (3): 264–282.

Thomson, J.J. (1971) A defense of abortion. *Philosophy and Public Affairs* 1: 47–66; reprinted, among other places, in Feinberg (1984, pp. 173–87).

Thomson, J.J. (1995). Abortion. *The Boston Review* XX (3) (Summer).

Tooley, M. (1983) *Abortion and Infanticide*. New York: Oxford University Press.

Warren, M.A. (1984) On the moral and legal status of abortion. In *The Problem of Abortion*, 2nd edn. J. Feinberg, ed., Belmont, CA: Wadsworth, pp. 102–119.

Further Reading

Bailey, R., Lee, P., and George, R.P. (2001). Are stem cells babies? reasononline.com at http://reason.com/archives/2001/07/11/are-stem-cells-babies (last accessed 6/17/13).

Beckwith, F. (1993) *Politically Correct Death: Answering the Arguments for Abortion Rights*. Grand Rapids, MI: Baker.

Beckwith, F. (2000) *Abortion and the Sanctity of Human Life*. Joplin, MO: College Press.

Chappell, T.D.J. (1998) *Understanding Human Goods: A Theory of Ethics*. Edinburgh: Edinburgh University Press.

Feinberg, J., ed. (1984) *The Problem of Abortion*, 2nd edn. Belmont, CA: Wadsworth.

Finnis, J. (1999) Abortion and health care ethics. In *Bioethics: An Anthology*, ed. H. Kuhse and P. Singer, pp. 13–20. London: Blackwell.

Finnis, J. (2001). Abortion and cloning: some new evasions. http://lifeissues.net/writers/fin/fin_01aborcloneevasions.html (last accessed 6/17/13). George, R. (2001) We should not kill human embryos – for any reason. In *The Clash of Orthodoxies: Law, Religion, and Morality in Crisis*. Wilmington, DL: ISI Books, pp. 317–323.

Grisez, G. (1990) When do people begin? *Proceedings of the American Catholic Philosophical Association* 63: 27–47.

Lee, P. (2004) The pro-life argument from substantial identity: a defense. *Bioethics* 18 (3): 249–263.

Marquis, D. (1989) Why abortion is immoral. *The Journal of Philosophy* 86: 183–202.

Oderberg, D. (2000) *Applied Ethics: A Non-Consequentialist Approach*. New York: Oxford University Press.

Pavlischek, K. (1993) Abortion logic and paternal responsibilities: one more look at Judith Thomson's "Defense of abortion". *Public Affairs Quarterly* 7: 341–361.

Schwarz, S. (1990) *The Moral Question of Abortion*. Chicago, IL: Loyola University Press.

Stone, J. (1987) Why potentiality matters. *Journal of Social Philosophy* 26: 815–830.

Stretton, D. (2000) The argument from intrinsic value: a critique. *Bioethics* 14: 228–239.

The President's Council on Bioethics (2002) *Human Cloning and Human Dignity: the Report of the President's Council on Bioethics*. New York: Public Affairs.

CHAPTER THREE

The Moral Permissibility of Abortion

Margaret Olivia Little

Introduction

When a woman or girl finds herself pregnant, is it morally permissible for her to end that pregnancy? One dominant tradition says "no"; its close cousin says "rarely" – exceptions may be made where the burdens on the individual girl or woman are exceptionally dire, or, for some, when the pregnancy results from rape. On both views, though, there is an enormous presumption against aborting, for abortion involves the destruction of something we have no right to destroy. Those who reject this claim, it is said, do so by denying the dignity of early human life – and imperiling their own.[1]

I think these views are deeply flawed. They are, I believe, based on a problematic conception of how we should value early human life; more than that, they are based on a profoundly misleading view of gestation and a deontically crude picture of morality. I believe that early abortion is fully permissible, widely decent, and, indeed, can be honorable. This is not, though, because I regard burgeoning human life as "mere tissue": on the contrary, I think it has a value worthy of special respect. It is, rather, because I believe that the right *way* to value early human life, and the right way to value what is involved in and at stake with its development, lead to a view that regards abortion as both morally sober and morally permissible. Abortion at later stages of pregnancy becomes, for reasons I shall outline, multiply more complicated; but it is early abortions – say, abortions in the first half of pregnancy – that are most at stake for women.

The Moral Status of Embryos and Early Fetuses

According to one tradition, the moral case against abortion is easily stated: abortion is morally impermissible because it is murder. The fetus, it is claimed, is a *person* – not just

Contemporary Debates in Applied Ethics, Second Edition. Edited by Andrew I. Cohen and Christopher Heath Wellman.

a life (a frog is a life), or an organism worthy of special regard, but a creature of full moral status imbued with fundamental rights. Abortion, in turn, constitutes a gross violation of one of that person's central-most such rights: namely, its right to life.

Now, for a great many people, the idea of a 2-week blastocyst, or 6-week embryo, or 12-week fetus counting as an equivalent rights-bearer to more usual persons is just an enormous stretch. It makes puzzles of widely shared intuitions, including the greater sense of loss most feel at later rather than earlier miscarriages, or again the greater priority we place on preventing childhood diseases than on preventing miscarriages. However else we may think such life worthy of regard, an embryo or early fetus is so far removed from our paradigmatic notion of a person that regarding it as such seems an extreme view.

The question is why some feel pushed to such an extreme. It is, in part, a reflection of just how inadequate our usual theories are when they bump up against reproduction. Surely part of the urge to cast a blastocyst as a full-fledged person, for instance, is a by-product of the impoverished resources our inherited theory has for valuing germinating human life: if the only category of moral status one has is a person or rights-holder, then the only way to capture our sense of the kind of respect or honor that embryos might deserve (the only way to capture the loss many feel at early miscarriage, for instance, or the queasiness over certain aspects of human embryo research) is to insist on fetal personhood from the moment of conception. The alternative, of course, is to challenge the assumption: instead of making the fetus match those terms of moral status, we ask what our theory of value should look like to accommodate the value of an entity like the fetus.

Or again, part of the urge to cast the embryo as a person is the worry that drawing subsequent distinctions in moral status over the course of fetal development would be fatally ad hoc. But such a worry already presupposes a certain metaphysics: it is only if one believes that discrete events and steady states are the fundamental explanatory classifications that distinctions of stages will feel troublingly arbitrary. A metaphysics that accommodates *becoming* or *continua* as fundamental explanatory classifications will be more likely to regard the distinction between zygote and matured person as inherently graduated. It would not expect to find – because it would not think to need – any distinction between discrete properties adequate to the job.

This is not to say that everything about moral status is degreed. But if we expand our moral categories beyond *rights* to notions of *value*, and accept *continua* as everyday phenomena rather than special puzzles, the road is paved for a picture of burgeoning human life that accords far better with the intuitions of so many: burgeoning human life has a status and worth that deepen as its development progresses.

But, it will be said, such an account misses something crucial. Unlike other inherently gradualist processes – the building of a house, say – there is here something already extant that should ground full moral status to the embryo: namely, a potential or telos for personhood. The only gradualist element in the picture is its unfolding. This, it will be urged, is what really grounds the moral standing of early human life: it is not because the embryo or early fetus *is* a person, but because the right way to value potential persons is to regard them as deserving the same deference *as* persons.

Now, I think there is a very important sense in which we should regard human embryos and fetuses as potential persons. We are in part biological animals, and biology classifies organisms as the types of creature they are by giving explanatory primacy to certain trajectories over others. While there are an infinite number of trajectories that fish eggs, for instance, could take – from developing into fish, to being eaten as caviar, to being infused with sheep DNA and becoming a sheep – they are understood as the kind of biological organism they are by privileging the first as their "matured state" and expressive of their "nature." It is in this sense that a fish egg is a potential fish, while a salamander egg – which could in principle be turned into a fish with enough laboratory machinations – is only thereby a possible one. Similarly, a human embryo is understood biologically as the kind of organism it is by giving explanatory primacy to the trajectory of its developing into a matured human, that is a person – something that cannot be said of a given sperm or egg.[2]

Lest we hang too much on this point, though, we need to remember that biology is not the only rubric that matters here. There is no direct isomorphism from the idea of a biological potential to a normative end – something that should or must be realized. Indeed, on one view, biological potential is only a candidate for normative upshot for creatures who independently count as having moral standing – a view that grounds moral status in potentiality turns out to have things exactly backwards. More deeply, though, the particular classification at issue here carries an intrinsic tension. For the trajectory in virtue of which we connect this sort of organism with that further state is a trajectory that depends on what *another person* – the pregnant woman – is able and willing to do. That is, *unlike* most biological organisms, the trajectory we privilege as the fetus's "natural" development – against which we classify its "potential" and measure when its existence is "truncated" – depends on the actions and resources of an autonomous *agent*, not the events and conditions of a *habitat*. Knowing what to think of the fetus thus requires assessing moves that have their home in biology (classifying organisms based on privileging certain environmental counterfactuals) applied when the biological "environment" is, at one and the same time, an autonomous agent subsumable under normative, not just biological, categories.

If this is easy to miss, it is in part because of how human gestation itself tends to get depicted. Metaphors abound of passive carriage; the pregnancy is a project of nature's. The woman is, perhaps, an especially close witness to that project, or again its setting, but the project is not her own. Her agency is thus noticed when she cuts off the pregnancy but passes unnoticed when she continues it. If, though, gestation belongs to the *woman* – if its essential resources are hers – her blood, her hormones, her energy, all resources that could be going to other of her bodily projects – then the concept of potential person is a hybrid concept from the start, not something we can read off from the neutral lessons of biology. In an important sense, then, talk of the fetus as potential person is dangerously misleading. For it encourages us to think of the embryo's development as mere *unfolding* – as though all that's needed other than the passage of time is already intrinsically there, or at least there independently of the woman.

In my own view, the biological capacities of early human life provide, once again, a degreed basis for according regard. Such biological potential marks out early human life as specially *respect-worthy* – which is why we should try to avoid conception where children are not what is sought (or again, why we do not think we should tack up human embryos on the wall for art, or provide them for children to dissect at school if fertilized chicken eggs get too pricey). To say that such life is respect-worthy, though, is not the same as claiming we are charged to defer as we would those with moral status.

Abortion and Gestational Assistance

Thus far, I have argued that morally restrictive views of abortion ride atop a problematic view of how we should value early human life. I now want to argue that they also ride atop a problematic misconception of the act of aborting itself. Let me illustrate first by returning to the claim that, if the fetus *were* a person, abortion would be a violation of its right to life.

We noted above that, while certain metaphors depict gestation as passive carriage (as though the fetus were simply occupying a room until it is born), the truth is of course far different. One who is gestating is providing the fetus with sustenance – donating nourishment, creating blood, delivering oxygen, providing hormonal triggers for development – without which it could not live. For a fetus, as the phrase goes, to live *is* to be receiving aid. And whether the assistance is delivered by way of intentional activity (as when the woman eats or takes her prenatal vitamins) or by way of biological mechanism, assistance it plainly is. But this has crucial implications for abortion's alleged status as murder. To put it simply, the right to life, as Judith Thomson famously put it, does not include the right to have all assistance needed to maintain that life (Thomson, 1971). Ending gestation will, at early stages at least, certainly lead to the fetus's demise, but that does not mean that doing so would violate its right to life.

Now Thomson herself illustrated the point with an (in)famous thought experiment in which one person is kidnapped and used as life support for another: staying connected to the Famous Violinist, she points out, may be the kind thing to do, but disconnecting oneself does not violate the Violinist's rights. The details of this rather esoteric example have led to widespread charges that Thomson's point ignores the distinction between killing and letting die, and would apply at any rate only to cases in which the woman was not responsible for procreation occurring. In fact, though, I think the central insight here is broader than the example, or Thomson's own analysis, indicates.[3]

As Frances Kamm's work points out (Kamm, 1992), in the usual case of a killing – if you stab a person on the street, for instance – you interfere with the trajectory the person had independently of you. She faced a happy enough future, we will say; your action changed that, taking away from her something she would have had but for your action. In ending gestation, though, what you are taking away from this person is something she would not have had to begin with, without your aid. She comes to you with a downward trajectory, as it were: but for you she would already be dead. In removing that assistance, you are not violating the person's right to life, judged in the traditional terms of a right against interference. While all killings are tragedies, then, not all are alike: some killings, as Kamm puts it, share the crucial "formal" feature of letting die, which is that they leave the person no worse off than before she encountered you. Of course, if one *could* end the assistance without effecting death, then, absent extraordinary circumstances, one should. (Part of the debate over so-called partial birth abortions is whether and when we encounter such circumstances.)[4]

The argument is not some crude utilitarian one, according to which you get to kill the person because you saved her life (as though, having given you a nice lamp for your birthday, I may therefore later steal it with impunity). The point, rather, is that where I am still in the process of saving – or sustaining or enabling – your life, and that life cannot be thusly saved or sustained by anyone else, ending that assistance, even by active means, does not violate your right to life.

Some, of course, will argue that matters change when the woman is causally responsible for procreation. In such cases, it will be said, she is responsible for introducing the person's need. She is not like someone happening by an accident on the highway who knows CPR; she is like the person who *caused* the accident. Her actions introduced a set of vulnerabilities or needs, and we have a special duty to lessen vulnerabilities and repair harms we have inflicted on others.

But there is a deep disanalogy between causing the accident and procreating. The fact of causing a crash itself introduces a harm to surrounding drivers: they are in a worse position for having encountered that driver. But the simple act of procreating does not worsen the fetus's position: without procreation, the fetus would not exist at all; and the mere fact of being brought into existence is not a bad thing. To be sure, creating a human is creating someone who comes with needs. But this, crucially, is not the same as inflicting a need *on to* someone (see Silverstein, 1987). It is not as though the fetus already existed with one level of needs and the woman added a new one (as does happen, for instance, if a woman takes a drug after conception that increases the fetus's vulnerability to, say, certain cancers). The woman is (partially) responsible for creating a life, and it is a life that necessarily includes needs, but that is not the same as being responsible for the person being needy rather than not. The pregnant woman has not made the fetus more vulnerable than it would otherwise have been: absent her procreative actions, it would not have existed at all.

Even if the fetus were a person, then, abortion would not be murder. More broadly, abortion is not a species of *wrongful interference*. This is not to say that abortion is thereby necessarily unproblematic. It is to argue, instead, that the crucial moral issue needs to be relocated to the question of what, if any, positive obligations pregnant women have to continue gestational assistance. The question abortion really asks us to address is a question about the *ethics of gestation*. But this is a question that takes us into far richer, and far more interesting, territory than that occupied by discussions of murder. In particular, it requires us to discuss and assess claimed grounds of obligation, and to assess the very specific kinds of burdens and sacrifice involved in rendering *this* type of assistance.

I have argued elsewhere that if or when the fetus is a person, then the question of when a woman might have some obligation to provide use of her body to save its life turns out to be a fascinatingly deep matter, and one that is ultimately deeply contextual (Little, forthcoming). The issue I want to turn my attention to here is what picture we get when we join the two views I have outlined: a view that regards burgeoning human life as respect-worthy but not endowed with substantial moral status, and a view that recognizes abortion as the ending of gestational support. Abortion, I want to argue, is both permissible and widely decent, for reasons involving what we might call *authorship* and *stewardship*. Let me take each one in turn.

Intimacy, Pregnancy, and Motherhood

When people first ask what is at stake in asking a woman to continue a pregnancy, what usually get emphasized are the physical and medical risks. And indeed, they are important to emphasize. While many pregnancies go smoothly, many do not; and the neutral language of an obstetrics text hardly captures the lived reality. I think of a friend I visited who had been put in lock-down on the psychiatric ward from pregnancy-related psychosis (and whose physician would not discuss inducing at 39 weeks because there was no "obstetrical indication"). Or my sister, whose two-trimester "morning sickness" – actually gut-wrenching dry heaves every 20 minutes and three hospitalizations – was the equal of many an experience of chemotherapy. Or another acquaintance, whose sudden onset of eclampsia during delivery brought her so close to dying that it left us all breathless. Asking women to take on the *ex ante* medical risks of pregnancy is asking a lot.

Then there are the social risks pregnancy can represent for some women – risks it is very hard for those of us in more comfortable lives to fully appreciate. Pregnancy is a marker for increased domestic violence. It leads, for many, to abandonment by family and community, even as it can lead the woman to feel tied to a relationship she would otherwise leave.

All of these burdens are important to appreciate. But there is something incomplete in such renditions of pregnancy's stakes. For a great many women, it is another set of issues that motivate the desire to end a pregnancy – issues having to do with the extraordinarily *personal* nature of gestation.

To be pregnant is to allow another living creature to live in and off of one's body for nine months. It is to have one's every physical system shaped by its needs, rather than one's own. It is to share one's body in an extraordinarily intimate and extensive – and often radically unpredictable – way. Then there is the aftermath of the nine months: for gestation does not just turn cells into a person; it turns the woman into a mother. One of the most common reasons women give for wanting to abort is that they do not want to become a mother – now, ever, again, with this partner, or no reliable partner, with these few resources, or these many that are now, after so many years of mothering, slated finally to another cause. Not because motherhood would bring with it such burdens – though it can – but because motherhood would so thoroughly change what we might call one's fundamental practical identity. The enterprise of mothering restructures the self – changing the shape of one's heart, the primary commitments by which one lives one's life, the terms by which one judges one's life a success or a failure. If the enterprise is eschewed and one decides to give the child over to another, the identity of mother still changes the normative facts that are true of one, as there is now someone by whom one does well or poorly (Ross, 1982). And either way – whether one rears the child or lets it go – to continue a pregnancy means that a piece of one's heart, as the saying goes, will forever walk outside one's body.

Gestation, in short, is not just any activity. It involves sharing one's very body. It brings with it an emotional intertwinement that can reshape one's entire life. It brings another person into one's family. Deciding whether to continue a pregnancy is not like being asked to write a check for charity, however large; it is an enormous undertaking that has reverberations for an entire lifetime. To argue that women may permissibly decline this need not trade on a view that grants no value to early life; it is, in essence, to argue about the right way to value *pregnancy* and *parenthood*. It is to recognize a level of moral prerogative based not just on the concretely understood burdens of the activity in question, but also on its deep connection to authoring a life. To illustrate, consider the following.

56   **Margaret Olivia Little**

Imagine that the partner of your family's dreams is wildly in love with you and asks for your hand in marriage. As it turns out, substantial utility would accrue by your accepting him: his connections would seal your father's bid for political office, raise the family profile yet higher, and add nicely to its coffers just as your eldest brother faces expensive restoration of the family estate. It would also, and not incidentally, keep the fellow himself from falling into a pit of despair, as it is clear you are the only one for him.

All of this utility notwithstanding, many will believe that you do not thereby have a moral *obligation* – even a prima facie one – to accept the proposal. You might have a responsibility to give the proposal serious thought; but if, on reflection, you realize that marriage to this man – or to any man – is not what you want, then there we are. And this, even if we stipulate that marriage would not be a setback to your happiness: the utility function you would enjoy following acceptance might, indeed, surpass the one that would follow refusal. This, even if we think that the needs presented would have coalesced to form a duty if the assistance required had been burdensome (say, writing a big check) rather than intimate.[5] Nor, finally, need we think the resistance must trace to a conviction that it would be morally wrong to accept the proposal – that it would in some way transgress the norms governing marriage. It is, we will imagine, quite obvious to you that you would come to have an enduring love if you accept; he understands this and relishes the prospective courtship. It is not that you would *use* him if you accept; it is that you do not *want* to have an enduring love with him, now, or at all.

Or again, imagine that your providing sexual service would help comfort and inspire the soldiers readying for battle. Many will believe this does not ground a requirement, even prima facie, to offer intercourse. This, even if you are the only one around capable of offering such service, and even if doing so would not actually be distressful to you. Such an intuition, again, need not trade on thinking that it would be wrong to give sex for such a purpose. Those with more permissive views of sexuality might well think someone who authentically and with full self-respect wanted to share her body for this purpose would be doing something generous and fine. One just does not want to make doing so the subject of obligation.

Now not all agree to these intuitions. If Victorian novels are to be believed, the upper classes of Regency England believed both that marriage and sex were fair candidates for obligation (especially when the family estate was at stake). But for many, there is something about marriage as a relationship, and sexual intercourse as a bodily connection, that makes them deserving of some special kind of deference when assessing moral obligation. The deference is doubtless limited: one need be absolutist here no more than elsewhere. But the defense, crucially, is not merely a function of plain utility considerations; it is the intimacy, not just the concrete welfare, that matters.

An important part of being a self is that the boundaries of one's self – the borders and use of one's body, the identity by which one knows oneself as oneself – are matters over which one deserves special moral deference. We might say it is on pain of imposing alienation. But the point is not to urge some fetishism about the evil of alienation (morality, after all, does not give a whit if you feel alienated when returning the borrowed library book), but to insist that some activities can have a sufficiently tight connection to self that alienation with respect to *them* is specially problematic to maintaining our status *as* selves. One's self is not always implicated in sexual intercourse and marriage; where it is, one may not care. But where it is, and you do, that fact is worthy of a deference or protection in a way that caring about how one's garden grows is not.

Gestation, like sex, is a bodily intimacy of the first order. Motherhood, like marriage, is a relational intimacy of the first order. If one believes that decisions about whether to continue a pregnancy are deserving of moral prerogative, it need not be because one believes early human life has no value – any more than assigning prerogatives over sex and marriage denies the value of one's family, the boys in fighting blue, or the relationship of marriage. Such views instead stem from the conviction that the proper way to value the relationship of motherhood and the bodily connection of pregnancy is to view them as intimacies deserving of special deference. Even if continuing a pregnancy represents *no* welfare setback to the woman, classically construed, we should recognize a strong moral prerogative over whether to continue that pregnancy.

This is not a claim that any reason to abort is a good one. Human life, even in nascent forms, should not to be extinguished lightly; one who decides to end a pregnancy because she wants to fit into a party dress, say, is getting wrong the value of burgeoning human life. To abort for such reasons is to act indecently. But this does not mean that such a woman now has an obligation to continue the pregnancy. What it means, in the first instance, is that she should not regard such a reason as adequate for the conclusion; not that the conclusion is not available to her.

It is not that decency is some optional ideal. Quite to the contrary: if one realizes that an action is indecent, one must not do it. But the "it" in question is, as Barbara Herman (1993, p. 147) puts it, an action–reason pair – it is, though it makes our deliberations sound more formal than they are – a piece of practical syllogism. To say that a practical syllogism is indecent means one should discard it, but that does not yet comment on what action one should do. More specifically, it does not mean that one cannot decently arrive at its conclusion, for there may well be decent reasons waiting in the wings.

Take a standard example. A soldier, we might well decide, does not have an obligation to risk death by falling on the grenade that threatens his comrades. Nonetheless, if the reason he declines has nothing to do with wanting to live and everything with wanting his hated comrades to die, his refusal is indecent. He betrays a dreadful understanding of what is here at stake; he should not refuse on that basis. But this does not mean he thereby faces now an obligation or imperative to fall on the grenade. For there is extant a reason the soldier can deploy as an honorable basis for declining – namely, that doing so would sacrifice his life.

Or again, to return to our fanciful examples, if the reason you decide not to marry the suitor is not because you do not want to enter such a commitment at this stage of your life, but because you do not like the wart on his big toe, or the color of the drawing-room walls in his mansion, or if you decline sexual intercourse for racist reasons, your behavior is indecent. To think these acceptable reasons – to think them adequate premises to support a practical conclusion of declining – is to fundamentally misappreciate the various values here implicated. But we do not thereby conclude that the person is now under a requirement to accept (as though it is the woman with the dreadful reasons who now has an obligation to have sex). For there is extant a reason that would be honorable to deploy as a basis for declining – that one does not want to have sex, or enter marriage. Similarly, the fact that a given woman might deploy a genuinely trivial or offensive basis for aborting does not mean she is now obliged to continue the pregnancy. For there are available reasons – about sharing her body and entering motherhood – she may deploy as a basis for honorably declining.

Norms of Responsible Creation

Now some will urge that those who are (at least jointly) responsible for procreation thereby have a heightened obligation to continue gestating. People, of course, disagree over what it takes to count as "responsible" here – whether voluntary but contracepted intercourse is different from intercourse without use of birth control, and again from intentionally deciding to become pregnant at the IVF clinic. But those who satisfy the relevant criteria, it is often said, must thereby face greater duty to "see the pregnancy through." Unease is expressed at the thought of heterosexual intercourse conducted in callous disregard of procreative potential, of creating only to let wither. If you are going to allow a new life to begin, it is thought, you had better see it through to fruition.

I think these intuitions point to important issues, but not the ones usually thought. Let us start with that notion of sexual irresponsibility. For many people, there is something troubling about the idea of couples engaging in heterosexual intercourse in complete disregard of contraception – say, when one is highly fertile and birth control is just an arm's reach away. Such a view points to an important set of intuitions about another layer of respect, namely, respect for creation itself. Respect for burgeoning human life carries implications, not just for the accommodation we might owe such life once extant, but for the conditions under which we should undertake activities with procreative potential in the first place. To regard something as a value sometimes enjoins us to make more of it, and sometimes, as with people, to take care about the conditions under which we make any.[6]

There are, as we might put it, norms of responsible creation. Such a view seems exactly right to me. Part of what I imagine teaching my own children about sexuality is that human life as such deserves respect (whatever the metaphysical details), and respect requires that one not treat one's procreative capacities in a cavalier way. But none of this means that one has a special responsibility to gestate if one *does* get pregnant. For one thing, these norms, while very important (and far too little emphasized in our current culture), are norms about the activities that can lead to procreation, not what one owes should procreation take place. They specify, as it were, the good faith conditions one should meet for engaging in certain activities. Even if the norms are broached – one has sex in callous disregard to its potential to lead to new human life – that does not itself imply that one now (as punishment?) must gestate: it says one should not have had that sort of sex. Indeed, for many of us, the thought that negligence here means one should continue a pregnancy has an internal disconnect: that one had irresponsible sex is no reason at all to bring a new person into the world.

This last point begins to point to a very different approach to the ethics of creation. The salience of responsibility for procreation to the responsibilities of gestation is not just complex: decisions about abortion are often located *within* the norms of responsible creation. Let me explain.

Many people have deeply felt convictions about the circumstances under which they feel it right for them to bring a child into the world – can it be brought into a decent world, an intact family, a society that can minimally respect its agency? These considerations can persist even after conception has taken place; for while the embryo has already been created, a person has not. Some women decide to abort, that is, not because they do not *want* the resulting child – indeed, they may yearn for nothing more, and desperately wish that their circumstances were otherwise – but because they do not think bringing a child into the world the right thing for them to do.

As Barbara Katz Rothman (1989) puts it, decisions to abort often represent not a decision to destroy, but a refusal to create. These are abortions marked by moral language. A woman wants to abort because she knows she could not give up a child for adoption but feels she could not give the child the sort of life, or be the sort of parent, she thinks a child *deserves*; a woman who would have to give up the child thinks it would be unfair to bring a child into existence already burdened by rejection, however well grounded its reasons; a woman living in a country marked by poverty and gender apartheid wants to abort because she decides it would be wrong for her to bear a daughter whose life, like hers, would be filled with so much injustice and hardship.

Some have thought that such decisions betray a simple fallacy: unless the child's life were literally going to be worse than non-existence, how can one abort out of concern for the future child? But the worry here is not that one would be imposing a *harm* on the child by bringing it into existence (as though children who are in the situations mentioned have lives that are not worth living). The claim is that bringing about a person's life in these circumstances would do violence to her ideals of creating and parenthood. She does not want to bring into existence a daughter she cannot love and care for; she does not want to bring into existence a person whose life will be marked by disrespect or rejection. In struggling with these issues, the worry is not that the child would have been better off never to have been born – as though children who are in the situations just mentioned have lives that are not worth living;[7] it is that continuing a pregnancy in such circumstances would violate the woman's commitments of respectful creation.

Nor does the claim imply judgment on women who *do* continue pregnancies in similar circumstances – as though there were here an obligation to abort. For the norms in question need not be impersonally authoritative moral claims. Like ideals of good parenting, they mark out considerations all should be sensitive to, perhaps, but equally reasonable people may adhere to different variations and weightings. Still, they are normative for those who do have them; far from expressing mere matters of taste, the ideals one does accept carry an important kind of categoricity, issuing imperatives whose authority is not reducible to mere desire. These are, at root, issues about *integrity*, and the importance of maintaining integrity over one's participation in this enterprise precisely because it is so normatively weighty.

Some will protest the thought of our deciding such matters. We have no dominion, it will be said, to pick and chose the conditions under which human life, once started, proceeds. On what we might call a "stewardship" view of creation, in contrast, this dominion is precisely part of the responsibility involved in creation. It is a grave matter to end a developing human life by not nurturing it; but it can be an equally grave decision to continue a process that will result in the creation of a person. The present case, note, is thus importantly different from the other area of controversy over dominion over life, namely, actions intending to hasten death. Whatever one thinks of that matter, it diverges in a key respect from abortion. When we stand by rather than hasten death, we are allowing a trajectory independent of us to proceed without our influence. Not to abort, though, *is* to do something else – namely, to create a person.

Gestation is itself a creative endeavor. Not in the sense that its constitutive activities are each or mostly intentional (as if the issue were whether the pregnant woman, like an athlete, deserves credit for the bodily activity involved). But if personhood emerges through pregnancy, and one has choices about whether to continue pregnancy, then decisions to do so themselves involve norms of respect. And not all norms of respect for creation, it turn outs, tell in favor of continuing.

None of this is to say that abortion is morally neutral. Abortion involves loss. Not just loss of the hope various parties have invested in the pregnancy, but loss of something valuable in its own right. Abortion is thus a sober matter, an occasion, often, for moral emotions such as grief and regret. Given the value at stake, it is only fitting to feel grief – a sorrow that life begun is now ended – or to feel moral regret – that the actions needed to help these cells develop into a person would have compromised too significantly the life of someone who already was one. Such regret, that is, can signal appreciation of the fact, not that the action was indecent, but that decent actions sometimes involve loss.

It takes enormous investment to develop early human life into a human being. Understanding the morality of early abortion involves assessing not just welfare, but intimacy, not just destruction, but creation. As profound as the respect we should have for burgeoning human life, we should acknowledge moral prerogatives over associations such as having another inhabit and use one's body in such an extraordinarily enmeshed way, over identity-constituting commitments and enterprises as profound as motherhood, and over the weighty responsibility of bringing a new person into the world.

Notes

1   Portions of this chapter draw on my essay, "Abortion" (Little, 2003).
2   At least, one of a couple of weeks' standing: earlier blastocysts' trajectories turn out to be fascinatingly underdetermined. There is, for instance, no fact of the matter internal to its own cellular information as to whether a one-week blastocyst will be one person or more; and at very early stages there is no fact of the matter as to which cells will become the fetus and which will become the placenta.
3   RU-486, which essentially interrupts the production of progesterone needed to maintain a placenta, provides a good example of an abortion method that is more straightforwardly a "letting die" than an active killing.
4   Later abortions are thus multiply complicated: fetal status increases even as its dependencies decline. On the one hand, later fetuses are much closer to, and at some stage likely count as, persons; on the other hand, they are no longer solely and fully dependent on gestational assistance for life, hence enlarging possibilities for removing assistance without effecting death.
5   That is, the action is not simply a token that falls under an imperfect duty. It is a fascinating question how to parse the structure of imperfect duties, a question I here leave aside.
6   Of course, just how much "care" one must exert to avoid conception will be heartily contested. Those, like myself, who value spontaneity in sexual relations and have mild views about the value of burgeoning human life will advance something quite modest – urging, say, good faith attempts to use birth control if it is safe, easily obtained, and immediately convenient. Others will advance stringent principles indeed, requiring, say, that one not have sex at all until one is prepared to parent.
7   My thanks to Adrienne Asche for this way of putting the point.

References

Herman, B. (1993) *The Practice of Moral Judgment*. Cambridge, MA: Harvard University Press.

Kamm, F.M. (1992) *Creation and Abortion: A Study in Moral and Legal Philosophy*. New York: Oxford University Press.

Little, M. (forthcoming) *Intimate Duties: Re-thinking Abortion, the Law, and Morality*. Oxford: Oxford University Press.

Little, M.O. (2003) Abortion. In *A Companion to Applied Ethics*, ed. R.G. Frey and C.H. Wellman, pp. 313–325. Oxford: Blackwell.

Ross, S. (1982) Abortion and the death of the fetus. *Philosophy and Public Affairs* 11: 232–245.

Rothman, B.K. (1989) *Recreating Motherhood: Ideology and Technology in a Patriarchal Society*. New York: Norton.

Silverstein, H.S. (1987) On a woman's "responsibility" for the fetus. *Social Theory and Practice* 13: 103–119.

Thomson, J.J. (1971) A defense of abortion. *Philosophy and Public Affairs* 1: 47–66.

Further Reading

Callahan, J.C. (April 11, 1986) The fetus and fundamental rights. *Commonweal* 13: 203–209.

Crittenden, A. (2001) *The Price of Motherhood*. New York: Henry Holt and Co.

Denes, M. (1976) *In Necessity and Sorrow: Life and Death in an Abortion Hospital*. New York: Basic Books, Inc.

Dworkin, R. (1993) *Life's Dominion: An Argument About Abortion, Euthanasia, and Individual Freedom*. New York: Alfred A. Knopf.

Dwyer, S. and Feinberg, J., eds (1997) *The Problem of Abortion*, 3rd edn. Belmont, CA: Wadsworth Publishers.

Feinberg, J. (1992) Abortion. In *Freedom and Fulfillment: Philosophical Essays*, ed. J. Feinberg, p. 37. Princeton, NJ: Princeton University Press.

Hursthouse, R. (1987) *Beginning Lives*. Oxford: Open University.

MacKinnon, C.A. (1991) Reflections on sex equality under law. *The Yale Law Journal* 100 (5): 1281–1328.

McDonaugh, E. (1996) *Breaking the Abortion Deadlock: From Choice to Consent*. New York: Oxford University Press.

Quinn, W. (1993) Abortion: identity and loss. In *Morality and Action*, ed. W. Quinn. New York: Cambridge University Press.

Steinbock, B. (1992) *Life Before Birth: The Moral and Legal Status of Embryos and Fetuses*. New York: Oxford University Press.

Wertheimer, R. (1974) Understanding the abortion argument. In *The Rights and Wrongs of Abortion*, ed. M. Cohen, T. Nagel, and T. Scanlon, pp. 23–51. Princeton, NJ: Princeton University Press.

West, R. (1993) Jurisprudence and gender. In *Feminist Legal Theory: Foundations*, ed. D.K. Weisberg, pp. 75–98. Philadelphia: Temple University Press.

Euthanasia

CHAPTER FOUR

In Defense of Voluntary Active Euthanasia and Assisted Suicide

Michael Tooley

In this chapter I shall defend the following two claims: first, given appropriate circumstances, neither voluntary active euthanasia nor assisting someone to commit suicide is in any way morally wrong; secondly, there should be no laws prohibiting such actions, in the relevant cases.

The discussion is organized as follows. First, I set out some preliminary concepts and distinctions. Then, in the next two sections, I offer two arguments in support of the thesis that assisted suicide and voluntary active euthanasia are not morally wrong. Finally, I ask whether there is any reason for thinking that, even if, as I have argued, voluntary active euthanasia and assisted suicide are not morally wrong, they should, nevertheless, not be legally permitted – and I argue that this is not the case.

Important Concepts and Distinctions

Writers on this topic define the term "euthanasia" in quite different ways. In the following discussion, I shall use the term "euthanasia" to refer to any action where a person is intentionally killed or allowed to die because it is believed that the individual would be better off dead than alive – or else, as when one is in an irreversible coma, at least no worse off. So understood, under what conditions, if any, is euthanasia morally acceptable, and should it ever be legally permitted?

Two familiar distinctions are important here. First, there is the threefold distinction involving voluntary euthanasia, non-voluntary euthanasia, and involuntary euthanasia. Thus, euthanasia is voluntary if the person who undergoes it has requested it. It is non-voluntary if the person is unable to indicate whether or not he or she wants to undergo euthanasia. (This will include, for example, cases involving infants, and adults

Contemporary Debates in Applied Ethics, Second Edition. Edited by Andrew I. Cohen and Christopher Heath Wellman.
© 2014 John Wiley & Sons, Inc. Published 2014 by John Wiley & Sons, Inc.

who have permanently lost consciousness.) Finally, it is involuntary if the person in question wants to go on living.

The second important distinction is between active euthanasia and passive euthanasia. How this distinction is best drawn is controversial, and there are two slightly different ways of doing so, depending upon how cases involving the withdrawal of life-support systems are classified. Thus, one way of drawing the distinction is in terms of the contrast between acting and doing nothing at all: it is active euthanasia whenever anything at all is done – including the withdrawal of a life-support system – that facilitates the person's death, and passive euthanasia only if nothing is done that brings about the person's death.

A different way of drawing the distinction is in terms of whether what might be called the "primary cause" of death is some human action, or, instead, an injury or disease: one has a case of active euthanasia whenever the primary cause of death is human action, and a case of passive euthanasia whenever the primary cause of death is some injury or disease.

Precisely where the line should be drawn between active euthanasia and passive euthanasia is important if one holds, as a significant number of people do, that passive euthanasia is morally permissible, but that active euthanasia is not. Here, however, we can ignore this issue, given that my goal is to argue that voluntary *active* euthanasia is morally permissible.

Before turning to a defense of assisted suicide and voluntary active euthanasia, it should be noted that some opponents of voluntary active euthanasia and assisted suicide define the term "euthanasia" much more narrowly than I have done – indeed, often very narrowly indeed. This is especially so in the case of writers who are defending the Roman Catholic view on these issues. Thus, for example, Daniel Callahan (Chapter 5 in this volume) offers the following definition: "By euthanasia I mean the direct killing of a patient by a doctor, ordinarily by means of a lethal injection" (2014, p. 92, n.1).

Notice that such a definition is narrower than what I have offered in three ways. First, cases where one allows a person to die do not get classified as euthanasia, even if one's intention is precisely the same as when one kills a person to enable that person to escape from the suffering that he or she is undergoing. Second, cases where, for example, a doctor administers a dose of morphine that it is known will cause death via respiratory failure do not get classified as cases of euthanasia, since it is held that the killing is not "direct": the doctor's intention is, it is said, merely to relieve the pain, not to kill, even though the doctor knows that the action will kill the patient. Finally, by incorporating the restriction to terminally ill persons, cases where a person is not terminally ill, but is suffering greatly from pain that cannot be relieved, are being defined as lying outside the scope of euthanasia.

Such a definition of "euthanasia" seems to me ill-advised in the extreme. In the first place, one is deprived of crisp and very useful expressions – such as "passive euthanasia" – for referring to cases where a terminally ill person is allowed to die. Second, and more seriously, the person who identifies euthanasia with the direct killing of a terminally ill person typically does so because he or she views the indirect killing of a terminally ill person as morally unproblematic, and similarly for an action of merely allowing a terminally ill person to die. If one holds, however, that such actions are morally permissible, but that the direct killing of a terminally ill person is morally wrong, then among the most crucial issues that one needs to address are, first, why the direct versus indirect distinction has such moral significance, and, second, why the same is true in the case of the distinction between killing and letting die. If one defines euthanasia broadly, as I have done, those issues are immediately in front of one. By contrast, a narrow definition of euthanasia makes it very easy to pass over those crucial questions without even any comment, let alone careful discussion and argument.

A Fundamental Defense of Assisted Suicide and Voluntary Active Euthanasia

The argument

A very plausible argument in support of the claim that voluntary active euthanasia and assisted suicide are not morally wrong in themselves is as follows:

(1) If a person is suffering considerable pain due to an incurable illness, then in some cases that person's death is in his or her own interest.

(2) If a person's death is in that person's own interest, then committing suicide is also in that person's own interest.

(3) Therefore, if a person is suffering considerable pain due to an incurable illness, then in some cases committing suicide is in that person's own interest. (From (1) and (2).)

(4) A person's committing suicide in such circumstances may very well also satisfy the following two conditions:
 (a) it neither violates anyone else's rights, nor wrongs anyone;
 (b) it does not make the world a worse place.

(5) An action that satisfies conditions (a) and (b), and that is not contrary to one's own interest, cannot be morally wrong.

(6) Therefore, a person's committing suicide when all of above conditions obtain would not be morally wrong. (From (3), (4), and (5).)

(7) It could be morally wrong to assist a person in committing suicide only if (i) it was morally wrong for that person to commit suicide, or (ii) committing suicide was contrary to the person's own interest, or (iii) assisting the person to commit suicide violated an obligation one had to someone else.

(8) Circumstances may very well be such that neither assisting a person to commit suicide nor performing voluntary active euthanasia violates any obligations that one has to others.

(9) Therefore, it would not be wrong to assist a person in committing suicide in the circumstances described above. (From (3), (6), (7), and (8).)

(10) Whenever assisting a person in committing suicide is justified, voluntary active euthanasia is also justified, provided the latter action does not violate any obligation that one has to anyone else.

(11) Therefore, voluntary active euthanasia would not be morally wrong in the circumstances in question. (From (8), (9), and (10).)

This argument, progressing from suicide, through assisted suicide, and on to voluntary active euthanasia, is a very natural one, and the assumptions involved seem quite modest. But is the argument sound? Next, I shall argue that it is.

The soundness of the argument

Anyone who holds that assisted suicide and voluntary active euthanasia are never *in themselves* morally permissible must hold that the above argument is unsound. Can that contention be sustained? I shall argue that it cannot.

An argument can be unsound in two different ways. First, it may involve fallacious reasoning. Second, it may contain one or more false premises. Anyone who wishes to reject the conclusion of the above argument needs to show, therefore, that it is defective in one (or both) of these ways.

As regards the first possible shortcoming, the fundamental way of determining whether an argument contains any fallacious reasoning is to formulate the argument in a logically rigorous way, and then to determine whether each step in the reasoning is in accordance with some truth-preserving rule of inference. But one can also go back to the definition of validity, according to which a given inference is deductively valid if it is logically impossible for the conclusion to be false if all of the premises are true. In setting out the above argument, I have indicated, for each step in the reasoning, what earlier statements the conclusion is supposed to follow from. Readers can therefore ask themselves, in each case, whether the conclusion drawn could possibly be false if the relevant premises were true, and I suggest that, when this is done, it will be seen that the argument is deductively valid.

If this is right, then the argument can only be unsound if at least one of the premises is false. So let us consider whether any good reason can be offered for rejecting any of the premises.

The starting point of the argument is the following claim:

(1) If a person is suffering considerable pain due to an incurable illness, then that person's death may very well be in his or her own interest.

This claim is, I suggest, very plausible indeed. For one thing, the level of suffering that people undergo in connection with some incurable illnesses is such that they come to hope that death will occur sooner rather than later. In addition, when death does come in such cases, those who loved the individual who has died welcome death, and view it as in the interest of the individual in question.

Let us consider, then, the second premise:

(2) If a person's death is in that person's own interest, then committing suicide is also in that person's own interest.

Some would argue that this premise is false. In particular, Roman Catholic philosophers who accept the teachings of their church would argue that even if one is in a situation where one would be better off dead than alive, it is not in one's interest to *bring about* one's own death, since suicide is a mortal sin, and this means that someone who makes a fully informed decision to commit suicide will wind up much worse off, since they will suffer eternal torment in Hell.

A full answer to this question would require a major detour through the philosophy of religion. A brief response, however, is as follows. The Catholic Church holds that many things, beside suicide, are mortal sins – including masturbation, any type of premarital sexual activity, homosexual sex, and the use of contraceptives within marriage. Anyone who wishes to appeal to the authoritative teachings of the Catholic Church in order to object to the second premise needs to be prepared, accordingly, to argue that the Catholic Church is right in holding that the other actions just mentioned also place one at serious risk of spending eternity in Hell. I would suggest that the chances of successfully doing this are not very great.

The third premise of my argument was this:

(4) A person's committing suicide in such circumstances may very well also satisfy the following two conditions:
  (a) it neither violates anyone else's rights, nor wrongs anyone;
  (b) it does not make the world a worse place.

This premise is, I suggest, very plausible. For while it is true that many people who are thus suffering have obligations to others – especially their husbands or wives, and their children – the obligations in question are typically ones that they could not possibly meet, given that they are in a state of extreme pain. In addition, most obligations are not of such a nature that one is morally obliged regardless of the cost to oneself, so that even if one could meet certain prima facie obligations by soldiering on in the face of extreme pain, it will rarely be the case that one acts wrongly if, in those circumstances, one does not meet the prima facie obligation. Finally, the ending of one's life, in such circumstances, will not only end one's own suffering; it will also end the emotional suffering experienced by those who love one. So, in general, the ending of a person's life in such circumstances will make the world a better place, not a worse one.

Some opponents of euthanasia would object, however, that although suicide may very well not violate the rights of other humans, it does not follow that condition (4a) is satisfied. Moreover, that condition, they would contend, is in fact never satisfied, since all lives belong to God, and so the destruction of anyone's life – including destruction by the person in question – violates God's right of ownership.

This "divine ownership" objection is unsound for at least three reasons. First, it can be shown that many bad things that are present in the world, such as undeserved suffering, make it very unlikely that God, understood as an all-powerful, all-knowing, and perfectly good being, exists. Second, persons cannot be the property of others, since autonomy is a right that persons possess by virtue of their nature as beings capable of conscious experience, thought, and rational choice. Third, consider sentient beings that are not persons. Such beings can be owned, but ownership does not make it permissible to compel such beings to suffer. Similarly, if, contrary to the second point, persons could be owned by others, that would still not render it permissible to prohibit persons from committing suicide when that was in their rational self-interest, and not morally wrong.

The fourth premise of the argument was this:

(5) An action that satisfies conditions (a) and (b), and that is not contrary to one's own interest, cannot be morally wrong.

The claim that this premise is plausible can be supported as follows. First of all, it initially seems plausible that for an action to be wrong, there must be some individual – either a person, or a sentient being that is not a person – who is wronged by the action.

But if condition (a) is satisfied, then no one else is wronged, and so the only possibility is that in ending one's own life, one is wronging oneself. We are considering, however, a case where suicide is, by hypothesis, in one's own interest. But if an action is in one's own interest, how can one do wrong to oneself by performing that action? Surely one cannot. If so, then the upshot is that no one – either oneself or anyone else – is wronged by the action.

So far, so good. However, reflections concerning future generations have convinced many philosophers that an action may be wrong even if it wrongs no one (Parfit, 1984, pp. 357–361). For consider two actions, one of which will lead to future generations that enjoy an extremely high quality of life, and the other of which will result in future generations that have lives worth living, but only barely so. Other things being equal, would not the second action be morally wrong? But notice that there may be no one who is worse off if the second action is performed, since it may be that none of the people who have lives worth living, but only barely so, when the second action is performed, would have existed if the first action had been performed, while the people who *would* have enjoyed lives of very high quality if the second action had been performed are not worse off, since they never exist. So it would seem that no one is wronged if the second action is performed, since no one is worse off.

What is true, however, is that the world is a worse place given the second action than it would have been if the first action had been performed. So if, as is generally thought, the second action is wrong, then a natural conclusion is that actions can be wrong if, even though they wrong no one, they make the world a worse place than it would otherwise be.

The reason for including condition (b) in statement (5), accordingly, is to address this possibility. This having been done, it would seem, then, that the fourth premise, thus formulated, is very plausible.

The fifth premise of my argument was this:

(7)   It could be morally wrong to assist a person in committing suicide only if, (i) it was morally wrong for that person to commit suicide, or (ii) committing suicide was contrary to the person's own interest, or (iii) assisting the person to commit suicide violated an obligation one had to someone else.

Here, the supporting line of thought is this. Suppose that someone is considering performing an action that is not morally wrong. How could it be wrong to help them to perform that action? Two possibilities come to mind. First, it could be that while it was not morally wrong for the other person to perform the action, it was an action that was very seriously contrary to that person's own best interests, and that, because of this, it would be wrong for one to provide the person with assistance in performing the action. Second, it could be that one has obligations to someone else that one would violate if one helped the person to perform the action in question. One might, for example, belong to a religious group where it is a condition of membership that one does not provide assistance to someone in committing suicide.

In the absence of either of these circumstances, however, is there any way in which it could be wrong to help a person to commit suicide? It is, I suggest, very hard to see any other possibility here. It would certainly seem, then, that the fifth premise is justified.

70   **Michael Tooley**

The sixth premise was this:

(8)   Circumstances may very well be such that neither assisting a person to commit suicide nor performing voluntary active euthanasia violates any obligations that one has to others.

The ground for accepting this premise is simply that, while one might have obligations to others that would make it wrong for one to assist someone to commit suicide, or for one to perform voluntary active euthanasia – obligations that arose, for example, from membership in some religious group, or professional union, that prohibited such actions – it will not in general be true that one has such obligations.

Finally, the concluding premise of my argument was this:

(11)   Whenever assisting a person in committing suicide is justified, voluntary active euthanasia is also justified, provided the latter action does not violate any obligation that one has to anyone else.

Here the thought is simply this. Provided that one does not have any obligations to others that would make it wrong for one to provide someone with voluntary active euthanasia, then the difference between helping someone to end his or her life, and doing it for that person, cannot be morally significant. So the final assumption in the argument is justified.

Voluntary Passive Euthanasia versus Voluntary Active Euthanasia

The argument

My second argument in support of the thesis that voluntary active euthanasia and assisted suicide are not morally wrong in themselves focuses upon the relationship between active and passive euthanasia. To arrive at that argument, consider the following closely related, well-known argument:

(1)   Voluntary passive euthanasia is not morally wrong in itself.
(2)   Intentionally killing a person and intentionally letting a person die are, in themselves, morally on a par.
(3)   The only intrinsic difference between voluntary active euthanasia and voluntary passive euthanasia is that the former is a case of killing, and the latter a case of letting die.
(4)   Therefore, voluntary active euthanasia is not morally wrong in itself. (From (1), (2), and (3).)

Given that (3) is true by definition, and that few think that (1) is mistaken, the crucial premise in the argument appears to be (2). Is it true, then, that killing and letting die are morally on a par? The answer is not entirely clear. On the one hand, a number of philosophers have argued that intentionally killing and intentionally letting die have precisely the same moral status (Rachels, 1975; Tooley, 1980; Oddie, 1997, 1998). One very interesting way of attempting to establish this conclusion, for example, is by means of a "Bare Difference Argument," where the basic idea is to focus upon two cases, each involving a person's death, that differ only in that one is a case of killing, and the other a case of letting die, and where there does not appear to be any morally significant difference between the two cases. (See, for example, Rachels, 1975, p. 79.) If there are such cases, must it not follow that there is no intrinsic moral difference between killing and letting die?

The status of Bare Difference Arguments has been disputed, with many philosophers holding that this form of argument is sound (Rachels, 1979; Malm, 1992; Oddie, 1997), and others holding that it is not sound (Beauchamp, 1977; Foot, 1977, pp. 101–102; Kagan, 1988). On the face of it, the argument certainly appears sound. The problem, however, is that there are cases where the intuitions of most non-consequentialists are that killing and letting die are not morally equivalent. One of the most famous cases, discussed at length by Harris (1975), involves the possibility of killing a healthy person in order to use that person's organs to save two people who need transplants if they are to survive. If killing and letting die are morally on a par, should not killing one person to save two be not only permissible, but also commendable, and perhaps obligatory? Many people, however, feel that that is not so.

I think it can be shown that the Bare Difference Argument is sound. What I shall do here, however, is argue instead that one can avoid this controversial question by shifting from the above argument to a slightly different one.

To see how this can be done, consider the following, asymmetry principle:

(A) Both the property of killing a person and the property of allowing a person to die are wrong-making properties of actions, but the former is a weightier wrong-making property than the latter.

If this principle were correct, then statement (2) in the argument above would be false, and the argument itself would fail. But principle (A) is not sound. The reason is that, as David Boonin (2000, pp. 160–161) has contended, any grounds for holding that there is a moral difference between killing and letting die must also be grounds for holding that a certain much more general principle is correct – the principle, namely, that intentionally causing a given harm is intrinsically more wrong than intentionally allowing that harm to occur. Or, to put it in terms of wrong-making properties:

(B) Both the property of intentionally *causing* a harm, and the property of intentionally *allowing* a harm to occur, are wrong-making properties of actions, but the former is a weightier wrong-making property than the latter.

But if this is right, then to the extent that the killing versus letting die distinction is morally significant, it is so *precisely because* it is just an instance of the more general distinction between intentionally causing harm and intentionally allowing harm to happen. But then the original asymmetry principle stated above cannot be an accurate formulation of what may be true in the killing versus letting die case, since it fails to distinguish between cases where killing and letting die are *harms* and cases where they are *benefits*. What is needed, then, is not (A), but the following, modified asymmetry principle:

(C)   Both the property of killing a person, *when the killing harms the person*, and the property of allowing a person to die, *when allowing the person to die harms the person*, are wrong-making properties of actions, but the former is a weightier wrong-making property than the latter.

Next, given (B), the question naturally arises as to whether there is a corresponding principle dealing with *benefits*, and, in response, I would suggest that if (B) is plausible, then the following principle must also be plausible:

(D)   Both the property of intentionally *causing* a benefit, and the property of intentionally *allowing* a benefit to occur, are right-making properties of actions, but the former is a weightier right-making property than the latter.

Or, at the very least, if (B) is plausible, then surely the following more modest variant on (D) must also be plausible:

(E)   Both the property of intentionally causing a benefit, and the property of intentionally allowing a benefit to occur, are right-making properties of actions, but the former is *at least as weighty* a right-making property as the latter.

But then, finally, if (E) is plausible, then surely the following principle must also be acceptable:

(F)   Both the property of killing a person, *when the killing benefits the person*, and the property of allowing a person to die, *when allowing the person to die benefits the person*, are right-making properties of actions, and the former is *at least as weighty* a right-making property as the latter.

Given principle (F), the argument that I want to advance is then as follows:

(1)   Voluntary passive euthanasia is not morally wrong in itself.
(2)   Both the property of killing a person, *when the killing benefits the person*, and the property of allowing a person to die, *when allowing the person to die benefits the person*, are right-making properties of actions, and the former is *at least as weighty* a right-making property as the latter.
(3)   The only intrinsic difference between voluntary active euthanasia and voluntary passive euthanasia is that the former is a case of killing, and the latter a case of letting die.
(4)   Therefore, voluntary active euthanasia cannot be morally worse in itself than voluntary passive euthanasia. (From (2) and (3).)
(5)   Therefore, voluntary active euthanasia is not morally wrong in itself. (From (1) and (4).)

In Defense of Voluntary Active Euthanasia and Assisted Suicide

An evaluation of the second argument

The argument just set out starts from the following premise:

(1)   Voluntary passive euthanasia is not morally wrong in itself.

This is a claim that few would challenge, as the view that voluntary passive euthanasia is, in general, morally permissible is very widely accepted indeed. But it is not really a claim that should be taken for granted, especially given that many arguments offered against voluntary active euthanasia are in fact arguments against voluntary passive euthanasia as well (Tooley, 1995).

How, then, should one defend this premise? My own approach would be to defend this by the same line of argument that I used earlier to defend the view that suicide is not morally wrong, at least in certain circumstances. Those who hold that suicide is morally wrong would need, of course, to argue along different lines, but that is not something that we need to consider here.

The second premise of the argument is this:

(2)   Both the property of killing a person, when the killing benefits the person, and the property of allowing a person to die, when allowing the person to die benefits the person, are right-making properties of actions, and the former is *at least as weighty* a right-making property as the latter.

How might this premise be challenged? The only challenge, I think, that deserves serious consideration is one that argues that the property of killing a person, when the killing benefits the person, cannot be a right-making property of actions, since the *direct killing* of an innocent person is always wrong in itself.

The proper response to this challenge to the second premise is, I suggest, to ask what basis can be offered for the claim that the direct killing of innocent persons is always morally wrong in itself. One possibility would be an axiological underpinning, according to which the existence of innocent persons is *valuable*, in the sense of making the world a better place. But this way of attempting to explain the principle in question is open to two serious objections. The first is that if this explanation of the wrongness of killing innocent persons were correct, then intentionally refraining from bringing innocent persons into existence would also be morally wrong, and to the very same degree, since the failure to create an object that would have a certain value makes precisely the same difference with regard to the overall value of the world as the destruction of an already existing object of the same sort, other things being equal. But the failure to bring an innocent person into existence is not morally on a par with destroying an innocent person who already exists. So the principle that the direct killing of innocent persons is always wrong in itself cannot be explained axiologically.

The second objection is this. Consider an innocent person who is suffering terribly from an incurable illness, and who would prefer to be killed, rather than to go on living. If one holds that killing such a person would be wrong because one would thereby be destroying something of value, then one should also hold that one would make the world a better place by creating an additional innocent person who one knew would suffer to the same degree as a result of the same incurable disease. But the latter, surely, is very implausible.

A second way of attempting to defend the claim that the direct killing of an innocent person is always wrong in itself is by appealing to the idea of rights, and by holding that such an action is wrong because an innocent person has a right to life. But this account is also open to at least two objections. The first is that people can, in general, waive their rights. Thus, for example, the fact that one has a right to some object does not mean that one does something wrong if one destroys it, or gives it to someone. Why, then, should the situation be any different with regard to the right to life? Why should it not be permissible, for example, to commit suicide? Why should the right to life not be a right that, like other rights, one can waive?

A second and deeper objection involves asking how rights function. A plausible view, I suggest, is that rights function in two ways. First, they function to protect an individual's interests. Second, they function to provide individuals with the freedom to make decisions concerning what they will do with their lives. But if this is correct, then when a person wants to die, helping her to do so will further that person's autonomy, while if it is in the person's interest to be dead, killing that person will further that person's interests. So if rights function to protect interests and autonomy, then, when both the relevant conditions are satisfied, so that it is in the person's interest to be killed, and the person asks to be killed, granting that request will not be contrary to either of the things that rights function to protect. Accordingly, the claim that the direct killing of an innocent person is always wrong in itself cannot be defended by appealing to the idea of rights. The conclusion that will be supported by an appeal to the idea of a right to life will be, at most, the much more limited one that the killing of an innocent person is wrong in itself if that action is either contrary to what the person really wants, or contrary to the person's interest.

The third and final premise of the second argument is this:

(3)   The only intrinsic difference between voluntary active euthanasia and voluntary passive euthanasia is that the former is a case of killing, and the latter a case of letting die.

But this is unproblematic, since it follows from the relevant definitions.

In conclusion, then, the second argument also appears to provide a sound justification for the claim that voluntary active euthanasia is morally acceptable. It is then a straightforward matter to argue that the same is true of assisting a person to commit suicide, in the appropriate circumstances.

Should Assisted Suicide and Voluntary Active Euthanasia Be Legal?

If assisted suicide and voluntary active euthanasia are morally permissible, what should their legal status be? Certainly, the fact that an action is not morally wrong constitutes strong prima facie grounds for concluding that it should not be illegal. It is possible, however, for actions that are not morally wrong, nevertheless, to be such as should be prohibited, on the ground that allowing the actions in question would give rise to other actions that would harm individuals, or violate their rights. Further consideration is therefore necessary, and so, in this final section, I shall consider three important objections to the legalization of assisted suicide and voluntary active euthanasia.

The first argument, put forward by Yale Kamisar (1958), focuses upon possible harm to those who choose to undergo euthanasia. The thrust of Kamisar's argument is that if voluntary active euthanasia is available, some people will choose to be killed in circumstances where being killed is contrary to their interest.

Kamisar's argument is problematic in two ways. In the first place, if a person were tempted to choose euthanasia in a situation where that was contrary to that person's own interest, and where the person was not emotionally disturbed, it is hard to see why, if the person were presented with the reasons why it would be better to go on living, he or she would be unable to appreciate the force of those reasons.

In the second place, it can be shown (Tooley, 1995) that the possibilities for irrational choice that Kamisar proposes, whatever weight they have, generally have precisely as much weight in the case of voluntary *passive* euthanasia. So one cannot (as Kamisar does), hold that such possibilities constitute grounds for not legalizing active euthanasia without equally holding that they are also grounds for not allowing passive euthanasia.

The other two arguments against legalization deserve more careful consideration. First, there are what are often referred to as "wedge" or "slippery slope" arguments against voluntary active euthanasia. These come in two forms. Both maintain that legalizing active euthanasia would be a mistake because doing so would be likely to lead to undesirable consequences involving the legalization of other things. According to one version of the argument, these consequences would follow by virtue of a logical relation: if one legalizes voluntary active euthanasia, then logical consistency requires that one also legalize, for example, involuntary euthanasia (Sullivan, 1975, p. 24). The second version of the wedge argument, by contrast, maintains that the undesirable consequences would follow simply due to certain facts about human psychology.

The problem with the "logical consistency" version of the argument will be clear from the discussion in the previous section. For the present argument can be seen to rest upon the assumption that the relevant basic moral principle involved here is something along the lines of:

(1)   The *direct killing* of an innocent person (or, alternatively, an innocent human being) is always wrong in itself.

But, as we in effect saw earlier, such a principle is not correct. It needs to be replaced, instead, by a principle such as:

(2)   Innocent persons have a right to life.

But then the point is that, while (2) does support a claim such as:

(3)   It is prima facie wrong to kill an innocent person if it is in that person's interest to go on living, or if the person has not given permission to have his or her life terminated.

76   **Michael Tooley**

It does not support a claim such as:

(4)   It is prima facie wrong to kill an innocent person who has a fixed and rational desire to be dead, and who has given permission to have his or her life terminated.

Thus, once the unsound claim that the direct killing of an innocent person is always wrong in itself is replaced by the sound principle that innocent persons have a right to life, the present argument collapses, since one can consistently hold both that voluntary active euthanasia is morally permissible and that involuntary euthanasia is not.

The second form of the wedge argument, by contrast, need not involve any unsound assumption about the relevant moral principles. For here it is granted, at least for the sake of argument, that voluntary active euthanasia is not wrong *in itself*. It is then argued, however, that acceptance of voluntary active euthanasia may lead to the acceptance of actions that *are* wrong in themselves – such as involuntary active euthanasia.

But what reasons are there for thinking that this will take place? Kamisar, in advancing this version of the wedge argument, offers three reasons. First, he claims that advocates of the legalization of voluntary active euthanasia often seem to hold that the case for legalizing certain types of non-voluntary euthanasia is at least as compelling as legalizing voluntary active euthanasia (1958: 1027–8). Second, he cites a poll that measured the amount of public support for, on the one hand, euthanasia for defective infants and, on the other hand, euthanasia for incurably and painfully ill adults, and where the result was that more people approved of the former than of the latter (45 percent versus 37.3 percent) (1958, 1029). Finally, Kamisar appeals (1958, pp. 1031–1032) to what happened under the Nazis, citing the description offered by Leo Alexander (1949; emphasis in original):

> The beginnings at first were merely a subtle shift in emphasis in the basic attitude of the physicians. *It started with the acceptance of the attitude, basic in the euthanasia movement, that there is such a thing as life not worthy to be lived.* This attitude in its early stages concerned itself merely with the severely and chronically sick. Gradually the sphere of those to be included in this category was enlarged to encompass the socially unproductive, the ideologically unwanted, the racially unwanted and finally all non-Germans.
>
> But it is important to realize that the infinitely small wedged-in lever from which this entire trend of mind received its impetus was the attitude toward the non-rehabilitable sick.

How strong are the considerations offered by Kamisar? The problem with the first two types of support that he offers is that they concern attitudes toward *nonvoluntary* euthanasia, and so they are not relevant to the claim that one is in danger of sliding down a slope that leads from voluntary active euthanasia to things that are morally wrong unless one holds that non-voluntary euthanasia – as contrasted with *involuntary* euthanasia – is morally wrong. But, in the first place, this is a deeply controversial claim, as is shown by one of the very facts that Kamisar cites – namely, that more Americans in the poll that he referred to approved of euthanasia in the case of "defective infants" than in the case of "incurably and painfully ill adults." Second, there are strong arguments that can be offered in support of the moral acceptability of euthanasia in the case of severely defective infants – arguments that Kamisar does not even address.

This leaves Kamisar's appeal to the case of Nazi Germany. Here there are at least two questions that need to be asked. The first is whether the claim that is advanced in the above passage is in fact correct. For some writers – such as Marvin Kohl (1975) and Joseph Fletcher (1973) – have argued that the Nazi mass murders, rather than growing out of attitudes toward the non-rehabilitable sick, were based upon the idea of the protection and purification of the Aryan stock, and upon an intense anti-Semitism – and one that was long established in Europe (Hay, 1951). Moreover, if one examines Hitler's *Mein Kampf*, there appears to be very strong evidence for that view, and against Leo Alexander's claim. Consider, for example, the following passages:

> What we must fight for is to safeguard the existence and reproduction of our race and our people, the sustenance of our children and the purity of our blood, the freedom and independence of the fatherland, so that our people may mature for the fulfillment of the mission, allotted it by the creator of the universe. (Hitler, 1971, p. 214)
>
> The Jewish doctrine of Marxism . . . withdraws from humanity the premise of its existence and culture. As a foundation of the universe, this doctrine would bring about the end of any order intellectually conceivable to man. And as, in this greatest of all recognizable organisms [humans], the result of an application of such a law could only be chaos, on earth it could only be destruction for the inhabitants of this planet.
>
> If, with the help of his Marxist creed, the Jew is victorious over the other peoples of the world, his crown will be the funeral wreath of humanity and this planet will, as it did thousands of years ago, move through the ether devoid of men. (Hitler, 1971, p. 65)

In the light of passages such as these, Leo Alexander's claim that the starting point of the Holocaust was "with the acceptance of the attitude, basic in the euthanasia movement, that there is such a thing as life not worthy to be lived" seems clearly untenable.

Secondly, even if Leo Alexander were right, one would still need to go on to ask to what extent the Nazi experience, which occurred in a dictatorship, is a good indicator of what is likely to happen in a democratic society such as the United States.

The answer, surely, is that it is not: if someone in the United States were to advocate such a program, the opposition would be overwhelming.

Finally, it is also possible to offer empirical evidence against the wedge argument, as is done, for example, by Rachels (1993, p. 62). He argues that there is "historical and anthropological evidence that approval of killing in one context does not necessarily lead to killing in different circumstances," and cites, as illustrations, the killing of defective infants in various societies, and the killing of people in self-defense in our own society. So there is good reason for thinking that people are perfectly capable of drawing clear and firm moral lines, and therefore are not in danger of sliding down what are claimed to be slippery slopes.

In addition, however, one can now offer empirical evidence of a very direct sort, since there is a society where voluntary active euthanasia has, for the past few years, been permitted – namely, the Netherlands. For while some who oppose legalization of voluntary active euthanasia have claimed that the Dutch experiment provides support for the slippery slope argument – on the grounds that there have been cases of involuntary euthanasia in the Netherlands – in fact the opposite is the case, as emerges if one compares the situation in the Netherlands with what obtains in societies where voluntary active euthanasia is not permitted. In particular, if one compares the results of surveys carried out in the Netherlands in 1990 and 1995, and in Australia in 1995–16, the following facts emerge. First, in the Netherlands, the percentage of active terminations without the patient's explicit consent fell from 0.8 percent to 0.7 percent over the period from 1990 to 1995, whereas, in Australia, the percentage of such cases in 1995–6 was 3.5 percent – that is, five times higher than in the Netherlands. Second, in the Netherlands, in 1995, 13.5 percent of all deaths involved a decision to withhold or withdraw treatment, whereas in Australia in 1995–6, this occurred in 30.5 percent of cases. Moreover, in Australia, in 22.5 percent of the cases, the decision to withhold or withdraw treatment was done without the patient's explicit consent (Kuhse et al., 1997; Oddie, 1998). The conclusion, accordingly, is that the rights of individuals are more likely to be violated when voluntary active euthanasia is illegal, than when it is permitted.

This brings me to the final objection that I shall consider to the legalization of voluntary active euthanasia. The thrust of this objection is that there are serious problems about how to implement the legalization of euthanasia. Should there be no laws at all concerning voluntary active euthanasia? That surely would lead to significant abuse. But if laws are needed, what form should they take? If the laws introduced complex and stringent procedures, then relatively few people who would benefit from voluntary active euthanasia might wind up being able to do so. On the other hand, if the procedures were very relaxed ones, would the likelihood of abuses not re-emerge?

In response to this problem, Rachels (1993, pp. 63–65) has suggested that one can bypass the problem of writing difficult and detailed legislation dealing with when voluntary active euthanasia is permissible by instead introducing a rule to the effect that, just as the fact that a killing has been done in self-defense may serve as a defense against a charge of homicide, so the fact that a killing was one of voluntary euthanasia could function in the same way – that is, as a satisfactory defense against a charge of homicide.

Rachels' proposal is an interesting one, and it would certainly appear to be a desirable change. It is unclear, however, how much access people would have to voluntary active euthanasia as a result. But perhaps one could combine Rachels' suggestion with the introduction of rather conservative legislation that would prescribe procedures under which voluntary active euthanasia, in certain clear-cut types of case, would be legally permissible. The combination of these two approaches might then both provide access to voluntary active euthanasia for those in need, while at the same time minimizing the likelihood of abuse, since anyone committing euthanasia in a borderline case would need to be prepared to prove that it was indeed a case of voluntary euthanasia.

One final important issue is this. It is usually assumed that if voluntary active euthanasia or assisted suicide were to be legalized, then such actions would be carried out by doctors. This assumption has led to strong opposition to legalization on the part of the American Medical Association, which has held that, in view of the basic orientation of the practice of medicine toward the saving of lives, doctors should not perform active euthanasia. Advocates of legalization have tended to respond by challenging the latter view – arguing, for example, that assisted suicide and voluntary active euthanasia are not really contrary to the Hippocratic Oath. But, even if this is so, one might very well ask whether it might not in fact be better if doctors were not involved, and if, instead, both the relevant counseling, and the carrying out of the actions in question, were in the hands of other trained professionals. For, in the first place, it may very well be psychologically difficult, for many people, to shift from attempting to do everything that can be done to save a person's life, to doing something to end that person's life. In the second place, would it not be better for euthanasia and assisted suicide to be carried out by people who have been specially trained to do this, people who are willing to step in when doctors have done all that they can, who are knowledgeable about the needs and the psychology of those who are dying, and who are therefore better able to provide the support and comfort that is needed at such a time?

References

Alexander, L. (1949) Medical science under dictatorship. *New England Journal of Medicine* 241: 39–47.

Beauchamp, T.L. (1977) A reply to Rachels on active and passive euthanasia. In *Social Ethics: Morality and Social Policy*, ed. T. Mappes and J. Zembaty, pp. 67–76. New York: McGraw-Hill.

Boonin, D. (2000) How to argue against active euthanasia. *Journal of Applied Philosophy* 17 (2): 157–168.

Callahan, D. (2014) A case against euthanasia. In *Contemporary Debates in Applied Ethics*, 2nd edn, ed. A.I. Cohen and C.H. Wellman, pp. 82–92. Malden, MA: Wiley-Blackwell.

Fletcher, J. (1973) Ethics and euthanasia. In *To Live and to Die*, ed. R.H. Williams, pp. 113–122. New York: Springer Verlag. (Reprinted in Fletcher, J. (1979) *Humankind: Essays in Biomedical Ethics* (pp. 149–58). Buffalo, MA: Prometheus Books.).

Foot, P. (1977) Euthanasia. *Philosophy & Public Affairs* 6 (2): 85–112.

Harris, J. (1975) The survival lottery. *Philosophy* 50: 81–87.

Hay, M. (1951) *The Foot of Pride*. Boston: Beacon Press. (Reprinted in 1960 as *Europe and the Jews*. Boston, NJ: Beacon Press.).

Hitler, A. (1971 [1925–7]) *Mein Kampf*, trans. R. Manheim. Boston, MA: Houghton Mifflin Company.

Kagan, S. (1988) The additive fallacy. *Ethics* 99: 5–31.

Kamisar, Y. (1958) Some nonreligious views against proposed "mercy-killing" legislation. *Minnesota Law Review* 42 (6): 969–1042.

Kohl, M. (1975) Voluntary beneficent euthanasia. In *Beneficent Euthanasia*, ed. M. Kohl, pp. 130–141. Buffalo, NY: Prometheus Books.

Kuhse, H., Singer, P., Baume, P. *et al.* (1997) End-of-life decisions in Australian medical practice. *Medical Journal of Australia* 166: 191–196.

Malm, H. (1992) In defense of the contrast strategy. In *Ethics: Problems and Principles*, ed. J.M. Fischer and M. Ravizza, pp. 272–277. New York: Harcourt Brace Jovanovich.

Oddie, G. (1997) Killing and letting-die: from bare differences to clear differences. *Philosophical Studies* 88: 267–287.

Oddie, G. (1998) The moral case for the legalization of voluntary euthanasia. *Victoria University of Wellington Law Review* 28: 207–224.

Parfit, D. (1984) *Reasons and Persons*. Oxford: Oxford University Press.

Rachels, J. (1975) Active and passive euthanasia. *The New England Journal of Medicine* 292(2): 78–80.

Rachels, J. (1979) Euthanasia, killing, and letting die. In *Ethical Issues Relating to Life and Death*, ed. J. Ladd, pp. 146–163. Oxford: Oxford University Press.

Rachels, J. (1993) Euthanasia. In *Matters of Life and Death*, 3rd edn, ed. T. Regan, pp. 30–68. New York: McGraw Hill.

Sullivan, J.V. (1975) The immorality of euthanasia. In *Beneficent Euthanasia*, ed. M. Kohl, pp. 12–33. Buffalo, NY: Prometheus Books.

Tooley, M. (1980) An irrelevant consideration: killing versus letting die. In *Killing and Letting Die*, ed. B. Steinbock, pp. 56–62. Englewood Cliffs, NJ: Prentice-Hall.

Tooley, M. (1995) Voluntary euthanasia: active versus passive, and the question of consistency. *Revue Internationale de Philosophie* 49 (3): 305–322.

Further Reading

"Ad Hoc" Committee of the Harvard Medical School (1968) A definition of irreversible coma: report of the "ad hoc" committee of the Harvard Medical School to examine the definition of brain death. *Journal of the American Medical Association* 205 (6): 337–340.

Den Hartogh, G. (1998) The slippery slope argument. In *A Companion to Bioethics*, ed. H. Kuhse and P. Singer. Oxford: Blackwell.

Gay-Williams, J. (1979) The wrongfulness of euthanasia. In *Intervention and Reflection: Basic Issues in Medical Ethics*, ed. R. Munson, pp. 141–143. Belmont, CA: Wadsworth Publishing Company.

Hume, D. (1985 [1777]) Of suicide. Reprinted In *Essays Moral, Political and Literary*, ed. E.F. Miller, pp. 577–589. Indianapolis: Liberty Fund.

Kuhse, H. and Singer, P., eds (1998) *A Companion to Bioethics*. Oxford: Blackwell.

Lamb, D. (1988) *Down the Slippery Slope*. London: Croom Helm.

Maguire, D.C. (1975) A Catholic view of mercy killing. In *Beneficent Euthanasia*, ed. M. Kohl, pp. 34–43. Buffalo, NY: Prometheus Books.

Sacred Congregation for the Doctrine of Faith (1980). *Declaration on Euthanasia.* http://www.vatican.va/roman_curia/congregations/cfaith/documents/rc_con_cfaith_doc_19800505_euthanasia_en.html (last accessed 6/17/13).

Tooley, M. (1979) Decisions to terminate life and the concept of a person. In *Ethical Issues Relating to Life and Death*, ed. J. Ladd, pp. 62–92. Oxford: Oxford University Press.

Walton, D. (1992) *Slippery Slope Arguments*. Oxford: Oxford University Press.

Williams, G. (1958) "Mercy-killing" legislation – a rejoinder. *Minnesota Law Review* 43 (1): 1–12.

CHAPTER FIVE

A Case Against Euthanasia

Daniel Callahan

Consider what I take to be a mystery. Life presents all of us with many miseries, sick or well. Why is it then that so few people choose to end their own lives in response to them? Why is it that when someone does commit suicide – even for reasons that seem understandable – the common reaction (at least in my experience) is one of sorrow, a feeling of pity that someone was driven to such a desperate extreme, particularly when most others in a similar situation do not do likewise? I ask these questions because, behind the movement and arguments in favor of euthanasia or physician-assisted suicide (PAS) – and I consider euthanasia a form of suicide – lies an effort to make the deliberate ending of one's life something morally acceptable and justifiable; and which looks as well to the help of government and the medical profession to move that cause along.[1]

It goes against the grain, I believe, of reason, emotion, and tradition, and all at the same time. If not utterly irrational, it is at least unreasonable – that is, it is not a sensible way to deal with the tribulations of life, of which a poor death is only one of life's horrible possibilities. Suicide generally provokes a negative emotional response in people, even if they can grasp the motive behind it. That response does not prove it is wrong, but it is an important signal of a moral problem. As for tradition, the doctor is being asked by a patient to go against the deep historical convictions of his discipline, to use his or her skills to take life rather than to preserve it, and to lend to the practice of euthanasia the blessing of the medical profession. I understand all of this to be opening the door to new forms of killing in our society, not a good development.

There have been, in Western culture, only three generally accepted reasons for taking the life of another, which is what euthanasia amounts to: self-defense when one's life is threatened, warfare when the cause is serious and just, and capital punishment, the ultimate sanction against the worst crimes. The movement to empower physicians legally to take the life of a patient, or help the patient take his own life, would then legitimate a form of suicide, but would also add still another reason by calling on medical skills to end a person's life.

Contemporary Debates in Applied Ethics, Second Edition. Edited by Andrew I. Cohen and Christopher Heath Wellman.

Suicide: The Way (Rarely) Taken

Let me return to the first of my two questions. Why do comparatively few people turn to suicide as a way of dealing with awful lives? People die miserable deaths all the time, from a wide range of lethal diseases and other causes. While it may cross their minds from time to time, few seem to want euthanasia or PAS as a way out. Millions of people have been brutally treated in concentration camps, with many of them ultimately to die – and yet suicide has never been common in such camps. Many millions of others have undergone all kinds of personal tragedy – the death of children or a spouse, the end of marriage or a deep romance, failures in their work or profession – but most of them do not turn to suicide either. The disabled have been long known to have a lower suicide rate than able-bodied people.

Euthanasia is often presented as a "rational" choice for someone in great pain and whose prospects are hopeless. And yet rationality implies some predictability of behavior, that is, some reasonable certainty that people will act in a consistent and foreseeable way under certain familiar circumstances. Yet it is almost impossible, save for severe depression, to predict whether someone suffering from a lethal illness is likely to turn to suicide. It is far more predictable that, when faced with even the worst horrors of life, most people will *not* turn to suicide. It is no less predictable that, when gripped by pain and suffering, they will want relief, but not to the extent of ending their lives to get it.

We may of course say that people fear ending their own lives, lacking the nerve to do so, or that religious beliefs have made suicide a taboo, or that it has hitherto been difficult to find expert assistance in ending one's life. Those are possible explanations, but since some people do in fact commit suicide, we know that it is hardly impossible to overcome those deterrents. Moreover, to say that most of the great religions and moral traditions of the world have condemned suicide does not in the end explain much at all. *Why* have they done so, even when, at the same time, they usually do not condemn laying down one's life to save another? In the same vein, why has the Western medical tradition for some 2,500 years, going back to Hippocrates, prohibited physicians from helping patients to commit suicide?

My guess is that the answer to the first of those two questions is that suicide is seen as a particularly bad way to handle misery and suffering, even when they are overwhelming – and the behavior of most people in turning away from suicide suggests they share that perception. It is bad because human life is better, even nobler, when we human beings put up with the pain and travail that come our way. Life is full of pain, stress, tragedy, and travail, and we ought not to want to tempt others to see suicide as a way of dealing with it. We would fail ourselves and, by our witness, our neighbor as well, who will know what we did and be led to do so themselves some day.

I began by asking at the outset why most suicides are treated as unhappy events, even when they obviously relieved someone's misery, which we would ordinarily consider valuable. Those readers who have been to the funerals of suicides will know how rarely those at such funerals feel relief that the misery of the life leading up to them has now been relieved. They almost always wish the life could have ended differently, that the suffering could have been borne. My surmise is that those of us who are bystanders or spectators to such deaths know that a fundamental kind of taboo of a rational kind has been broken, some deep commitment to life violated, and that no relief of pain and suffering can justify that. To say this is by no means to condemn those who do so. We can often well enough comprehend why they were driven to that extreme. Nor do I want to imply that they must have been clinically depressed. I am only saying that it is very hard to feel good about suicide or to rejoice that it was the way chosen to get out of a burdensome life.

I present these considerations about suicide as speculations only, not as some kind of decisive arguments against euthanasia. But I think it important to see what sense can be made of a common revulsion against suicide, and sadness when it happens, that has marked generations of people in most parts of the world. Moreover, as I will develop more fully below, it turns out that the experience with the Dutch euthanasia laws and practice, as well as with the Oregon experience with PAS, indicates that it is not misery, pain, and suffering in any ordinary sense that are the motivation for the desire to put an end to one's life. It is instead in great part a function of a certain kind of patient with a certain kind of personality and outlook upon the world.

It is, I believe, important that we try to make sense of these background experiences and reactions. They tell us something about ourselves, our traditions, and our human nature. They offer an enriched perspective when considering the most common arguments in favor of euthanasia. On the surface those arguments are meant to seem timely, in tune with our mainstream values, commonsensical and compassionate, and of no potential harm to our medical practice or our civic lives together. I would like to show that they are indeed in tune with many of our mainstream values, but that they are misapplied in this case, harmful to ourselves and others if we accept them.

Three Arguments in Favor of Euthanasia

I want now to turn to the main arguments in favor of euthanasia, and to indicate why I think they are weak and unpersuasive. I will follow that with a discussion of the legal problem of euthanasia and PAS, and conclude with some comments on the experience with euthanasia and PAS in the Netherlands and the state of Oregon.

Three moral arguments have been most prominent in the national debate. One of them is that we ought, if we are competent, to have the right to control our body as we see fit and to end our life if we choose to do so. This is often called the right of self-determination. Another is that we owe it to each other, in the name of beneficence or charity, to relieve suffering when we can do so. Still another is that there is no serious or logical difference between terminating the treatment of a dying patient, allowing the patient to die, and directly killing a patient by euthanasia. I will look at each of these arguments in turn.

If there is any fundamental American value, it is that of freedom and particularly the freedom to live our own lives in light of our own values. The only limit to that value is that, in the name of freedom, we may not do harm to others. At least a hundred years ago the value of freedom was extended to the inviolability of our bodies – that is, our right not to have our bodies invaded, abused, or used without our consent. Even to put our hands on another without their permission can lead to our being charged with assault and battery. That principle was extended to participation in medical research and the notion of informed consent: no individual can use your body for medical research without your specific informed consent granting them permission to do so. In later years, many construed earlier bans on abortion as an interference with the right of a woman to make her own choices about her body and the continuance of a pregnancy.

It seemed, then, only a small and logical step to extend the concept of freedom and self-determination to the end of our life. If you believe that your pain and suffering are insupportable, and if there is no hope that medicine can cure you of a fatal disease, why should you not have the right to ask a physician directly to end your life (euthanasia) or to provide you with the means of doing so (PAS)? After all, it is your body, your suffering, and if there is no reason to believe others will be harmed by your desire to see your life come to an end, what grounds are there for denying you that final act of self-determination? As I suggested above, it is precisely because the claim of self-determination in this context seems so much in tune with our traditional value of liberty that it seems hard to find a reason to reject it.

But we should reject it, and for a variety of considerations, three of which seem most important. The first is that euthanasia is mistakenly understood as a personal and private matter only of self-determination. Suicide, once a punishable crime, was removed from the law some decades ago in the United States. But it is one thing not to prosecute a person for attempting suicide and quite another to think that euthanasia is a private act, impacting on no other lives. On the contrary, with euthanasia as its means, it becomes a social act by virtue of calling upon the physician to take part in it. Legalizing it would also provide an important social sanction and legitimation of those practices. They would require regulation and legal oversight.

Most critically, it would add to the acceptable range of killing in our society, noted above – one more occasion for the taking of life. To do so would be to reverse the long-developing trend to limit the occasions of socially sanctioned killing, too often marked by abuse. Euthanasia would also reinstate what I would call "private killing," by which I mean a situation where the agreement of one person to kill another is ratified in private by the individuals themselves, not by public authorities (even if it is made legal and supposed safeguards put in place). Dueling as a way of settling differences was once accepted, a form of private killing, something between the duelists only. But it was finally rejected as socially harmful and is nowhere now accepted in civilized society. The contention that it was *their* bodies at stake, *their* private lives, was rejected as a good moral reason to legally accept dueling.

Euthanasia as a Social, not Private, Act

A closely related objection is that what makes euthanasia and PAS social, and not individual, matters is that, by definition, they require the assistance of a physician. Two points are worth considering here. The first is whether we want to sanction the private killing that is euthanasia by allowing physicians to be one of the parties to euthanasia agreements. Since the doctor–patient relationship is protected by the long-standing principle of confidentiality – what goes on between doctor and patient may not be legally revealed to any third party – that gives doctors enormous power over patients.

Whatever the law might be, there will be no way of knowing whether doctors are obeying regulations allowing for euthanasia or PAS, no way of knowing whether they are influencing patient decisions in wrongful ways, no way of knowing whether they are acting with professional integrity. As Sir Charles Allbutt, a British physician, nicely put the problem a century ago:

> If all professions have their safeguards they also have their temptations, and ours is no exception. . . . Unfortunately the game of medicine is played with the cards under the table . . . who is there to note the significant glance, the shrug, the hardly expressed innuendo of our brethren. . . . Thus we work not in the light of public opinion but in the secrecy of the chamber. (Cited in Scarlett, 1991, pp. 24–25)

To give physicians the power to kill patients, or assist in their suicide, when their actions are clothed in confidentiality is to run a considerable risk, one hard to spot and one hard to act upon. As will be noted below, the Dutch experience with euthanasia makes clear how easy it was for doctors to violate the court-established rules for euthanasia and to do so with impunity. There is just no way, in the end, for outsiders to know exactly what doctors do behind the veil of confidentiality; that in itself is a threat.

The second consideration is that the tradition of medicine has, for centuries, opposed the use of medical knowledge and skill to end life. Every important Western medical code of ethics has rejected euthanasia – and rejected it even in those eras when there were many fewer ways of relieving pain than are now available. That could hardly have been because earlier generations of doctors knew less about, or were more indifferent to, pain and suffering. Their relief was at the very heart of the doctor's professional obligation.

There was surely another reason. The medical tradition knew something of great importance: doctors are all too skilled in knowing how to kill to be entrusted with the power to deliberately use that skill. This is not to say that physicians are corrupt, prone to misuse their power; not at all. It is only to say, on the one hand, that the very nature of their profession is to save and protect life, not end it; and that they also, on the other hand, become inured much more than the rest of us to death. Ordinary prudence suggests that the temptation to take life should be kept from them as far as possible. To move in any other direction is to risk the corruption of medicine and to threaten the doctor–patient relationship.

But what of the duty to relieve suffering, to act out of compassion for another? Did the moral strictures against euthanasia and PAS in effect simply forget about, or ignore, that duty? Not at all, but the duty to relieve suffering has never been an absolute duty, overriding all moral objections. No country now allows, or has ever allowed, euthanasia without patient consent even if the patient is incompetent and obviously suffering. Nor are patients' families authorized to request euthanasia under those circumstances. Moreover, as time has gone on, the ability of physicians to relieve patients of just about all pain and suffering through good palliative care has shown that most suffering can be relieved without the ultimate solution of killing the patient. In any event, any alleged duty to relieve suffering has historically always given way to the considerations, noted above, about the nature of medicine as a profession whose principal duty is construed as the saving not the taking of life, and not even when the life cannot medically be saved.

The third argument against euthanasia I want to consider is based on the belief that there is no inherent moral difference between killing a patient directly by euthanasia and allowing a patient to die by deliberately terminating a patient's life-supporting treatment (by turning off a ventilator, for example). Since physicians are allowed to do the latter, it is said that they should be allowed to do the former as well – and indeed that it may be more merciful to carry out euthanasia than to stop treatment, perhaps increasing and prolonging the suffering before the patient actually dies. In effect, the argument goes, terminating treatment will foreseeably end the life of the patient, a death hastened by the physician's act; and that is no different, in its logic or outcome, from killing the patient directly by euthanasia (Rachels, 1975).

There are some mistakes in this argument. One of them is a failure to remember that patients with truly lethal, fatal diseases cannot be saved in the long run. The most that can be accomplished is, by aggressive treatment, to delay their death. At some point, typically, a physician will legitimately decide that treatment cannot bring the patient back to good health and cannot reverse the downhill course of the illness. The disease is in control at that point and, when the physician stops treatment, the disease takes over and kills the patient. It has been long accepted that, in cases of that kind, the cause of death is the disease, not the physician's action.

Moreover, how can it be said that a physician has "hastened" a patient's death by ending life-saving treatment? After all, but for the doctor's action in keeping the patient alive in the first place and then continuing the life-sustaining treatment, the patient would have died much earlier. Put another way, the doctor saves the patient's life at one point in time, sustains the patient's life through a passage of time, and then allows the patient to die at still another time. Since no physician has the power to stay indefinitely the hand of death, at some point or other, in any case, the physician's patient will be irreversibly on the way to death; that is, at some point, life-sustaining treatment will be futile. To think that doctors "kill" patients by terminating treatment is tantamount at that point to saying that doctors have abolished lethal disease and that they now die only because of a physician's actions. It would be lovely if doctors have achieved that kind of power over nature, with death solely in their hands. It has not happened, and is not likely ever to happen. To say this is not to deny that physicians can misuse their power to terminate treatment wrongly: they can stop treatment when it could still do some good, or when a competent patient wants it continued. In that case, however, the physician is blameworthy. It is still the underlying disease that does the killing, but the physician is culpable for allowing that to happen when it ought not to have happened.

Euthanasia and the Law

I have provided some reasons why, ethically speaking, euthanasia and PAS cannot be well defended. But what of the law? If we claim to live in a free country, and believe in pluralism, should not the law leave it up to us as individuals to decide how our lives should end? Many people will reject my arguments against euthanasia, and public opinion polls have consistently shown a majority of Americans to be in favor of it. A law that simply allowed those practices, but coerces no one to embrace them, would seem the most reasonable position.

Not necessarily. I noted earlier that the moral acceptance of euthanasia would have the effect of legitimating the role of the physician as someone now empowered to end life. It would also bring an enormous change in the role of the physician, changing the very notion of what it means to be one (Kass, 2004). Seen in that light, a law permitting euthanasia would have social implications far beyond simply giving patients the right to choose how their lives end. As in so many other matters, what on the surface looks like a narrowly private decision turns out, with legalization, to send much wider ripples through society in general and the practice of medicine in particular. It has been said that, in addition to its regulatory functions, the law is a teacher, providing a picture of the way we think people should live together. Legalized euthanasia would teach the wrong kind of lesson.

The actual enforcement of a law on euthanasia would be enormously difficult to carry out. The privacy of the doctor–patient relationship means that there is an area that the law cannot enter. Whatever conditions the law might set for legal euthanasia, there is in the end no good way to know whether it is being obeyed. Short of having a policeman sitting in on every encounter between a doctor and a patient, what they agree to will remain unavailable to the rest of us. All laws are subject to abuse, particularly when they are controversial in the first place. Not everyone will agree with the law as written, and we can be sure that some will bend it or ignore it if they can get away with it. But in most cases it is possible to detect the violation. We know when our goods have been stolen, just as we can know when someone has been brutally beaten.

It would be far more difficult to detect abuses with euthanasia. For one thing, two of the main reasons offered in favor of euthanasia – self-determination and the relief of suffering – do not readily lend themselves to the limits of law. Why should a right of self-determination be limited to those in a terminal state, which is what is commonly proposed and is required in Holland and Belgium? The Dutch law, which does not require a terminal illness, but only unbearable suffering, is in that respect much more perceptive about the logical and legal implications of the usual moral arguments in favor of euthanasia, which is why it rejected a terminal illness requirement.

The Dutch realized that the open-ended logic of the moral reasons behind euthanasia do not lend themselves well to artificial, legal barriers. Impending death is not the only horrible thing in life and, if an individual's body is her own, why should any interference with her choice be tolerated? The requirement of an impending death seems arbitrary in the extreme. As for the relief of suffering, why should someone have to be competent and able to give consent, as if the suffering of those lacking such capacities counts for less? In short, the main reasons given for the legalization of euthanasia seem, logically, to resist the kinds of limit built into the Oregon and Belgian laws. That reality opens the way to abuse of the law. All it requires is a physician who finds the law too narrow, the deed too easy, and a desperate patient all too eager to die.

The Dutch Experience

This is not speculation. The Netherlands offers a case study of how it happens. For many decades, until a formal change in the law only recently, the Dutch courts had permitted euthanasia if certain conditions were met: a free choice, a considered and persistent request, unacceptable suffering, consultation with another physician, and accurate reporting on the cause of death. Throughout the 1970s and 1980s euthanasia (and occasionally PAS) was carried out, with many assurances that the conditions were being met. But, curious to find out about the actual practice, the Dutch government established a Commission on Euthanasia in 1990 to carry out an anonymous survey of Dutch physicians (Van der Maas, 1992).

The survey encompassed a sample of 406 physicians, and two other studies, which, taken together, were eye-opening. The official results showed that, based on their sample, out of a total of 129,000 deaths there were some 2,300 cases of voluntary ("free choice") euthanasia and 400 cases of assisted suicide. In addition, most strikingly, there were some 1,000 cases of intentional termination of life without explicit request, what the Dutch called "non-voluntary euthanasia." In sum, out of 3,300 euthanasia deaths, nearly one-third were non-voluntary. Less than 50 percent of the euthanasia cases were reported as euthanasia: another violation of the court rules. Worst of all, some 10 percent of the non-voluntary cases were instances of euthanasia with competent patients who were not asked for their consent.

None of that was supposed to be happening, and was a clear abuse of the court-established rules. A number of doctors had obviously taken it upon themselves to unilaterally end the lives of many patients. If that could happen there, it could happen here. Since that time, the Dutch have officially established a legal right to euthanasia (replacing the early court-established guidelines), but the government there has recently found that only 50 percent of the physicians who carry it out report doing so, and that there continue to be 400 cases a year of voluntary euthanasia (Sheldon, 2003).

The American state of Oregon, which legalized PAS in 1994, but whose actual implementation was delayed until 1997 by a number of court challenges, offers a variety of further insights into the practice. To the surprise of many, the actual number of people to take advantage of the new law has been small. The number of prescriptions for PAS, written for the first four years, beginning in 1998, has been 24 (1998), 33 (1999), 39 (2000) and 44 (2001) – and the actual number of deaths from their use has been 16 (1998), 27 (1999), 27 (2000), and 21 (2001) (Oregon Death with Dignity, 2002, pp. 24–25). While it is not wholly clear why there have been so few, or why all of those for whom lethal drugs were prescribed did not in the end use them, a possible reason is the high quality of pain relief care available in Oregon. Palliative care (that part of medicine that aims to reduce pain and suffering) has been particularly strong in Oregon, in great part because both proponents and opponents of the law worked hard to improve palliative care at the end of life, making it the most effective state in the country for doing so.

Has there been any abuse of the law in Oregon, as was the case in the Netherlands? One group supportive of the law, the Oregon Death with Dignity Legal Defense and Education Center (2002), has flatly stated that "after four full years of legalized death with dignity in Oregon, there have been no missteps, abuses or coercive tendencies" (Oregon Death with Dignity, 2002, pp. 24–25). That may possibly be true, but they cannot possibly *know* it to be true. The experience of Holland, which uncovered abuses only after careful anonymous surveys of physicians, ought to raise a cautionary flag. Before that survey, Dutch euthanasia supporters issued equally confident statements about the purity of their practice. The underlying question, in the absence of such anonymous information, is whether there still exist incentives for physicians to violate the Oregon law. My surmise is that there are, simply because not all patients who want PAS will meet the legal standards and because not all physicians will agree with those standards. That is what happened in Holland and may quite possibly be happening in Oregon as well. We will only know after an anonymous survey.

Not Pain but Loss of Control

But for me the most interesting information to come out of the Oregon situation are the reasons given for wanting PAS. The standard argument, used frequently in getting the Oregon law changed to permit PAS, was that of the relief of unbearable pain and suffering. But as it has turned out, only a minority of those who availed themselves of the law cited inadequate pain relief. The pattern for 2001 was similar to that of the preceding year: the major motive (for about 80 percent of those wanting PAS) was "loss of autonomy and a diminished ability to participate in activities that make life enjoyable," to use the language of a major organizational supporter of the law (Oregon Death with Dignity, 2002, pp. 24–25). No doubt that is a form of suffering for those for whom "loss of autonomy" is a grievous affliction. But of course that is one of the results of lethal disease and old age. We will all, eventually in our lives, lose our autonomy and see a reduction of our ability to do that which makes life enjoyable. Our bodies just give out at some point.

If that is our human fate – forestalled a bit by modern medicine, but not nullified – why is it that only a tiny fraction of the population wants euthanasia as the solution to the medical miseries of their lives even if they might let the law accept it with others? To refine that question a bit: why is it that PAS in Oregon seemed to attract people with a heavy focus on autonomy in their lives? From the clinical evidence, there seems to be no fixed response of human beings to suffering, and certainly no probability that people will typically see suicide as the way out. Much, if not everything, seems to depend on individual differences in values, not in bodily responses to pain or impending death. In reporting on the first year of the Oregon law, the state Oregon Health Commission noted that a majority of the 16 reported cases involved people with a particular fear of a "loss of control or the fear of loss of general control, and a loss of bodily function" (Chin et al., 1999, pp. 580, 582). It was not the unbearable and unrelievable physical pain so often and luridly emphasized in the efforts to legalize PAS, or a fear of abandonment, or dependency on others (though some mentioned that), or a feeling of meaninglessness in suffering.

Worry about such a loss represents a particular set of personal (and idiosyncratic) values, by no means a widely distributed set. This was well brought out in the official state report. What the state officials did was to match those who received PAS (called the "case" group) with a group (called the "control" group) of patients with "similar underlying illnesses," and matched as well for age and date of death (Chin et al., 1999, p. 578). Their findings were striking: the PAS group was much more concerned about autonomy and control than the other group. Even more provocative was the fact that the PAS group was far more able to function physically than the control group: "21 percent of the case patients, as compared with 84 percent of the control patients . . . were completely disabled" (Chin et al., 1999, p. 580). In other words, the PAS group was far better off physically than the control group. It was their personal values that led them in one direction rather than another, not the objective intensity of their incapacities. Or to put it in terms we used earlier, PAS represents a legitimation of suicide for those who have a particular conception of the optimum life and its management, one of complete control.

Catering to a Small Minority

If it turns out, then, that PAS heavily attracts a particular kind of person, one very different from most terminally ill people, then much of the public policy argument on its behalf fails. It is not a general problem requiring drastic changes in law, tradition, and medical practice. Just as suicide in general, whatever the level of misery, is not the way most people seek to deal with it, so also are euthanasia and PAS the desire of a tiny minority. These results, it should be added, are much the same as those found in the Netherlands. At a 1991 conference there with the leaders of the Dutch euthanasia movement, I asked the physicians how it was possible reliably to diagnose "unbearable" or "untreatable" suffering as a medical condition and thus suitable for their euthanasia or PAS ministrations. They conceded that there is no reliable medical diagnosis, no way of really knowing what was going on within the mind and emotions of the patient, and – consistent with the findings of the Oregon state study – no correlation *whatever* between a patient's actual medical condition and the reported suffering.

Perhaps euthanasia is not, as many would like to put it, simply a logical extension of the physician's duty to relieve pain and suffering, an old obligation in a new garment. Perhaps it is just part of the drift toward the medicalization of the woes of life, particularly that version of life that regards the loss of control as the greatest of human indignities. Not only that, but even the fear of a loss of control is for many tantamount to its actual loss. I wonder if the voters of Oregon, and all of those who believe euthanasia a needed progressive move, mean to empower unto death that special, and small, subclass of patients uncommonly bent on the control of their lives and eager to have the help of doctors to do so. Somehow I doubt it, but it looks as if that may be what they got.

Underlying much of what I have written here are two assumptions, which need some defense. One of them is that good palliative care, a rapidly growing medical specialty, can relieve most pain and suffering. Some cases, I readily concede, may not be helped, or not enough, by even the best palliative care, but the overwhelming majority can be. My second assumption is this: it is bad public policy to abandon long-standing legal prohibitions, with important reasons and traditions behind them, for the sake of a very small minority, and particularly when the consequences open the way for abuse and a fundamental change in medical values. The fact, for so it seems, that the small minority reflects not some general human response to pain and suffering but a personal, and generally idiosyncratic, view of suffering is all the more reason to hesitate before legally blessing euthanasia. Human beings, in their lives and in their deaths, have long been able to see their lives come to an end without feeling some special necessity to have it ended of them, directly by euthanasia or self-inflicted by PAS.

What about the notion of "death with dignity," a phrase much used by euthanasia supporters? It is a misleading, obfuscating phrase. Death is no indiginity, even if accompanied by pain and a loss of control. Death is a fundamental fact of human biology, as fundamental as any other part of human life. If that human life has dignity as human life, it cannot be lost because death brings it to an end, even if in a disorderly, unpleasant fashion. It takes more than that to erase our dignity. Human beings in concentration camps did not lose their essential human value and dignity by being tortured, humiliated, and degraded. Euthanasia confers no dignity on the process of dying; it only creates the illusion of dignity for those who, mistakenly, believe a loss of control is not to be endured. It can be, and most human beings have endured it. No one would say that the newborn baby, unable to talk, incontinent, utterly unable to control her situation, and unable to interact with others, lacks dignity. Neither does the dying older person, even if displaying exactly the same traits. Dignity is not so easily taken from human beings. Nor can euthanasia confer it on someone.

Note

1   Unless there is a need to deal with the difference between euthanasia and physician-assisted suicide (PAS), I will hereafter refer only to euthanasia. By euthanasia I mean the direct killing of a patient by a doctor, ordinarily by means of a lethal injection. By PAS I mean the act of killing oneself by means of lethal drugs provided by a physician.

References

Chin, A.E., Hedberg, K., Higginson, G.K., and Fleming, D.W. (1999) Legalized physician-assisted suicide in Oregon – the first year's experience. *The New England Journal of Medicine* 340: 577–583.

Kass, L.R. (2004) "I will give no deadly drugs": why doctors must not kill. In *The Case against Physician-assisted Suicide: For the Right to End-of-life Care*, ed. K. Foley and E. Hendin. Baltimore, MD: Johns Hopkins University Press.

Oregon Death with Dignity Legal Defense and Education Center (2002) *Oregon Death with Dignity*. Portland, OR: Oregon Death with Dignity Legal Defense and Education Center.

Rachels, J. (1975) Active and passive euthanasia. *The New England Journal of Medicine* 292: 78–80.

Scarlett, E. (1991) What is a profession? In *On Doctoring: Stories, Poems, Essays*, ed. B.R. Reynolds and J. Stone, pp. 124–125. New York: Simon & Schuster.

Sheldon, T. (2003) Only half of Dutch doctors report euthanasia. *British Medical Journal* 326: 1164.

Van der Maas, P.J. (1992) *Euthanasia and other Decisions at the End of Life*. Amsterdam: Elsevier.

Further Reading

Emanuel, L.L., ed. (1998) *Regulating How We Die: The Ethical, Medical, and Legal Issues Surrounding Physician-assisted Suicide*. Cambridge, MA: Harvard University Press.

New York State Task Force on Life and the Law (1994) *When Death is Sought: Assisted Suicide and Euthanasia in the Medical Context*. New York: York State Task Force on Life and the Law.

Quill, T.E. (1996) *A Midwife Through the Dying Process: Stories of Healing and Hard Choices at the End of Life*. Baltimore, MD: Johns Hopkins University Press.

Weir, R.F., ed. (1997) *Physician-assisted Suicide*. Bloomington, IN: Indiana University Press.

Animals

CHAPTER SIX

Empty Cages: Animal Rights and Vivisection

Tom Regan

Animals are used in laboratories for three main purposes: education, product safety testing, and experimentation – medical research in particular. Unless otherwise indicated, my discussion is limited to their use in harmful, non-therapeutic medical research (which, for simplicity, I sometimes refer to as "vivisection"). Experimentation of this kind differs from therapeutic experimentation, where the intention is to benefit the subjects on whom the experiments are conducted. In harmful, non-therapeutic experimentation, by contrast, subjects are harmed, often seriously, or put at risk of serious harm, in the absence of any intended benefit for them; instead, the intention is to obtain information that might ultimately benefit others.

Human beings, not only nonhuman animals, have been used in harmful, non-therapeutic experimentation. In fact, the history of medical research contains numerous examples of human vivisection, and it is doubtful whether the ethics of animal vivisection can be fully appreciated apart from the ethics of human vivisection. Unless otherwise indicated, however, the current discussion of vivisection and my use of the term are limited to harmful, non-therapeutic experimentation using nonhuman animals.

The Benefits Argument

There is only one serious moral defense of vivisection.[1] That defense proceeds as follows. Human beings are better off because of vivisection. Indeed, we are (we are told) much better off because of it. If not all, at least the majority of the most important improvements in human health and longevity are indebted to vivisection. Included among the advances often cited are open heart surgery, vaccines (for polio and smallpox, for example), cataract and hip-replacement surgery, and advances in rehabilitation

Contemporary Debates in Applied Ethics, Second Edition. Edited by Andrew I. Cohen and Christopher Heath Wellman.
© 2014 John Wiley & Sons, Inc. Published 2014 by John Wiley & Sons, Inc.

techniques for victims of spinal cord injuries and strokes. Without these and the many other advances attributable to vivisection, proponents of the Benefits Argument maintain, the incidence of human disease, permanent disability, and premature death would be far greater than it is today.

Defenders of the Benefits Argument are not indifferent to how animals are treated. They agree that animals used in vivisection sometimes suffer both during the research itself, and because of the restrictive conditions of their life in the laboratory. That the research can harm animals, no reasonable person will deny. Experimental procedures include drowning, suffocating, starving, and burning; blinding animals and destroying their hearing; damaging their brains, severing their limbs, crushing their organs; inducing heart attacks, ulcers, paralysis, seizures; forcing them to inhale tobacco smoke, drink alcohol, and ingest various drugs, such as heroin and cocaine (Diner, 1985).

These harms are regrettable, vivisection's defenders acknowledge, and everything possible should be done to minimize animal suffering. For example, to lessen the stress caused by overcrowding, animals should be housed in larger cages. But, so the argument goes, there is no other way to secure the important human health benefits vivisection yields so abundantly, benefits that greatly exceed any harms endured by animals.

What the Benefits Argument Omits

Any argument that rests on comparing benefits and harms must not only state the benefits accurately, it must also do the same for the relevant harms. Advocates of the Benefits Argument fail on both counts. Independent of their tendency to minimize the harms done to animals and their fixed resolve to marginalize non-animal alternatives,[2] advocates overestimate the human benefits attributable to vivisection and ignore the massive human harms that are an essential part of vivisection's legacy. Even more fundamentally, they uniformly fail to provide an intelligible methodology for comparing benefits and harms across species. I address each of these three failures in turn.

The overestimation of human benefits

Proponents of the Benefits Argument would have us believe that most of the truly important improvements in human health could not have been achieved without vivisection. The facts tell a different story. Public health scholars have shown that animal experimentation has made, at best, only a modest contribution to public health. As a matter of fact, the vast majority of the most important health advances have resulted from improvements in living conditions (in sanitation, for example) and changes in personal hygiene and lifestyle, none of which has anything to do with animal experimentation (LaFollette and Shanks, 1996; Greek and Greek, 2000, 2002).

The underestimation of human harms

Advocates of the Benefits Argument conveniently ignore the hundreds of millions of deaths and the uncounted illnesses and disabilities that are attributable to reliance on the "animal model" in research. Sometimes the harms result from what reliance on vivisection makes available; sometimes they result from what reliance on vivisection prevents. The deleterious effects of prescription medicines are examples of the former.

Prescription drugs are first tested extensively on animals before being made available to consumers. As is well known, there are problems involved in extrapolating results obtained from studies on animals to humans. In particular, many medicines that are not toxic for test animals prove to be highly toxic for human beings. In fact, it is estimated that one hundred thousand Americans die and some two million are hospitalized annually because of the harmful effects of the prescription drugs they are taking (US General Accounting Office, 1990). That makes prescription drugs the fourth leading cause of death in America, behind only heart disease, cancer, and stroke – a fact that, without exception, goes unmentioned by the advocates of the Benefits Argument.

Worse, the Food and Drug Administration, the federal agency charged with regulating prescription drugs, estimates that physicians report only 1 percent of adverse drug reactions. In other words, for every adverse drug response reported, 99 are not. Clearly, before vivisection's defenders can reasonably claim that human benefits greatly exceed human harms, they must acknowledge how often and how much reliance on this model leads to prescribed therapies that cause massive human harm (Kessler, 1993).

Massive harm to humans is also attributable to what reliance on vivisection prevents. The role of cigarette smoking in the incidence of cancer is a case in point. As early as the 1950s, human epidemiological studies revealed a causal link between cigarette smoking and lung cancer. Nevertheless, repeated efforts, made over more than 50 years, rarely succeeded in inducing tobacco-related cancers in animals. Despite the alarm sounded by public health advocates, governments around the world for decades refused to mount an educational campaign informing smokers about the grave risks they were running. Today, the costs to the US economy from smoking are nearly $200 million, including substantial medical costs and loss of productivity (American Cancer Society, 2013).

How much of this massive human harm could have been prevented if the results of vivisection had not directed government healthcare policy? It is not clear that anyone knows the answer, beyond saying, "A great deal. More than we will ever know." One thing we do know, however: advocates of the Benefits Argument contravene the logic of their argument when they fail to include these harms in their defense of vivisection.

Comparisons across species

Not to go unmentioned, finally, is the universal failure of vivisection's defenders to explain how we are to weigh benefits and harms across species. Before we can judge that vivisection's benefits for humans greatly exceed its harms to other animals, someone needs to explain how to make the relevant comparisons. For example: how much animal pain equals how much human relief from a drug that was tested on animals? It does not suffice to say – to quote the American philosopher Carl Cohen – that "the suffering of our species does seem somehow to be more important than the suffering of other species" (Cohen and Regan, 2001, p. 291). Not only does this fail to explain how much more important our suffering is supposed to be, it offers no reason why anyone should think that it is.

Until those who support the Benefits Argument offer an intelligible methodology for comparing benefits and harms across species, the claim that human benefits derived from vivisection greatly exceed the harms done to animals is more in the nature of unsupported ideology than demonstrated fact.

Human Vivisection and Human Rights

The Benefits Argument suffers from an even more fundamental defect. Despite appearances to the contrary, the argument begs all the most important moral questions; in particular, it fails to address the role that moral rights play in assessing harmful, non-therapeutic research on animals. The best way to understand its failure in this regard is to position the argument against the backdrop of human vivisection and human rights.

Human beings have been used in harmful, non-therapeutic experiments for thousands of years (Lansbury, 1985, chs 1–4; Annas and Grodin, 1992, chs 1–7, and 11; Jones, 1993; Lederer, 1995, chs 2, 4, and 5; Homblum, 1999). Not surprisingly, most human "guinea pigs" have not come from the wealthy and educated, not from the dominant race, not from those with the power to assert and enforce their rights. No, most victims of human vivisection have been coercively conscripted from, for example, the ranks of young children (especially orphans), elderly, severely developmentally disabled, insane, poor, illiterate, members of "inferior" races, homosexuals, military personnel, prisoners of war, and convicted criminals.

The scientific rationale behind vivisecting human beings needs little explanation. Using human subjects in research overcomes the difficulty of transposing results from another species to our species. If "benefits for humans" establishes the morality of animal vivisection, should we favor human vivisection instead? After all, vivisection that uses members of our own species promises even greater benefits.

No serious advocate of human rights can support such research. This judgment is not capricious or arbitrary; it is a necessary consequence of the logic of basic moral rights, including our rights to bodily integrity and to life. This logic has two key components (Regan, 1983, 2004b).

First, possession of these rights confers a unique moral status. Those who possess these rights have a kind of protective moral shield – an invisible "No Trespassing" sign, so to speak – that prohibits others from injuring their bodies, taking their life, or putting them at risk of serious harm, including death (Nozick, 1974). When people violate our rights, when they "trespass on our moral property," they do something wrong to us directly.

This does not mean that it must be wrong to hurt someone or even to take his life. When terrorists exceed their rights by violating ours, we act within our rights if we respond in ways that can cause serious harm to the violators. Still, what we are free to do when someone violates our rights does not translate into the freedom to override their rights without justifiable cause.

Second, the obligation to respect others' rights to bodily integrity and to life trumps any obligation we have to benefit others (Dworkin, 1977). Even if society in general were to benefit if the rights of a few people were violated, that would not make violating their rights morally acceptable to any serious defender of human rights. The rights of the individual are not to be sacrificed in the name of promoting the general welfare. This is what it means to affirm our rights. It is also why the basic moral rights we possess, as the individuals we are, have the moral importance that they do.

Why the Benefits Argument Begs the Question

Once we understand why, given the logic of moral rights, respect for the rights of individuals takes priority over any obligation we might have to benefit others, we can understand why the Benefits Argument fails to justify vivisection on nonhuman animals. Clearly, all that the Benefits Argument *can* show is that vivisection on nonhuman animals benefits human beings. What this argument *cannot* show is that vivisecting animals for this purpose is morally justified. And it cannot show this because the benefits humans derive from vivisection are irrelevant to the question of animals' rights. We cannot show, for example, that animals have no right to life because we benefit from using them in research in which they are killed.

It does not suffice that advocates of the Benefits Argument insist that "there are no alternatives" to vivisection that will yield as many human benefits for two reasons. First, this reply is disingenuous. The greatest impediment to developing new scientifically valid non-animal alternatives and to using those that already exist is the hold that the ideology of vivisection currently has on medical researchers and those who fund them. Second, whether animals have rights is not a question that can be answered by saying how much vivisection benefits human beings. No matter how great the human benefits might be, the practice is morally wrong if animals have rights that vivisection violates.

But *do* animals have any rights? The best way to answer this question is to begin with an actual case of human vivisection.

The Children of Willowbrook

Now closed, Willowbrook State Hospital was a mental hospital located in Staten Island, one of New York City's five boroughs. For 15 years, from 1956 to 1971, under the leadership of New York University Professor Saul Krugman, hospital staff conducted a series of viral hepatitis experiments on thousands of the hospital's severely retarded children, some as young as three years old. Among the research questions asked was: "Could injections of gamma globulin (a complex protein extracted from blood serum) produce long-term immunity to the hepatitis virus?" (Rothman and Rothman, 1984).

What better way to find the answer, Dr Krugman decided, than to separate the children in one of his experiments into two groups. In one, children were fed the live hepatitis virus and given an injection of gamma globulin, which Dr Krugman believed would produce immunity; in the other, children were fed the virus but received no injection. In both cases, the virus was obtained from the feces of other Willowbrook children who suffered from the disease. Parents or guardians were asked to sign a release form that would permit their children to be "given the benefit of this new preventive."

The results of the experiment were instrumental in leading Dr Krugman to conclude that hepatitis is not a single disease transmitted by a single virus; there are, he confirmed, at least two distinct viruses that transmit the disease, what today we know as hepatitis A and hepatitis B, the latter of which is the more severe of the two. Early symptoms include fatigue, loss of appetite, malaise, abdominal pain, vomiting, head-ache, and intermittent fever; then the patient becomes jaundiced, the urine darkens, the liver swells, and enzymes normally stored in the liver enter the blood. Death results in 1–10 percent of cases.

Everyone agrees that many people have benefited from this knowledge and the thera-pies that Dr Krugman's research made possible. Some question the necessity of his research, citing the comparable findings that Baruch Blumberg made by analyzing blood antigens in his laboratory, where no children were harmed or put at risk of griev-ous harm. But even if we assume that Dr Krugman's results could not have been achieved without experimenting on his uncomprehending subjects, what he did was wrong.

The purpose of his research, after all, was not to benefit each of the children. If that was his objective, he would not have withheld injections of gamma globulin from half of them. *Those* children certainly could not be counted among the intended beneficiar-ies. (Thus the misleading nature of the release form: not *all* the children were "given the benefit of this new preventive.")

Moreover, it is a perverse moral logic that says, "The children who received the injec-tions of gamma globulin but who did not contract hepatitis – they were the real benefi-ciaries." Granted, if these children already had the hepatitis virus and failed to develop the disease because of the injections, it would make sense to say that they benefited from Dr Klugman's experiment. But these children did not already have the virus; they were given the virus by Dr Klugman and his associates. How can they be described as "beneficiaries"? If I hide a time bomb armed with an experimental device that I think will defuse the bomb before it is set to go off under your bed, and if the device works, you would not shake my hand and thank me because you benefited from my experi-ment. You would wring my neck for placing you in grave danger.

No serious advocate of human rights can accept the moral propriety of Dr Krugman's actions. By intentionally infecting all the children in his experiment, he put each of them at risk of serious harm. And by withholding the suspected means of preventing the disease from half the children, he violated their rights twice over: first, by willfully placing them at risk of serious physical illness; second, by risking their very lives. This grievous breach of ethics finds no justification in the benefits derived by others. To violate the moral rights of the few is never justified by adding the benefits for the many.

The Basis of Human Rights

Those who deny that animals have rights frequently emphasize the uniqueness of human beings. We not only write poetry and compose symphonies, read history and solve math problems, but we also understand our own mortality and make moral choices. Other animals do none of these things. That is why we have rights and they do not.

This way of thinking overlooks the fact that many human beings do not read history or solve math problems, do not understand their own mortality or make moral choices.

The profoundly retarded children used by Dr Krugman in his research are a case in point. If possession of the moral rights to bodily integrity and life depended on understanding one's mortality or making moral choices, for example, then those children lacked these rights. In their case, therefore, there would be no protective moral shield, no invisible "No Trespassing" sign that limited what others were free to do to them. Lacking the protection of rights, there would not have been anything about the moral status of the children themselves that prohibited Dr Krugman from injuring their bodies, taking their life, or putting them at risk of serious harm. Lacking the protection of rights, Dr Krugman did not – indeed, he could not – have done anything wrong to the children. Again, this is not a position any serious advocate of human rights can accept.

But what is there about those of us reading these words, on the one hand, and the children of Willowbrook, on the other, that can help us understand how they can have the same rights we claim for ourselves? Where will we find the basis of our moral equality? Not in the ability to write poetry, make moral choices, and the like; not in human biology, including facts about the genetic make-up humans share. All humans are (in some sense) biologically the same. However, biological facts are indifferent to moral truths. Who has what genes has no moral relevance to who has what rights. Whatever else is in doubt, this we know.

But if not in some advanced cognitive capacity or genetic similarity, then where might we find the basis of our equality? Any plausible answer must begin with the obvious: the differences between the children of Willowbrook and those who read these words are many and varied. We do not denigrate these children when we say that our lives have a richness that theirs lacked. Few among us would trade our life for theirs, even if we could.

Still, as important as these differences are, they should not obscure the similarities. For, like us, these children were the subjects-of-a-life, *their* life, a life that was experientially better or worse for the child whose life it was. Like us, each child was a unique somebody, not a replaceable something. True, they lacked the ability to read and to make moral choices; nevertheless, what was done to these children – both what they experienced and what they were deprived of – mattered to them as the individuals they were, just as surely as what is done to us, when we are harmed, matters to us.

In this respect, as the subjects-of-a-life, we and the children of Willowbrook are the same; we are equal. Only in this case, our sameness – our equality – is morally important. Logically, we cannot claim that harms done to us matter morally, but that harms done to these children do not. Relevantly similar cases must be judged similarly. This is among the first principles of rational thought – a principle that has immediate application here. Logically, we cannot claim our rights to bodily integrity and to life, then deny these same rights in the case of the children. Without a doubt, if we have rights, so too did the children of Willowbrook.

Why Animals Have Rights

We routinely divide the world into animals, vegetables, and minerals. Amoebae and paramecia are not vegetables or minerals; they are animals. No one engaged in the vivisection debate thinks that the use of such simple animals poses a vexing moral question. By contrast, everyone engaged in the debate recognizes that using nonhuman primates must be assessed morally. All parties to the debate, therefore, must "draw a line" somewhere between the simplest forms of animate life and the most complex, a line that marks the boundary between those animals that do, and those that do not, clearly matter morally.

One way to avoid some of the controversies in this quarter is to follow Charles Darwin's lead. When he compares "the Mental Powers of Man and the Lower Animals," Darwin restricts his explicit comparisons to humans and other mammals (Darwin, 1976).

His reasons for doing so depend in part on structural considerations. In all essential respects, these animals are physiologically like us, and we, like them. Now, in our case, an intact, functioning central nervous system is associated with our capacity for subjective experience. For example, injuries to our brain or spinal cord can diminish our sense of sight or touch, or impair our ability to feel pain or remember. By analogy, Darwin thinks it is reasonable to infer that the same is true of animals that are most physiologically similar to us. Because our central nervous system provides the physical basis for our subjective awareness of the world, and because the central nervous system of other mammals resembles ours in all the relevant respects, it is reasonable to believe that their central nervous systems provide the physical basis for their subjective awareness.

Of course, if attributing subjective awareness to nonhuman mammals clashes with common sense, makes their behavior inexplicable, or is at odds with our best science, Darwin's position should be abandoned. But just the opposite is true. Every person of common sense agrees with Darwin. All of us understand that dogs and pigs, cats and chimps enjoy some things and find others painful. Not surprisingly, they act accordingly, seeking to find the former and avoid the latter. In addition, both humans and other mammals share a family of cognitive abilities (we both are able to learn from experience, remember the past, and anticipate the future) as well as a variety of emotions (Darwin (1976) lists fear, jealousy, and sadness). Not surprisingly, again, these mental capacities affect their behavior. For example, other mammals will behave one way rather than another because they remember which ways of acting had pleasant outcomes in the past, or because they are afraid or sad.

Moreover, that these animals are subjectively present in the world, Darwin understands, is required by evolutionary theory.[3] The mental complexity we find in humans did not arise from nothing. It is the culmination of a long evolutionary process. We should not be surprised, therefore, when Darwin summarizes his general outlook in these terms: "The differences between the mental faculties of humans and the higher animals, great as it is, is one of degree and not of kind" (Darwin, 1976, p. 80).

The psychological complexity of mammals (henceforth "animals," unless otherwise indicated) plays an important role in arguing for their rights. As in our case, so in theirs: they are the subjects-of-a-life, *their* life, a life that is experientially better or worse for the one whose life it is. Each is a unique somebody, not a replaceable something. True, like the children of Willowbrook, they lack the ability to read, write, or make moral choices; nevertheless, what is done to animals – both what they experience and those things of which they are deprived – matters to them, as the individuals they are, just as what was done to the children of Willowbrook, when they were harmed, mattered to them.

In this respect, as the subjects-of-a-life, other mammals are our equals. And in this case, our sameness, our equality, is important morally. Logically, we cannot maintain that harms done to us matter morally, but that harms done to these animals do not. Relevantly similar cases must be judged similarly. As was noted earlier, this is among the first principles of rational thought, and one that again has immediate application here. Logically, we cannot claim our rights to bodily integrity and life, or claim these same rights for the children of Willowbrook, and deny them when it comes to other mammals. Without a doubt, if humans have rights, so too do these animals.

Challenging Human and Animal Equality: Speciesism

The argument for animal rights sketched above implies that humans and other animals are equal in morally relevant respects. Some philosophers repudiate any form of species egalitarianism. According to Cohen (Cohen and Regan, 2001), whereas humans are equal in morally relevant respects, regardless of our race, gender, or ethnicity, humans and other animals are not morally equal in any respect, not even when it comes to suffering. Here are a few examples that will clarify Cohen's position.

First, imagine that a boy and girl suffer equally. If someone assigns greater moral weight to the boy's suffering because he is a white male from Ireland, and less moral weight to the girl's suffering because she is a black female from Kenya, Cohen would protest – and rightly so. Human racial, gender, and ethnic differences are not morally relevant differences. The situation differs, however, when it comes to differences in species. Imagine that a cat and dog both suffer as much as the boy and girl. For Cohen, there is nothing morally prejudicial, nothing morally arbitrary in assigning greater importance to the suffering of the children, because they are human, than to the equal suffering of the animals, because they are not.

Proponents of animal rights deny this. We believe that views like Cohen's reflect a moral prejudice against animals that is fully analogous to moral prejudices, like sexism and racism, that humans often have against one another. We call this prejudice speciesism (Ryder, 1975).

For his part, Cohen affirms speciesism (human suffering does "somehow" count for more than the equal suffering of animals) but denies its prejudicial status. Why? Because according to him, while there are no morally relevant differences between human men and women, or between whites and blacks, "the morally relevant differences [between humans and other animals] are enormous" (Cohen and Regan, 2001, p. 62). In particular, human beings but not other animals are "morally autonomous"; we can, but they cannot, make moral choices for which we are morally responsible.

This defense of speciesism is no defense at all. Not only does it overlook the fact that a very large percentage of the human population (children up through many years of their life, for example) are not morally autonomous, but moral autonomy is not relevant to the issues at hand. An example will help explain why.

Imagine someone says that Jack is smarter than Jill because Jack lives in Syracuse, Jill in San Francisco. Where the two live is different, certainly, and where different people live sometimes is a relevant consideration – for example, when a census is being taken or taxes are levied. But everyone will recognize that where Jack and Jill live has no logical bearing on whether Jack is smarter. To think otherwise is to commit a fallacy of irrelevance familiar to anyone who has taken a course in elementary logic.

The same is no less true when a speciesist says that Toto's suffering counts for less than the equal suffering of Dorothy because Dorothy, but not Toto, is morally autonomous. If the question we are being asked is whether Jack is smarter than Jill, we are given no relevant reason for thinking one way or the other if we are told that Jack and Jill live in different cities. Similarly, if the question we are being asked is, "Does Toto's pain count as much as Dorothy's?" we are given no relevant reason for thinking one way or the other, even if we are told that Dorothy is morally autonomous, and Toto is not.

This is not because the capacity for moral autonomy is never relevant to our moral thinking about humans and other animals; sometimes it is. If Jack and Jill have this capacity, then they, but not Toto, will have an interest in being free to act as their conscience dictates. In this sense, the difference between Jack and Jill, on the one hand, and Toto, on the other, *is* morally relevant. But just because moral autonomy is morally relevant to the moral assessment of *some* cases, it does not follow that it is relevant in *all* cases. And one case in which it is not relevant is the moral assessment of pain. Logically, to discount Toto's pain because Toto is not morally autonomous is fully analogous to discounting Jill's intelligence because she does not live in Syracuse.

The question, then, is: can any relevant defensible reason be offered in support of the speciesist judgment that the moral importance of human and animal pain, equal in other respects, should always weigh in favor of the human being over the animal? To this question, neither Cohen nor any other philosopher, to my knowledge, offers a logically relevant answer. To persist in judging human pains (I note that the same applies to equal pleasures, benefits, harms, and so on, throughout all similar cases) as being more important than the like pains of other animals, because they are human pains, is not rationally defensible. Speciesism is a moral prejudice. Contrary to Cohen's assurances otherwise, it is wrong.

Other Objections, Other Replies

Not everyone who denies rights to animals is a speciesist. Some critics agree that human and nonhuman animals are equal in some morally relevant respects; for example, if a man and a mouse suffer equally, then their suffering should count the same, when judged morally. These critics simply draw the line when it comes to moral rights. Humans have them, other animals do not. Why this difference? The answers vary. Here, briefly, is a summary statement of some of the most common objections to animal rights together with my replies.[4] It is to be recalled that the rights in question are the moral rights to bodily integrity and life.

Objection: Animals do not understand what rights are. Therefore, they have no rights.
Reply: The children of Willowbrook, and all young children for that matter, do not understand what rights are. Yet we do not deny rights in their case, for this reason. To be consistent, we cannot deny animals rights.
Objection: Animals do not respect our rights. For example, lions sometimes kill innocent people. Therefore, they have no rights.
Reply: Children sometimes kill innocent people. Yet we do not deny rights in their case, for this reason. To be consistent, we cannot deny animals rights.

Objection: Animals do not respect the rights of other animals. For example, lions kill wildebeests. Therefore, they have no rights.

Reply: Children do not always respect the rights of other children; sometimes they kill them. Yet we do not deny rights in their case, for this reason. To be consistent, we cannot deny animals rights.

Objection: If animals have rights, they should be allowed to vote, marry, file for divorce, and immigrate, for example, which is absurd. Therefore, animals have no rights.

Reply: Yes, permitting animals to do these things is absurd. But these absurdities do not follow from claiming rights to life and bodily integrity, either in the case of animals or in that of the children of Willowbrook.

Objection: If animals have rights, then mosquitoes and roaches have rights, which is absurd. Therefore, animals have no rights.

Reply: Not all forms of animate life must have rights because some animals do. In particular, neither mosquitoes nor roaches have the kind of physiological complexity associated with being the subject-of-a-life. In their case, therefore, we have no good reason to believe that they have rights, even while we have abundantly good reason to believe that other animals, mammals in particular, do.

Objection: If animals have rights, then so do plants, which is absurd. Therefore, animals have no rights.

Reply: "Plant rights" do not follow from animal rights. We have no reason to believe, and abundant reason to deny, that carrots and cabbages are subjects-of-a-life. We have abundantly good reason to believe, and no good reason to deny, that mammals are. In claiming rights for animals, therefore, we are not committed to claiming rights for plants.

Objection: Human beings are closer to us than animals; we have a special relation to them. Therefore, animals have no rights.

Reply: Yes, we have relations to humans that we do not have to other animals. However, we also have special relations to our family and friends that we do not have to other human beings. But we do not conclude that other humans have no rights, for this reason. To be consistent, we cannot deny animals rights.

Objection: Only human beings live in a moral community in which rights are understood. Therefore, all human beings, and only human beings, have rights.[5]

Reply: Yes, at least among terrestrial forms of life, only human beings live in such a moral community. But it does not follow that only human beings have rights. Only human beings live in a scientific community in which genes are understood. From this we do not conclude that only human beings have genes. Neither should we conclude, using analogous reasoning, that only human beings have rights.

Objection: Humans have rights, and animals do not, because God gave rights to us but withheld rights from them.

Reply: No passage in any sacred book states, "I (God) give rights to humans. And I (God) withhold them from animals." We simply do not find such declarations in the Old Testament, the New Testament, the Torah, or the Koran, for example (Regan, 2004a, ch. 8; 1991, pp. 143–158).

Objection: Animals have some rights to bodily integrity and life, but the rights they have are not equal to human rights. Therefore, human vivisection is wrong, but animal vivisection is not.

Reply: This objection begs the question; it does not answer it. What morally relevant reason is there for thinking that humans have greater rights than animals? Certainly it cannot be any of the reasons examined in the objections above. But if not in any of them, then where? The objection does not say.

The objections just reviewed have been considered because they are among the most important, not because they are the least convincing. Their failure, individually and collectively, goes some way towards suggesting the logical inadequacy of the anti-animal rights position. Morality is certainly not mathematics. In morality, there are no proofs like those we find in geometry. What we can find, and what we must live with, are principles and values that have the best reasons, the best arguments on their side. The principles and values that pass this test, whether most people accept them or not, are the ones that should guide our lives. Given this reasonable standard, the principles and values of animal rights should guide our lives.

Conclusion

As was noted at the outset, animals are used in laboratories for three main purposes: education, product safety testing, and experimentation, harmful non-therapeutic experimentation in particular. Of the three, the latter has been the object of special consideration. However, the implications for the remaining purposes should be obvious (Regan, 2004b, ch. 10). It is wrong when any animal's rights are violated in pursuit of benefits for others. It is conceivable, however, that some uses of animals for educational purposes – for example, having students observe the behavior of injured animals when they are returned to their natural habitat – may be justified. By contrast, it is not conceivable that using animals in product testing can be. Harming animals to establish what brands of cosmetics or combinations of chemicals are safe for humans is an exercise in power, not morality. In the moral universe, animals are not our tasters, and we are not their kings.

The implications of animal rights for vivisection are both clear and uncompromising. Vivisection is morally wrong. It should never have begun and, like all great speciesist evils, it ought to end, the sooner, the better. To reply that "there are no alternatives" not only misses the point, it is false. It misses the point because it assumes that the benefits humans derive from vivisection are derived morally when they are not, and it is false because, apart from using already existing and developing new non-animal research techniques, there is another, more fundamental alternative to vivisection. This is to stop doing it. When all is said and done, the only adequate moral response to vivisection is empty cages, not larger ones.[6]

Notes

1   One could attempt to justify animal vivisection by arguing that it is interesting, challenging, and yields knowledge, which is intrinsically good even when it is not useful. However, a defender of human vivisection could make the same claims, and no one (one hopes) would think that this settles any moral question in that case. Logically, there is no reason to judge

animal vivisection any differently. Even if it is interesting and challenging, and even if it yields knowledge (which is intrinsically good), that would not make it right.

2 The philosopher Carl Cohen, the most strident defender of the Benefits Argument, is guilty on both counts. The most he will admit is that "some" animals "sometimes" are caused "some pain"; as for alternatives, he dismisses their validity as "specious." See his contribution (and my rejoinder) in Cohen and Regan (2001). I discuss his ideas more pointedly in the sequel.

3 Many people of good will do not believe in evolution. They believe that human existence is the result of a special creation by God, something that took place approximately 10,000 years ago. For these people, the evidence for animal minds provided by evolutionary theory is no evidence at all. Despite first impressions, the rejection of evolution need not undermine the main conclusions summarized in the previous paragraph. All of the world's religions speak with one voice when it comes to the question before us. Read the Bible, the Torah, or the Koran; study Confucianism, Buddhism, Hinduism, or Native American spiritual writings. The message is everywhere the same; mammals *most certainly* are psychologically present in the world. These animals *most certainly* have both preference and welfare interests. In these respects, all the world's religions teach the same thing. Thus, while the argument I have given appeals to the implications of evolutionary theory, the conclusions I reach are entirely consistent with the religiously based convictions of people who do not believe in evolution. And for those who believe both in God and in evolution? Well, these people have reasons of both kinds for recognizing the minds of other animals with whom we share a common habitat: the Earth.

4 I address a number of more philosophical objections in Regan (2001, pp. 39–65).

5 Cohen (1997) favors this argument. I reply more fully in Cohen and Regan (2001, pp. 281–284).

6 This chapter adapts material from my chapter in Cohen and Regan (2001) and Regan (2002).

References

American Cancer Society (2013) Questions about smoking, tobacco, and health. http://www .cancer.org/cancer/cancercauses/tobaccocancer/questionsaboutsmokingtobaccoandhealth/ questions-about-smoking-tobacco-and-health-toc (last accessed 6/17/13).

Annas, G.J. and Grodin, M.A., eds (1992) *The Nazi Doctors and the Nuremberg Code: Human Rights in Human Experimentation*. New York: Oxford University Press.

Cohen, C. (1997) Do animals have rights? *Ethics and Behavior* 7: 91–102.

Cohen, C. and Regan, T. (2001) *The Animal Rights Debate*. Lanham, MD: Rowman & Littlefield.

Darwin, C. (1976) Comparison of the mental powers of man and the lower animals. In *Animal Rights and Human Obligations*, ed. T. Regan and P. Singer, pp. 72–81. Englewood-Cliffs, NJ: Prentice Hall.

Diner, J. (1985) *Behind the Laboratory Door*. Washington, DC: Animal Welfare Institute.

Dworkin, R. (1977) *Taking Rights Seriously*. Cambridge, MA: Harvard University Press.

Greek, C.R. and Greek, J.S. (2000) *Sacred Cows and Golden Geese: The Human Costs of Experiments on Animals*. New York: Continuum.

Greek, C.R. and Greek, J.S. (2002) *Specious Science: How Genetics and Evolution Reveal Why Medical Research on Animals Harms Humans*. New York: Continuum.

Homblum, A.M. (1999) *Acres of Skin*. London: Routledge.

Jones, J. (1993) *Bad Blood: The Tuskegee Syphilis Experiment*. New York: Free Press.

Kessler, D.A. (1993) Introducing MedWatch: a new approach to reporting medication and adverse effects and product problems. *Journal of the American Medical Association* 269: 2765–2768.

LaFollette, H. and Shanks, N. (1996) *Brute Science: Dilemmas of Animal Experimentation*. London: Routledge.

Lansbury, C. (1985) *The Old Brown Dog: Women, Workers, and Vivisection in Edwardian England*. Madison, WI: University of Wisconsin Press.

Lederer, S.E. (1995) *Subjected to Science: Human Experimentation in America Before the Second World War*. Baltimore, MD: John's Hopkins University Press.

Nozick, R. (1974) *Anarchy, State, and Utopia*. New York: Basic Books.

Regan, T. (1983) *The Case for Animal Rights*. Berkeley: University of California Press.

Regan, T. (1991) *The Thee Generation: Reflections on the Coming Revolution*. Philadelphia, PA: Temple University Press.

Regan, T. (2001) *Defending Animal Rights*. Champaigne, IL: University of Illinois Press.

Regan, T. (2002) Empty cages: animals rights and vivisection. In *Animal Experimentation: Good or Bad?* ed. T. Gilland, pp. 19–36. London: Hodder & Stoughton.

Regan, T. (2004a) *Animal Rights, Human Wrongs: An Introduction to Moral Philosophy*. Lanham, MD: Rowman & Littlefield.

Regan, T. (2004b) *Empty Cages: Facing the Challenge of Animal Rights*. Lanham, MD: Rowman & Littlefield.

Rothman, D. and Rothman, S. (1984) *The Willowbrook Wars*. New York: Harper & Row.

Ryder, R. (1975) *Victims of Science: The Use of Animals in Science*. London: David-Poynter.

US General Accounting Office (1990) *Report to the Chairman, Subcommittee on Human Resources and Intergovernmental Relations Committee on Government Operations, House of Representatives, FDA Drug Review, Postapproval Risk, 1976–1985*. Washington, DC: US Government Printing Office.

Further Reading

Americans for Medical Progress (2013). http://www.ampef.org/ (last accessed 6/17/13).

Physicians Committee for Responsible Medicine (2013). http://www.pcrm.org/ (last accessed 6/17/13).

CHAPTER SEVEN

Animals and Their Medical Use

R.G. Frey

May we use animals in medicine, in order to enhance and to extend our lives?[1] That we do so is commonplace, and their numbers, especially given developments in genetic engineering, xenotransplantation, cloning, and the like have increased, even as questions have been raised today about their continued use. Thus (e.g., in the search for "designer" mice that exhibit just those features that we are breeding them to exhibit), vast numbers of mice as by-products are produced along the way. Again, we need only to imagine a series of successful xenotransplants to believe that a wholesale effort to produce human organs in animals would take root with a passion. Moreover, genetic engineering and cloning continue to take place in animal models, before being attempted in humans, and trial and error in this regard is likely to result in increasing numbers of animals created for these essentially human ends. In any event, it remains true today that millions of animals continue to be used in medical experiments, even if we ignore those animals used in countries about which we lack adequate information or which effectively hide their research projects from prying eyes.

Are we justified in using animals in these ways? This question must not be thought applicable only because some of our efforts at the moment result in failure. For it would apply even if, as in the case of the development of Salk vaccine for polio, we were eventually proved to be successful in eradicating a disease. If, for example, we were successful through genetic engineering and eventually gene therapy in eliminating Huntington's disease, would we have been justified in reaching this happy outcome through using animals in order to do so?

It is tempting to see this question as one pitting life enhancement and extension in humans against suffering in and the very lives of animals – tempting, in other words, to see the issue as one in which we have to decide whether it is permissible to use animal suffering and lives in order to benefit humans. Can we give a principled justification of this use (Frey, 1989, 2003)?

Contemporary Debates in Applied Ethics, Second Edition. Edited by Andrew I. Cohen and Christopher Heath Wellman.
© 2014 John Wiley & Sons, Inc. Published 2014 by John Wiley & Sons, Inc.

I believe that we can give a principled justification of this use, but I also believe that it is not easy to do so and that the kind of justification to be given exacts from us a cost that many people will not be prepared to pay. To this extent, I think the case for anti-vivisectionism is far stronger than most people suppose.

I do not have space here to give any very detailed account of how I think this case for animal experimentation goes and so for how we are to choose between animal and human lives, but I certainly can give an indication of some of the important issues that bear upon this choice.

The way I have put the central issue pits human and animal lives against each other. For it is surely wrong to maintain that the bulk of medical experimentation takes place for the benefit of animals themselves, even though it may be true, through the incorporation of discoveries into veterinary practice, that animals may indeed at times benefit from the experiments of which they are a part. Seen in this light, two obvious positions suggest themselves, namely, abolitionism on the one hand, or the view that it is always impermissible to use animals for human benefit, and, on the other, anything goes, or the view that it is always permissible so to use animals.

In the case of abolitionism, all experiments involving animals, whether invasive or not, however far advanced, whatever the likelihood of imminent or eventual success, must be stopped at once. In the case of anything goes we may do whatever we like to animals, short perhaps of excessive cruelty and wanton slaughter, in the name of medical advance, most especially if what is proposed figures in the research protocol that is subject to peer review and if it is carried out in accordance with what counts as usual levels of standard of care for the animals in question.

These two positions are, I think, too extreme. For different reasons, they strike me as objectionable; the second is objectionable in ways that take us to the very core of the choice between human and animal lives. Some more middle position strikes me as preferable, and I here set out what I take to be the first steps towards that middle position. Obviously, as with any middle position, it will be exposed to attack from the two extremes, and I will try to show how it might try to deal with some of those attacks.

The Abolitionist Appeal to Animal Rights

Abolitionism fails because the vehicle by which the case in favor is to be made cannot bear the weight that is put upon it. In the main today this vehicle is moral rights, but not just moral rights under any conception. For under most conceptions it will not follow that a case for human use of animals in experimentation will be barred. There will be merely a prima facie right on the animal's part, and such rights can have countervailing concerns arrayed against them and so possibly be outweighed. So, the theorist must come up with a conception of a right that bars precisely this effect.

Most mainstream rights theorists today either do not confer rights upon animals or do so only in some attenuated sense. Tom Regan, on the other hand, wants to confer upon animals rights in the sense of a trump, much along the lines of Ronald Dworkin's sense of certain moral rights as trumps in the human case (Regan, 2001). That is, Regan conceptualizes rights as trumps to considerations of the general welfare. Giving animals rights in this sense disallows appeals to human benefit as a justification for the use of animals in medical research, since that would amount to using appeals to the general welfare to justify an infringement of an animal's right, say, to life. And what, other than appeal to human benefit, is animal research all about?

In Regan's picture, then, we are left with no way to raise the issue of animal research. For the only way moral perplexity registers at any deep level in Regan's picture is if some countervailing right comes into the matter, such that it then poses a conflict with the animal's right to life; then, one is on the familiar, though nonetheless difficult, terrain of a rights-theorist having to deal with a conflict of rights. Conflicts, of course, pose problems, and their resolution is not always easy. But in the case of medical experimentation there is no countervailing right: we do not have a right to use animals merely in order to benefit ourselves. Our convenience battles their right, and we lose. There is no way, then, to register the moral perplexity people feel between weighing and balancing human and animal lives and seeing whether there can be a case for using animals.

In Regan's picture, rights are powerful things to have, and, if animals have them, they have them in the full sense that human beings or persons do. Nothing contends against a right except another right, since anything else is not sufficiently weighty to contend; all medical research has is gain in human benefit. Thus, there is no way to portray the effects of polio vaccine as eliminating one of the scourges of human life in order to justify using monkeys in the research. All invasive (or, for that matter, non-invasive) medical research that is for our mere benefit must be stopped at once, for benefit can never trump rights. And that, basically, covers all the medical research that exists.

The problem here is that Regan has set out to endow animals with rights in so powerful a sense that nothing is able to contend with them. Certainly no argument grounded in human benefit can. Thus, to cite benefit as a ground for an argument in support of animal experimentation fails to appreciate the force of the rights that animals possess – rights that do not accept benefits to others as a reason for their infringement. It is not possible even to state the pro-research position, since all such statements inevitably run through human benefit and thus fail to grasp that, in Regan's eyes, animals have rights that trump our attempts to achieve that very benefit. It seems odd not even to be able to state the pro-research position, even if it ultimately turns out to be mistaken, which, I suspect, is one important reason why mainstream rights theorists continue to resist endowing animals with rights as trumps. But the matter seems worse than odd: to bestow upon the animal a right so strong that one thereby ensures that no case from benefit can even register and then to turn around and point to the fact that no case from benefit can overcome an animal's right (to life) seems to achieve the desired result by cooking the broth.

The "Anything Goes" View on Animals

If the abolition of animal experiments and the forgoing of all benefits in terms of the removal of illnesses and the prolongation of life that animal research confers or promises constitute an extreme position, so too does the "anything goes" position. On this position, anything we might do to animals appears justified, provided only that the benefit obtained, actually or potentially, is significant enough to offset massive animal suffering and deaths. (Obviously, what counts as "significant enough" here is contentious, and it can often appear, even to sympathetic observers to the research cause, that the benefit gained is trivial compared to the cost exacted.) This position also strikes me as extreme: to hold it on plausible philosophical grounds requires one, I think, to argue that animals do not matter morally in the sense that they are not members of the moral community. This strikes me as mistaken.

What is it about animals that does not warrant our moral concern? The usual answers are their pain and suffering and their lives. As for the former, everywhere today the medical research community has presented guidelines governing animal pain and suffering that insist that these be controlled, limited, mitigated where feasible, and justified in the research protocol and actual experiment, and the very care that researchers bestow upon their animals shows that they take animal suffering seriously, as does the insistence that animals be euthanized before they recover from certain painful experiments. If this level of care should be absent, government and funding oversight committees can challenge – indeed, close down – research projects.

On the other hand, to take seriously or to count morally animal suffering, but not animal lives, is implausible, since so much of the worry over suffering, whether in our case or theirs, is precisely owing to the way it can blight, impair, and destroy a life. If animal lives have no value, why should we care about ruining them? Why, in medical research, do we go to such great lengths to justify animal sacrifice? Why do we demand that such sacrifice be directly related to the achievement of the protocol's results? If, however, animal lives have some value, then we need to justify their destruction and the intentional diminution of their quality of life.

At bottom, adherents to the anything goes position must hold that there is a genuine moral difference between the human and animal cases, where pain and suffering and/or the destruction of valuable lives are concerned. But what is the genuine moral difference between burning a man and burning a baboon? Between infecting a man with a certain disorder and genetically engineering a baboon to be subject to that disorder? Between killing both man and baboon? What is at issue is not the claim that it is worse to do these things to the man, but that, according to the anything goes view, doing these things to the baboon is of no moral concern whatever, even though – as in the man's case – suffering occurs, the quality of life is drastically lowered, and killing takes place. If done to the man, these things are wrong; if done to the baboon, they are not. How can species membership make this difference? For it is not easy to see how species membership can constitute a moral difference between two relevantly similar acts of killing or lowering of quality of life; in the case of pain and suffering, I cannot see how they constitute a moral difference at all. Nor is my view any different if we substitute a rat for a baboon. If, in other words, we use something other than a primate by way of contrast, for, as will be seen below, my views of moral standing and the comparative value of lives assigns both to the lives of rats.

It should be obvious that these issues involved in the anything goes position take us to the very center of the debate on the choice between human and animal lives. I will now elaborate on some of these issues, with an eye towards indicating how, if we must choose to use certain creatures in medical experimentation (since fully developed alternatives are not yet in existence), we are to choose those creatures (Frey, 1996b, 1997a, 1997b, 2002).

In my view, moral standing or considerability turns upon whether a creature is an experiential subject with an unfolding series of experiences that, depending upon their quality, can make that creature's life go well or badly. Such a creature has a welfare that can be positively and negatively affected; with a welfare that can be enhanced and diminished, a creature has a quality of life. In this guise, rodents and baboons are experiential subjects, with a welfare and a quality of life that our actions can affect, and this is so whether or not they are agents (which we think they are not) and whether or not they are the bearers of rights (which most of us think they are not). (Thus, agency and rights to my mind are irrelevant to the issue of moral standing.) Such creatures have lives that consist in the unfolding of experiences and so have a welfare and a quality of life, and while there may be some creatures about which I am uncertain of these things, the usual experimental subjects in laboratories are not among them. Thus, to my mind, these laboratory creatures have moral standing, and are, therefore, part of the moral community on the same basis that we are.

I reject, then, the central claims of the anything goes position. I see no reason to deny that rats and baboons feel pain, and I can see no moral difference between burning a man and burning a rat or a baboon. Pain is pain, and species strikes me as irrelevant; what matters is that a creature is an experiential one, and pain typically represents an evil in the lives of all such creatures, if only instrumentally, with respect to quality of life. But if pain and suffering count morally, it is hard to see why animal lives do not; as what concerns us so much about pain and suffering in our case is how these things can impair and significantly diminish the quality of life, they can also, it seems reasonable to believe, in the cases of all creatures who can experience them. No one takes intense pain or prolonged suffering, other things equal, to indicate a high or desirable quality of life, and animals, just as we ourselves, are living creatures with experiential lives, and thus beings with a quality of life. For these reasons, I think animal lives have value.

Thus, I reject abolitionism and the anything goes position. And I am not a speciesist. I do not think we can justify animal experimentation by citing species as a morally relevant reason for using animals in experiments. Nor do I deny that animals are members of the moral community; they are. So how is mine a position that can support some animal research?

The Value of Lives and Quality of Life

In my view, not all members of the moral community have lives of equal value, and where sacrificing life is concerned the threshold for taking lives of lesser value is lower than it is for taking lives of higher value.

It is deeply unpalatable to many to think that some lives are less valuable than others; they would dearly love it to be true that at least all human lives are equally valuable. But when I speak of not all lives being equally valuable I am not referring only to the difference between animal and normal adult human lives; I refer also to human lives themselves. A quality of life view of the value of a life makes the value of a life a function of its quality, and it is commonplace in the medical world today that not all human lives are of equal quality. Indeed, some people lead lives of such a quality that even they themselves seek release from them, as some cases involving a right to die and physician-assisted suicide make clear, and it seems somewhat bizarre to tell such people that, after all, according to some abstraction or other that one happens to believe, they really do have lives as valuable as normal adult human lives. There are some lives we would not wish upon even our worst enemies, and it seems mere pretense to claim that these are as valuable as normal adult human lives. Of course, no one can deny that some may find comfort in such abstractions that substitute for, or, indeed, may even reflect the old adage that all lives are equal in the eyes of God; but I take it to be equally obvious today that many people no longer find this venerable adage comforting.

What is at issue, then, is the comparative value of human and animal life. If we think that not all human lives have the same value, and if we think about the depths to which human life can tragically plummet, then it may well turn out that some animal lives have a higher quality than some human lives. And if we have to use lives in experiments (if, I emphasize, we *have* to), then surely we are here also to use the life of lower quality in preference to the life of higher quality. (Here, I allude only to the logic of the position, not to any side effects that might easily bar one from acting on that logic.)

My account of how we are to decide the comparative value of human and animal life must be subject to scrutiny; that is, at the very least, I must have something to say, in addition to trying to assess the comparative value of these lives, for going about assessing it in the way I do.

One of the strengths of my position on the value of human and animal life, I think, is that it coheres nicely with recent discussions of the value of life in medical ethics and allied areas. In a word, what matters is not life but quality of life. The value of a life is a function of its quality, its quality of its richness, and its richness of its capacities and scope for enrichment; it matters, then, what a creature's capacities for a rich life are. The question is not, say, whether a rat's life has value; I agree that it does. The rat has an unfolding series of experiences and can suffer, and it is perfectly capable of living out a life appropriate to its species. The question is whether the rat's life approaches normal adult human life in quality (and so value) given its capacities and the life that is appropriate to its species, and this is a matter of the comparative value of such lives. Here, the claim is that normal adult human life is more valuable than animal life, based on greater richness and greater potentialities for enrichment. Autonomy or agency can help augment that value. How?

The claim is not that autonomy will inevitably or certainly enhance the value of a life; rather, it is that autonomy can be used for that purpose. In my view, autonomy is instrumentally, not intrinsically, valuable; its value depends upon the uses made of it, and, in the case, at least, of normal adult humans, those possible usages significantly enrich a life. To direct one's own life to secure what one wants; to make one's own choices in the significant affairs of life; to assume responsibility over a domain of one's life and so acquire a certain sense of freedom to act; to decide how one will live, and to mold and shape one's life accordingly: these are the sorts of thing that open up areas of enrichment in a life with consequent effect upon that life's quality and value. Equally, however, it is possible that nothing of the sort will issue from the exercise of one's autonomy: just because a life's value can be augmented through the exercise of auton-omy in no way shows that it is inevitably or always so augmented. The point behind all this, of course, is that these ways of augmenting the value of our lives are, arguably, not available to animals. It does not follow that animal life has no value (indeed, exactly the opposite is my view) or that an animal life cannot have greater value than some human life (again, it is my view that it can). Rather, what is centrally at issue is the comparative value of normal adult human life and animal life and how we go about deciding the matter.

Certainly, were we to adopt some Eastern religion or some form of quasi-religious metaphysic, it is possible that we might come to have a different view of animals and of how we stand to them. Indeed, we might come to take a different view of our relations to the animal kingdom (and to the inanimate environment), without any specifically religious impulses at all. This much is clear through poetry, through cultural differences we encounter among the individuals that make up our society, and through exposure to the art of different ages and cultures. From these different possible views of our relations to animals different possible accounts of the comparative value of human and animal lives may flow. But from the mere fact that different possible accounts of this comparative value may arise nothing follows per se about the adequacy of any single one. Argument must establish the soundness of such accounts, and if, for example, one's claims about comparative value turn upon one's adoption of an Eastern religion, some religious metaphysic, or some abstraction (such as the claim that all life, whatever its quality, has the same "intrinsic" or "inherent" or "innate" value), then it is that religion, metaphysic, or abstraction that must be subjected to scrutiny.

As I have indicated, since not all human lives have the same richness and potentialities for enrichment, not all human lives are equally valuable. In fact, some human lives can be so blighted, with no, or so little, prospect for enrichment that the quality of such lives can fall well below that of ordinary, healthy animals.

It might be claimed that we can know nothing of the richness of animal lives, but ethologists and animal behaviorists, including some sympathetic to the "animal rights" cause, certainly think otherwise. How else, for example, could the claim that certain rearing practices blight animal lives be sustained? That we cannot know everything about the inner lives of animals, of course, in no way implies that we cannot know a good deal.

Quality of life views turn upon richness, and if we are to answer the question of the comparative value of human and animal life we must inquire after the richness of their respective lives. Intra-species comparisons are sometimes difficult, as we know in our own cases in, say, medical ethics; but such comparisons are not completely beyond us. Inter-species comparisons of richness and quality of life are likely to be even more difficult, though again not impossible. To be sure, as we descend from the "higher" animals, we lose behavioral correlates that we use to gain access to animals' interior states. Yet scientific work that gives us a glimpse into animal lives continues to appear, though it is hard as yet to make out much of a case for extensive richness (or so it appears).

Again, we must not simply think that criteria for assessing the richness of human lives apply straightforwardly to animals. Rather, we must use all that we know about animals, especially those closest to us, to try to gauge the quality of their lives in terms appropriate to their species. Then we must try to gauge what a rich, full life looks like to an animal of that species, and, subsequently, try to gauge the extent to which this approaches what we should mean in the human case when we say of someone that they had led a rich, full life. A rich, full life for the rat, science seems to suggest, does not approach a rich, full life for a human; the difference in capacities is just too great. However, if one is going to suggest otherwise, we need evidence of what in the rat's case compensates for its apparent lack of certain capacities, since by its behavior alone we do not normally judge it to have comparable richness.

In order to adopt a quality of life view of the value of a life of a rat, we must try to place ourselves in the rat's position, adopting the capacities and life of the rat. This may be difficult, but it does not appear impossible, and in the case of primates, or animals closer to ourselves, we may well be able to overcome many difficulties that impede our doing this with rats, chickens, or birds.

Can one drop the provision that quality be determined by richness and so avoid the judgment of reduced richness and quality in the animal case? However, richness does not determine quality of life (i.e., by the extent, variety, and quality of experiences), so what else can determine it?

Of course, one might just want to claim that humans and animals have different capacities and lives, and that each leads a rich and full though different life. But this makes it appear that we are barred from comparative judgments, when, in fact, the central ingredients of the respective lives – namely, experiences – appear remarkably alike. Surely I can know something of the lives of animals? Ethologists and animal behaviorists support this. But I have no reason to believe that the rat's life possesses anything like the variety and depth of ways of enrichment that normal adult human life possesses, and I need evidence to make me believe that, for example, one of the rat's capacities so enriches its life that it approaches normal adult human life in richness. The rat has a keener sense of smell, but how does this fact transform the richness of its life to approximate the richness that all the variety and depth of human capacities typically confer upon us? We need evidence to think this.

Two Senses of Moral Community

The ultimate problem over vivisection should now be obvious: we cannot be sure that human life will always and in every case be of higher quality than animal life. And if we are to use the life of lower quality in preference to the life of higher quality, assuming that some life or other has to be used, then we seem committed to using the human life of lower quality in this case. To be sure, this thought might (or will) outrage people, and adverse side effects might (or will) make us choose otherwise. And I am not advocating that we use humans. What I am doing is trying to point out the logic of the position. Today, in medical ethics, we appeal constantly to concerns of quality of life, and we treat quality as if it determines the value of a life. What can be cited that guarantees that human life will be more valuable than animal life? I cannot think of any such thing.

Well, it might be said, this just goes to show the problem with using quality of life talk in this kind of context; perhaps. But it seems a peculiar reason to, say, believe in God or to endorse some abstraction about the (greater) inherent value of all human life because one has thought through the logic of the position on animal experimentation and can find no argument that enables us to continue to only use animals for our benefit.

Finally, I think we must draw a distinction between two different senses of the moral community and show how this distinction fits the earlier discussion.

No one will deny that the patient in the final throes of Alzheimer's disease or the severely mentally enfeebled are members of the moral community in the sense of having moral standing, since they remain experiential subjects with a welfare and quality of life that can be augmented or diminished by what we do to them. This is true of all kinds of human beings who presently, as the result of disease or illness, have had the quality of their lives radically diminished, from those seriously in the grip of amyotrophic lateral sclerosis to those with Huntington's disease. All kinds of human beings presently live lives of massively reduced quality – reduced from the quality of life we find in healthy, normal, adult humans. Yet, they remain members of the moral community.

On the other hand, patients in a permanently vegetative state or anencephalic infants are more problematic candidates for membership in the moral community in this first sense. For though what happens to them may well affect the welfare and quality of life of other people, such as their parents, it is not obvious that they have experiential states that would include them as members of the moral community in their own right.

This first sense of moral community, then, is that in which the creatures that figure within it are all those who are morally considerable in their own right. I have indicated how I think (the "higher") animals get into this sense of the moral community, but it does not matter if one thinks that sentience in one of its senses encompasses animals, and so admits them. For all accounts of moral community in this first sense are accompanied by disclaimers that animals are not moral agents – are incapable, in the sense of agency that matters for the assessment of moral responsibility, of acting for and weighing reasons. They are not morally responsible for what they do, not because they fall outside the moral community in this first sense, but because they do not weigh reasons for action.

Animals, then, are morally considerable; what befalls them as patient or as the object of actions on our part that affect their welfare counts morally. This is no mean consideration, since, heretofore, many have insisted that animals are not morally considerable. But all the creatures that fall within the class of morally considerable beings are not alike: some are included as agents, some as patients, and there is a (further) sense of moral community in the case of the former that is not present in the case of the latter.

In this second sense of moral community, members have duties to each other, reciprocity of action occurs, standards for the assessment of conduct figure, reasons for action – especially where deviation from standards occurs – are appropriately offered and received. The absence of agency – the absence of the proffering and receiving of standards and reasons – matters because those who cannot do these things are not appropriately regarded as moral beings in the sense of being held accountable for their actions. To be accountable for what one does in a community of others who are accountable for what they do is not the same thing as being considerable in one's own right.

Plainly, some humans are not members of the moral community in this second sense: they are incapable of adducing standards for the evaluation of conduct, of conforming their conduct to those standards, and of receiving and weighing reasons for action. Disease and illness, for example, can undo agency in this sense. Equally, perfectly normal children and many of the very severely mentally enfeebled are not members of the moral community in this second sense. In this sense, many more humans can fall outside the moral community as a community of agents than fall outside the moral community as a community of morally considerable beings.

Some humans, such as those in permanently vegetative states and anencephalic infants, fall outside the moral community in both senses. (Hence, much of the controversy about, say, whether the former may permissibly be removed from respirators or whether the latter may permissibly be used as organ donors.) On the other hand, while a great many, if not all, animals arguably fall outside the moral community in the second sense, a great many fall inside the moral community in the first sense. Thus, there are some humans outside the moral community altogether, even while some animals are within the moral community in the first sense, and if one were going to select a creature upon which to experiment, this consideration, at least to morally serious beings, would seem to be relevant.

(I do not here address the question of whether creatures that fall outside the scope of both senses of moral community can be more easily killed than creatures that fall without the second but within the first sense. In fact, given some argument from potentiality that encompasses children, if it works, I do think the threshold for killing is lower in the cases of those in a permanently vegetative state and anencephalic infants. But the facts that affect this case for a lower threshold are too numerous and complex to go into on the present occasion.)

Membership in the moral community in this second sense has nothing to do with whether a being is morally considerable. Agency construed as acting and weighing reasons for action in the light of proffered standards is not required in order to be morally considerable in one's own right. So to what is it relevant? The answer, I believe, is that it *can* be relevant – note, the absence of any necessity in the matter – to augmenting and helping to determine the value of a life.

On a quality of life view of the value of a life, being a member of the moral community in the second sense can enrich one's life and, therefore, enhance its quality. It does this by informing the relations in which we stand to others and thus affecting how we live and judge our lives (Frey, unpublished).[2]

The moral relations in which we stand to each other are part of the defining characteristics of who we are. We are husbands, fathers, sons, brothers, friends, and so on. These are important roles we play in life, and they are informed by a view of the moral burdens and duties they impose on us, as well as the opportunities for action they allow us. Seeing ourselves in these relations is often integral to whom we take ourselves to be, from the point of view of the son as well as that of the father. In these relations, we come to count on others to entwine ourselves with the fate of at least some others, to be moved by what befalls these others, and to be motivated to affect the fates of these others to the extent that we can. Our lives, and how we live them, are affected in corresponding ways. Though there is no necessity in any of this, being a functioning member of a unit of this kind can be one of the great goods of life – enriching the very texture of the life one lives.

Again, binding ourselves to others, pledging ourselves to perform within the moral relations in which we stand to others, and holding ourselves responsible for shortcomings in this regard are all part of what we mean by being a functioning member of a moral community, within which we live our lives with other members. We come to count on others and they on us: the reciprocity of action and regard so characteristic of fully functioning moral communities find their root in these moral relations.

Part of the richness conferred on our lives by being a functioning member of a community characterized by these moral relations is that we come to take certain reasons for action almost for granted. We come to take the standards to be at least prima facie ones that it is appropriate by which to judge our own and others' actions. Again, there is no necessity about any of this, for we can come to reject the standards implied in the usual understanding of these moral relations in our societies in favor of others. But that these standards take a normative form by which we can evaluate reasons and actions, whatever their substance, is the crucial point; it is a normative understanding of these roles that seems crucial (1) to how we see ourselves within them, (2) to how we live our lives and judge many of our actions within those lives, and so (3) to how we judge how well or badly those lives are going.

Participation in such a community can enrich our lives; even in a minimal form, it achieves this by enabling us to cooperate over extensive areas of our lives with at least some others to achieve those of our ends that can only be achieved through cooperation. Put differently, the relations in which we stand to each other aid us in the pursuit of our ends and projects, many of which require the cooperation of others to achieve, and the pursuit of these ends and projects. The pursuit, as some philosophers would have it, of one's own conception of the good, adds enormously to how well we take our lives to be going. Since our welfare is, to a significant extent, bound up in these kinds of pursuits, to ignore this fact is to give a radically impoverished account of a "characteristically" human life. Since all these ends and projects can vary between persons, there is, in this sense, no life "appropriate" to our species, no single way of living to which every one of us "has" to conform. Agency, of course, enables us to select different ends and projects, to mold and shape our lives differently, and to achieve and accomplish different things.

But, beyond any such minimal form, we should note that the very way we live our lives as, for instance, fathers and sons – in order to fulfill what we see as our obligations within these moral relations – forms part of the texture and richness of our lives. We often cannot explain who we are and what we take some of our prized ends in life to be except in terms of these relationships in a moral community, and we often find it difficult to explain why we did something that obviously was at great cost to ourselves except through citing how we see ourselves linked to certain others. Thus, the fact that most humans are members of the moral community in the second sense is a powerful and important feature of their lives: they can live out lives of their own choosing, molded and shaped in the ways they want, in order to reflect and capture the ends and projects they want to pursue. More than this, they can live out these lives in a normative understanding of, for example, the relationships that characterize their interactions with others, relationships in terms of which they see themselves as linked to these others. Here, also, the normative understanding of these relationships enables us to see our lives as going well or badly depending upon how these relationships are affected by what we do to others and by what they do to us. The reciprocity so characteristic of a fully functioning moral community is not the mere reciprocity of action; it is the reciprocity of judging actions from a normative point of view that sees something like enhancing the welfare of another as a reason for action.

Of course, much that we do mirrors the animal case, but agency enables us to fashion a life for ourselves, to live a life molded and shaped by choices that are of our own making and reflect, presumably, how we want to live. Achievement or accomplishment of ends so chosen in this regard is one of the great goods of human life and is one of the factors that can – again, there is no necessity in the matter – enrich individual human lives.

Conclusion

When we seek to compare human and animal life, then, in order to make judgments about the comparative richness of lives, account must be taken of this fashioning of a life for ourselves in a community of shared moral relations. Nothing I have said has implied that there may not be to animal lives features that enable them to make up in richness what agency can confer on ours. We should need evidence of this, of course, thus we have added reason to take seriously the subjective experiences of animals. All I am claiming is that agency can enable normal adult humans to enhance the quality and value of their lives in ways that no account of the activities that we share with animals comprehends (as best we know), and in seeking to give some account of the comparative value of human and animal lives, this kind of difference is obviously both relevant and important.

Notes

1 For a sample of my other writings on this issue, which I draw upon in this chapter, see Frey (1996a, 1997a, 1997b, 1998, 2001, 2003).
2 What follows draws upon material referenced earlier, especially material from Frey (1997b, 2003, unpublished manuscript).

References

Frey, R.G. (1989) Vivisection, medicine, and morals. In *Animal Rights and Human Obligations*, ed. T. Regan and P. Singer, pp. 223–226. Englewood Cliffs, NJ: Prentice-Hall.

Frey, R.G. (1996a) Medicine, animal experimentation, and the moral problem of unfortunate humans. *Social Philosophy and Policy* 13: 181–211.

Frey, R.G. (1996b) Autonomy, animals, and conceptions of the good life. *Between the Species* 12: 8–14.

Frey, R.G. (1997a) Moral community and animal research in medicine. *Ethics and Behavior* 7: 123–136.

Frey, R.G. (1997b) Moral standing, the value of lives, and speciesism. In *Ethics in Practice*, ed. H. LaFollette, pp. 139–152. Oxford: Blackwell.

Frey, R.G. (1998) Organs for transplant: animals, moral standing, and one view of the ethics of xenotransplantation. In *Animal Biotechnology and Ethics*, ed. A. Holland and A. Johnson, pp. 190–208. London: Chapman & Hall.

Frey, R.G. (2001) Justifying animal experimentation: the starting point. In *Why Animal Experimentation Matters*, ed. E.F. Paul and J. Paul, pp. 197–214. New Brunswick, NJ: Transaction Publishers.

Frey, R.G. (2002) Ethics, animals, and scientific inquiry. In *Applied Ethics in Animal Research*, ed. J. Gluck, T. DiPasquale, and F.B. Orlans, pp. 13–24. West Lafayette, IN: Purdue University Press.

Frey, R.G. (2003) Animals. In *Oxford Handbook to Practical Ethics*, ed. H. LaFollette, pp. 151–186. New York: Oxford University Press.

Frey, R.G. (unpublished manuscript) Lives within the moral community.

Regan, T. (2001) *Defending Animal Rights*. Urbana: University of Illinois Press.

ISSUES IN JUSTICE

Affirmative action

CHAPTER EIGHT

A Defense of Affirmative Action

Albert Mosley

Introduction

For over 300 years in what is now the United States of America, it was socially and legally acceptable to discriminate on the basis of race. Religion and science were used to justify enslaving African Americans, and after slavery was abolished, to justify excluding them from educational, employment, and investment opportunities provided to other Americans. Since the landmark Supreme Court decisions of the 1950s declaring segregation unconstitutional, the federal government has taken the lead in guaranteeing an end to racial and sexual discrimination. In publicly available education, accommodations, employment, and investment opportunities, overt discrimination against individuals on the basis of race, sex, religion, or ethnicity in the award of public goods has been legally prohibited.

But legal prohibitions against racial and sexual discrimination have not been sufficient to erase the effects of centuries of bias. Racist and sexist stereotypes, in conjunction with long-established habits and networks, continue to exclude minorities and women from educational, employment, and investment opportunities. To address this, executive orders, legislative statutes, and judicial rulings have mandated not only that discrimination cease, but that "affirmative action" be taken to end the legacies of racism and sexism.[1] Institutions doing business with or receiving payments or grants from the federal government have been required to show a good faith effort to address racial and sexual disparities in the award of educational, employment, and investment opportunities. Affirmative action is a broad set of policies that public and private institutions have evolved in response to the need to end not just the practice but also the legacy of racial and sexual discrimination. The aim of these policies is to provide women and minorities with access to positions they otherwise would be unlikely to get because of the continuing effect of historical oppression (Patterson, 1998, p. 10).

Contemporary Debates in Applied Ethics, Second Edition. Edited by Andrew I. Cohen and Christopher Heath Wellman.

Affirmative Action as a Remedy for Past Injustices

Affirmative action utilizes procedures designed to reach out to women and minorities to ensure that they are informed of opportunities and are given fair consideration for those opportunities. It is a way of recognizing that in the past, many employment, educational, and investment opportunities were not made known to the public at large, but were discussed by word of mouth and awarded through personal networks. Thus, admission to select educational institutions was often on the basis of recommendations from faculty, staff, or alumni; employment opportunities and union memberships were obtained by referral from individuals already employed by the firm or already a member of the relevant union; and business opportunities were made known and awarded on the basis of connections to the right people.

As a result of such networks and practices, members of groups excluded through state-sanctioned action in the past are more likely to be excluded in the present, even when the explicit basis of exclusion is not race or sex. Like laws that allowed one to vote if one's grandfather had voted, networks and procedures established by past practices constitute neutral ways of perpetuating exclusions based on race and sex.

Affirmative action policies mandate taking extra steps to ensure that women and minorities are made aware of opportunities by public advertising and extensive searches. Nonetheless, many continue to assume that minorities and women are more naturally suited for menial positions because that is where most are found. Affirmative action has been a principal means of assuring that selection and evaluation procedures are not tainted by unnecessary qualifications and unconscious biases.

But many who support outreach and fairness measures designed to eliminate discriminatory practices oppose stronger affirmative action measures that take race and sex into account as a means of increasing the representation of minorities and women. While sexual differences may seem relevant in choosing applicants for many types of positions, using the race of an applicant as a relevant factor has proven to be more controversial. For many, if it was wrong to deny a person an opportunity because of his or her race, then it should be wrong to award a person an opportunity because of his or her race. If it was wrong for white people to get preferential treatment then it ought to be wrong for black people to get it.

But such reasoning, while appealing in its simplicity, is ahistorical and ignores the lingering effect of the past on the present. The historical fact is that when slavery was protected by the constitution of the United States, a black person could be enslaved but a white person could not be. Treated as property like horses and dogs, black people were denied the benefits of their labor, denied the right to accumulate wealth, to share it with their families, or to bequeath it to their progeny. Slavery was justified on the grounds that black people were morally and cognitively incapable of acting as responsible agents, and required the direction provided by their masters. Most whites of that era who opposed slavery did so not because they believed black people were their moral and cognitive peers, but because slave labor undermined the viability of free labor. Even after slavery, most continued to believe that black people were incapable of satisfying the duties of democratic citizenship. Such views have not disappeared (Levin, 1997; Kershnar, 2000, 2003).

After the abolition of slavery, legal segregation sought to insulate whites from contact with blacks, except where the latter provided services to the former. The intent was to guarantee that blacks received educational, employment, and investment opportunities commensurate with their inferior status. Individuals considered to be members of the inferior races of Europe were able to escape their status by immigrating to America and identifying generically as white. This, in turn, gave them the privilege to displace and exclude the progeny of slaves wherever opportunities were to be had (Ignatiev, 1995; Jacobson, 1998). The enforced inferiority of Africans and their descendants justified the assumption that they were innately less competent, and continues to be used to justify their over-representation among the least well off and under-representation among the most well off. Consider some sociological data: the incarceration rate of black men in America is six times higher than the incarceration rate of black men in South Africa at the height of apartheid (Guinier and Torres, 2002, p. 263); black Americans make up 12 percent of the population but over 30 percent of the poor (Appiah and Gutmann, 1996, p. 147); in 2000 the unemployment rate was 3.5 percent for whites but 7.6 percent for blacks and 5.7 percent for Hispanics; in 2000, 7.5 percent of non-Hispanic whites, 22 percent of blacks, and 21 percent of Hispanics were living in poverty (US Census Bureau, 2002, pp. 291, 368). More: schools and housing are becoming increasingly segregated, minorities are hired less often than whites with similar qualifications, earn less with similar responsibilities, and are charged more often for similar products and services (Oppenheimer, 1996).

Even when it is admitted that slavery and segregation were unjust, opponents of measures that take race into consideration in awarding opportunities emphasize that people living today were neither slaves nor slaveholders. They argue that descendants of European immigrants should not be punished for something they had nothing to do with, just as the descendants of African slaves should not be rewarded for suffering they did not experience. Some go further and argue that even if the descendants of European immigrants benefited from the sins of state-sponsored slavery and segregation, nothing is owed to the descendants of slaves for the disadvantages they have inherited. Even if the immediate ancestors of contemporary whites did commit injustices against the immediate ancestors of contemporary blacks, it does not follow that contemporary whites owe contemporary blacks. The fact that x benefits from a wrong done to y does not imply that x owes y compensation. To illustrate this point, Stephen Kershnar presents the following scenario:

> Jim, a white American, is the second best tennis player in the world, second only to a Chinese-American, Frank. As a result of Frank's superiority, Jim makes only one-third the money that Frank makes. One weekend, however, Frank is out on the town with his girl-friend, and is viciously beaten and stabbed by a racist Brooklyn mob. This mob has no connection to Jim. Jim, now freed of competition from Frank, wins more tennis tournaments and as a result his income triples. Jim has thus directly benefited from an injustice done to Frank. (Kershnar, 1997, p. 354)

This example is meant to illustrate how a person may benefit from a racial injustice yet be neither morally nor legally obligated to compensate the innocent victim. As Kershnar concludes, "Merely benefiting from an unjust act is not a sufficient condition to obligate payment on the basis of compensatory justice" (1997, p. 355).

But the simplicity of Kershnar's example begs the question. If Frank was only a random victim of the mob, then Frank's bad luck is merely Jim's good luck, much as if a car fleeing a robbery had struck Frank. That the mob was racist might be as irrelevant as that the bank robbers were racist. But if the mob's intent was to compromise Frank's ability to compete so that a minority player would not be #1 and a white player would be, then Jim's good fortune is not the result of mere chance but is morally compromised. If Jim colluded with the mob, then he is culpable for the harm suffered by Frank and should be forced to relinquish his position. If Jim had no involvement in the mob's attack on Frank but the attack was nonetheless done with the intent of benefiting Jim – and Jim comes to know this – then I believe Jim *is* morally obligated to condemn the attack and to relinquish in some way some of the benefits of his ill-gotten gains as a way of discouraging such possibilities in the future.

Like Kershnar, Louis Pojman uses a common-sense example in arguing that the innocent beneficiary of unjust acts need not assume the liabilities caused by those acts. Suppose Albert's parents buy a growth hormone for Albert, hoping he will become a great basketball star. However, Michael's parents steal the hormone, and give it to Michael, who, instead of Albert, grows to be 6 foot 10 inches and makes millions playing basketball. Both Albert's parents and Michael's parents die. Does Michael owe Albert anything? (Pojman, 1992, p. 195; 1998, p. 102). In Pojman's estimation, Michael does not owe Albert anything, either morally or legally. And the coach, upon hearing of the incident, is not obligated to compensate Albert by giving him Michael's position on the basketball team. Pojman concludes: "If minimal qualifications are not adequate to override excellence in basketball, even when the minimality [that is, the possession of minimal qualifications] is a consequence of wrongdoing, why should they be adequate in other areas?" (1992, p. 195).

Pojman's remarks suggest that what is true of athletes should be equally true of pilots, military leaders, business executives, and university professors. That their skills were acquired at the cost of injustices to others may be unfortunate, but this is nevertheless morally irrelevant. For both Pojman and Kershnar, individuals can legitimately inherit the benefits of unjust acts, so long as they themselves were not complicit in the performance of those acts.

But such a position ignores the fact that the agent of injustice is benefited indirectly, because the injustice furthers the agent's aim – one of which is to provide those who inherit the agent's estate with wealth they otherwise would not likely have. This position increases the probability that mobs might engage in acts that transfer wealth to those they identify with, even if that wealth does not benefit members of the mob directly. It encourages acts of injustice by tolerating them, so long as the perpetrator is not the direct beneficiary. And it makes considerations of justice less important than effectiveness, efficiency, and utility.

Affirmative Action as a Form of Compensatory Justice

By construing persons as atomized individuals, critics of restitution ignore how the prospect of benefiting those one identifies with is often a greater source of motivation than benefiting oneself. A person may commit a great injustice and be prepared to bear the personal sacrifice it entails if it is likely that his family and progeny may benefit.

If this option is not discouraged, then acquiring and bequeathing unjust benefits will be sanctioned as a morally and legally permissible strategy. But human beings are not atomized, self-serving entities. Rather, human beings typically conceive themselves as having distinct family lines and group identities, and are, more often than not, as concerned with providing benefits to those with whom they identify as they are concerned with benefiting themselves (Ridley, 1995, pp. 253–266).

Some argue that selection procedures that take race into consideration in the awarding of opportunities are wrong because they do more harm than good; they especially harm those blacks who are provided with such consideration by reinforcing the public's belief that blacks cannot compete on a fair basis. Moreover, this argument continues, using race as a plus factor rewards members of such groups who are most qualified and, therefore, least harmed by past injustices. The end result is that society as a whole is harmed because the best-qualified candidates are not chosen, increasing the likelihood of ineptitude and inefficiency.

Such objections play on the fear that candidates whose race or sex is a factor in the award of an opportunity are likely to be less productive, if not unqualified, for the position they attain. To extend the scenario introduced by Kershnar in arguing against restitution, if Frank's arm is broken by the racist mob, he should not be given the #1 tennis ranking he probably would have retained had the mob attack not occurred. Likewise, Albert should not be given the position he is more likely to have had had Michael's parents not stolen his growth hormone. But these are not objections to the moral duty to provide restitution. At best, they are objections against providing restitution of a particular kind. If certain persons are rendered unable to perform the duties of a position they otherwise are likely to have occupied had they not been unjustly injured, restitution is not achieved by putting them where they are expected to do what they cannot do. This merely adds insult to injury.

Where possible, one of the aims of restitution is to put the injured party in the position he or she would have attained had the unjust injury not occurred. Thus, suppose Frank and Jim are playing a championship match, Jim wins by having Frank's water doped, and this is subsequently made public. Then we would expect Jim's title to be invalidated, and the title awarded instead to Frank. In this way, Frank is granted what he otherwise would probably have achieved had the doping not occurred. But where the injury renders the victim incapable of fulfilling the duties he or she likely would have been capable of, an alternative aim of restitution is to provide appropriate substitutes so that the disadvantages suffered by an injured party are minimized. Thus, if Frank's arm were broken before his match with Jim, it would do no good to offer Frank the opportunity to play Jim that he otherwise would have had. On the other hand, it would be pernicious to allow Jim to gain the title by forfeit, especially if the intent of breaking Frank's arm was so that Jim would win. Even if Jim is not complicit in causing Frank's arm to be broken, he becomes complicit if, upon learning that Frank was injured in order to enrich him, he does nothing to rectify the injuries done to Frank.

One of the central concerns of compensatory justice focuses not on the costs to the victim, but on the possible rewards to the perpetrator of the injury. Consider the following scenario: Jim is the Great White Hope of boxing and knows he can make $10,000,000 in one year if he becomes the new champion. Frank, being the typical black boxing champion, only expects to make $1,000,000 in the subsequent year. Jim discovers a dope that can only be detected at least one year after its use, has it administered to Frank during the fight, Jim wins, and his duplicity is discovered a year and a half later. Should he only be obligated to forfeit the title and the $1,000,000 Frank expected to make? Should Jim be allowed to keep the other $9,000,000 so long as it goes to his estate but not to him? I believe most people would be uneasy with a morality that tolerated injustice for the sake of innocent beneficiaries (Sher, 1981, pp. 10, 17; Ridley, 1995). Imposing fines, penalties, and other damages that exceed the cost to the injured is one way of guaranteeing that the injuring party does not benefit or pass on benefits from the unjust injury.

Taking race or sex into consideration is not simply reversing the historical discrimination against women and people of color, for it does not affect the ability of white males to perform in positions of status and power. Rather, taking race and sex into consideration is a practical acknowledgment that prejudices and historical practices have unjustly limited the opportunities of qualified women and people of color, and that exclusion will be maintained in many areas unless directly addressed.

Many Americans resent being asked to apologize and provide restitution for injuries they had no part in. But there are many situations in which we are expected to assume moral responsibility for actions we did not do personally. Suppose A makes B a gift of $100,000 to get started in a business. But unbeknownst to B, A has robbed C of a million dollars. If B becomes aware of the robbery, but nonetheless refuses to accept any responsibility for C's fate, then B becomes complicit in the original act and continues the injury of that act (Marino, 1998). We should not be surprised that B, acting in self-interest, would explain C's injuries in such a way as to minimize the effect and the injustice of A's assault, while disavowing any personal inclination to inflict similar harms. In a similar fashion, many whites disavow any personal inclination to deny any person opportunities on the basis of race alone, but also believe that being black is highly correlated with having lower intelligence, lower morals, lower motivation, and so on. By avoiding overt racist justifications and opposing the use of racial categories altogether, it is possible to condemn racial oppression while maintaining the effects of state supported racial exclusion.

Standardized Tests and Race

Some who oppose using race as a factor in the selection of candidates for opportunities argue that affirmative action should only guarantee the right to compete, not the right to succeed (Wolf-Devine, 1997, p. 183). But the very right to compete is compromised when selection procedures are biased. This was clearly true before the Civil Rights revolution, when being of European origin was a necessary condition to be selected for the most prestigious institutions and offices. It is also true, though less clear, that selection based on the results of standardized tests is biased as well. One of the most important factors in selecting applicants for admission to select postsecondary and professional schools is their score on the ACT, SAT, GRE, LSAT, and so on, all of which are highly correlated with standard IQ tests. Typically, black, Hispanic, and Native American applicants have average scores on these tests that are lower on the average score of white applicants (Rosser, 1989; Nisbett, 1998). This has reinforced the claim that less-qualified minority applicants are replacing more-qualified white and Asian candidates.

130   **Albert Mosley**

Such claims resonate with benign justifications of slavery and segregation which held that, because Africans were less intelligent, they were prone to immoral acts and irrational beliefs, and it was the white man's burden to help save them from themselves. While few contemporary whites are prone to advocate slavery or segregation as a solution to the presence of Africans in the United States, a substantial proportion of whites continue to believe that blacks are less intelligent than whites. And this is not merely a belief of the uneducated. More than half the educational psychologists in the top US universities believe that the difference in average IQ score between blacks and whites is due to genetic factors that are inherited, and which are resistant to social and environmental changes (Synderman and Rothman, 1987, pp. 137–144; Patterson, 1998, p. 61).

In *The Bell Curve*, Herrnstein and Murray (1994), for instance, attribute average socioeconomic class differences to average differences in intelligence capacity (IQ), and differences in average intelligence between races to differences in genetic makeup. Because genetic information is resistant to somatic influences, they suggest that changes in social and physical environment can affect genetic differences only minimally. They acknowledge that in particular cases, a less intelligent person may be more successful than a more intelligent person, but the evidence they present suggests this is not what we should expect on the average. Similarly, one may on particular occasions find a black person who is more intelligent than a white person, but this is not what we should expect on the average. The status quo is the way it is because of innate differences between species and between races. Such a point of view appears to receive scientific support from aggregate test results that show that, even when blacks achieve a middle-class status, the average IQ of their children remains below the average IQ of the children of the lower-class whites (Hacker, 1992, p. 146). But such facts conceal as much as they reveal. Income parity does not mean that a middle-income black family is alike in all relevant respects to a middle-income white family. In fact, a middle-income black family has fewer assets than a lower-class white family, their children attend schools that are less well endowed, and they and their children are more likely to be denied employment and convicted of a crime (Brooks, 1990, p. 65; Oliver and Shapiro, 1995, pp. 101, 111). And many recent Asian immigrants are highly educated, but accept low incomes in order to gain a foothold in America.

Like IQ tests today, in the earlier part of the twentieth century, the cephalic index was considered a reliable measure of intelligence capacity. The cephalic index measured skull shape and capacity, and was considered to be genetically determined. However, Franz Boas measured average cephalic indices for immigrants from "lower European races," and showed that averages changed dramatically between descendants born in Europe and those born in America from the same parents. Such sudden changes could not be accounted for by changes in the distribution of genes between generations. Some of the most damaging evidence to the claim that IQ differences are fixed from birth reflects the earlier work of Boas regarding the claim that the intellectual potential of lower European races was limited by genetic factors fixed from birth (Boas, 1912).

More than five years before *The Bell Curve* was published, James Flynn released data which shows that, from generation to generation, IQ scores have been rising at a faster pace than can be explained by genetic changes (Flynn, 1987, pp. 171–191). Herrnstein and Murray acknowledge Flynn's results, and conclude that, "on the average, whites today may differ in IQ from whites, say, two generations ago as much as whites today differ from blacks today. Given their size and speed, the shifts in time necessarily have been due more to changes in the environment than to changes in the genes" (Herrnstein and Murray, 1994, pp. 307–308; see also Swain, 2002, ch. 8). Many other examples from public health show how improved nutrition, health, educational opportunities, and smaller family sizes have produced dramatic changes in attributes otherwise believed to be fixed and permanent features of a group's racial essence.

There is also much evidence that intelligence and aptitude tests are culturally biased. Critics of paper-and-pencil intelligence tests have pointed to numerous assumptions built into the test and the test-taking environment that create barriers for otherwise qualified candidates. Even the manner in which questions are posed on a test has been shown to differentially influence the performance of blacks and whites (Freedle, 2003). Moreover, IQ tests are not good predictors of who will be most academically successful, and academic success is not a good predictor of professional success (Rosser, 1989). Nonetheless, a long history of racist and sexist arguments makes it easy to ignore this and other evidence of the extent to which differences in test scores are products of the social and physical environment.

Presumably, written tests eliminate selection based on birth, family connections, class, and other considerations; their role is to provide an objective yardstick for measuring qualifications, one that affords each individual an equal opportunity to demonstrate his or her individual merit. Those who score highest on the tests believe it is an indication that they have more merit and are entitled to the opportunity in question. African, Native, and Hispanic Americans are considered less qualified compared with European and Asian Americans because they tend to have lower test scores and GPAs.

But we should not enshrine test-taking scores and GPAs as the principal criteria for selection. Bowen and Bok (1998) show that students admitted to our most prestigious schools under affirmative action programs are typically as successful, and are more civically involved, than those not admitted under affirmative action. Indeed, those who scored highest on admissions tests often gave least back to their communities in terms of involvement in civic affairs. A study of Harvard Law School graduates showed an inverse relationship between entering LSAT scores and postgraduate income, community involvement, and professional satisfaction (Lempert et al., 2000, p. 468). If one of the objectives of higher education is to contribute to the practical good of human communities, those admitted under affirmative action have often given back as much or more than their higher scoring counterparts.

Because of their importance in admissions decisions, there is a growing trend for students to study to pass tests such as the ACT, SAT, LSAT, and GRE, and for parents who can afford it to purchase expensive test-preparation programs that guarantee higher scores on standardized tests for their children. But SAT scores are little better than chance in predicting college performance after the first year of study. Instead of providing an objective way of predicting who will do best in college and afterwards, SAT scores correlate significantly more with parental income than with success in college. Thus, it is questionable whether using SAT scores as a major factor in admissions improves a college's ability to admit those candidates who are most likely to be successful in fulfilling the mission of higher education (Crouse and Trusheim, 1988, p. 128). To the extent that admission to select institutions is based on scores on standardized tests, the selection procedures will exclude those whose families have traditionally been denied the opportunity to accumulate wealth.

In short, we should be wary of assuming that timed paper-and-pencil tests provide a reliable estimate of who will do best as students, workers, and citizens. Some suggest that people who do best on standardized tests may often be least prepared for real-life situations involving competing perspectives and ineliminable uncertainties. Paper-and-pencil test have been notoriously inadequate in their ability to predict an individual's capacity for creative choices and collaborative involvement (Guinier and Sturm, 2001). Limiting educational opportunities at our most select institutions by the use of paper-and-pencil tests limits participation in the workplace and in civic activities at the highest levels.

Even where professional success is our primary criterion, a survey of top executives of Fortune 1000 companies revealed that most people considered qualities such as creativity, drive, and leadership to be more important than SAT scores. Many cautioned that multiple-choice pencil-and-paper tests were poor measures of the attributes most important for success in corporate America. Only 4 percent of the executives interviewed considered standardized tests such as the SAT and the ACT to be important for long-term success, and only 20 percent cited grades in college or graduate school as good predictors of success. The core attributes considered important for success were: integrity, will to succeed, determination, hard work, ability to motivate, and ability to overcome obstacles (National Urban League, 2005).

A more diverse classroom and workplace help to counteract habitual thinking about race and gender differences (Guinier and Sturm, 2001, p. 9). The fact that in the past women and minorities have not been leaders in science, commerce, and the arts should not be our guide to the choice of future leaders. This requires eliminating both the discriminatory intent of individuals and the disparate impact of institutional barriers. Selection procedures need to integrate the insight that many people learn on the job in ways that are not replicated by paper-and-pencil tests. As our world evolves and institutions change, past procedures may not be reliable guides to success in the future. Affirmative action has provided the occasion for us critically to examine the extent to which aptitude and intelligence tests are good predictors of academic and practical success. Instead of estimating capacity in order to determine who participates, it acknowledges participation as necessary for developing capacity (Guinier and Sturm , 2001).

Affirmative Action and Equal Protection

The use of race in deciding to include some rather than others for benefits is severely limited by the Fourteenth Amendment and the 1964 Civil Rights Act to situations where doing so is necessary to achieve a compelling state interest. Thus, the Supreme Court has sanctioned the use of race in cases where an agency has continued a documented practice of invidious racial discrimination and is mandated, as part of a settlement, to include members of the formerly excluded group. In such a case, race is not irrelevant to achieving the designated goal of dismantling a culture of racial exclusion.

According to Justice Powell's position in *Regents of the University of California* v. *Bakke* (1978), the state has a different but equally compelling interest in producing a diverse learning environment for its future leaders. Powell argued that, for the purpose of achieving diversity, the use of race as a factor in the selection process is constitutionally valid. However, that purpose is forward-looking, not remedial, and seeks to ensure that future leaders have been exposed to diverse points of view. The goal is to assemble the optimum learning environment. For Powell, "The Nation's future depends upon leaders trained through wide exposure to . . . ideas" (*Regents of the University of California* v. *Bakke*, 1978, p. 312). The First Amendment protection accorded the free exchange of ideas recognizes the importance of providing an arena for the exploration of different points of view. Democratic governance requires an informed citizenry that is able to explore and choose from a cross-section of ideas.

According to Justice Powell, the kind of creative play and experimentation that vitalizes higher education is best achieved with a diverse student body: "The atmosphere of speculation, experiment and creation – so essential to the quality of higher education – is widely believed to be supported by a diverse student body" (*Regents of the University of California* v. *Bakke*, 1978, p. 312). We would expect the farm boy from North Dakota to bring different perspectives from those of the prep school graduate from New England, and we would expect a black student from a middle-class family in Clinton, Maryland to offer a different perspective from that of a white student from a middle-class family in Chevy Chase, Maryland.

Nonetheless, any legally sanctioned use of race is suspect, and prohibited unless a strong case can be made that it is necessary to achieve a compelling state interest. Thus, Cheryl Hopwood did not sue the University of Texas Law School because she was rejected even though she had higher scores than more than 100 other white applicants who were admitted. She sued because certain students were admitted whose race was used as a factor in their assessment. She argued that her rights had been violated because the Fourteenth Amendment and the Civil Rights Bills (of, e.g., 1964, 1971, etc.) explicitly prohibit state support for institutions that make choices using racial differences a significant factor.

In making its decision in this case, the Fifth Circuit Court of Appeals explicitly rejected the claim that race may be used as a factor in choosing between applicants in order to foster diversity:

> Within the general principles of the 14th Amendment, the use of race in admissions for diversity in higher education contradicts, rather than furthers, the aim of equal protection. Diversity fosters, rather than minimizes, race [as a discriminatory factor]. It treats minorities as a group, rather than as individuals. It may further remedial purposes [the only permissible rationale in the court's view] but, just as likely, may promote improper racial stereotypes, thus fueling racial hostility. (*Hopwood* v. *State of Texas*, 1996, p. 945)

But just because two individuals are treated differently because of race does not mean that one of them has been treated unfairly. Fairness and equal consideration are not always achieved by identical treatment and color blindness (Appiah and Gutmann, 1996, p. 109). Fairness too often is construed as identical treatment. But in fact it is unfair to treat people with significantly different histories and capacities as if they were identical. To take an extreme example, treating a paraplegic as one would a normally ambient person is not being fair, and it is not fair because they are being treated identically. Fairness is best construed as providing equal concern, not identical treatment. Different people may have different needs and different potentials for producing effective solutions. Using physical strength as a measure of potential for effective policing is unfair to women because it fails to consider other capacities that may be as, or more, effective in resolving conflicts and defusing volatile situations (Sturm and Guinier, 2008).

The backward-looking justification for affirmative action contends that centuries of white supremacy impose historical liabilities that put people of color at greater risk. When attempting to redress a wrong or achieve diversity, a qualified candidate may be given preference over other qualified candidates because of the possession of an attribute that is connected to righting the wrong or introducing an important perspective. Equal consideration of relevant differences instead of identical treatment makes equality of opportunity a reality rather than merely an abstract principle.

We do not live in a time and place where skin color makes no difference. We are not color-blind. Color-consciousness has been an important part of American history and continues to influence our perceptual judgments. To act as if color made no difference would be to ignore the facts (Appiah and Gutmann, 1996, pp. 110, 125; Guinier and Torres, 2002, pp. 274–275). Color may be as important a qualification for a school with few or no black members as being from the south-west may be for a school whose members would otherwise all be from the north-east.

But being color-conscious does not commit one to the position that skin color is a biological sign of predictable physical, cognitive, and behavioral differences. Like all concepts, racial categories evolve, and being identified as a person of color can change from being a mark of inferiority to being a locus of historical oppression and resistance. Most of the categories we use originated in the past, but we do not always continue to use them with their original meanings (Mosley, 1997). We continue to use the terms "sunrise" and "sunset" to distinguish our perceptual awareness of the relative motion of the earth and sun, though we no longer believe that the sun is itself moving above and then below the horizon. Likewise, continuing to use racial categories to distinguish human beings does not preclude giving those categories new and more appropriate meanings.

Racialization and racism have changed over time and appear differently in different historical eras. Before World War II, the notion of European sub-races coexisted with the distinction between European, African, Asian, and Native American races. American racism created a generic white race that enabled European immigrants to displace African, Asian, and Native Americans from employment, education, and investment opportunities. The operation of racism within European populations has been displaced from view by the focus of attention on people of non-European origin. Many whites suppress the historical experience of their own racialization, while continuing to view "poor white trash" as a race apart, often immorally conceived and genetically marred.

Hitler and the Nazis brought general discredit to race theories of the past, giving rise to the view that race has no biological validity and is purely a social construct. Such views have been used to support the demand for race-blind policies and procedures: if the concept of race has no valid biological meaning, then it was a mistake to have used it to exclude individuals from opportunities, and it is equally a mistake to use it to include individuals for opportunities sought by the public at large. Using racial notions with benign intentions, it is argued, is just as ill-conceived as using racial notions invidiously. If there are no races, it becomes difficult to see how there can be such a thing as a meaningful quest for racial diversity.[2]

A Defense of Affirmative Action

In this way, critics of affirmative action argue that eliminating all uses of racial categories is a legitimate way of banishing racism. Delegitimizing race deters individuals from banding together as members of the same or different races. People disadvantaged by state-sponsored racism in the past are dissuaded from coming together around notions of race, and are often persuaded that any reference to a racial affiliation is illegitimate.

Intended originally to protect blacks, the Equal Protection clause of the Fourteenth Amendment is now being used to protect advantages others have gained from past acts of exclusion. For some philosophers, it is better to let traditional victims bear the primary costs of the past, rather than extend those costs to innocent beneficiaries (Kekes, 1998, p. 886).

On the other hand, color-conscious policies acknowledge the continuing effect of slavery and segregation, and ask that those who benefit from the unjust acts of the past relinquish some of those benefits. White applicants are not asked to bear the burden of past racial and gender injustice alone, but to relinquish the increased odds of success made possible through the inheritance of unjust benefits.

Given the prospect of losing benefits, there is little wonder that many whites are motivated to believe that blacks are not as intelligent as whites, are lazier, more violent, and prefer welfare to work (Sniderman and Piazza, 1993; Guinier and Torres, 2002, p. 261; Swain, 2002, p. 149). Such beliefs resonate with beliefs of generations past, and provide comfortable explanations of why higher proportions of blacks than whites are incarcerated, undereducated, impoverished, sick, injured, and likely to die younger. Explanations from the past reappear in a new guise and repeat habits of thought that maintain the practical effects of an era of white supremacy. Bringing more blacks into the professional mainstream provides more opportunities to challenge such ideas and explore solutions that take all sides into consideration.

Conclusions

Opponents of affirmative action programs that take race and gender into consideration agree that it is important that we learn to interact with individuals from diverse backgrounds. But they do not take the high percentage of blacks among the least well off and low percentage of blacks among the most well off to be primarily the products of slavery and segregation. Even if there had been no slavery and segregation, it does not follow that women and minorities would be represented in all areas in proportion to their presence in the general population. To assume that any disproportionate representation is the result of an unjust act is, they argue, overbroad. Natural and cultural differences between groups of people may predispose them to different professions and to different proportional representations within professions – without this being the effect of systemic injustices.

Louis Pojman points out that African American men are over-represented and Asian American men are under-represented in professional basketball. Should our quest for diversity lead us to insist that Asian Americans be hired until parity is reached with African Americans? Would this improve the quality of professional basketball? On the other hand, Asian Americans are over-represented in the sciences. Should we require that their numbers be limited so that African Americans can be integrated into those disciplines? Would this improve the quality of science? (Pojman, 1998, p. 106).

Basketball is used as an example of an arena in which anti-discrimination is sufficient to allow talent to exhibit itself, and where affirmative action could be little more
than an artificial attempt to achieve proportional representation. But basketball is a
bad analogy because there is no history of excluding Asian Americans in basketball. If
there were such a history, we might well suspect that the low proportion of Asian
American players was the result of persistent attempts to eliminate them from the
competition. Because we know that the participation of women and minorities in the
sciences has been historically restricted, we should be concerned whether effects of that
past might not be contributing to a continuing injustice.

Carl Cohen, like so many who oppose explicit attempts to increase minority
enrollments, commits himself to addressing the evil of racism. He is even prepared to
accept policies that use race as a factor in determining admissions, as long as race is
not "dispositive." By this, he means that between two equally qualified candidates, race
cannot be used as a "tie-breaker." Cohen (1997) accused the University of Michigan of
using race in this fashion and of maintaining separate tract systems. Of applicants to
the law school with similar GPA and LSAT scores, 85 percent of the minority applicants
but only 5 percent of the non-minorities were admitted. And in the undergraduate
school, of applicants with similar GPA and SAT scores, 11.5 percent of the non-
minority applicants and 100 percent of the minority applicants were admitted. Cohen
believes the difference in admission rates between minorities and non-minorities (17
to 1 in the law school admissions process and 9 to 1 in undergraduate admissions)
proves that race is not just a factor, but also a "dispositive" factor, and is accorded more
weight than is fair.

What Cohen in fact shows is how misleading percentages can be without attention
to the actual number of cases involved. In the law school example, 6 out of 124 white
applicants with GPAs and LSAT scores in the low range were accepted, while 17 out of
20 black applicants in this range were accepted. If treated identically, only 1 black
applicant would have been admitted and an additional 16 higher-scoring non-minority
applicants. Likewise, in undergraduate admissions, whites with the lower-range scores
were admitted 11.5 percent of the time but black students with similar scores were
admitted 100 percent of the time. For applicants with this range of test scores and
GPAs, minorities were about 9 times more likely to be accepted than non-minorities.
But in terms of actual numbers, if minorities were accepted at the identical rate as
non-minorities, 5 minorities would have been admitted and 56 additional majority
students. Taking into consideration the small number of qualified minority applicants
helps make clear how a description in terms of percentages merely distort the real situation. Contrary to the intent of equal consideration, identical treatment is more likely
to perpetuate than eliminate socially determined educational disparities.

For Cohen, utilizing racial categories even for benign purposes is akin to using an
evil means to achieve a good end.[3] In contrast, I have argued that the constitution does
not prohibit the use of race in order to disassemble a pattern and culture of racial
exclusion. Nor need we assume that the use of racial categories commits us to the
meanings and theories originally attached to those terms. Anti-discrimination policies
have helped us to recognize that race-blind descriptions and procedures may distribute
costs and benefits selectively between blacks and whites without using racial terminology at all. Many procedures that make no mention of racial categories have nonetheless
been shown to have a disparate impact on a historically excluded group.

A Defense of Affirmative Action   137

Affirmative action works to sever the link between skin color and social destiny by placing qualified people of color in positions they otherwise would be unlikely to achieve. Many institutions see it as part of their mission to help counter the lingering effects of a racist past by accepting qualified blacks for stereotypically white positions in greater numbers than would normally be expected. As Appiah and Gutmann put it, "By hiring qualified blacks for stereotypically white positions in greater numbers than blacks would be hired by color blind employers the US will move farther and faster in the direction of providing fair opportunity to all its citizens" (1996: 131). But without an appreciation of the wrong perpetrated on people of color, and a commitment to correct that wrong, it is debatable why a concern with diversity should give special attention to the "racial" variety.

The question is whether and how America addresses the continuing effect of an era of exclusion on the basis of race. I believe all Americans – white, black, red, and yellow – have an obligation not to allow certain groups to constitute the primary victims of history, while certain other groups are its primary beneficiaries. Victims and beneficiaries of past unjust acts must be reconciled, not by banishing reference to racial ills, but by addressing them openly and directly. It remains to be seen whether the most controversial part of affirmative action, the use of race as a factor in addressing the lingering effects of state-enforced racism, will be subject to such "strict scrutiny" that racial exclusion is condemned in theory but maintained in practice.

Notes

1 Executive Orders include: Executive Order 8802 by Franklin Roosevelt (1941); Executive Order 10952 by John F. Kennedy (1961) ("take affirmative action to insure that persons are hired without regard to race, color, or creed); Executive Order 11246 by Lyndon Johnson (1965) (establishing Office of Federal Contract Compliance); and the 1970 Philadelphia Plan. Legislative statutes include: 1964 Civil Rights Act, Equal Employment Opportunity Act of 1972, and the 1990 Civil Rights Bill. Supreme Court decisions include *Brown*, *Griggs* v. *Duke Power Co.*, *Bakke*, *United Steelworkers* v. *Weber*, *Sheetmetal Workers Union* v. *EEOC*, *Richmond* v. *Crosson* (1989), *Adarand* (1995), as well as many other rulings at different levels of the judiciary.

2 The very use of racial categories is considered by some to continue a racist agenda. People who trace their disadvantages to racial injustice come to be viewed like people who blame witches for their misfortune. There are no witches, and there are no races. It is possible for certain individuals to have been harmed by the false belief in the existence of races, just as individuals have been harmed by the false belief in witches.

3 "We all aspire one day to transcend the racism that has so long pervaded American life. Difficult to achieve, that goal will certainly not be advanced by the continued reliance upon the very evil we seek to eradicate" (Cohen, 1997). We may ask how this passage is consistent with Cohen's claim that race can be used as a factor, so long as it is not "dispositive." (Thanks to Ernie Alleva for this point.)

References

Appiah, K.A. and Gutmann, A. (1996) *Color Conscious: The Political Morality of Race*. Princeton, NJ: Princeton University Press.

Boas, F. (1912) *Changes in the Bodily Form of Immigrants*. Senate Document No. 208, 61st Congress, May 1912.

Bowen, W. and Bok, D. (1998) *The Shape of the River: Long Term Consequences of Considering Race in College and University Admissions*. Princeton, NJ: Princeton University Press.

Brooks, R. (1990) *Rethinking the American Race Problem*. Berkeley, CA: University of California Press.

Cohen, C. (1997) Admissions policy lawsuit: affirmative action debate – Letter from Carl Cohen, to members of the University community 10/22/97. http://www.umich.edu/~rescoll/AffActDebate/affirmx2.html (last accessed 6/17/13).

Crouse, J. and Trusheim, D. (1988) *The Case Against the SAT*. Chicago, IL: University of Chicago Press.

Flynn, J.R. (1987) Massive IQ gains in 14 nations: what IQ tests really measure. *Psychological Bulletin* 101: 171–191.

Freedle, R.O. (2003) Correcting the SAT's ethnic and social-class bias: a method for re-estimating SAT scores. *Harvard Educational Review* 73: 1–43.

Guinier, L. and Sturm, S. (2001) *Who's Qualified?* Boston, MA: Beacon Press.

Guinier, L. and Torres, G. (2002) *The Miner's Canary*. Cambridge, MA: Harvard University Press.

Hacker, A. (1992) *Two Nations*. New York: Random House.

Herrnstein, R. and Murray, C. (1994) *The Bell Curve: Intelligence and Class Structure in American Life*. New York: Free Press.

Hopwood v. State of Texas (1996) 78 F.3d 932 (3rd Cir. 1996).

Ignatiev, N. (1995) *How the Irish Became White*. New York: Routledge.

Jacobson, M.F. (1998) *Whiteness of a Different Color*. Cambridge, MA: Harvard University Press.

Kekes, J. (1998) The injustice of affirmative action involving preferential treatment. In *Ethics: History, Theory, and Contemporary Issues*, ed. S. Cahn and P. Markie, pp. 879–887. New York: Oxford University Press.

Kershnar, S. (1997) Strong affirmative action programs at state educational institutions cannot be justified via compensatory justice. *Public Affairs Quarterly* 11: 345–364.

Kershnar, S. (2000) Intrinsic moral value and racial differences. *Public Affairs Quarterly* 14: 205–224.

Kershnar, S. (2003) Experiential diversity and *Grutter*. *Public Affairs Quarterly* 17: 159–170.

Lempert, R., Chambers, D., and Adams, T. (2000) The river runs through law school. *Law and Social Inquiry* 25: 395–506.

Levin, M. (1997) Natural subordination, Aristotle on. *Philosophy* 72 (280): 241–257.

Marino, G. (1998) Apologize for slavery facing up to the living past. *Commonweal* 25: 11–14.

Mosley, A. (1997) Are racial categories racist? *Research in African Literatures* 28: 101–111.

National Urban League (2005) *Spotting talent and potential in the business world: lessons from Corporate America for College Admissions*. http://web.archive.org/web/20050514142026/http://cgi.nul.org/studyresults.html (last accessed 6/19/13).

Nisbett, R. (1998) Race, genetics, and IQ. In *The Black–White Test Score Gap*, ed. C. Jencks and M. Philips, pp. 86–102. Washington, DC: Brookings Institution Press.

Oliver, M. and Shapiro, T. (1995) *Black Wealth/White Wealth*. New York: Routledge.

Oppenheimer, D. (1996) Understanding affirmative action. *Hastings Constitutional Law Quarterly* 23: 948–949.

Patterson, O. (1998) *The Ordeal of Integration*. Washington, DC: Civitas/Counterpoint.

Pojman, L. (1992) Equal human worth: a critique of contemporary egalitarianism. *Public Affairs Quarterly* 6: 181–206.

Pojman, L. (1998) The case against affirmative action. *International Journal of Applied Philosophy* 12: 97–115.

Regents of the University of California v. Bakke (1978) 438 US 265 (1978).

Ridley, A. (1995) Ill-gotten gains: on the use of results from unethical experiments in medicine. *Public Affairs Quarterly* 9: 253–266.

Rosser, P. (1989) *The SAT Gender Gap: Identifying the Underlying Causes.* Washington DC: Center for Women's Policy Studies.

Sher, G. (1981) Ancient wrongs and modern rights. *Philosophy and Public Affairs* 10: 3–17.

Sniderman, P. and Piazza, T. (1993) *The Scar of Race.* Cambridge, MA: Harvard University Press.

Sturm, S. and Guinier, L. (2008) The future of affirmative action. *The Boston Review* 25: 6. http://web.archive.org/web/20080513042057/http://bostonreview.net/BR25.6/sturm .html (last accessed 6/19/13).

Swain, C. (2002) *The New White Nationalism in America.* New York: Cambridge University Press.

Synderman, M. and Rothman, S. (1987) Survey of expert opinion on intelligence and aptitude testing. *American Psychologist* 42: 137–144.

US Census Bureau (2002) *Statistical abstract of the United States.* http://www.census.gov/prod/ www/statistical-abstract-02.html (no longer available).

Wolf-Devine, C. (1997) *Diversity and Community in the Academy.* Lanham, MD: Rowman & Littlefield.

CHAPTER NINE

Preferential Policies Have Become Toxic

Celia Wolf-Devine

The debate over affirmative action in the United States has long been a bitter one, and the parties to the debate show no signs of drawing closer together. Many people have looked to the law for clear guidance, but in vain. The law of affirmative action is lacking in coherence, since there are three different standards deriving from different sources, and they do not fit well together.[1] The tide has turned against it, but in the recent cases involving the University of Michigan, the Supreme Court has, somewhat unexpectedly, stabilized around a consensus in favor of diversity, at least in higher education.[2] The important thing to keep in mind, however, is that virtually all affirmative action pro-grams are currently being undertaken voluntarily.[3] What the law permits is not for that reason required; law and morality are not coextensive. As philosophers, our concern is not with how to apply existing laws or how to construct a brief for our client, but with helping people understand what the issues are so they can make up their minds about what ought to be done.

What is keeping such programs in place in spite of growing opposition to them is largely a combination of institutional inertia and a feeling on the part of many people that to favor affirmative action for women and people of color is to be on the side of the angels.[4] I do not think this is true. The arguments in favor of preferential policies do not withstand critical examination; such policies are unjust, and they are having and can be expected to continue to have very bad consequences.

Framing the Issue

The main thing Albert Mosley and I disagree about is whether we ought to engage in what I call "preferential" affirmative action (he calls this "strong" affirmative action). While "procedural" affirmative action seeks only to ensure that members of target

Contemporary Debates in Applied Ethics, Second Edition. Edited by Andrew I. Cohen and Christopher Heath Wellman.
© 2014 John Wiley & Sons, Inc. Published 2014 by John Wiley & Sons, Inc.

groups are encouraged to apply and receive fair consideration for jobs, preferential affirmative action, by contrast, involves selecting a woman or person of color who appears to be less well qualified by the usual criteria than some white male applicant.[5] To tell whether preferential affirmative action has occurred, ask yourself: "If another black person had applied whose credentials matched those of the rejected white candidate, would that person have gotten the job over the black candidate who was in fact chosen?" If the answer is "yes," then preferential affirmative action is at work.[6] What I am opposing is preferential affirmative action.

Preferential policies are often described in misleading ways. Even the label "affirmative action" suggests that its proponents are in favor of doing something positive, while opponents of such policies are presumed to favor doing nothing (perhaps trusting the market to correct for discrimination). This is a false dichotomy; we are *not* forced to choose between affirmative action policies in their current form and doing nothing. There is any number of things we might do to make our society more just and to help the disadvantaged other than adopting preferential policies, and I believe we ought to undertake a number of such reforms.

Calling preferential policies "policies of inclusion" blurs the distinction between procedural and preferential affirmative action (Mosley, 1998, pp. 161–168). It is one thing to encourage members of groups who have until now been scarce in certain work environments to apply, to welcome them, to tell them that we value their talents and will treat them in every way as members of the community. It is another to include them by giving them positions for which they are less well qualified than some white male applicant. Since preferential policies are zero sum (they do not create jobs but only redistribute them), the *inclusion* of one person necessarily involves the *exclusion* of another.

Finally, affirmative action is often presented as a continuation of the great civil rights movement of the 1950s and 1960s. But while affirmative action did indeed have its roots in the 1960s, such policies have undergone a significant change in the 40 years since the Civil Rights Act of 1964, which was designed primarily to provide black citizens the same rights white people had, and enable them to participate fully in the social, economic, and political life of their communities. The large federal agencies set up in the 1960s to oversee employment discrimination moved quickly and aggressively beyond simply trying to root out discrimination to promoting proportional representation of target groups. A number of groups emerged as beneficiaries from these policies, including women, Hispanics, Orientals, American Indians, and Asians and Pacific Islanders. For an in-depth discussion of the history of affirmative action policies in their legal, economic, and political contexts, as well as detailed analyses of the arguments for the various target groups, see my book (Wolf-Devine, 1997, pp. 5–46).

Instituting temporary programs (such as scholarships, low-interest loans for black-owned businesses or mortgages for black people, job-training programs, and the creation of jobs for young, unemployed black men) to be financed out of tax revenues and designed to help black people move forward would at least have made sense and been doable in the political and economic climate of the 1960s. What we have now makes *no* sense, and should not be allowed to feed off the moral capital of the civil rights movement – which I believe was one of our high points as a nation – when we were beginning to really try to take our own American ideals seriously.[7]

Disentangling Race and Sex

Most defenders of preferential policies, including Mosley, develop arguments tailored specifically to the situation of black people and then throw in women and other beneficiaries with little or no argument. Political action has, perhaps, required banding together to advance common interests, but the situations of the groups commonly accorded preferences differ in important ways. Hispanics, for example, are almost all recent immigrants or children of immigrants. They are disproportionably young and many do not speak fluent English. So while we should help them acquire the skills they need to enter the mainstream of American life as previous immigrant groups have done, they are not entitled to preferences over poor non-Hispanic white people whose families have been here for generations.

The situation of women is radically different from that of either Hispanics or black people. The way in which sex and race have become interwoven in the United States is largely the result of historical accident. The type of feminism prevalent in the United States since the 1970s was started by women who were active in the civil rights movement, and the notion of "sexism" was consciously modeled on "racism." Politically active women in other countries do not think about the problems faced by women in this sort of way (Hewlett, 1987, pp. 164–167). On the legal front also, racial and sexual discrimination became linked by something of a fluke. A Southern opponent of black civil rights added women in an amendment to Title VII at the very last minute, hoping perhaps to secure the defeat of the Civil Rights Act itself, or at least throw a monkey wrench into its enforcement by bringing in a whole new protected class.

Women themselves were bitterly divided over the advisability of adding women to Title VII. Affluent and careerist Republican women favored this, while Democratic women active in the blue-collar and pink-collar women's unions opposed it because it was too general, and preferred "specific bills for specific ills" (such as the Equal Pay Act of 1963). Some of the reasons they opposed it were that, first, it ran together discrimination against women and that against black people, which they thought should be addressed separately because they involved very different problems; second, it would endanger the women's protective legislation which they deemed essential to protect working-class women; and, third, it would divert attention and resources away from the more pressing needs of black people.

The strongest arguments in favor of preferential treatment of black people do not hold water for women. Being female is not passed on from generation to generation; women have fathers, brothers, and sons as well as daughters. They do not live in segregated communities; their lives are closely interwoven with men. They share the social class of their fathers while growing up, and that of their husbands when married. For this reason, disadvantage is not inherited in the way it is for black people. Simple non-discrimination would be enough to ensure that they were able to attain positions commensurate with their abilities. They should of course receive equal pay for equal work,[8] but this is an entirely different issue from whether they should be given preferences over better-qualified men at the hiring level.

The state has no compelling interest in promoting proportional representation of women in all the professions. If the great majority of firefighters or mathematicians continue to be male, what difference would this make so long as the jobs are done well and those women with the motivation and ability to do the job well are not excluded unfairly? In fact, if the tests firefighters have to pass had to be changed so that more women could pass them (say, by lessening the amount of weight the firefighter must be able to carry), it would be to our *dis*advantage to have women proportionally represented among firefighters.

Unlike racial differences, the biological differences between men and women are extensive, so the likelihood that these will have some impact on their behavior and capacities is far higher. Sex-based job preferences do *not* level the playing field. They tilt it in a way that makes it harder for couples that would prefer to have the man be the primary breadwinner. I believe that much of the push toward proportional representation of women comes from those who want to eliminate traditional sex roles. But the way couples arrange their domestic lives should be up to them.

Giving women preferences has created a new minority – namely, young white men who are not well established in their careers and who are asked to bear all the burden of such policies. Finally, it has cushioned the class impact of affirmative action programs, since the job lost by a middle-class white man might go to his wife with no net gain to the worst off.

Affirmative Action for Black People: Evaluating the Arguments

When doing social philosophy, although it is important to get a sense of what sort of world we live in, caution is required when relying on statistics or polls. People can put different spins on the same statistics. For example, a widely cited Federal Reserve study of racial disparities in mortgage loan approval rates indicated that black and Hispanic applicants were rejected much more often than white ones.[9] But the study did not control for net worth, the credit histories, or the existing debts of the applicants. Subsequent studies brought to light other important considerations such as the fact that minority applicants generally had greater debt burdens, poorer credit histories, sought loans covering a larger percentage of the value of the property and were also more likely to seek to finance multiple-dwelling units. It turned out that if you looked at *approved* borrowers, the minority borrowers were approved with incomes only three-quarters as high as the approved whites, and assets worth less than half the value of the assets of the whites (Sowell, 2002, pp. 175–176). Statistics, thus, can be very tricky to interpret properly, and polls are notoriously manipulable. I therefore encourage you to reflect about your own experiences and to question your parents and grandparents about the sorts of change they have seen in their lifetimes and use this as a check on what you read.

The compensatory (or backward-looking) argument

This argument relies on the straightforward principle that the one who wrongs another owes the other. The underlying model is that of tort law; defendant has wronged plaintiff and the court must try to restore plaintiff to the position he or she would have been

in had defendant not wronged him or her. Trying to apply this model to wrongs span-
ning several generations, however, generates unmanageable problems.

What makes black people's claim for compensation stronger than that of others who
have suffered discrimination is the fact that their ancestors were brought here as slaves.
Buying and selling a human being as property violates that person's human dignity,
and as a result of slavery and the Jim Crow laws subsequently instituted to keep black
people in subordinate positions, those so treated were both wronged and harmed.

There is a problem, however, for those who defend racial preferences now – that is,
early in the twenty-first century. They must establish a strong connection between a
current black candidate and harms inflicted on the black community by slavery and
Jim Crow laws so as to justify what at least *looks like* an unfair employment practice –
namely disfavoring the other candidate because he or she is white. And it is not enough
to show that the black candidate deserves compensation; it must also be shown that
this particular way of compensating him or her is just.

When discussing entry-level jobs we are talking mainly about people born in the
1980s or late 1970s. Thus the black candidate has not been directly harmed by (or
the white candidate benefited by) slavery, Jim Crow laws, or, for the most part, overtly
racial exclusions of any kind. (Recent black immigrants are not entitled to preferences
on compensatory grounds.) Attempts have been made to show that the current generation of white people has benefited materially from the unpaid labor of slaves appropri-
ated by their owners because this was passed down through the white community
(Boxill, 1972, p. 120). But slavery was only marginally profitable economically, only a
small number of white people benefited from it (some were actually harmed), and most
of those who did benefit were ruined by the war. And the enormous amount of immigration that has occurred since the abolition of slavery makes any claim that all presently living white people can be supposed to have benefited from slavery indefensible.

Even if we focus on material gains the current white job candidate's parents or
grandparents might have obtained because of Jim Crow laws or overtly racial exclusions (young white men at the start of their careers are no more likely to have benefited
than others, and probably less likely), we do not *know* in any given case that this
occurred, and such exclusions are falling rapidly into the past, in any case. The law
does not allow people to collect for wrongs done to their parents or grandparents (with
a few exceptions such as wrongful death of a parent, or requiring the heir of a thief to
return stolen property – where the victim, thief, and object stolen can be identified
clearly). History is full of injustices of every kind (not just racial ones), so it would
appear unfair to compensate some but not others. At some point we just need to pick
up and go on if we do not want to become like the Middle East where the old angers
fester for millennia.

People have different moral intuitions about the innocent beneficiaries of past
wrongs. Even if many currently living white people are better off than they would have
been in the absence of racism, I do not think that this obligates them to compensate
black people. If you know about a wrong in advance and could have done something
to prevent it, or if it was committed specifically to benefit *you*, then you owe compensation, but not otherwise. I also object to the procedure of projecting moral intuitions
that concern one-on-one interactions onto a large and complex society.

Perhaps instead of focusing on material damages, we might understand the harm
inflicted by slavery and Jim Crow laws on the current generation of American black

people as a form of cultural damage. Being a slave is not conducive to a strong work ethic, habits of deferring gratification, saving and planning for the future, taking initiatives, and so on, and black family structure was also adversely affected in various ways by slavery. To the extent that black culture in the United States has been shaped by slave experience, this is likely to make it harder for black people to hold down jobs and maintain stable families. And whatever problems immigrants faced, they did not have the cultural baggage of slavery to contend with. There are two problems, however, with using the cultural damage argument to support preferences.

First, a culture that maximizes members' capacity to succeed in an individualistic and competitive society is not necessarily better than one that does not (more strongly communal cultures may afford a more humane quality of life, for example), and features of black culture that hold them back are likely to be connected with other features that they rightly want to retain. Second, the cultural damage argument undermines itself. If black people claim that their culture has inculcated traits in them that disable them from performing certain sorts of jobs successfully, it is unreasonable to then turn around and ask to be preferentially appointed to such jobs.

Finally, there are two underlying problems with the whole compensatory project. The first is that if it were not for slavery, the current generation of American black people would not exist, since their ancestors would have remained in Africa and married different people. Mosley concedes this, but suggests that had it not been for slavery and racism, some Africans might have come here like other immigrant groups, and been successful in the same way they were (Mosley and Capaldi, 1996, p. 34). You can write hypothetical history any way you want, but we can only act in the present, and we are the people our biology and history have made us.

The compensatory argument also requires some sort of clear standard to determine when justice has been achieved. Otherwise preferential policies open the door to endless turf war. In practice, defenders of preferences have fallen back on the assumption that in the absence of unjust discrimination, black people would be proportionally represented in all the various professions. Mosley concedes that it is not possible to tell what level of representation they would have achieved, but says that proportional representation is "the only *fair* assumption to make" (Mosley and Capaldi, 1996, p. 28).[10] But there are any number of reasons other than discrimination that might cause different racial and ethnic groups to clump together in certain occupations and be absent from others – the most important being cultural differences.

Cultures vary widely in the character traits they admire and strive to inculcate among members, the professions they regard as most prestigious, and the sort of family life they aspire to and achieve. Cultural differences may not be ineradicable, but they go very deep, change very slowly, and often persist in an ethnic group over hundreds of years even when they are scattered all over the globe. Chance or the environment they originated in also affect the occupations that members of an ethnic group enter (Sowell, 2002).

There are thus enough reasons not to expect to find members of various ethnic and racial groups proportionally represented at all levels in all occupations without any appeal at all to genetic factors – which have been and still are fiercely controversial. I am skeptical about whether there are genetically based differences in capacities between different racial groups. But Mosley says, "The possibility of a selective distribution of behavioral traits causally determined by race cannot be ruled out as impossible," and that, "the concept of race has both a biological and historical legitimacy" (Mosley, 1984, pp. 226, 234). On such premises, there would be even less reason to suppose that different racial groups would be distributed throughout the professions at all levels in a random manner in the absence of injustice.

Corrective argument

The corrective argument defends preferences not to compensate for past wrongs, but to counteract *existing* bias in the hiring process. This argument has an advantage over the compensatory argument in that it does not paint the beneficiaries in a demeaning light as victims. Preferences are viewed as a way of selecting the candidates who are in fact best qualified by correcting for the bias against them (Rachels, 1993, p. 220). The bias, they argue, is located either in the prejudices of those making hiring decisions, or in the standards by which candidates are evaluated.

If we restrict ourselves to specific cases where there is clear evidence of bias at work (and not just statistical disparities), then there is a good case for taking some sort of corrective action, but when the corrective argument is generalized it becomes pernicious, because bias is presumed rather than shown. White people are taken to be so infected with racism (conscious or unconscious) that they cannot fairly judge the qualifications of black people, and this stigmatizes them. Or else all the criteria used to evaluate candidates are taken to be biased against black people whenever they do not yield the proper racial mix. Since there is usually at least a reasonable fit between the sorts of test employed to screen applicants and the skills needed to perform successfully, the push to eliminate all standards that do not yield the "right" results is likely to result in increased incompetence on the job.

Forward-looking arguments

Defenders of preferential policies sometimes argue for them on consequentialist grounds by pointing to the desirable results that are hoped will be produced by them, such as providing role models for other members of the group, or creating a more diverse work force or student body.

Role models Putting black people in desirable positions, it is argued, gives other black people the message that such positions are open also to them and this will encourage them to work harder and aspire to succeed. In spite of the influence this hypothesis has had upon policy-makers, surprisingly little empirical evidence has been supplied to support it.[11] Employing *preferences* in order to provide role models, in any case, sends a mixed signal. Certainly it is a good thing that there are some highly visible, successful black people, but a few really top-notch ones in different fields are enough to send the message that the field is open to black people.

Diversity and representation "Diversity" is not so much an argument as it is a kind of umbrella under which a variety of quite different programs take shelter, so if you find this argument confusing, the problem is not just with you. Compensatory or corrective arguments are often disguised as appeals to diversity to evade legal restrictions. And all too often what is going on under the surface of demands for greater diversity is either sheer politics – "more of us; less of you" – or else an attempt to advance some ideological agenda such as feminism or multiculturalism.

Diversity is a mixed good. Deep differences of outlook between people often generate conflict (even bloody wars), so diversity must be balanced by shared values or goals that hold the group together. In practice, no one advocates limitless diversity, of course, and the idea that we should simply "celebrate diversity" is silly. Some individuals are pedophiles or racists, and some of the diverse cultures the world has seen include features we rightly find morally horrifying, such as infanticide or the routine torture of prisoners of war.

Advocates of diversity employ a mixture of aesthetic and political rhetoric to move us to accept the type of diversity they value on other grounds. Aesthetic metaphors include things like the rainbow in which a variety of colors contribute to an overall beautiful appearance, or a stew in which the different ingredients each add their distinctive flavor. Such metaphors are highly subjective; some prefer blended soups, some prefer chunky ones, some are purists who prefer to savor the taste of each food separately. Another problem is that aesthetic metaphors treat members of the groups as interchangeable. Onions, after all, are supposed to add a distinctive onion flavor to the stew.

A black person in an elite position is often said to "represent" other black people (and so on for other groups). But, a lawyer represents clients and a senator represents voters because they are hired or elected by those whose interests they purport to represent, and can be removed if they fail to do so satisfactorily. No such mechanisms are in place for college professors, accountants, CEOs, or students. Why should we suppose that they represent anyone but themselves?

In contexts where people's ideas matter (academia, for example), using this sort of rhetoric improperly puts pressure on members of the groups in question to conform to what is taken to be the official position of their group instead of being given the same right white men have to make up their own minds. (For an in-depth discussion of ways in which the corrective, role model, mentor, and diversity arguments play themselves out in the university context, see Wolf-Devine, 1997, chs 3 and 4.)

Which groups get special consideration is all too often a function of bureaucratic inertia. Since affirmative action policies are already in place for women and certain minority groups, it is easy to favor those same groups, regardless of whether this particular sort of diversity makes sense in relation to the activity in question and its goals.[12] Some press for special consideration of their own group in order to advance their interests and those of their friends. Some support preferences as a way to work for social justice in the broader society by bringing in more members of disadvantaged groups, or compensating those who have been victims of injustice. Since so many Americans confuse class and race, they may see improving the situation of black people relative to white people as a way of breaking the cycle of poverty. Universities may want to include those with perspectives that we think will contribute to a better environment for learning, or enable future citizens to learn to deal with those different from them. Sometimes it is just party politics, as when students agitating for the appointment of a Hispanic professor at a prestigious law school objected to a candidate on the grounds that he was a Republican!

Assessing the Arguments

I will argue that when we disentangle the different types of motivation involved, each of the goals sought could be obtained in a better way, and further that racial preferences are exacerbating the problems they are supposed to be helping solve.

Of course if one favors a group just because it is one's own group, preferences may be the only way to go. But if black people are entitled to favor their own, there is no reason why white people may not do the same. If one's goal is breaking the cycle of poverty, this is better attained by policies directly targeting the poor. Race is not an adequate proxy for poverty; only 27.7 percent of the poor are black, 20.1 percent Hispanic, and 48.1 percent are white. Since black people are disproportionately poor, they would benefit more from such programs, but focusing on poverty rather than race has the advantage of being fair to the white poor. If one's goal is compensating black people for past injustices, one needs to invest resources and thought into black community development, healthcare, and education. Black students at age 17 are four years behind white students in reading and five years behind them in math (Thernstrom and Thernstrom, 2003, p. 13). Preferential admission to college does not fix this problem.

If we are concerned about students coming into contact with those different from themselves so that they will be better able to function as citizens of a pluralistic democracy, racial diversity is only one of many sorts of diversity, and there is no reason to suppose it is the most important one. In terms of cultural differences, Asian Americans clearly bring the greatest diversity, and one that might help students function better in a global economy. The South is unknown territory to most Northerners, conservative Bible Christians are very scarce on elite campuses, and being able to deal with people from a different class background than one's own is an important skill. Studying texts from other historical periods provides another important sort of diversity.

Racial preferences confirm negative racial stereotypes. Most people form their opinions about members of groups other than their own on the basis of their own experience. When there were anti-Jewish quotas at elite colleges, Jewish students had to be brighter than gentiles to be admitted, and as a result, Jews got a reputation for being especially brainy. But under a regime of racial preferences, black students will be, on average, less well qualified than their white classmates. This artificially contrived situation will reinforce the perception that black people are less able academically, and sow seeds of self-doubt in black students.[13] The strongest advocates of racial preferences themselves seem to have a rather dim view of the capacities of black students. Bowen and Bok (1998, p. 39), for example, compare race-sensitive admissions policies with parking places for the handicapped.

In fact, black students in the Bowen and Bok study did underperform, winding up in the 23rd percentile of their classes (and that includes those who would have been admitted without preferences). Mosley is critical of SATs (although he suggests no alternative). They do, however, have some predictive value and in fact they over-predict the performance of black students (who did not do as well as others with the same SAT scores) (Thernstrom and Thernstrom, 1997, ch. 13, 2001). But opponents of racial preferences in admissions are not committed to their being the best, let alone the only, criterion for selecting students. The point is merely that all students should be judged by the same criteria, whatever they are.

If race were taken into account in a loose sort of way in the same manner regional diversity is, this would probably be relatively benign (especially if adequate remediation was provided for those who need it). The problem is that this is not what has been happening. The amount of preference accorded black students is extremely large. Harvard may be able to employ such preferences and still get reasonably competent students, but schools with less prestige will not.

One hidden cost of racial preferences is the drop-out rate of black students, which is characteristically three times that of white students. Another one shows up later when they have to take a race-blind test. A recent study found that 43 percent of the black students admitted to law school on the basis of race either dropped out or failed to pass a bar exam, and in 1988 51 percent of black medical students failed the required Part I exam given by the National Board of Medical Examiners (Thernstrom and Thernstrom, 2001, pp. 195–196).

The worry that motivates those advocating racial preferences in admissions, of course, is that race-neutral policies would sharply diminish the number of black students getting college educations, thus keeping them in socially inferior positions. I do not believe that the consequences of race-neutral admissions policies would be, on balance and in the long run, bad. The number of black students admitted to the most selective colleges would diminish, at least initially, but certainly not to the vanishing point. At the University of California (UC), for example, the initial decrease in black enrollments on the Berkeley campus (they are on the rise again now) was offset by increases at the other UC campuses, and already minority students who could only get into community colleges are increasing their transfer rate to the UC campuses. It is not even clear that fewer black students will graduate from the Berkeley campus under race-neutral admissions policies if ending preferences has the effect of reducing their drop-out rate.

If there were 3.5 percent or 4 percent of black students on some elite campuses for a while, rather than 7 percent or 8 percent, there would still be opportunities for racial interaction, and the interaction would be on a healthier footing in that all students would know the others had been admitted on their merits. Bowen and Bok overestimate the importance of attending an elite school. A careful examination of the schools attended by those in elite positions (white as well as black) indicates only a small fraction of them attended such institutions. There are a number of different pathways to satisfying and lucrative careers. Black people made some of their strongest progress economically before preferences went into effect.[14] To suppose that the progress they have made in recent years is a result of preferences would seem to imply either that they are not competent enough to attain decent jobs on their own or that white people are so hopelessly racist that they would not have hired them without pressure.

If one is concerned about underserved minority communities, and if there is evidence that those admitted under affirmative action programs practice law or medicine more in such communities than those not (there is dispute over whether this is true), communities could be served as well or better by giving scholarships or other incentives to any student who would make a commitment to serve there after graduation for a certain number of years.

Finally, one good result that eliminating preferences might have is that such elimination would force people to confront the root problems instead of being able to paper them over. Some good and creative programs are already being instituted in those states where racial preferences in admissions have been eliminated. As Fullinwider and Lichtenberg note:

> It was not until the thumb on the scale was removed that universities in Florida, Texas, and California intensified their intervention programs, and that state legislatures opened their purses and began to put real money behind intervention. Perhaps the abolition of affirmative action is required to motivate institutions of higher learning and state legislatures to address causes, not just symptoms.[15]

Conclusion

In Lewis Carroll's *Alice in Wonderland*, Alice asks the Cheshire Cat: "Would you tell me, please, which way I ought to walk from here?" and the cat replies: "That depends a good deal on where you want to get to" (1946, p. 64). At this point, then, I will step back and say a few things about where we are now, where I would like to see us go from here, and why preferential policies are not the way to get there. Preferential policies are flawed in three ways. They do not address the root problems our society is facing. They are divisive at a time when we desperately need programs that can bring people together. And they re-entrench racial categories just when it is beginning to seem possible that they may really begin to fade in importance.

We have some very serious problems now that we did not have in the 1960s. Economically, we are faced with the continued growth of poverty (in spite of our high per capita gross domestic product), an extraordinary increase over the last 30 years in the polarization between rich and poor (Phillips, 1990, 2002), increasing job insecurity caused by downsizing, and the fact that globalization has led many businesses to close plants here and move their operations to countries where labor is cheaper. Politically, the big problems are severe distrust of government and loss of a sense of the common good.

Preferential policies do not address these root problems. As Mosley admits:

> Strong affirmative action is a conservative response to racial injustice. It does not seek to eliminate the growing gap between rich and poor. Rather, it seeks to eliminate the overrepresentation of Blacks among the least well off and their underrepresentation among the most well off . . . It does not create new jobs. Rather, it addresses how jobs already created shall be distributed. (Mosley and Capaldi, 1996, p. 59)

Some argue that preferential policies are better than nothing. But racial, gender, and ethnic preferences are a step in the wrong direction. They are inherently divisive because they are zero sum. They were put in place by executive orders and by large federal agencies rather than by democratic means, so continuing them alienates people further from the government (this is especially true of the white working class). Those who lose out economically are likely to suspect that affirmative action was the cause, whether or not it actually was, which fuels racial hostilities.

My suggestion is that we should focus on the plight of America's children, of whom 20 percent live below the poverty level; the earlier you intervene to break the cycle of poverty, the more successful your efforts will be. Middle-class children are also at risk in a number of ways, and family-friendly policies could do a lot to help them flourish

(Hewlett, 1991). Children are our future, and programs designed to give them a chance to develop their talents and contribute to society at least stand a chance to win popular support and bring people together. Targeting poor children would help black people disproportionately since they are disproportionately poor and young.

Finally, racial categories are entrenched when important benefits and burdens are distributed on the basis of race-based preferences, and when large bureaucracies are set up to oversee such programs. Indeed, such programs have a strong emotional and material interest in preserving racial categories; if people were to stop regarding skin color as all that important, they would be out of a job. But there is considerable evidence that racial barriers are finally beginning to break down.[16] My own experience – and that of people I know from a variety of backgrounds – is that anti-black racial prejudice among whites has decreased significantly. For example, my first cousin's daughter married an African American, and although her grandparents were upset, none of the younger members of the family has shown any discomfort with this.

Perhaps the deepest difference between Albert Mosley and me is that he wants to see racial categories preserved, and I do not. He wants them preserved because he wants those who have been harmed through racial categories to receive compensation through these same categories, and because he believes that self-identification of black people as members of the black race can help build "a positive sense of self-identity and self pride" (Mosley, 1997, pp. 108–109). This conjunction links black identity and being a victim to being an individual whom something is owed. To the degree that black identity is linked to a culture of "oppression and resistance," this will reinforce one of the major things that may be holding black people back – namely, the tendency to regard working hard and doing well in school as "acting white." There is no reason why black people cannot move into the mainstream of American life while still having pride in their own culture, much as Irish or Italian Americans have done.

Racial attitudes in this country are complex, ambiguous, and shifting. Given that things are delicately poised, it is important to be careful not to make race relations worse by generating reactive racism. Reactive racism occurs when people are told that members of the other race dislike them or think they are inferior, stupid, racists, bigots, or whatever. If someone tells me that another person likes me, this will dispose me to take a liking to him or her, whereas if I am told that someone dislikes me or thinks badly of me, I will approach that person differently. Expectations tend to be self-fulfilling. So, I conclude with a plea for seeking common ground and treating one another with respect as individuals.

Notes

I wish to express my gratitude to the Earhart Foundation, whose generous assistance enabled me to take time off teaching to work on this project. I also wish to thank my husband Phil Devine for helping me talk out ideas and for commenting on several drafts of this chapter.

1   For a good, clear discussion of the law of affirmative action by a noted discrimination lawyer, see Rutherglen (1997). I am indebted to subsequent informal conversations with Rutherglen for some of the points I make concerning the law.

2   The recent decisions reinforce the court's support for Justice Powell's position in *Bakke*. Astute discussions of the *Bakke* case are found in the essays by Carl Cohen and Ronald

Dworkin in Cahn (2002b), and in Fullinwider and Lichtenberg (2004, ch. 9). An amicus brief filed by the National Association of Scholars critiquing the Gurin report, which the University of Michigan relied on in its defense of the educational value of diversity, can be found at ⟨www.nas.org⟩. Another valuable amicus brief on the same subject can be found at ⟨www/iwf.org⟩. The Supreme Court rejected a system employed by the undergraduate school in which applicants were given extra points for being black, but permitted the law school's more loose and indeterminate use of race as one plus factor among many so long as it is not the determining factor. In practice, however, the distinction between being *a* factor and being *the* factor is almost impossible to draw. I am inclined to think the court was swayed by briefs from former military officers and corporations that have affirmative action policies in place, and is worried that discontinuing preferences would dry up the supply of available black candidates, and that diversity really has nothing to do with the case.

3 Legally, courts may order employers to institute affirmative action programs to remedy *their own* prior acts of discrimination, and have done so in several particularly egregious cases. But voluntarily undertaking preferential hiring programs designed to counteract general societal discrimination is not permissible because they involve racial discrimination against white people. People often undertake such programs as a defense against a possible suit for racial or gender discrimination, but in fact there is very little exposure to the threat of litigation so long as employers have at least *some* members of target groups on their staff. The one area where racial, gender, and ethnic preferences still have real clout is in the government contracts program, where there are "set-asides" for female and minority contractors. ("Set asides" in hiring are illegal, so I find it puzzling that the government employs them in contracting.)

4 On the political front, the major setback to such policies has been laws along the lines of Proposition 209 (the California Civil Rights Initiative) which stipulates: "The state shall not discriminate against, or grant preferential treatment to, any individual or group on the basis of race, sex, color, ethnicity, or national origin in the operation of public employment, public education or public contracting." Similar laws have subsequently been passed also in Texas, Florida and Washington State.

5 I am indebted for this distinction to Steven M. Cahn (2002a, pp. 71–80).

6 I owe this particularly clear way of determining when preference is at work in a hiring decision to Thomas Nagel (1979, ch. 7).

7 The ideals of the civil rights movement grew out of a shared religious heritage – Protestant, Catholic, and Jewish – that grounded demands for universal human rights and respect for the dignity of each person on a common human nature, and this foundation has been called into question by those who claim to be its heirs. For an excellent account of the disintegration of the early New Left, see Gitlin (1995).

8 See Furchtgott-Roth and Stolba (1999) for a sophisticated analysis of data on women's income and progress in the professions. They argue that the wage gap between men and women is rapidly disappearing.

9 See Sowell (2002, p. 175) for a discussion of "Expanded HMDA data on residential lending: one year late," Federal Reserve Bulletin, November 1992.

10 To be fair, he goes on to say that he does not believe they should be *maintained* in such positions. I do not know what to make of this. I cannot believe that he is seriously proposing that we engineer proportional representation of all the current beneficiaries of preferences and then see what happens. The amount of governmental intrusion into people's lives required to attain this would be well beyond what Americans would (or should) tolerate. And if the desired proportions had been attained in this way, the result would be highly unstable, since those preferentially hired would have been set up to fail. The point about being set up to fail is made by Pojman (1998, p. 171).

Preferential Policies Have Become Toxic   153

11 For one thing, it is difficult to state the hypothesis in a way that makes it actually testable, and for another one needs to specify what counts as a "same kind" role model. I am aware of only a few empirical studies of the importance of role models (all in educational settings), and these have generally failed to find any statistically significant correlation between having a teacher of one's own race or sex and student performance (see Wolf-Devine, 1997, pp. 81–86 for discussion and references). The only exception is a study conducted in Tennessee in the 1980s that found a small but statistically significant correlation between K-3 students' performance and their having a teacher of their own race (Thernstrom and Thernstrom, 2003, pp. 201–202).

12 For one thing, these are the groups about which statistics are collected. I am sure, for example, that Pentecostal Christians have lower average incomes than Episcopalians or Jews, but the census does not collect data on religious affiliation.

13 I am indebted throughout this section to the Thernstroms' critique of Bowen and Bok (Thernstrom and Thernstrom, 2001, pp. 169–231) for arguments, statistics, and for the point about how Jewish quotas gave Jews a reputation for being especially smart, as well as to Thernstrom and Thernstrom (1997, ch. 14).

14 The 1940s, 1950s and 1960s was a period of growth in the black middle class. The black–white income gap hit its lowest point in 1972, and then increased again because the heavy manufacturing industries in the Midwest, where many black people were employed, began decline. A lot of the variance between racial groups, I suspect, is a function of the way in which major structural economic factors impact the regions where they live and the industries in which they are employed, rather than of racial prejudice. For excellent historical discussions of the way underlying economic trends differentially affected black and white workers, see Edsall and Edsall (1991, chs 6 and 11) and Madrick (1995).

15 I am quoting here from a draft of Fullinwider and Lichtenberg (2004, ch. 11), which the authors sent me in the spring of 2003. These sentences will not appear in the published version.

16 Polls reveal enormous shifts in racial patterns and attitudes over the past 40 years both among black people and among white people, and this is especially marked if you compare the responses of adults and teens within each group. Valuable discussions of poll data are found in Everett Ladd's essay in Thernstrom and Thernstrom (2002) and in Thernstrom and Thernstrom (1997, ch. 17).

References

Bowen, W. and Bok, D. (1998) *The Shape of the River: Long-Term Consequences of Considering Race in College and University Admissions*. Princeton, NJ: Princeton University Press.

Boxill, B. (1972) The morality of reparation. *Social Theory and Practice* 2 (1): 113–122.

Cahn, S.M. (2002a) Two concepts of affirmative action. In *Puzzles and Perplexities: Collected Essays*, ed. S.M. Cahn, pp. 71–80. Lanham, MD: Rowman & Littlefield.

Cahn, S.M. (2002b) *The Affirmative Action Debate*, 2nd edn. New York: Routledge.

Carroll, L. (1946) *Alice in Wonderland*. Kingsport, TN: Kingsport Press for Grosset & Dunlap.

Edsall, T.B. and Edsall, M.D. (1991) *Chain Reaction: The Impact of Race, Rights and Taxes on American Politics*. New York: W.W. Norton & Company.

Fullinwider, R. and Lichtenberg, J. (2004) *Leveling the Playing Field: Justice, Politics, and College Admissions*. Lanham, MD: Rowman & Littlefield.

Furchtgott-Roth, D. and Stolba, C. (1999) *Women's Figures: An Illustrated Guide to Women's Progress in America*. Washington, DC: American Enterprise Institute.

Gitlin, T. (1995) *The Twilight of Common Dreams: Why America is Wracked by Culture Wars*. New York: Metropolitan Books.

Hewlett, S. (1987) *A Lesser Life: The Myth of Women's Liberation in America*. New York: Warner Books.

Hewlett, S. (1991) *When the Bough Breaks: The Cost of Neglecting Our Children*. New York: Basic Books.

Madrick, J. (1995) *The End of Affluence: The Causes and Consequences of America's Economic Dilemmas*. New York: Random House.

Mosley, A.G. (1984) Negritude, nationalism and nativism: racists or racialists? In *African Philosophy: Selected Readings*, ed. A. Mosley, pp. 216–235. Englewood Cliffs, NJ: Prentice Hall.

Mosley, A.G. (1997) Are racial categories racist? *Research in African Literatures* 26 (4): 101–111.

Mosley, A.G. (1998) Policies of straw or policies of inclusion? A review of Pojman's 'Case against affirmative action'. *The International Journal of Applied Philosophy* 12 (2): 161–168.

Mosley, A.G. and Capaldi, N. (1996) *Affirmative Action: Social Justice or Unfair Preference?* Lanham, MD: Rowman & Littlefield.

Nagel, T. (1979) *Mortal Questions*. Cambridge: Cambridge University Press.

Phillips, K. (1990) *The Politics of Rich and Poor: Wealth and the American Electorate in the Reagan Aftermath*. New York: Random House.

Phillips, K. (2002) *Wealth and Democracy: A Political History of the American rich*. New York: Broadway Books.

Pojman, L. (1998) The case against affirmative action. *The International Journal of Applied Philosophy* 12 (1): 97–115.

Rachels, J. (1993) Are quotas sometimes justified? In *Affirmative Action and the University: A Philosophical Inquiry*, ed. S.M. Cahn. Philadelphia, PA: Temple University Press.

Rutherglen, G. (1997) Affirmative action in faculty appointments: a guide to the perplexed. Appendix to Wolf-Devine (1997: 181–204).

Sowell, T. (2002) Discrimination, economics and culture. In Thernstrom and Thernstrom (2002: 167–80).

Thernstrom, A. and Thernstrom, S., eds (2002) *Beyond the Color Line: New Perspectives on Race and Ethnicity in America*. Stanford, CA: The Hoover Institution Press.

Thernstrom, A. and Thernstrom, S. (2003) *No Excuses: Closing the Racial Gap in Learning*. New York: Simon and Schuster.

Thernstrom, S. and Thernstrom, A. (1997) *America in Black and White*. New York: Touchstone.

Thernstrom, S. and Thernstrom, A. (2001) Racial preferences in higher education: an assessment of the evidence. In *One America*, ed. S.A. Renshon, pp. 169–231. Washington, DC: Georgetown University Press. (An earlier version of this paper can be found in UCLA Law Review, 46/5 (June 1999), under the title "Reflections on the shape of the river.").

Wolf-Devine, C. (1997) *Diversity and Community in the Academy: Affirmative Action in Faculty Appointments*. Lanham, MD: Rowman & Littlefield.

Further Reading

Cahn, S.M., ed. (1993) *Affirmative Action and the University: A Philosophical Inquiry*. Philadelphia, PA: Temple University Press.

Cohen, C. and Sterba, J. (2003) *Affirmative Action and Racial Preference: A Debate*. New York: Oxford University Press.

Steele, S. (1990) *The Content of Our Character: A New Vision of Race in America*. New York: St. Martin's Press.

Steele, S. (1998) *A Dream Deferred: The Second Betrayal of Black Freedom in America*. New York: HarperCollins Publishers Inc.

Capital punishment

CHAPTER TEN

A Defense of the Death Penalty

Louis P. Pojman

> *Who so sheddeth man's blood, by man shall his blood be shed.*
>
> (Genesis 9: 6)

There is an ancient tradition, going back to biblical times but endorsed by the mainstream of philosophers, from Plato to Thomas Aquinas, from Thomas Hobbes to Immanuel Kant, Thomas Jefferson, John Stuart Mill, and C.S. Lewis, that a fitting punishment for murder is the execution of the murderer. One prong of this tradition, the *backward-looking* or deontological position, epitomized in Aquinas and Kant, holds that because human beings, as rational agents, have dignity, one who with malice aforethought kills a human being, forfeits his or her right to life and deserves to die. The other, the *forward-looking* or consequentialist tradition, exemplified by Jeremy Bentham, Mill, and Ernest van den Haag, holds that punishment ought to serve as a deterrent, and that capital punishment is an adequate deterrent to prospective murderers. Abolitionists such as Hugo Adam Bedau (1980, 1982) and Jeffrey Reiman (1998) deny both prongs of the traditional case for the death penalty. They hold that long prison sentences are a sufficient retributive response to murder and that the death penalty probably does not serve as a deterrent. I will argue that both traditional defenses are sound and together they make a strong case for retaining the death penalty. That is, I hold a combined theory of punishment: a backward-looking judgment that the criminal has committed a heinous crime plus a forward-looking judgment that a harsh punishment will deter would-be murderers are sufficient to justify the death penalty. I turn first to the retributivist theory in favor of capital punishment. Then I will examine the deterrence theory. Finally, I will present four of the major objections to the death penalty along with the retributivist's response to each of them.

Contemporary Debates in Applied Ethics, Second Edition. Edited by Andrew I. Cohen and Christopher Heath Wellman.
© 2014 John Wiley & Sons, Inc. Published 2014 by John Wiley & Sons, Inc.

In Favor of the Death Penalty

Retribution

The small crowd that gathered outside the prison to protest the execution of Steven Judy softly sang: "We Shall Overcome." But it didn't seem quite the same hearing it sung out of concern for someone who, on finding a woman with a flat tire, raped and murdered her and drowned her three small children, then said that he hadn't been "losing any sleep" over his crimes.

I remember the grocer's wife. She was a plump, happy woman who enjoyed the long workday she shared with her husband in their ma-and-pa store. One evening, two young men came in and showed guns, and the grocer gave them everything in the cash register.

For no reason, almost as an afterthought, one of the men shot the grocer in the face. The woman stood only a few feet from her husband when he was turned into a dead, bloody mess.

She was about 50 when it happened. In a few years her mind was almost gone, and she looked 80. They might as well have killed her too.

Then there was the woman I got to know after her daughter was killed by a wolf-pack gang during a motoring trip. The mother called me occasionally, but nothing that I said could ease her torment. It ended when she took her own life.

A couple of years ago I spent a long evening with the husband, sister and parents of a fine young woman who had been forced into the trunk of a car in a hospital parking lot. The degenerate who kidnapped her kept her in the trunk, like an ant in a jar, until he got tired of the game. Then he killed her.[1]

Human beings have dignity as self-conscious rational agents who are able to act morally. One could maintain that it is precisely their moral goodness or innocence that bestows dignity and a right to life on them. Intentionally taking the life of an innocent human being is so evil that the perpetrator forfeits his own right to life. He or she deserves to die.

The retributivist holds three propositions: (1) that all the guilty deserve to be punished; (2) that only the guilty deserve to be punished; and (3) that the guilty deserve to be punished in proportion to the severity of their crime. Thomas Jefferson supported such a system of proportionality of punishment to crime:

> Whosoever shall be guilty of rape, polygamy, sodomy with man or woman, shall be punished, if a man, by castration, if a woman by cutting through the cartilage of her nose a hole of one half inch in diameter at the least. [And] whosoever shall maim another, or shall disfigure him . . . shall be maimed, or disfigured in the like sort: or if that cannot be, for want of some part, then as nearly as may be, in some other part of at least equal value. (Quoted in van den Haag, 1975, p. 193)

One need not accept Jefferson's specific penalties to concur with his central point of some equivalent harm coming to the criminal.

Criminals such as Steven Judy, Timothy McVeigh, Ted Bundy (who is reported to have raped and murdered more than 100 women), and the two men who gunned down the grocer (mentioned in the quotation by Royko, above) have committed capital offenses and deserve nothing less than capital punishment. No doubt malicious acts like the ones committed by these criminals deserve a worse punishment than death, but at a minimum, the death penalty seems warranted.

People often confuse retribution with revenge. While moral people will feel outrage at acts of heinous crimes, such as those described above by Royko, the moral justification of punishment is not vengeance, but desert. Vengeance signifies inflicting harm on the offender out of anger because of what he has done. Retribution is the rationally supported theory that the criminal deserves a punishment fitting to the gravity of his crime.

The nineteenth-century British philosopher James Fitzjames Stephens (1967, p. 152) thought vengeance was a justification for punishment, arguing that punishment should be inflicted "for the sake of ratifying the feeling of hatred – call it revenge, resentment, or what you will – which the contemplation of such [offensive] conduct excites in healthily constituted minds". But retributivism is not based on hatred for the criminal (though a feeling of vengeance may accompany the punishment). Retributivism is the theory that the criminal *deserves* to be punished and deserves to be punished in proportion to the gravity of his or her crime – whether or not the victim or anyone else desires it. We may all deeply regret having to carry out the punishment, but consider it warranted.

On the other hand, people do have a sense of outrage and passion for taking revenge on criminals for their crimes. Stephens (1863, p. 80) was correct in asserting that "[t]he criminal law stands to the passion for revenge in much the same relation as marriage to the sexual appetite". Failure to punish would no more lessen our sense of vengeance than the elimination of marriage would lessen our sexual appetite. When a society fails to punish criminals in a way thought to be proportionate to the gravity of the crime, the danger arises that the public would take the law into its own hands, resulting in vigilante justice, lynch mobs, and private acts of retribution. The outcome is likely to be an anarchistic, insecure state of injustice. As such, legal retribution stands as a safeguard for an orderly application of punitive desert.

Our natural instinct is for vengeance, but civilization demands that we restrain our anger and go through a legal process, letting the outcome determine whether, and to what degree, to punish the accused. Civilization demands that we not take the law into our own hands, but the laws should also satisfy our deepest instincts when they are consonant with reason. Our instincts tell us that some crimes, such as McVeigh's, Judy's, and Bundy's, should be severely punished, but we refrain from personally carrying out those punishments, committing ourselves to the legal processes. The death penalty is supported by our gut animal instincts as well as our sense of justice as desert.

The death penalty reminds us that there are consequences to our actions, and that we are responsible for what we do, so that dire consequences for immoral actions are eminently appropriate. The death penalty is such a fitting response to evil.

Deterrence

The second tradition justifying the death penalty is the forward-looking utilitarian theory of deterrence. This holds that by executing convicted murderers we will deter would-be murderers from killing innocent people. The evidence for deterrence is controversial. Some scholars, such as Sellin (1967) and Bedau, argue that the death penalty is not such a superior deterrent of homicides as long-term imprisonment.

Others, such as Ehrlich (1975), make a case for the death penalty as a significant deterrent. Granted, the evidence is ambiguous and honest scholars can differ on the results. However, one often hears abolitionists claiming that the evidence shows that the death penalty fails to deter homicide. This is too strong a claim. The sociological evidence does not show either that the death penalty deters or that it fails to deter. The evidence is simply inconclusive. But a common-sense case can be made for deterrence.

Imagine that every time someone intentionally killed an innocent person he was immediately struck down by lightning. When mugger Mike slashed his knife into the neck of the elderly pensioner, lightning struck, killing Mike. His fellow muggers witnessed the sequence of events. When burglar Bob pulled his pistol out and shot the bank teller through her breast, a bolt leveled Bob, and his compatriots beheld the spectacle. Soon men with their guns lying next to them were found all across the world in proximity to the corpses of their presumed victims. Do you think that the evidence of cosmic retribution would go unheeded?

We can imagine the murder rate in the United States and everywhere else plummeting. The close correlation between murder and cosmic retribution would surely serve as a deterrent to would-be-murderers. If this thought-experiment is sound, we have a prima facie argument for the deterrent effect of capital punishment. In its ideal, prompt performance, the death penalty would likely deter most rational, criminally minded people from committing murder. The question then becomes: how do we institute the death penalty in a manner that would have the maximal deterrent effect without violating the rights of the accused?

The accused would have to be brought to trial more quickly, and the appeals process of those found guilty "beyond reasonable doubt" limited. Having DNA evidence should make this more feasible than hitherto. Furthermore, public executions of the convicted murderer would serve as a reminder that crime does not pay. Public executions of criminals seem an efficient way to communicate the message that if you shed innocent blood, you will pay a high price. Hentoff (2001, p. 31) advocated that Timothy McVeigh be executed in public so that the public themselves would take responsibility for such executions. I agree with Hentoff on the matter of accountability, especially if such publicity would serve to deter homicide.

Abolitionists sometimes argue that because the statistical evidence in favor of the deterrent effect of capital punishment is indecisive, we have no basis for concluding that it is a better deterrent than long prison sentences. If I understand these abolitionists, their argument presents us with an exclusive disjunct. Either we must have conclusive statistical evidence (i.e., a proof) for the deterrent effect of the death penalty, or we have no grounds for supposing that the death penalty deters. Many people accept this argument. Recently, a colleague said to me, "There is no statistical evidence that the death penalty deters," as if to dismiss the argument from deterrence altogether. This confuses the proposition "there is no statistical proof for the deterrence-effect" with the proposition "there is statistical proof against the deterrence-effect." This is a fallacious inference, for it erroneously supposes that only two opposites are possible. There is a middle position that holds that while we cannot prove conclusively that the death penalty deters, the weight of evidence supports its deterrent effect. Furthermore, I think there are too many variables to hold constant for us to prove via statistics the deterrence hypothesis, and even if the requisite statistics were available, we could question whether they were cases of mere correlation versus causation. On the other hand, common-sense or anecdotal evidence may provide insight into the psychology of human motivation, providing evidence that fear of the death penalty deters some types of would-be criminals from committing murder. Granted, people are sometimes deceived about their motivation. But usually they are not deceived, and, as a rule, we should presume that they know their motives until we have evidence to the contrary. The general common-sense argument goes like this:

1   What people (including potential criminals) fear more will have a greater deter-rent effect on them.
2   People (including potential criminals) fear death more than they do any other humane punishment.
3   The death penalty is a humane punishment.
4   Therefore, people (including criminals) will be deterred more by the death penalty than by any other humane punishment.

Since the purpose of this argument is to show that the death penalty very likely deters more than long-term prison sentences, I am assuming it is humane – that is, acceptable to the moral sensitivities of the majority in our society. Torture might deter even more, but it is not considered humane.

Common sense informs us that most people would prefer to remain out of jail, that the threat of public humiliation is enough to deter some people, that a sentence of 20 years will deter most people more than a sentence of two years, and that a life sentence will deter most would-be criminals more than a sentence of 20 years. I think that we have common-sense evidence that the death penalty is a better deterrent than long prison sentences. For one thing, as Wilson and Herrnstein (1986) have argued, a great deal of crime is committed on a cost–benefit schema, wherein the criminal engages in some form of risk assessment as to his or her chances of getting caught and punished in some manner. If he or she estimates the punishment to be mild, the crime becomes inversely attractive, and vice versa. The fact that those who are condemned to death generally do everything in their power to get their sentences postponed or reduced to long-term prison sentences, in the way lifers do not, shows that they fear death more than life in prison.

The point is this: imprisonment constitutes one evil, the loss of freedom, but the death penalty imposes a more severe loss, that of life itself. If you lock me up, I may work for a parole or pardon. I may learn to live stoically with diminished freedom, and I can plan for the day when my freedom has been restored. But if I believe that my crime may lead to death, or loss of freedom followed by death, then I have more to fear than mere imprisonment. I am faced with a great evil plus an even greater evil. I fear death more than imprisonment because it alone takes from me all future possibility.

I am not claiming that the fear of legal punishment is all that keeps us from criminal behavior. Moral character, good habit, fear of being shamed, peer pressure, fear of authority, or the fear of divine retribution may have a greater influence on some people. However, many people will be deterred from crime, including murder, by the threat of severe punishment. The abolitionist points out that many would-be murderers simply do not believe they will be caught. Perhaps this is true for some. While the fantastic egoist has delusions of getting away with his crime, many would-be criminals are not so bold or delusionary.

A Defense of the Death Penalty   163

Former Prosecuting Attorney for the State of Florida, Richard Gernstein, has set forth the common-sense case for deterrence. First of all, he claims, the death penalty certainly deters the murderer from any further murders, including those he or she might commit within the prison where he is confined. Second, statistics cannot tell us how many potential criminals have refrained from taking another's life through fear of the death penalty. He quotes Judge Hyman Barshay of New York: "The death penalty is a warning, just like a lighthouse throwing its beams out to sea. We hear about ship-wrecks, but we do not hear about the ships the lighthouse guides safely on their way. We do not have proof of the number of ships its saves, but we do not tear the lighthouse down" (Gernstein, 1960, p. 253).

Some of the common-sense evidence is anecdotal, as the following quotation shows. British Member of Parliament Arthur Lewis explains how he was converted from an abolitionist to a supporter of the death penalty:

> One reason that has stuck in my mind, and which has proved [deterrence] to me beyond question, is that there was once a professional burglar in [my] constituency who consist-ently boasted of the fact that he had spent about one-third of his life in prison. . . . He said to me "I am a professional burglar. Before we go out on a job we plan it down to every detail. Before we go into the boozer to have a drink we say 'Don't forget, no shooters' – shooters being guns." He adds: "We did our job and didn't have shooters because at that time there was capital punishment. Our wives, girlfriends and our mums said, 'Whatever you do, do not carry a shooter because if you are caught you might be topped [executed].' If you do away with capital punishment they will all be carrying shooters." (British Par-liamentary Debates, 1982)

It is difficult to know how widespread this reasoning is. My own experience corrobo-rates this testimony. Growing up in the infamous Cicero, Illinois, home of Al Capone and the Mafia, I had friends, including a brother, who drifted into crime, mainly bur-glary and larceny. It was common knowledge that one stopped short of killing in the act of robbery. A prison sentence could be dealt with – especially with a good lawyer – but being convicted of murder, which at that time included a reasonable chance of being electrocuted, was an altogether different matter. No doubt exists in my mind that the threat of the electric chair saved the lives of some of those who were robbed in my town. No doubt some crimes are committed in the heat of passion or by the temporally (or permanently) insane, but many are committed through a process of risk assessment. Burglars, kidnappers, traitors, and vindictive people will sometimes be restrained by the threat of death. We simply do not know how much capital punish-ment deters, but this sort of common-sense, anecdotal evidence must be taken into account in assessing the institution of capital punishment.

John Stuart Mill admitted that capital punishment does not inspire terror in hard-ened criminals, but it may well make an impression on prospective murderers:

> As for what is called the failure of the death punishment, who is able to be judge of that? We partly know who those are whom it has not deterred; but who is there who knows whom it has deterred, or how many human beings it has saved who would have lived to be murderers if that awful association had not been thrown round the idea of murder from their earliest infancy. (Mill, 1986, pp. 97–104)

Mill's points are well taken: first, not everyone will be deterred by the death penalty, but some will; second, the potential criminal need not consciously calculate a cost–benefit analysis regarding his crime to be deterred by the threat. The idea of the threat may have become a subconscious datum "from their earliest infancy." The repeated announcement and regular exercise of capital punishment may have deep causal influence.

Gernstein quotes the British Royal Commission on Capital Punishment (1949–53), which is one of the most thorough studies on the subject and which concluded that there was evidence that the death penalty has some deterrent effect on normal human beings. Some of its evidence in favor of the deterrence effect includes:

1 Criminals who have committed an offense punishable by life imprisonment, when faced with capture, refrained from killing their captor though by killing, escape seemed probable. When asked why they refrained from the homicide, quick responses indicated a willingness to serve life sentence, but not risk the death penalty.
2 Criminals about to commit certain offenses refrained from carrying deadly weapons. Upon apprehension, answers to questions concerning absence of such weapons indicated a desire to avoid more serious punishment by carrying a deadly weapon, and also to avoid use of the weapon which could result in imposition of the death penalty.
3 Victims have been removed [by criminals] from a capital-punishment State to a non-capital-punishment State to allow the murderer opportunity for homicide without threat to his own life. This in itself demonstrates that the death penalty is considered by some would-be-killers. (Gernstein, 1960, p. 253)

Gernstein then quotes former District Attorney of New York, Frank S. Hogan, representing himself and his associates:

We are satisfied from our experience that the deterrent effect is both real and substantial . . . for example, from time to time accomplices in felony murder state with apparent truthfulness that in the planning of the felony they strongly urged the killer not to resort to violence. From the context of these utterances, it is apparent that they were led to these warnings to the killer by fear of the death penalty that they realized might follow the taking of life. Moreover, victims of hold-ups have occasionally reported that one of the robbers expressed a desire to kill them and was dissuaded from so doing by a confederate. Once again, we think it not unreasonable to suggest that fear of the death penalty played a role in some of these intercessions.

On a number of occasions, defendants being questioned in connection with homicide have shown a striking terror of the death penalty. While these persons have in fact perpetrated homicide, we think that their terror of the death penalty must be symptomatic of the attitude of many others of their type, as a result of which many lives have been spared. (Gernstein, 1960, pp. 253–254)

It seems likely that the death penalty does not deter as much as it could do, because of its inconsistent and rare use. For example, in 1949, out of an estimated 23,370 cases of murder, non-negligent manslaughter, and rape, there were only 119 executions carried out in the United States. In 1953, out of 27,000 murder cases, only 62 executions for those crimes took place. Few executions were carried out in the 1960s and none at all from 1967 to 1977. Gernstein (1960, p. 254) points out that at that rate a criminal's chances of escaping execution are better than 100 to 1. Actually, since

Gernstein's report, the figures have become even more weighted against the chances of the death penalty. In 1993, there were 24,526 cases of murder and non-negligent manslaughter and only 56 executions, while in 1994 there were 23,305 cases of murder and non-negligent manslaughter and only 31 executions – a ratio of more than 750 to 1 in favor of the criminal. The average length of stay for a prisoner executed in 1994 was ten years and two months. If potential murderers perceived the death penalty as a highly probable outcome of murder, would they not be more reluctant to kill? Gernstein notes:

> The commissioner of Police of London, England, in his evidence before the Royal Commission on Capital Punishment, told of a gang of armed robbers who continued operations after one of their members was sentenced to death and his sentence commuted to penal servitude, but the same gang disbanded and disappeared when, on a later occasion, two others were convicted of murder and hanged. (Gernstein, 1960, p. 254)

Gernstein sums up his data:

> Surely it is a common-sense argument, based on what is known of human nature, that the death penalty has a deterrent effect particularly for certain kinds of murderers. Furthermore, as the Royal Commission opined, the death penalty helps to educate the conscience of the whole community, and it arouses among many people a quasi-religious sense of awe. In the mind of the public there remains a strong association between murder and the penalty of death. Certainly one of the factors which restrains some people from murder is fear of punishment and surely, since people fear death more than anything else, the death penalty is the most effective deterrent. (Gernstein, 1960, p. 254)

A retentionist is someone who advocates retaining the death penalty as a mode of punishment for some crimes. Given the retributivist argument for the death penalty based on desert, the retentionist does not have to prove that the death penalty deters *better* than long-prison sentences, but if the death penalty is deemed at least as effective as its major alternative, it would be justified. If evidence existed that life imprisonment were a *more effective* deterrent, the retentionist might be hard-pressed to defend it on retributivist lines alone. My view is that the desert argument plus the common-sense evidence – being bolstered by the following argument, the Best Bet Argument – strongly supports retention of the death penalty.

Ernest van den Haag (1968) set forth what he calls the Best Bet Argument. He argues that even though we do not know for certain whether the death penalty deters or prevents other murders, we should bet that it does. Indeed, due to our ignorance, any social policy we take is a gamble. Not to choose capital punishment for first-degree murder is as much a bet that capital punishment does not deter as choosing the policy is a bet that it does. There is a significant difference in the betting, however, in that to bet against capital punishment is to bet against the innocent and for the murderer, while to bet for it is to bet against the murderer and for the innocent.

The point is this: we are accountable for what we let happen, as well as for what we actually do. If I fail to bring up my children properly, so that they are a menace to society, I am to some extent responsible for their bad behavior. I could have caused it to be somewhat better. If I have good evidence that a bomb will blow up the building you are working in and fail to notify you (assuming I can), I am partly responsible for your death, if and when the bomb explodes. So we are responsible for what we omit doing, as well as for what we do. Purposefully to refrain from a lesser evil which we know will allow a greater evil to occur is to be at least partially responsible for the greater evil. This responsibility for our omissions underlies van den Haag's argument, to which we now return.

Suppose that we choose a policy of capital punishment for capital crimes. In this case we are betting that the death of some murderers will be more than compensated for by the lives of some innocents not being murdered (either by these murderers or by others who would have murdered). If we are right, we have saved the lives of the innocent. If we are wrong, we have, unfortunately, sacrificed the lives of some murderers. But say we choose not to have a social policy of capital punishment. If capital punishment does not work as a deterrent, we have come out ahead, but if it does work, then we have missed an opportunity to save innocent lives. If we value the saving of innocent lives more highly than we do the loss of the guilty, then to bet on a policy of capital punishment turns out to be rational. Since the innocent have a greater right to life than the guilty, it is our moral duty to adopt a policy that has a chance of protecting them from potential murderers.

It is noteworthy that prominent abolitionists, such as Charles Black, Hugo Adam Bedau, Ramsey Clark, and Henry Schwartzchild, have admitted to Ernest van den Haag that even if every execution were to deter 100 murders, they would oppose it, from which van den Haag concludes: "to these abolitionist leaders, the life of every murderer is more valuable than the lives of a hundred prospective victims, for these abolitionists would spare the murderer, even if doing so will cost a hundred future victims their lives." Black and Bedau said they would favor abolishing the death penalty even if they knew that doing so would increase the homicide rate by 1,000 percent.[2] This response of abolitionists is puzzling, since one of Bedau's arguments against the death penalty is that it does not bring back the dead: "We cannot do anything for the dead victims of crime. (How many of those who oppose the death penalty would continue to do so if, *mirabile dictu*, executing the murderer might bring the victim back to life?)" (Bedau, 1989, p. 190). Apparently, he would support the death penalty if it brought a dead victim back to life, but not if it prevented 100 innocent victims from being murdered.

If the Best Bet Argument is sound, or if the death penalty does deter would-be murderers, as common sense suggests, then we should support some uses of the death penalty. It should be used for those who commit first-degree murder, for whom no mitigating factors are present, and especially for those who murder police officers, prison guards, and political leaders. Many states rightly favor it for those who murder while committing another crime, for example, burglary or rape. It should also be used for treason and terrorist bombings. It should also be considered for egregious white-collar crimes such as for bank managers who embezzle the savings of the public. The Savings & Loan scandals of the 1980s and the corporate scandals of 2002, involving wealthy bank officials and CEOs engaging in fraudulent business behavior, ruined the lives of many people, while providing the perpetrators with golden parachutes. This gross violation of the public trust may well warrant the electric chair.

Objections to the Death Penalty

Finally, let us examine four of the major objections to death penalty, as well as the retentionist's responses to those objections.

Objection 1

Capital punishment is a morally unacceptable thirst for revenge. As former British Prime Minister Edward Heath put it:

> The real point that is emphasized to me by many constituents is that even if the death penalty is not a deterrent, murderers deserve to die. This is the question of revenge. Again, this will be a matter of moral judgment for each of us. I do not believe in revenge. If I were to become the victim of terrorists, I would not wish them to be hanged or killed in any other way for revenge. All that would do is deepen the bitterness that already tragically exists in the conflicts we experience in society, particularly in Northern Ireland. (British Parliamentary Debates, 1982)

Response Retributivism, as I argued above, is not the same thing as revenge, although the two attitudes are often intermixed in practice. Revenge is a personal response to a perpetrator for an injury. Retribution is an impartial and impersonal response to an offender for an offense done against someone. You cannot desire revenge for the harm of someone to whom you are indifferent. Revenge always involves personal concern for the victim. Retribution is not personal but is based on objective factors: the criminal has deliberately harmed an innocent party and so deserves to be punished, whether I wish it or not. I would agree that I or my son or daughter deserves to be punished for our crimes, but I do not wish any vengeance on myself or my son or daughter.

Furthermore, while revenge often leads us to exact more suffering from the offender than the offense warrants, retribution stipulates that the offender be punished in proportion to the gravity of the offense. In this sense, the *lex talionis* that we find in the Old Testament is actually a progressive rule, where retribution replaces revenge as the mode of punishment. It says that there are limits to what one may do to the offender. Revenge demands a life for an eye or a tooth, but Moses provides a rule that exacts a penalty equal to the harm done by the offender.

Objection 2

Perhaps the murderer does deserve to die, but by what authority does the state execute him or her? Both the Old and New Testament say, "'Vengeance is mine, I will repay,' says the Lord" (Deut. 32: 35 and Romans 12: 19). You need special authority to justify taking the life of a human being.

Response The objector fails to note that the New Testament passage continues with a support of the right of the state to execute criminals in the name of God: "Let every person be subjected to the governing authorities. For there is no authority except from God, and those that exist have been instituted by God. Therefore he who resists what God has appointed, and those who resist will incur judgment. . . . If you do wrong, be afraid, for [the authority] does not bear the sword in vain; he is the servant of God to execute his wrath on the wrongdoer" (Romans 13: 1–4). So, according to the Bible, the authority to punish, which presumably includes the death penalty, comes from God.

But we need not appeal to a religious justification for capital punishment. We can cite the state's role in dispensing justice. Just as the state has the authority (and duty) to act justly in allocating scarce resources, in meeting the minimal needs of its (deserving) citizens, in defending its citizens from violence and crime, and in not waging unjust wars, so too it has the authority, flowing from its mission to promote justice and the good of its people, to punish the criminal. If the criminal, as one who has forfeited a right to life, deserves to be executed, especially if it will likely deter would-be murderers, the state has a duty to execute those convicted of first-degree murder.

Objection 3

Miscarriages of justice occur. Capital punishment is to be rejected because of human fallibility in convicting innocent parties and sentencing them to death. In a survey done in 1985, Bedau and Radelet found that 25 of the 7,000 persons executed in the United States between 1900 and 1985 were innocent of capital crimes (quoted in van den Haag, 1986, p. 1664). While some compensation is available to those unjustly imprisoned, the death sentence is irrevocable. We cannot compensate the dead. As John Maxton, a British Member of Parliament puts it, "If we allow one innocent person to be executed, morally we are committing the same, or, in some ways, a worse crime than the person who committed the murder" (British Parliamentary Debates, 1982).

Response Mr Maxton is incorrect in saying that mistaken judicial execution is morally the same or worse than murder, for a deliberate intention to kill the innocent occurs in a murder, whereas no such intention occurs in wrongful capital punishment.

Sometimes this objection is framed as follows. It is better to let ten criminals go free than to execute one innocent person. If this dictum is a call for safeguards, then it is well taken; but somewhere there seems to be a limit on the tolerance of society towards capital offenses. Would these abolitionists argue that it is better that 50 or 100 or 1,000 murderers go free than that one guilty person be executed? Society has a right to protect itself from capital offenses even if this means taking a tiny chance of executing an innocent person. If the basic activity or process is justified, then it is regrettable, but morally acceptable, that some mistakes are made. Fire trucks occasionally kill innocent pedestrians while racing to fires, but we accept these losses as justified by the greater good of the activity of using fire trucks. We judge the use of automobiles to be acceptable, even though such use causes an average of 50,000 traffic fatalities each year. We accept the morality of a defensive war even though it will result in our troops accidentally or mistakenly killing innocent people.

The fact that we can err in applying the death penalty should give us pause and cause us to build a better appeals process into the judicial system. Such a process is already in place in the American and British legal systems. That occasional error may be made, regrettable though this is, is not a sufficient reason for us to refuse to use the death penalty, if on balance it serves a just and useful function.

Furthermore, abolitionists are simply misguided in thinking that prison sentences are a satisfactory alternative here. It is not clear that we can always or typically compensate innocent parties who waste away in prison. Jacques Barzun has argued that a prison sentence can be worse than death and carries all the problems that the death penalty does regarding the impossibility of compensation.

> In the preface of his useful volume of cases, *Hanged in Error*, Mr Leslie Hale refers to the tardy recognition of a minor miscarriage of justice – one year in jail: "The prisoner emerged to find that his wife had died and that his children and his aged parents had been removed to the workhouse. By the time a small payment had been assessed as 'compensation' the victim was incurably insane." So far we are as indignant with the law as Mr Hale. But what comes next? He cites the famous Evans case, in which it is very probable that the wrong man was hanged, and he exclaims: "While such mistakes are possible, should society impose an irrevocable sentence?" Does Mr Hale really ask us to believe that the sentence passed on the first man, whose wife died and who went insane, was in any sense *revocable?* Would not any man rather be Evans dead than that other wretch "emerging" with his small compensation and his reason for living gone? (Barzun, 1962, pp. 188–189)

The abolitionist is incorrect in arguing that death is different from long-term prison sentences because it is irrevocable. Imprisonment also takes good things away from us that may never be returned. We cannot restore to the inmate the freedom or opportunities he or she has lost. Suppose an innocent 25-year-old man is given a life sentence for murder and 30 years later the error is discovered and he is set free. Suppose he values three years of freedom to every one year of life. That is, he would rather live 10 years as a free man than 30 as a prisoner. Given this man's values, the criminal justice system has taken the equivalent of 10 years of life from him. If he lives until he is 65, he has, as far as his estimation is concerned, lost 10 years, so that he may be said to have lived only 55 years.

The numbers in this example are arbitrary, but the basic point is sound. Most of us would prefer a shorter life of higher quality to a longer one of low quality. Death prevents all subsequent quality, but imprisonment also irrevocably harms one by diminishing the quality of life of the prisoner.

Objection 4

The death penalty is unjust because it discriminates against the poor and minorities, particularly African Americans, over against rich people and whites. Former Supreme Court Justice William Douglas wrote that "a law which reaches that [discriminatory] result in practice has no more sanctity than a law that in terms provides the same" (*Furman v. Georgia*, 1972). Stephen Nathanson (2001, p. 62) argues that, "in many cases, whether one is treated justly or not depends not only on what one deserves but on how other people are treated". He offers the example of unequal justice in a plagiarism case: "I tell the students in my class that anyone who plagiarizes will fail the course. Three students plagiarize papers, but I give only one a failing grade. The other two, in describing their motivation, win my sympathy, and I give them passing grades" (2001, pp. 62, 60). Arguing that this is patently unjust, he likens this case to the imposition of the death penalty and concludes that it too is unjust.

Response First of all, it is not true that a law that is applied in a discriminatory manner is unjust. Unequal justice is no less justice, however uneven its application. The discriminatory application, not the law itself, is unjust. A just law is still just even if it is not applied consistently. For example, a friend of mine once got two speeding tickets during a 100-mile trip (having borrowed my car). He complained to the police officer who gave him the second ticket that many drivers were driving faster than he was at the time. They had escaped detection, he argued, so it was not fair for him to get two tickets on one trip. The officer acknowledged the imperfections of the system but, justifiably, had no qualms about giving him the second ticket. Unequal justice is still justice, however regrettable. So Justice Douglas is wrong in asserting that discriminatory results invalidate the law itself. Discriminatory practices should be reformed, and in many cases they can be. But imperfect practices in themselves do not entail that the laws engendering these practices are themselves are unjust.

With regard to Nathanson's analogy with the plagiarism case, two things should be said against it. First, if the teacher is convinced that the motivational factors are mitigating factors, then he or she may be justified in passing two of the plagiarizing students. Suppose that the one student did no work whatsoever, showed no interest (Nathanson's motivation factor) in learning, and exhibited no remorse in cheating, whereas the other two spent long hours seriously studying the material and, upon apprehension, showed genuine remorse for their misdeeds. To be sure, they yielded to temptation at certain – though limited – sections of their long papers, but the vast majority of their papers represented their own diligent work. Suppose, as well, that all three had C averages at this point. The teacher gives the unremorseful, gross plagiarizer an F, but relents and gives the other two a D. Her actions parallel the judge's use of mitigating circumstances and cannot be construed as arbitrary, let alone unjust.

The second problem with Nathanson's analogy is that it would have disastrous consequences for all law and benevolent practices alike. If we concluded that we should abolish a rule or practice unless we treat everyone exactly by the same rules all the time, we would have to abolish, for example, traffic laws and laws against imprisonment for rape, theft, and even murder. Carried to its logical limits, we would also have to refrain from saving drowning victims if a number of people were drowning but we could only save a few of them. Imperfect justice is the best that we humans can attain. We should reform our practices as much as possible to eradicate unjust discrimination wherever we can, but if we are not allowed to have a law without perfect application, we will be forced to have no laws at all.

Nathanson (2001, p. 67) acknowledges this latter response, but argues that the case of death is different. "Because of its finality and extreme severity of the death penalty, we need to be more scrupulous in applying it as punishment than is necessary with any other punishment". The retentionist agrees that the death penalty is a severe punishment and that we need to be scrupulous in applying it. The difference between the abolitionist and the retentionist seems to lie in whether we are wise and committed enough as a nation to reform our institutions so that they approximate fairness. Apparently, Nathanson is pessimistic here, whereas I have faith in our ability to learn from our mistakes and reform our systems. If we cannot reform our legal system, what hope is there for us?[3]

More specifically, the charge that a higher percentage of blacks than whites are executed was once true, but is no longer so. Many states have made significant changes in sentencing procedures, with the result that, currently, whites convicted of first-degree murder are sentenced to death at a higher rate than blacks.[4]

One must be careful in reading too much into these statistics. While great disparities in statistics should cause us to examine our judicial procedures, they do not in themselves prove injustice. For example, more males than females are convicted of violent crimes (almost 90 percent of those convicted of violent crimes are males – a virtually universal statistic), but this is not strong evidence that the law is unfair, for there are biological/psychological explanations for the disparity in convictions. Males are on average and by nature more aggressive (usually linked to testosterone) than females; simply having a Y chromosome predisposes them to greater violence. Nevertheless, we hold male criminals responsible for their violence and expect them to control themselves. Likewise, there may be good explanations why people of one ethnic group commit more crimes than those of other groups, explanations that do not impugn the processes of the judicial system, nor absolve rational people of their moral responsibility.

At the time of writing, Governor Ryan of Illinois had just commuted the sentences of more than 167 death-row inmates. Abolitionists throughout the world celebrated this as a great victory. But they should have second thoughts. By summarily commuting the sentences of all of the condemned men, the Governor undermined the stability and integrity of the law as a viable institution in his state, overturning years of work by the police, prosecutors, judges, and juries, and turned his back on the right of the victims' families to see justice done. Apparently, some of those convicted were done so on insufficient evidence. If so, their sentences should have been commuted and the prisoners compensated. But such decisions should be taken on a case-by-case basis. Some of the convicts on death row were hardened unrepentant criminals, guilty of heinous crimes. If capital punishment is justified, its application should be confined to such clear cases in which the guilt of the criminal is "beyond reasonable doubt." But to overthrow the whole system because of a few possible miscarriages of justice is as unwarranted as it is a loss of faith in our system of criminal justice. No one would abolish the use of fire engines and ambulances because occasionally they kill innocent pedestrians while carrying out their mission.

The complaint is often made by abolitionists that only the poor get death sentences for murder. If their trials are fair, then they deserve the death penalty, but rich murderers may be equally deserving. At the moment, only first-degree murder and treason are crimes deemed worthy of the death penalty. Perhaps our notion of treason should be expanded to include those who betray the trust of the public, corporation executives who have the trust of ordinary people, but who, through selfish and dishonest practices, ruin their lives. My proposal is to broaden, not narrow, the scope of capital punishment, to include businessmen and women who unfairly and severely harm the public. As I have mentioned, the executives in the recent corporation scandals who bailed out with millions of dollars while they destroyed the pension plans of thousands of employees may deserve severe punishment and, if convicted, they should receive what they deserve. My guess is that the threat of the death sentence would have a deterrent effect in such cases. Whether it is feasible to apply the death penalty to horrendous white-collar crimes is debatable. But there is something to be said in its favor; it would certainly remove the impression that only the poor get executed.

Conclusion

While the abolitionist movement is gaining strength – due in part to the dedicated eloquence of opponents to the death penalty such as Hugo Adam Bedau, Stephen

Nathanson, and Jeffrey Reiman – a cogent case can be made for retaining the death penalty for serious crimes. The case primarily rests on a notion of justice as desert, but is strengthened by utilitarian arguments involving deterrence. It is not because retentionists disvalue life that we defend the use of the death penalty. Rather, it is because we value human life as highly as we do that we support its continued use. The combined argument based on both backward-looking and forward-looking considerations justify use of the death penalty.

The abolitionist points out the problems in applying the death penalty. We can concede that there are problems and that reform is constantly needed, but since the death penalty is justified in principle, we should seek to improve its application rather than abolish a just institution.[5] If civilized society can reduce racism and sexism and send people to the moon, surely it can reduce the injustices connected with the criminal justice system. We ought not to throw out the baby with the dirty bath water.

Notes

1 Mike Royko, quoted in Moore (1995, pp. 98–99).
2 Cited in Ernest van den Haag, "The Death Penalty Once More," unpublished manuscript. In "A Response to Bedau" (van den Haag, 1977, p. 798, n.5), van den Haag states that both Black and Bedau said that they would be in favor of abolishing the death penalty even if "they knew that its abolition (and replacement by life imprisonment) would increase the homicide rate by 10%, 20%, 50%, 100%, or 1000%. Both gentlemen continued to answer affirmatively." Bedau confirmed this in a letter to me (July 28, 1996).
3 An example might be the abolition of large numbers of institutions for the mentally ill in New York which began in the 1960s, sought by reformers because of documented abuses related to both inadequate treatment and due regard for patients' rights. It was argued that prevailing conditions could not be reformed, but large-scale release of long-institutionalized persons without adequate planning for their follow-up led to new problems, including visibly increased homelessness. In hindsight, many believe that more work should have been done to reform the institutions. Sometimes it is the lesser of two evils to keep an imperfect institution than to abolish it for an unknown effect.
4 The Department of Justice's Bureau of Justice Statistics Bulletin for 1994 reports that between 1977 and 1994, 2,336 (5%) of those arrested for murder were white, 1,838 (40%) were black, and 316 (7%) were Hispanic. Of the 257 who were executed, 140 (54%) were white, 98 (38%) were black, 17 (7%) were Hispanic, and 2 (1%) were other races. In 1994, 31 prisoners – 20 white men and 11 black men – were executed, although whites made up only 7,532 (41%) and blacks 9,906 (56%) of those arrested for murder. Of those sentenced to death in 1994, 158 were white men, 133 were black men, 25 were Hispanic men, 2 were Native American men, 2 were white women, and 3 were black women. Of those sentenced, relatively more blacks (72%) than whites (65%) or Hispanics (60%) had prior felony records. Overall, the criminal justice system does not seem to favor white criminals over black, though it does seem to favor rich defendants over poor ones.
5 I have discussed these problems in Pojman (1998).

References

Barzun, J. (1962) In favor of capital punishment. *The American Scholar* 31: 181–191.
Bedau, H.A. (1980) Capital punishment. In *Matters of Life and Death*, ed. T. Regan, pp. 148–182. New York: Random House.

Bedau, H.A. (1982) *The Death Penalty in America*. New York: Random House.

Bedau, H.A. (1989) How to argue about the death penalty. In *Facing the Death Penalty*, ed. M. Radelet, pp. 178–192. Philadelphia, PA: Temple University Press.

British Parliamentary Debates (1982) *Fifth Series*. vol. 23, issue 1243, House of Commons, 11 May 1982.

Ehrlich, I. (1975) The deterrent effect of capital punishment: a question of life and death. *The American Economic Review* 65(June): 397–417.

Furman v. Georgia (1972) 408 US 238.

Gernstein, R.E. (1960) A prosecutor looks at capital punishment. *The Journal of Criminal Law, Criminology, and Police Science* 51: 252–256.

Hentoff, N. (2001) The state closes our eyes as it kills. *The Village Voice* (May 1): 31. http://www.villagevoice.com/2001-04-24/news/the-state-closes-our-eyes-as-it-kills/1/ (last accessed 6/17/13).

Mill, J.S. (1986) *Parliamentary Debates. Third series*, April 21, 1868. Reprinted in *Applied Ethics*, ed. P. Singer, pp. 97–104. New York: Oxford University Press.

Moore, M. (1995) The moral worth of retributivism. In *Punishment and Rehabilitation*, ed. G. Murphy Jeffrie, pp. 94–130. Belmont, CA: Wadsworth.

Nathanson, S. (2001) *An Eye For An Eye: The Immorality of Punishing By Death*. Lanham, MD: Rowman & Littlefield.

Pojman, L.P. (1998) For the death penalty. In *The Death Penalty: For and Against*, ed. L.P. Pojman and J. Reiman, pp. 1–66. Lanham, MD: Rowman & Littlefield.

Reiman, J. (1998) Why the death penalty should be abolished in America. In *The Death Penalty: For and Against*, ed. L.P. Pojman and J. Reiman, pp. 67–133. Lanham, MD: Rowman & Littlefield.

Sellin, T. (1967) Effect of repeal and reintroduction of the death penalty on homicide rates. In *The Death Penalty in America*, ed. H. Bedau, pp. 339–343. Chicago, IL: Aldine Books.

Stephens, J.F. (1863) *A History of Criminal Law in England*. New York: Macmillan.

Stephens, J.F. (1967) *Liberty, Equality, Fraternity*. Cambridge: Cambridge University Press.

van den Haag, E. (1968) On deterrence and the death penalty. *Ethics* 78: 280–288.

van den Haag, E. (1975) *Punishing Criminals: Concerning a Very Old and Painful Question*. New York: Basic Books.

van den Haag, E. (1977) A response to Bedau. *Arizona State Law Journal* 4: 797–802.

van den Haag, E. (1986) The ultimate punishment: a defense. *Harvard Law Review* 99: 1662–1669.

Wilson, J.Q. and Herrnstein, R.J. (1986) *Crime and Human Nature*. New York: Simon and Schuster.

Further Reading

Davis, M. (1981) Death, deterrence, and the method of common sense. *Social Theory and Practice* 7: 145–178.

CHAPTER ELEVEN

Why We Should Put the Death Penalty to Rest

Stephen Nathanson

My aim in this chapter is to make the strongest case that I can to show that punishing people by death is an unjust and immoral practice. Although we often think of the death penalty debate as one of those eternal, irresolvable issues, I believe that the arguments for the death penalty are extremely weak and that the practice of punishing by death is morally indefensible.

I know, of course, that not everyone sees things this way. The laws of 38 of the 50 US states include death as a possible punishment, and public support in the United States for the death penalty since the 1970s has been very strong. This American consensus, however, is somewhat anomalous. The death penalty has been abolished in almost every modern, democratic country, and its abolition is now required of any country wanting to enter the European Union.

Still, in the United States, many people strongly support the death penalty. Why is this? While it could be that death penalty supporters simply want vengeance and do not care about morality, I doubt that this is true. I believe that most death penalty supporters are people of good will who think that the death penalty is right or necessary. While some political leaders use the death penalty for political gain, most people have no vested interest in the death penalty. If they are wrong about it, this is the result of honest mistakes. They either have mistaken factual beliefs or are confused in their moral thinking. If I can show that the factual and moral beliefs on which death penalty support depends are mistaken, this should lead them to see that it ought to be abolished where it is used and left to rest in peace where it has already been rejected.

Whether I can do this remains to be seen. One reason for optimism is that the death penalty debate differs from some other controversial issues. Some controversies are hard to resolve because people on opposing sides differ in their fundamental values. In such cases, it is hard to find values that can serve as a basis for reaching agreement. The death penalty debate is not like this. Both death penalty supporters and opponents

Contemporary Debates in Applied Ethics, Second Edition. Edited by Andrew I. Cohen and Christopher Heath Wellman.
© 2014 John Wiley & Sons, Inc. Published 2014 by John Wiley & Sons, Inc.

generally appeal to the same fundamental values: the pursuit of justice and respect for human life.

If I can show that a belief in the importance of justice and respect for human life is inconsistent with support for the death penalty, it would follow that people who hold these ideals and yet favor the death penalty are actually contradicting their own values. An argument that would show this would be very strong, both logically and psychologically. It would be logically strong because contradictory views are necessarily false, and psychologically strong because it appeals only to values that death penalty supporters themselves accept.

Of course, some people might reject the values of justice and respect for human life, but there would be a great cost to doing so. First, they would deprive themselves of some of the common arguments for capital punishment – such as that it is necessary for protecting human life and that it is a just punishment for murder. Indeed, they would be unable to say why murder is a serious crime since the condemnation of murder presupposes that human life has an especially high value. Second, rejecting these values would undermine their moral credibility. No one would listen to people who said that they were indifferent to justice and respect for human life because our society is publicly committed to these values (whether or not it actually takes the required steps to do achieve them).

We can assume, then, that all who support the death penalty and whose views we take seriously are committed to the values of justice and respect for human life. My goal is to show that the death penalty is inconsistent with these values. How can this be done?

My argument will proceed in two stages. First, a consideration of the death penalty in theory, followed by, second, an examination of the death penalty in practice. I will show that the principled bases for the death penalty are extremely weak. Then, I will show that even if the death penalty could be justified in theory, the actual practice of executing murderers violates both the values of justice and respect for human life. Even people who support the death penalty in theory should oppose it in practice.

The Death Penalty in Theory: Saving Lives and Doing Justice

The two basic arguments for the death penalty are: (1) that it is the best deterrent of murders and thus saves people's lives, and (2) that it is the punishment that justice requires for the crime of murder.

The argument from deterrence

According to this argument, the threat of execution is a more powerful deterrent than lesser punishments and therefore will lead to fewer deaths from homicide. If this is true, then anyone who values human life will be willing to support executing murderers because this will spare the lives of innocent people. Just as common-sense morality permits killing in self-defense or defense of others, so it permits the death penalty as a form of social self-defense, saving the lives of people who would otherwise be victims of murder.

In theory, this is a powerful argument. Anyone who values human life will want to diminish the number of people murdered, and if the death penalty is uniquely effective in preventing murders, then it cannot be dismissed as senseless violence or mere vengeance. Nor could it be said to be the same as murder, for while murders increase the number of innocent victims, the death penalty (according to this argument) diminishes the number of innocent victims.

The deterrence argument has been challenged on factual grounds, and the best evidence suggests that the death penalty is not a better deterrent than life imprisonment.[1] In general, countries and states that do not use the death penalty have lower homicide rates than countries that do. These are familiar points which I will not stress here because I want to consider the death penalty "in theory." I want to challenge the underlying moral principle, which is the idea that if a punishment deters more murders and thus saves more innocent lives, then it is justifiable. While this sounds plausible, it is false and thus fails to justify the death penalty. I offer two arguments to show this.

First, we can imagine punishments that have greater deterrent value than either the death penalty or imprisonment and yet would be wrong to inflict. Suppose we could deter more murders by executing not just the person who commits a murder but also the family or closest friends of such a person. If the idea behind the deterrence argument is that we deter more murders by threatening the most terrible punishments, then it is plausible to suppose that potential murderers who might be prepared to risk their own lives might be deterred by the loss of life to others that they care about – their husbands or wives, their children or parents, their closest friends. If, as it is often said, the death penalty is supposed to make potential murderers "think twice," this punishment would be likely to make them think three or four times. Its logic is the same as the argument used to support the death penalty over long-term imprisonment: the more terrible the punishment, the more powerful the deterrent.

Yet, even if this punishment succeeded, it would be an unjust, immoral punishment. It would save some innocent people's lives by killing other innocent people, and this is morally unacceptable.

Does this argument show that the death penalty is wrong? No, but it does show that the deterrence argument is not sufficient to justify it, even in theory. It shows that a punishment can be the best deterrent and still be morally wrong.

Someone might object to my use of a made-up example that involves a punishment that virtually no one supports. After all, they might say, the death penalty debate is about the execution of people who are guilty. No one defends the execution of innocent people to deter murders. But, of course, even in theory, death penalty supporters must acknowledge the risk that some innocent people will be executed and, even in theory, they must be willing to say that killing some innocent people is an acceptable price to pay for the saving of a greater number of innocent lives.[2] So my fanciful example shares an important feature with the actual death penalty for murderers: both involve a willingness to kill innocent people.

This is an issue that I will return to when considering the death penalty in fact. For now, I only want to show that even if the factual assumptions underlying the deterrence argument were correct, the argument by itself cannot justify the death penalty. If deterrence were all that mattered, then we would be logically committed to executing the family and friends of murderers as well as the murderers themselves. This is scarcely something that people committed to justice and respect for human life should support.

The argument from justice and desert

Many people support the death penalty for a different reason. They think it is the only truly just punishment for murder, and they often feel that anything less than death is morally unacceptable. Why is this?

In explaining this view, many people cite the expression "an eye for an eye." This familiar saying is probably the most influential basis for the death penalty. It benefits from both the authority of the Bible and from its surface plausibility as a fair rule for punishment. The "eye for an eye" principle tells us how to treat those who commit crimes. It says that if one person harms another, then the perpetrator should suffer the very same harm as the victim. This is how the "eye for an eye" principle is generally understood: the punishment should equal the crime.

For those in the know, this principle gains additional credibility from the fact that it is affirmed by the great philosopher Immanuel Kant in a very famous passage about punishment. He writes:

> What kind and degree of punishment does public legal justice adopt as its principle and standard? None other than the principle of equality . . . the principle of not treating one side more favorably than the other. Accordingly, any undeserved evil that you inflict on someone else . . . is one that you do to yourself. If you vilify him, you vilify yourself; if you steal from him, you steal from yourself; if you kill him, you kill yourself. Only the Law of retribution (*jus talionis*) can determine exactly the kind and degree of punishment. (Kant, 1965, p. 101)

And, he adds, if a person "has committed murder, he must die. In this case, there is no substitute that will satisfy the requirements of legal justice" (Kant, 1965, p. 102).

There is, no doubt, something appealing about the "eye for an eye" principle and if it provides a general criterion for determining the appropriate level of punishment for crimes, the death penalty will be justified because it satisfies the test of doing to the criminal what the criminal has done to the victim.

In spite of its surface plausibility, it is easy to see that the "eye for an eye" principle is defective and cannot provide a solid basis for the death penalty. People are doubly mistaken about the "eye for an eye" principle. They are mistaken in thinking that it is correct and mistaken in thinking that they actually accept it as an adequate guide to punishment. The only reason that people think they believe it is that they have not really thought about it.

There are three serious problems for "an eye for an eye." First, it requires unjust and barbaric punishments in cases where people have acted barbarically. Second, it conflicts with many of our beliefs about punishment and its justification. Third, in many cases, it provides no real guidance in determining the appropriate punishments.

Suppose that a person murders the entire family of someone that he regards as his enemy. If we describe his crime as "killing the family of his enemy," then the "eye for an eye" principle appears to require that we punish the killer by killing his entire family. Simply to execute the murderer alone would not satisfy the idea that the punishment should equal the crime. Yet no one would urge the death of the murderer's family as an appropriate punishment since they are innocent and should not suffer for the crimes of another.

Anyone who rejects this as a just punishment must reject the "eye for an eye" principle as well, and because virtually everyone would reject it, that shows that they do not really accept the "eye for an eye" principle. Even Kant, in spite of his strong affirmations of his version of "the eye for an eye" principle, departs from that principle for reasons very like the ones I have given. In a less famous passage than the one I quoted above, he asks: "[H]ow can this principle [of the equality of crime and punishment] be applied to punishments that do not allow reciprocation because they are either impossible in themselves or would themselves be punishable crimes against humanity in general?" (Kant, 1965, p. 132). Just by asking this question, Kant acknowledges that the "eye for an eye" principle cannot be applied to all cases and that it sometimes recommends punishments that would be immoral to inflict. He follows his question with three examples in which the "eye for an eye" principle would lead to an immoral punishment, and then he suggests an alternative. He writes: "Rape, pederasty, and bestiality are examples of the latter. For rape and pederasty, [the proper punishment is] castration . . . and for bestiality the punishment is expulsion forever from civil society since the criminal guilty of bestiality is unworthy of remaining in human society" (Kant, 1965, p. 132). Kant believes that raping the rapist, forcing the pederast to have homosexual sex, or forcing the person guilty of bestiality to have sex with animals would be "crimes against humanity." Instead, he proposes castration for the rapist and exile for the person guilty of sex with animals.

These may or may not be sensible suggestions, but they are clearly not instances of the "eye for an eye" principle. The lesson here is important. When we actually think about the implications of the "eye for an eye" principle, we quickly come upon cases in which it provides us either with defective guidance in the setting of punishments or with no guidance at all.

These problems are directly relevant to issues about the death penalty for murder. A common feature of the criminal law is that acts with similar effects are treated quite differently because of various facts about the crime. Yet the "eye for an eye" principle focuses on only one aspect of the crime: the harm caused to the victim. It says that the punishment should match the effect of the crime and from this it follows that the taking of a victim's life should be followed by the killing of the murderer.

Yet there are many different types of action that can result in the death of a victim, and we tend to think that these different types of actions should not all be punished in the same way. Consider the following:

- A hired killer lies in wait and shoots the intended victim.
- An argument degenerates into a fight in which one person strikes the other and kills him.
- A person sets fire to a building, thinking that it is empty; several people in the building die in the fire.
- A drunken driver kills a pedestrian.

While all of these actions have dire effects, most of us would view the hired killer as more culpable than the others. His action would generally be classified as first-degree murder, and in states with the death penalty, he might be sentenced to die.

In none of the other cases is death intended, even though the people involved engaged in dangerous actions. The second case, depending on the circumstances, might be classified either as second-degree murder (which is not punishable by death) or manslaughter, a lesser charge. If the arsonist had taken steps to ensure that there were no victims, he would most likely be charged with manslaughter, while the driver would be charged with vehicular homicide.

Most people agree that these actions should be dealt with differently. But anyone who believes this must reject the "eye for an eye" principle, since it requires us to treat them all in the same way. This supports both of the arguments I have put forward. First, because not all homicides should be punished in the same way, the "eye for an eye" principle is wrong. Second, if most people believe that we should treat homicides differently depending on the intentions and the circumstances, then those same people do not believe the "eye for an eye" principle. They might cite it in an argument, but they do not really believe it.

Whatever its customary force and initial plausibility, the "eye for an eye" principle is far from the last word on the appropriate punishment for particular crimes and cannot bear the burden of justifying death as a punishment for murder.

The Death Penalty in Practice

So far, I have considered the death penalty in theory. I have tried to show that the two most common reasons for supporting the death penalty – the "eye for an eye" principle and the argument from deterrence – are inadequate. Even in theory, the death penalty lacks a convincing moral justification.

But the death penalty debate is not merely an abstract moral issue. It is about actual institutions run by actual people in actual societies. Even if my arguments about the death penalty in theory had failed, there would still be a strong case against the death penalty in practice. To see this, consider the following: suppose that the deterrence argument worked in theory and that the "eye for an eye" principle provided an adequate principle for determining what punishment people deserve. It would still not follow that the death penalty should be adopted. Why? Because we need to consider how this punishment works out in practice.

Suppose that the death penalty deters more effectively than other punishments and suppose that murderers deserve to die. In addition, however, suppose that the legal institutions of a society that imposes the death penalty are not reliable. As a result, innocent people are often convicted of murder. Suppose, for example, that half of those convicted were innocent and that people know about the failings of the system. In this situation, even though the death penalty is justified in theory, it would be unjustified in practice. Indeed, it would be blatantly inconsistent for the death penalty to be retained by a society that is committed to the two values I emphasized at the start. It would not be respectful of human life because it would be killing innocent human beings, and it would not be consistent with a commitment to justice because it would be punishing innocent people. Any member of my imagined society who cares about justice and respect for human life should oppose the death penalty in that society even if they favor it in theory. To support the death penalty in that society would show that one had no genuine commitment to these values.

Whether the death penalty is justifiable or not, then, depends only partly on abstract beliefs about morality and justice. In addition, it depends on facts about a society and its institutions. Charles Black makes this point very effectively. He writes:

> We are not presently confronted, as a political society, with the question whether something called "the state" has some abstract right to kill "those who deserve to die." We are confronted by the single unitary question posed by reality: "Shall we kill those who are chosen to be killed by our legal process as it stands?" (Black, 1981, p. 166)

It is the practice of capital punishment – administered by real legal systems and real human beings – that kills people, and, we – as citizens of actual societies – have to decide if this practice should continue.

When I speak about an inconsistency in the pro-death penalty position, then, I mean an inconsistency between the values affirmed by death penalty supporters and the actual practice of capital punishment. My claim is that if death penalty supporters consider their own values, they will see that these values are violated by the institution of capital punishment, both as it exists now and as it is almost certain to exist for the foreseeable future.

Why the Death Penalty is Inconsistent with the Value of Justice

In order for the actual death penalty to be consistent with the value of justice, it would have to be true that people who are punished by death deserve the punishment. Since there is a widespread view that only some of those people who kill others deserve to die, a just system must be capable of two things: first, it must separate the guilty from the innocent and, second, it must be able to sort out the worst murderers – those who deserve to die according to the legal criteria of desert – from those who deserve a lesser punishment. If the system cannot do both of these reliably, then the results that it generates are unjust.

In fact, we know that the system in the United States is unreliable. We know this because a large amount of evidence shows that irrelevant features play a large role in determining the level of punishment that a person receives. In theory, the death penalty is imposed because of the terribleness of the specific crimes committed. In practice, actual death sentences are the result of arbitrary, irrelevant factors like race, socioeconomic status, and the quality of legal representation. I will briefly cite some facts about the influence of the irrelevant factors.

Race

One of the most widely studied influences on sentencing is race. A large body of research has shown that sentencing in capital cases is very much influenced by both the race of the offender and the race of the victim.

- Between 1976 and 1996, 83% of the people executed in the United States were charged with the murder of a white victim (Hood, 1998, p. 745).
- In the 20 years after *Gregg v. Georgia* (1976), the Supreme Court case that reinstated the death penalty, only 1 percent of executions were imposed on a white person who had killed a black victim (Hood, 1998, p. 745).

Socio-economic status

A person's social and economic status also plays a role in determining the sentence for the crime of murder. A comprehensive study of the death penalty in Georgia yielded the following result:

- In Georgia, defendants classified as having low socio-economic status were 2.3 times more likely to receive a death sentence than defendants seen as having higher status (Baldus *et al.*, 1990).

Quality of legal representation

Socioeconomic factors are related to the ability of people to hire competent lawyers. When people lack money, they must accept court-appointed lawyers who are often less competent. The same Georgia study showed:

- Defendants with court-appointed attorneys were 2.6 times more likely to receive a death sentence than defendants who could afford to hire lawyers (Baldus *et al.*, 1990, p. 158; Bright, 1997).

An investigation by the *Chicago Tribune* of 285 capital cases in Illinois concluded that the state's death penalty system was pervaded by "bias, error, and incompetence." It also cited poor legal representation, finding that 33 people sentenced to death had lawyers who were later disbarred or suspended (Armstrong and Mills, 2000).

A report by the American Bar Association (ABA) on the death penalty in the United States shows this to be a national problem. According to the ABA, court-appointed lawyers for defendants charged with first-degree murder often have no criminal trial experience and do not know the special rules and procedures for capital cases. They often have insufficient funds to cover the cost of preparing and investigating cases and frequently fail to make relevant objections during a trial so that they can be considered on appeal. In addition, they often fail to introduce mitigating factors during the part of the trial devoted to determining the sentence (American Bar Association, 2010, pp. 7–9).

The ABA report concluded that "in case after case, decisions about who will die and who will live turn not on the nature of the offense the defendant is charged with committing but rather on the nature of the legal representation the defendant receives" (American Bar Association, 2010, p. 6). This factual conclusion supports the following moral conclusion. If "decisions about who will die" do not depend on "the nature of the offense the defendant is charged with committing" but are determined by irrelevant factors such as race, social standing, and inadequate legal counsel, then the death penalty as it exists in our society cannot be relied on to produce just results.

Death penalty supporters claim that they want justice. Sometimes, in explaining why only some people guilty of homicide should be executed, they add that capital punishment should be restricted to people whose crimes are most terrible and whose culpability is greatest. What the evidence shows, however, is that the factors that determine whether people are executed or not differ from the factors cited in defense of the death penalty. Even if (in theory) justice would be achieved by executing the worst murderers, there is no reason to believe that this is what our system does.

Of course, these injustices would not support the abolition of capital punishment if the system could be reformed so as to eliminate the influence of these irrelevant factors. But there is no reason to believe that this can be done. The factors that interfere with the achievement of justice are too pervasive to be rooted out. Moreover, reforms have already been tried and have failed. In the United States, attempts to free capital sentencing from the influence of arbitrary factors have been ongoing since the 1972 case of *Furman v. Georgia*. Supreme Court Justice Harry Blackmun, who had supported the constitutionality of the death penalty in *Furman* and other cases, eventually argued that the defects in the system are unfixable. In the 1994 case *Callins v. Collins*, Blackmun announced:

> For more than 20 years I have endeavored . . . along with a majority of this Court to develop procedural and substantive rules that would lend more than the mere appearance of fairness to the death penalty endeavor. . . . I [now] feel morally and intellectually obligated to concede that the death penalty experiment has failed. (*Callins v. Collins*, 1994, p. 1145)

Anyone who is committed to the value of justice should follow Justice Blackmun's lead and reject the death penalty because of the injustices it has yielded in the past and is likely to yield in the future.

Why the Death Penalty is Inconsistent with Respect for the Value of Human Life

Having shown why support for the death penalty is inconsistent with a commitment to the value of justice, I will now show why it is inconsistent with respect for the value of human life and thus why it should be rejected by anyone who is committed to honoring that value.

When we examine how the death penalty system actually works, we see that it generates practices that show a callous disregard for human life. In fact, like the act of murder itself, the death penalty system embodies a lack of concern about the taking of human life.

In making this serious charge, I have a number of features of the death penalty system in mind. Consider the facts that I have cited about the injustice of the death penalty system. Since the quality of legal representation strongly influences the sentence imposed on a person, a system that tolerates inadequate representation for people who may be sentenced to death expresses indifference toward the value of these defendants' lives. There is no way that assigning incompetent lawyers to people in this position can be compatible with a commitment to take seriously the value of each person's life. Neither is the failure to provide court-appointed lawyers with the resources to investigate their clients' cases compatible with a commitment to take seriously the value of each person's life. Anyone concerned with the value of human life would be determined to ensure that executions occur only after the most exacting procedures have proved beyond a reasonable doubt that death is the proper punishment. There is an obvious inconsistency between affirming the value of human life and tolerating the current level of legal representation for people who face the possibility of death.

Problems with our system can lead to two kinds of mistaken judgment. The first is that a person who is guilty of a crime may receive a more severe punishment than would have been received had he or she been a member of a different race, had a higher social status, or had been able to hire a better lawyer. The second is that the poor quality of legal representation may result in innocent people being convicted of murder and sentenced to die. While there has been a widespread impression that the legal system in the United States bends over backwards to give defendants every conceivable advantage, the facts are quite otherwise. In fact, for many defendants, the system is stacked against them, and the results of the process are not reliable indicators of guilt or innocence. Consider the following facts (Armstrong and Mills, 2000) that set the stage for the moratorium on executions in Illinois:

- Between 1977 and 2000, the state of Illinois executed 12 people for murder and also released 13 people from death row because they were shown to be completely innocent.
- In some of the cases of wrongful convictions, police used coercive measures, including torture, to extract confessions from innocent persons.
- In at least 46 cases, convictions for murder were based on testimony from jailhouse informants; these informants often benefited from their testimony and in some cases had long records of lies and deceit.

These kinds of occurrence are not limited to Illinois. According to James McCloskey, while 226 people were executed in the United States between 1973 and 1995, 54 people were released from death row because of innocence. "This means," McCloskey (1996, p. 70) comments, "that during the last twenty years, for every five death row inmates executed, one has been released and exonerated. That points to a rather cracked system, one prone to serious and frequent mistakes".

At the national level, the causes of error resemble those in Illinois. Hugo Bedau and Michael Radelet identified 350 instances of wrongful convictions in capital cases and found that 82 of them resulted from questionable actions by police officers and prosecutors (Bedau and Radelet, 1987, pp. 56–59). This is consistent with a general pattern in the causes of wrongful convictions. One study of wrongful convictions in general (i.e., not simply in homicide cases) concluded: "If we had to isolate a single 'system dynamic' that pervades large numbers of these cases [of erroneous convictions], we would probably describe it as police and prosecutorial overzealousness" (Huff et al., 1996, p. 64). The chance of convicting and executing innocent persons is substantial, and misconduct by officials in the criminal justice system is a frequent source of error.[3]

While these practices and the resulting convictions of innocent people are dreadful in connection with any crimes and punishments, they are especially horrifying in the case of the death penalty, since they can result in the killing of people for crimes of which they are entirely innocent. Moreover, the death penalty makes corrections of errors impossible.[4]

In reply, death penalty supporters may argue that the fact that innocent people were exonerated and released shows that the system works. This reply, however, is inconsistent with the facts. In many cases, people have been spared from death only by chance or through the intervention of people outside the system. In Illinois, one person on death row was released through the work of students at Chicago-Kent College of Law, while three others were exonerated after investigations by journalism students at Northwestern University. One of these people, Anthony Porter, came within two days of being executed (Armstrong and Mills, 2000). Such down-to-the-wire cases that depend on the fortuitous intervention of outsiders are no evidence for the reliability or self-correcting nature of the legal system.

Moreover, when claims of innocence arise, officials are often resistant to them. As McCloskey notes: "Once wrongly convicted and sentenced to death, the criminal justice system treats you as a leper. No one wants to touch you. In my view, those in authority seem to be more interested in finality, expediency, speed, and administrative streamlining than in truth, justice, and fairness" (McCloskey, 1996, p. 70). McCloskey's claim about the true interests of those in the system is supported by the ABA report. It points out that the Supreme Court has ruled that "there is no constitutional right to counsel [i.e., representation by a lawyer] in post-conviction proceedings, even in capital cases" (American Bar Association, 2010, p. 9). As a result, people who have new evidence or justified procedural claims may lack the professional assistance that is required to assert claims in a legally credible way.

The lack of interest in correcting mistakes is nowhere more evident than in the time limits set by states for the submission of new evidence and in the Supreme Court's upholding of such limits. In *Herrera* v. *Collins* (1993) the Court ruled that new evidence in support of a claim of innocence could be disregarded because it had been submitted too late to meet the Texas 60-day deadline. In other words, the Court ruled that a person could be executed even though there was now evidence that he or she is innocent. Why? Because the evidence came in too late. In defending this shocking view, the Court majority noted that these deadlines were quite common. It noted:

> Texas is one of 17 States that requires a new trial motion based on newly discovered evidence to be made within 60 days of judgment. . . . Eighteen jurisdictions have time limits ranging between 1 and 3 years, with 10 States and the District of Columbia following the 2-year federal time limit. Only 15 States allow a new trial based on newly discovered evidence to be filed more than 3 years after conviction. . . . [Only] 9 States have no time limits. (*Herrera* v. *Collins*, 1993, pp. 409–411).

It is hard to see how the Court's judgment in this case or the state policies that are cited could be consistent with a commitment to respecting human life. What sort of commitment to the value of human life is shown by the 60-day deadline that Texas and 16 other states set for submitting new evidence of innocence? Or by the fact that only nine states place no limit on the time period for making sure that people are guilty before we execute them? What sort of attitude toward human life is exhibited by a Supreme Court that places respect for deadlines ahead of a concern about the death of innocent human beings?

It is hard to see how anyone who is committed to the ideal of respect for human life could approve of such practices. And yet, these practices are completely understandable. They reflect the desire of a legal bureaucracy to bring time-consuming appeals to a halt. They reflect the desire of officials who resist the exposure of errors because they do not want to be seen as incompetent, misled, or over-zealous. They reflect the desire of citizens who want lower taxes more than they want to pay for competent lawyers for people charged with murder. They reflect the fact that it is easier to respect the value of human life in words than to do so in deeds.

Death penalty supporters ought to acknowledge that even if in their ideal world the values of justice and human life would be affirmed by executing murderers, in our actual world the actual practice of capital punishment violates these very same values. If consistency with the values of justice and respect for human life are the appropriate criteria for deciding the issue, then people who understand the death penalty system should oppose the practice of punishing by death. Opposition to the death penalty is consistent with these values, while support for the death penalty violates them.

A Final Point

I have tried to show that the death penalty fails in theory and is inconsistent in practice with the values that death penalty supporters claim to support and I think I have succeeded in showing this. Nonetheless, I know that many people will not shift their view the next time they read about a terrible murder or about a particularly vicious criminal. They will think, "Surely, this person deserves to die. Surely, the death penalty is justified in this case." What they overlook is that the death penalty is not about the treatment of a particular individual. Rather, as has been clear in my discussion, the death penalty is a system. It is a system that empowers prosecutors to seek death as a punishment, judges and jurors to sentence people to death, and prison officials to impose death. To favor the death penalty is *not* to favor the execution of a particular person whom you or I believe deserves to die. Rather, it is to authorize many different people – whose motives and attitudes are unknown to us – to seek and authorize death as a punishment. These people may well make judgments that you or I would disagree with and yet when we say we favor the death penalty we are authorizing people to act on these judgments. Moreover, we are authorizing them to do so in the context of a system that we know to be unfair and unreliable.

Our views about the death penalty, then, are views about an institution, not about individual murders or murderers. It is quite possible to believe that some murderers deserve to die and yet to oppose the death penalty because one knows that others who do not deserve to die will be executed. And we do know this. We know that death sentences will result from racial prejudice, poor legal representation, and misdeeds by police and prosecutors. We know that evidence of innocence will be rejected by courts because deadlines are missed. We know that the practice of capital punishment as it actually exists violates the principles of justice and respect for human life. Anyone who genuinely cares about justice and the value of human life should conclude that the death penalty should be put to rest.

Notes

1 For a discussion of both common-sense and statistical evidence concerning the deterrence argument, see Nathanson (2001, ch. 2). For a survey of research on deterrence, see Baily and Peterson (1997).
2 For one example of the claim that killing innocent people is acceptable, see van den Haag (1975, pp. 219–221).
3 Official misconduct is not the only source of errors. For an analysis of a variety of sources of inaccuracy in criminal cases and proposals to increase accuracy, see Givelber (1997).
4 In a decision that was later overturned, Judge Jed Rakoff argued that the death penalty was unconstitutional because DNA evidence had definitively shown its imposition to be unreliable. For his decision, see *United States v. Quinones* (2002).

References

American Bar Association (2010) American bar association report on the death penalty. http://www.abanet.org/irr/rec107.html (last accessed 6/19/13).

Armstrong, K. and Mills, S. (2000) Ryan: until I can be sure – Illinois is first state to suspend death penalty. *Chicago Tribune* (February 1): 1.

Baily, W. and Peterson, R. (1997) Murder, capital punishment, and deterrence: a review of the literature. In *The Death Penalty in America: Current Controversies*, ed. H.A. Bedau, pp. 135–161. Oxford: Oxford University Press.

Baldus, D., Woodworth, G., and Pulaski, C., Jr (1990) *Equal Justice and the Death Penalty*. Boston, MA: Northeastern University Press.

Bedau, H.A. and Radelet, M. (1987) Miscarriages of justice in potentially capital cases. *Stanford Law Review* 40: 21–179.

Black, C., Jr (1981) *Capital Punishment: The Inevitability of Caprice and Mistake*, 2nd edn. New York: W.W. Norton.

Bright, S. (1997) Counsel for the poor: the death sentence not for the worst crime but for the worst lawyer. In *The Death Penalty in America: Current Controversies*, ed. H.A. Bedau, pp. 275–309. Oxford: Oxford University Press.

Callins v. Collins (1994) 510 US 1141.

Furman v. Georgia (1972) 408 US 238.

Givelber, D. (1997) Meaningless acquittals, meaningful convictions: do we reliably acquit the innocent? *Rutgers Law Review* 49: 1317–1396.

Gregg v. Georgia (1976) 428 US 153.

Herrera v. Collins (1993) 506 US 390.

Hood, R. (1998) Capital punishment. In *The Handbook of Crime and Punishment*, ed. M. Tonry, pp. 739–776. New York: Oxford University Press.

Huff, C.R., Rattner, A., and Sagarin, E. (1996) *Convicted But Innocent: Wrongful Convictions and Public Policy*. Thousand Oaks, CA: Sage Publications.

Kant, I. (1965 [1785]) *The Metaphysical Elements of Justice*, trans. J. Ladd. New York: Macmillan.

McCloskey, J. (1996) The death penalty: a personal view. *Criminal Justice Ethics* 15: 70–76.

Nathanson, S. (2001) *An Eye for an Eye? The Immorality of Punishing by Death*, 2nd edn. Lanham, MD: Rowman & Littlefield.

United States v. Quinones (2002) 196 F. Supp. 2d 416 (SDNY).

van den Haag, E. (1975) *Punishing Criminals*. New York: Basic Books.

Further Reading

ACLU (2012) *The Case Against the Death Penalty.* Washington, DC: American Civil Liberties Union. http://www.aclu.org/capital-punishment/case-against-death-penalty (last accessed 6/17/13).

Amnesty International (1987) *United States of America: The Death Penalty.* London: Amnesty International Publications.

Bedau, H.A. (1987) *Death is Different.* Boston, MA: Northeastern University Press.

Bedau, H.A. (1999) Abolishing the death penalty even for the worst murderers. In *The Killing State,* ed. A. Sarat, pp. 40–59. New York: Oxford University Press.

Bedau, H.A., Radelet, M., and Putnam, C. (1992) *Spite of Innocence.* Boston, MA: Northeastern University Press.

Bentele, U. (1998) Back to an international perspective on the death penalty as a cruel punishment: the case of South Africa. *Tulane Law Review* 73: 251–304.

Berns, W. (1979) *For Capital Punishment.* New York: Basic Books.

Bowers, W. (1984) *Legal Homicide: Death as Punishment in America, 1864–1982.* Boston, MA: Northeastern University Press.

Bowers, W. (1993) Capital punishment and contemporary values: people's misgivings and the Court's misperceptions. *Law & Society Review* 27: 165–186.

Davis, M. (1996) *Justice in the Shadow of Death.* Lanham, MD: Rowman & Littlefield.

Death Penalty Information Center. http://www.deathpenaltyinfo.org/.

McCleskey v. Kemp (1987) 481 US 279.

Nathanson, S. (1992) Is the death penalty what murderers deserve? In *The Moral Life,* 2nd edn, ed. S. Luper, pp. 380–389. New York: Harcourt Brace.

Nathanson, S. (1997) How (not) to think about the death penalty. *The International Journal of Applied Philosophy* 11: 7–10.

Nathanson, S. (1999) The death penalty as a peace issue. In *Institutional Violence,* ed. D. Curtin and R. Litke, pp. 53–59. Amsterdam: Rodopi.

Pojman, L. and Reiman, J. (1998) *The Death Penalty: For and Against.* Lanham, MD: Rowman & Littlefield.

Prejean, H. (1993) *Dead Man Walking: An Eyewitness Account of the Death Penalty in America.* New York: Random House.

Sarat, A., ed. (1999) *The Killing State.* New York: Oxford University Press.

Sorrell, T. (1987) *Moral Theory and Capital Punishment.* Oxford: Blackwell.

Steffens, L. (1998) *Executing Justice.* Cleveland: Pilgrim Press.

van den Haag, E. (1978a) In defense of the death penalty: a legal-practical-moral analysis. *Criminal Law Bulletin* 14: 51–68.

van den Haag, E. (1978b) The collapse of the case against capital punishment. *National Review* (March 31): 395–407.

Woodson v. North Carolina (1976) 428 U.S. 280.

Zimring, F. and Hawkins, G. (1986) *Capital Punishment and the American Agenda.* Cambridge: Cambridge University Press.

Reparations

CHAPTER TWELVE

Compensation and Past Injustice

Bernard Boxill

Misfortune makes its victims worse off than they would have been but for the misfortune. Compensation makes them no worse off than they would have been but for the misfortune (Nozick, 1974, p. 57). Why does justice require compensation? In an early discussion of corrective justice, Aristotle argued that when one individual treats another unjustly, corrective justice requires that the judge take the gain from the assailant and apply it to the sufferer thus reestablishing their equality (Aristotle, 1999, Book V.4). This may suggest that justice requires compensation because it restores a prior just distribution of goods that injustice distorted. This suggestion works smoothly if the injustice is theft and the judge acts promptly. In such a case restitution of what was stolen often leaves the victim of misfortune no worse off than she would have been but for the misfortune, and also corrects the distortion in the distribution caused by the theft. But restitution is often impossible, as, for example, when the stolen things have been lost or destroyed. And even when it is possible and may bring the victim of theft to the position she was in before the theft, it may still fail to leave her no worse off than she would have been but for the misfortune, and consequently may fail to compensate her. Further, the injustice that distorts a just distribution is often not theft. The victims of vandalism and assault suffer losses that will distort a just distribution, but vandals and assaulters often gain nothing comparable to the losses they cause. Consequently, although making them pay for those losses may compensate the victims it will not restore the original just distribution. We can try to fix the problem by simply defining the distribution that results following compensation as a just distribution. But that seems question begging. Finally, Aristotle was discussing something that he called "corrective" justice, which may not be the same thing as compensatory justice. And in any case when he spoke of corrective justice as restoring equality between the assailant and his victim he did not mean that corrective justice restores a prior just distribution because he did not believe that a just distribution was necessarily an equal distribution.

Contemporary Debates in Applied Ethics, Second Edition. Edited by Andrew I. Cohen and Christopher Heath Wellman.
© 2014 John Wiley & Sons, Inc. Published 2014 by John Wiley & Sons, Inc.

It would seem, then, that if justice demands that we compensate the victims of harmful injustice, it cannot demand this because compensation reinstates a former just distribution. A more recent discussion of the justice of compensation sets aside the issue of how compensation affects just distributions and focuses on how it affects the individuals who have suffered misfortune. On this view, justice requires compensation for misfortunes because people have a fundamental interest in exercising the capacity to have, to revise, and to pursue a conception of the good, and they can reasonably exercise that capacity only if they have some assurance that they will be compensated if they suffer unjust misfortune (Goodin, 1991, p. 152). I will also dispute this view. On my account, justice requires compensation because of the fidelity we owe to the principles of justice. If we are committed to the principles of justice then we are committed to undoing the evil results that often follow from their violation and *as far as reasonably possible* to making the world resemble what it would have been like had the violation never occurred – at least if this can be done without violating people's rights. Since misfortunes to individuals are among those evil results, our fidelity to the principles of justice requires that we see to it that these misfortunes are undone, which means that we must see to it that those individuals are compensated. It will be pointed out that violating just principles sometimes has good results, and then I will be asked whether I propose that these results should be undone too. My answer is yes, they should be undone at least if doing so is necessary to undo the evil effects of the injustice and does not violate fundamental rights.

Compensation, Misfortune, and Life Plans

Let us begin with the argument that the case for the justice of compensation for misfortune relies on the importance of our interest in being able to exercise the capacity to have, to revise, and to pursue a conception of the good. That interest is indeed among our fundamental interests. We are mostly rational forward-looking creatures who care about our future selves as well as future states of the world. Caring about our future selves and future states of the world, we therefore have conceptions of the good that we want to pursue and realize. Rawls defines a conception of the good as an "ordered family of ends and aims which specifies a person's conception of what is of value in human life" (Rawls, 2001, pp. 18, 19). Clearly, establishing such an ordering, as well as revising and pursuing it, require planning, which requires in turn a reliable and reasonably accurate estimation of our talents, opportunities, health, faculties, well-being and property. If such estimations are very likely to be thrown off because of misfortunes, we will become persuaded that there is not much point in making any plans. If justice guarantees our fundamental interests it must therefore protect us from such misfortunes, and when it cannot it must make up as much as possible for them by requiring that we be compensated.

But misfortunes tend to disrupt our life plans whether they are caused by injustice or accident, and those caused by accident can be as disruptive as those caused by injustice. If justice is concerned with protecting our interest in making and pursuing life plans and compensation protects that interest, justice must equally require compensation for those equally injured whether by accident, nature, or injustice. In other words, appealing solely to our interest in being able to have, revise, and pursue a conception of the good does not tell us why, or if, justice has an extra or special reason to require compensation for misfortunes caused by injustice that it does not have for requiring compensation for misfortunes caused by accident.

Of course it is often easier to actually compensate the victims of misfortunes caused by injustice than the victims of misfortunes caused by accident. Where misfortune is caused by injustice, we know who must pay to compensate the victims – the perpetrator of the injustice. Where misfortune is caused by accident or nature, we seem to have no very good reasons for imposing the costs of compensating its victims on anyone in particular. Indeed, if justice guarantees citizens' fundamental interest in exercising the capacity to have, to revise, and to pursue a conception of the good, justice may forbid imposing the costs of compensating misfortunes on people who did not unjustly cause the misfortunes. No one can reasonably make and follow through on life plans if he is aware that he is always in danger of being hauled in and made to bear the costs of compensating the victims of misfortunes he successfully took every precaution to avoid causing unjustly. Unfortunately, compensating the victims of misfortune caused by injustice can be as difficult as compensating the victims of misfortune caused by accident or nature. There may seem to be no one who must justly bear the cost of compensating the victims of misfortune caused by injustice because wrongdoers who cause misfortune often cannot compensate their victims. A destitute thug or a drunk driver may put dozens of people in wheelchairs for the rest of their lives. He can and should be punished for his crimes, of course, but he cannot be forced to make compensation to the victims of his crimes because no one can be forced to do what he or she cannot do. The difficulty is not simply that his crimes may not profit him enough to enable him to compensate them. The possible profits of crime may have nothing to do with the obligation of the criminal to compensate his victims. Criminals often profit from their crimes, but it is not their having thus profited that justifies compelling them to compensate their victims. The profits they make from crime may not be enough to compensate their victims. In that case, the criminal must compensate his victims from holdings that he did not come to possess as a result of any crime – if he has such holdings. Unfortunately he often does not.

There is a way to address this problem. A group of individuals may agree to establish themselves into an insurance system in which the whole group undertakes to compensate victims of misfortune whether it is the result of nature, accident, or injustice. At least three contingencies may make this arrangement attractive: the costs of compensating misfortunes are often greater than anyone can pay or is willing to pay; it is often difficult and expensive to establish fairly who unjustly caused the misfortune; and even the fairest procedures for establishing who unjustly caused the misfortune may result in innocent people being blamed for unjustly causing misfortunes that may be far beyond their means to pay to compensate. But this insurance scheme treats the misfortunes as if their causes, whether they are injustice or accident, are irrelevant to the case for compensating their victims. What matters is that misfortune has befallen someone and the promise of all to compensate him. Consequently, it fails to show why justice should be specifically concerned with the compensation of misfortunes caused by injustice.

A second problem with the argument for the justice of compensation under consideration is that it leads to the strange conclusion that sometimes justice does not require criminals to compensate their victims. Consider for example a successful swindler who has been able to hold on to his ill-gotten gains for so long that many of his victims have died or been induced to scuttle their old plans and to make new ones that do not depend on their having the money he stole from them. And suppose too that expecting to continue getting away with his crimes he has reasonably made a life plan that depends on his holding on to his loot. Emphasizing our fundamental interest in life planning could lead to the conclusion that we should allow him to keep his loot, let his victims proceed with their new plans, and let bygones be bygones (Waldron, 1992, p. 19). Even more shocking conclusions are possible. For suppose that though the swindler's victims are resigned to never regaining their property, they have made other life plans adjusted to their diminished resources but still remain inconsolable about their loss, and suppose also that driven by resentment they destroy what he stole from them simply to prevent him from having it and to frustrate his plans just as he frustrated theirs. The view that proposes to make the world safe for life planners might in this case require the swindler's victims to "compensate" him!

The proposal to make secure life planning the basis for just compensation runs into the above two problems because it depreciates the importance of past injustice. The swindler's life plans should not be interrupted because the past injustice of his swindling counts for nothing. For the same reason all victims of misfortunes have the same claim to be compensated; there is no special reason why the victims of unjustly caused misfortune be compensated. The view that justice requires compensation for misfortune is influenced by Rawls's view that justice aims to secure citizens' fundamental interests, among which is the interest in exercising the capacity to have, to revise, and to pursue a conception of the good. But I doubt that Rawls's theory of justice countenances the kind of depreciation of past injustice that the view in question seems to lead to. He does argue that justice must help guarantee our capacity to have and revise a conception of the good, but he also argues that justice must help guarantee our "capacity for a sense of justice" (Rawls, 2001, p. 112). Presumably then it must also be important that we follow the principles of political justice. *How* important? It would make no sense to answer that we must consider it a great obligation to follow the rules of justice, but when they are not followed, then we must consider the results as the product of nature or accident and do the best we can with them. One disaster may be the result of nature and a second the result of injustice. Both must be dealt with but not in the same way. In the first we look to the future and do what is best for everyone. In the second, justice demands that we look to the past. Debts must be paid, compensation made, and wrongdoers punished.

For example, the consensus among philosophers is that justice is so important that it must be followed even at the cost of benevolence. This must imply that no one is going to be allowed to benefit from unjust benevolence. If we say seriously that some principle must be followed, we imply that if it is not followed and this has consequences, then as far as possible and taking care not to violate rights or commit further injustice, we must return the world to what it would be like if the principle had been followed. Accordingly, if someone faced with a choice between acting justly and acting unjustly but benevolently were to ask philosophers for advice on what he should choose to do, they would impress on him that his duty is to act justly, warn him that if he chooses to act unjustly but benevolently that he is going to be punished and to be roundly condemned and reprimanded by all morally upright onlookers, that he will have to make full compensation to those who suffered misfortune as a result of his injustice, and that he would have acted in vain anyway for even the innocent beneficiaries of his injustice will not be allowed to keep the good things he gave them. And presumably they would give him the same advice if he informed them that he would follow the example of the successful swindler mentioned above by holding on to the loot until his victims give up on getting it back and make life plans that do not depend on their having it, and then make his own life plans as if the stolen money in his possession were his own money.

But suppose that after hearing and absorbing all this advice the individual decides to ignore it and succeeds in doing what he informed the philosophers he would do. On discovering what he has succeeded in doing the philosophers cannot then say, "Your theft was wrong, but dispossessing you of what you stole would now be wrong, so let us let sleeping dogs lie, let bygones be bygones, and let us get on with it." Such a response would make their earlier advice a pack of lies.

The Effects of Compensation

Since the rules of justice are not lies, they must require that we do what they say we will do when the rules are violated. And since they say that when the rules of justice are violated and people suffer misfortune as a result that we will compel the violators to compensate their victims, it follows that we must do exactly that, at least as far as we can. But let us get clearer on what this involves. Consider an example suggested by Gregory Kavka (1982): John had planned to go on a trip, but theft of a valuable item in his house persuades him to stay home. That night he and his wife Mary have intercourse and she conceives. The fetus would not have existed but for the theft, which was an injustice, and aborting it would remove one of the most salient consequences of the injustice and return the world to a condition that closely resembles what it would have been like had the injustice not occurred. But justice cannot require aborting the fetus for that reason. Following the logic of that reason, we would require abortions of all fetuses since it is true of all fetuses that they would not have been conceived but for some earlier injustice. Further, aborting the fetus is not relevant to the issue of the justice of compensation that the example raises. In the example, justice requires that we compensate for the misfortune that an injustice causes. The injustice was theft and the misfortune it caused was the loss of the valuable item. Justice requires the compensation of those who lost that item. Compensating them by making them no worse off that they would have been had they not lost the item would require at least (if possible) the return of the item. But returning the item to them may proceed without aborting the fetus although it would not have existed but for the theft. More generally, justice need not require compensation simply because an injustice has occurred. It requires compensation only when it causes misfortune but it need not always cause misfortune. On the contrary, sometimes it causes only good fortune. Consider for example a variation in the case just discussed. The theft was attempted but failed so that nothing was lost, though its attempt was enough to keep John home and this led to Mary becoming pregnant. In such a case there would be no one to compensate though the resulting pregnancy was the result of injustice. Only by confusing compensation with making the world resemble as much as possible what it would have been like if the injustice had not occurred could one suppose that compensation was an issue and would require that the fetus should be aborted.

In the above example I assumed that Mary's pregnancy was not a misfortune. But let us suppose that it was a misfortune; it was unwanted, would wreck the parents' plans, and perhaps was also a threat to Mary's life. In such a case does it follow that justice would require that the parents be compensated perhaps by terminating the pregnancy, supposing that this would be the best or only way to make the parents no worse off than what they would have been had the pregnancy never occurred? I doubt it. Since the pro-lifers deny that abortion is permissible, they would also deny that justice could require it even if the pregnancy is a misfortune. And though the free choicers hold that abortion is generally or at least often permissible, given their self-description they would deny that justice could *require* the abortion even if as in the case at hand the pregnancy it would end is a misfortune. This agreement between the sides in the abortion debate may tell us that in some cases justice cannot require compensation for misfortune if the compensation would involve another injustice. But it does not tell us that sometimes justice cannot require compensation for misfortunes that are *caused* by injustice. It seems true that the misfortune, in this case, the unwanted and dangerous pregnancy, would not have occurred but for the theft, which was of course an injustice. But it seems that the theft was merely a necessary condition for the misfortune rather than the proximate cause of it. So let us finally consider a pregnancy genuinely caused by an injustice, for example, rape. Rape is an injustice and it causes many misfortunes. Its victim is humiliated, insulted, debased, and probably physically and emotionally harmed. These are misfortunes caused by injustice and the victim must be compensated for them. What about the pregnancy the rape causes? Of course, pregnancy is often not a misfortune but in this case it seems most likely to be one. Indeed the victim will almost certainly consider her pregnancy to be a disaster – distinct from, in addition to, and perhaps worse than the disaster of the rape. If so, it seems that justice requires that she be compensated for that misfortune as well. What would constitute compensation in this case? What would make her – the victim of the misfortune of being pregnant as the result of a rape – no worse off than she would have been but for that misfortune? It may seem that here the answer is obviously that the fetus should be aborted. And we may suppose that she agrees, insisting that her compensation must at the very least begin with aborting the fetus. This seems acceptable if certain assumptions are taken to be true, for example, that a fetus has no right to life and that the victim of unjust misfortune has a decisive say in what will make her no worse off than she would have been had the misfortune not occurred; that is to say, the victim has a decisive say in what will compensate her for her misfortune. If those assumptions are true then aborting the fetus would be a necessary part of her compensation. But it does not follow that aborting the fetus would be justified. Many people believe that a fetus has a right to life. If they are right about this, its right to life could conceivably outweigh her right to compensation, and justice would forbid aborting the fetus even if doing this was the only thing that could be done to make the victim of the rape no worse off than she would have been had she not become pregnant as a result of that rape. But some philosophers maintain that a woman may permissibly abort a fetus attached to her even if it has a right to life; their argument is that its right to life, assuming it has one, does not give it a right to be attached to her. To handle this complication let us suppose that the fetus can be removed from her body and placed in some kind of incubator that enables it to live. But let us suppose that the victim still insists that as long as it lives she will be worse off than she would have been had it never been conceived. Even if she was insisting on the truth, justice would not require the death of the fetus, if we suppose that the fetus has a right to life. In that case its right to life would outweigh her specific right to compensation in the form of termination of the fetus. It would be a different matter if she insisted that as long as the fetus remains attached to her she will be worse off than she would have been had it never been conceived. In that case, justice could require that it be removed from her body, but not simply because doing so would compensate her. A further condition would be that its right to life does not give it a right to remain in her body.

To avoid the issue of fetal rights to life altogether let us suppose that the fetus comes to term and is now a six month old baby and the mother insists that as long as it lives she will be worse off than she would have been had she never been raped and impregnated. In this case justice cannot require and indeed would forbid killing the baby in order to compensate her. Suppose she rejects alternative compensations, large monetary payments for example, on the ground that they would not give her satisfaction, and suppose she is telling the truth. It would not matter. If only killing the child will make her no worse off than she would have been had she not been raped and impregnated as a result, then unfortunately she must remain uncompensated. No one would suppose that justice could require that a complete stranger be killed even if only his death could make her no worse off than she would have been had she never been raped and impregnated as a result. But the child is like the stranger in relevant respects. Like the stranger, it is guilty of no wrong, and although it owes its existence to an unjustly caused misfortune and consequently did profit from that misfortune, it did so innocently and thus forfeited none of its rights.

It follows that there are many things that justice will forbid that we do even if doing them is necessary to compensate the victims of the misfortunes of injustice. For example, since the child resulting from a rape did not forfeit its right to liberty it would be unjust and impermissible to force it to be her mother's slave – even if only this would make her mother no worse off than she would have been had she not been raped and impregnated. Or consider for example the infamous Tuskegee Syphilis Study and let us suppose that the study yielded information that led to better treatment of the disease. (I do not know if the study did in fact yield useful information. I am assuming that it did for the sake of argument.) Justice requires that those responsible for the study as well as those who those participated in it, knowing the injustice it involved, should be punished. And justice also requires that the victims of the study be compensated, here supposing that justice demanded that they get the treatment that was then available and that they were deceived and unjustly deprived of it. Finally, the costs of compensating them would be justly imposed on those who conducted the study, those who profited from it knowing its injustice, and those who knew about it but did nothing to stop it. But these costs cannot justly be imposed on those who were cured of syphilis as a result of the new treatments the study made possible and thus profited from its injustice. People have rights that must not be violated even if violating these rights would help compensate the injured victims of a prior injustice. It does not matter if they profited from the injustice at least if they did so innocently. The case considered earlier established this claim. We agreed that although a child conceived as a result of a rape profits from the injustice of the rape, for he would not exist if it had never occurred, he cannot therefore be justly enslaved to compensate the victim of the injustice, in this case his mother. Similarly, for the case of those cured of syphilis as a result of the study, it would be unjust to compel them to labor to pay for the compensation of the study's victims. Still less could justice require that the innocent beneficiaries of the study, those cured as a result of the research it permitted, be re-infected with syphilis on the grounds that doing so would make the world more closely resemble the condition it would have been in had the unjust study never been conducted. People have rights not to be deliberately infected with a very serious disease against their will even if they were cured of that same disease by treatment that was developed out of the information acquired as a result of injustice. Similarly, suppose for example that the swindler mentioned earlier used his ill-gotten gains to get his children the best possible educations. They went to the best (and most expensive) prep schools, the best (and most expensive) colleges, and the best (and again most expensive) medical schools, and as a result now command very high salaries. Nevertheless, if they did not participate in their father's fraud and were ignorant of the source of his largesse, being innocent of wrongdoing, and consequently not having forfeited their rights to work at the jobs of their own choice, it would violate their rights and be unjust to compel them to work at the high paying jobs in order to garnish their wages to help compensate their father's victims.

It would be a different story if the father had left his children a hefty monetary inheritance – even if the children inherited the money or goods in all innocence, believing that their father had acquired it honestly. Their inheritance could justly be taken from them in order to compensate their father's victims. Being innocent they have not forfeited any of their basic rights, including their rights to own property. But taking their inheritance from them does not violate their rights to own property. Having a right to own property does not imply that one has a right to whatever happens to be in one's possession. And they have no right to the money their father put in their possession. He had no rights to and therefore could not have given them any rights to it by simply taking it from one location – his bank account – to another location – their bank accounts. It would not make any difference that they had made life plans based on their continued possession of the money. A person does not have a right to things he needs in order to pursue his life plans just because he happens to have these things in his possession. He has to possess them justly, that is, as a result of the rules of justice being followed exactly. But the swindler's children did not come to possess the money in question as a result of the rules of justice being followed exactly. On the contrary, they came to possess the money in question as a result of the rules of justice being violated. They possessed the money but it did not belong to them. It belonged to the victims of the injustice. True, they may have made life plans based on having the money, and they may have to change their plans if the money were to be taken from them, but no one has a right not to have to change his life plans. Neither is a person to be identified with her life plans. She remains the same and her basic rights remain the same and need not be violated though her life plans change and change radically.

It may be objected that a person may have a right to what he needs to pursue his life plans just because he happens to possess those things even if he does not possess them as a result of the rules of justice being followed exactly. Though people may fail to possess the talents they were born or conceived with because someone violated their rights, people do not rightfully possess the talents they were born with just because no one violated their rights. A lack of talent is not necessarily a result of injustice, and justice does not require that people have the talents that they were born or conceived with as if had they not been born or conceived with those talents they would have been the victim of some injustice. Consider for example a person whose plan of life is to be a rocket scientist, and that one of his reasons for taking up that plan is that he has to a high degree the very special talents necessary to become a successful rocket scientist. He cannot say that he has his talents as the result of the rules of justice being followed exactly. In fact he may have his talents as a result of the rules of justice being violated, say for example, if his father, a gifted rocket scientist, raped his mother and he was conceived as a result. Although his existence and talents were the result of injustice, damaging his brain with drugs so as to deprive him of his talent would outrageously violate his rights. Indeed, I would add that it would also be unjust, and for the same reason, to use drugs to deprive the children of the swindler of the intellectual advantages that they acquired as a result of his injustice. But the reason why it would be unjust to do these things is not that the people in question need their talents or education to fulfill their life plans. It is because depriving them of their talents or education would require violating their right not to have their bodies and persons interfered with in the ways suggested. Taking away money or goods from the children of the swindler is not the same thing as taking away the results of their education. Innocent people have rights that their bodies not be invaded even if the invasion deprives them of things that they have only because of injustice. If someone steals a valuable diamond and I innocently swallow it and it gets stuck in my gut and only surgery can remove it so that it can be returned to its owner, performing the surgery on me without my consent would still be unjust. It would be a different matter if I swallowed the diamond in order to steal it. In that case, I may well forfeit my right not to be operated on without my consent. The rules regulating things outside our bodies are different from the rules regulating things inside our bodies. Since the things outside our bodies can be taken from us without violating our rights, we can justly hang on to them only if they did not come into our possession as a result of a violation of someone else's rights. If they came into our possession as a result of the violation of someone else's rights, they must be returned to the victims of the violation, however innocent we are and however much we covet them or want then in order to pursue our plan of life.

The point of following the principles of justice is to secure to people the things that the rules assign to them. If people lose some of these things as a result of a violation of the rules of justice, that is, if they suffer misfortune as a result of injustice, the appropriate response is therefore to return to them the things they lost; or if this is impossible to return them to a condition no worse than they would have been in had those things not been lost. This is what compensation attempts to do. It does not try to make the world resemble as much as possible what it would be like if the injustice had never occurred. We do not kill John's son because he would not have existed if his father had not been robbed, even if killing him might help to make the world resemble what it would have been like had the robbery never occurred. We compensate John for the loss of the item the robbery caused. Further, we do not compensate him if such compensation would require that we violate the fundamental rights of the innocent, rights they never forfeited. We do not kill the baby that is the result of a rape even if killing it is the only thing that will compensate her mother, making her no worse off than she would have been had she never suffered the misfortune of being impregnated as a result of a rape. In such cases the right to compensation conflicts with other rights that trump the right to be compensated for the misfortune in question. These rights were the rights to life and liberty. I assumed, reasonably, I believe, that these rights trump the right to receive monetary awards or the right to the satisfaction or peace of mind that comes from knowing that as far as possible all traces of one's unjustly caused misfortune are wiped clean. But there are cases where the right to compensation comes into conflict with no other rights or with rights that seem considerably less weighty than the rights to life and liberty. In the first case there is no reason why compensation should not proceed. But the second case is more complicated.

Closing Thoughts

As we have seen, the innocent beneficiaries of the Tuskegee syphilis study, those who were cured of the disease as a result of what was learned from the study, cannot justly be enslaved or forced to work at jobs they would not otherwise choose to work at in order to compensate the victims of the study; doing so would violate their rights to liberty. But what if they are already willingly working at high paying jobs? In that case, would it also be unjust to force them to give up a part of their wages in order to compensate those whose unjustly caused misfortunes made their cures possible? The fundamental right to liberty does not seem to be an issue; no one is suggesting that they be compelled to work at these jobs in order to compensate their benefactors. Further, holding on to all of one's high wages does not seem to be a fundamental right, or at least it does not seem to be as fundamental a right as the right to liberty. People earning high wages are often justly taxed although it would be unjust to enslave them. Thus it may be argued that while the right to compensation may be outweighed by the fundamental right to liberty, thus forbidding enslavement for the sake of compensation, it cannot be outweighed by the far less fundamental right to hold on to all of one's wages. And on this ground the further conclusion may be drawn that the very well-to-do innocent beneficiaries of the unjust misfortunes suffered by the victims of the Tuskegee study may be justly required to part with some of their earnings in order to compensate those from whose unjustly caused misfortunes they have innocently benefited.

But many will object to this. They will point out that the beneficiaries of the study are as innocent of wrongdoing as the other well-to-do members of their society, and consequently since they have forfeited none of their rights there can be no good reason to single them out to make compensation to the victims of the unjustly caused misfortunes – even if they are the only ones who benefited from those misfortunes. The objectors will contend that if the rights to compensation of the victims of the study are so weighty and simply must be satisfied, then *everyone* in the society should be required to contribute to satisfying those rights. This is a persuasive argument, but it has at least two disturbing features: first, it proposes to *increase* the number of innocent people whose rights have to be infringed to compensate the victims of the study; second, the innocent beneficiaries of the unjust study owe their good fortune to the unjust misfortunes of its victims – even if they do so innocently; the other innocent well-to-do members of the society do not. That is, the good fortune of the former is traceable to injustice; the good fortune of the latter is not. Of course, the world is such that inevitably, though those latter individuals do not owe their good fortune to the Tuskegee study, they owe at least some of their good fortune to some other injustice. But let us assume that they did so less directly and that the victims of that different injustice from whose misfortunes they have benefited have already been compensated or can no longer be identified. Given that assumption, although we can say with considerable assurance that they have benefited from injustice, we cannot actually trace their good fortune to any specific earlier injustice; the consequences of that injustice have been wiped clean by compensation or at least have been made undetectable to human capacities by the convergence and merging of the rush of events that constitute history. Many tributaries will flow into a great river and we cannot identify and trace the contribution that each tributary makes after it has entered the river. But it does not follow that we cannot identity the water in a tributary before it enters the great river. Fidelity to the principles of justice requires that in making compensation for injuries caused by injustice we should make the world look the way it would look if the unjustly caused injuries had not occurred, as long as doing so does not violate fundamental rights that have not been forfeited. By hypothesis, those responsible for the unjust Tuskegee study have been stripped of their holdings in order to compensate their victims. But also by hypothesis many of their victims still remain uncompensated. That is, the misfortunes caused by the unjust study have not been undone and reversed. Clearly discernible traces of the injustice remain. We cannot wipe them clean by compensation if compensation would require violating anyone's more fundamental rights. Since leaving the victims of the study uncompensated is a non-starter, we have two alternatives: compensate the victims by taxing the innocent and well-to-do who owe their good fortune to the unjust study; or, tax everybody, including those who do not owe their good fortune to the unjust study. Both involve taxing the innocent and compensating the victims of injustice. And both involve infringing the rights of the innocent, though these rights are outweighed by the more weighty rights of the victims to compensation. But the former has a signal advantage over the latter. It brings the world closer to what it would have been like if the injustice in question had never occurred. Fidelity to the principles of justice therefore requires that we choose the former.

Suppose for example that the swindler discussed earlier used his thefts wisely and ended up with far more than enough to compensate his victims – even if their compensation required far more than the sums he actually stole from them. Should he be allowed to keep the surplus after making compensation to his victims? Everyone will answer that he should not, and many will justify their answer by arguing that allowing him to keep the surplus will fail to deter future swindlers. But this forward-looking answer is vulnerable to the usual objections to similar utilitarian proposals; for example, what if the public will never know that the swindler was allowed to keep the surplus? Further, the deterrence of crime is the province of punishment, and punishing swindlers by taking away their surpluses after compensation would entail punishing successful swindlers more severely than unsuccessful ones, which would be unjust since punishment is unjust if its severity and extent varies from person to person and is decided after the crime. A better justification for refusing to allow the swindler to keep his surplus is fidelity to the principles of justice. This requires us, as far as reasonably possible, to make the world look the way it would look if the violation of the principles of justice had not occurred, provided that we compensate those unjustly injured and do not violate rights that are not outweighed by weightier rights to compensation. The swindler would not have the surplus in question had he never committed his injustices. To wipe out the traces of injustice, as far as we can without committing further injustices, we must deprive him of the surplus. Of course he might have earned an amount equal to the surplus had he worked honestly; and if he had he would be allowed to keep his earnings, *of course*. But in that case there would be no traces of injustice to wipe clean since his earnings would not be *traceable* to injustice.

References

Aristotle (1999) *Nichomachean Ethics*, trans. T. Irwin. Indianapolis, IN: Hackett.
Goodin, R.E. (1991) Compensation and redistribution. *Nomos* 33: 143–177.
Kavka, G. (1982) The paradox of future individuals. *Philosophy and Public Affairs* 11: 93–112.
Nozick, R. (1974) *Anarchy, State, and Utopia*. New York: Basic Books.
Rawls, J. (2001) *Justice as Fairness: A Restatement*. Cambridge: Harvard University Press.
Waldron, J. (1992) Superceding historic injustice. *Ethics* 103: 4–28.

CHAPTER THIRTEEN

Must We Provide Material Redress for Past Wrongs?

Nahshon Perez

This chapter aims to demonstrate several reasons why the plausible answer to this chapter's title is *rarely*. Owing to the nature of such a plural argument, "rarely" is an adequate conclusion. There is no single overwhelming reason to argue that material redress for past wrongs is unjustifiable, but there are numerous counter-arguments to such claims, which, in the aggregate, make it nearly implausible that there are convincing claims for redress following past wrongs. This volume does not allow for much elaboration, and only some of the arguments against redress will be considered here.[1]

This chapter begins with one normative assumption: the Rawlsian view that each person is separated from other persons (Rawls, 1999, p. 24).[2] Simply put, this assumption means that each person is different from other persons, and that each person should be treated with equal respect and concern. Specifically in the subject matter of intergenerational reparations, this assumption will act as a barrier versus various aggregative theories and policies. (More on this below.)

This chapter has two major sections: The first briefly presents some required definitions. The second raises the "non-identity problem" and examines two major attempts to answer this problem and justify intergenerational redress: the "identity" and "timing" arguments. These attempts to justify redressing past wrongs face several high hurdles and, by describing these hurdles, this chapter shows that few such claims will be able to overcome all of these challenges. It is better, succinctly put, to focus on the rights of existing people rather than to dwell on past wrongs.

Some Definitions: Past Wrongs, Nozickian Compensation, Restitution

Past wrongs of the kinds discussed here are cases of substantial past injustices in which all the wrongdoers and victims have since died. I shall assume that the wrong has ended

Contemporary Debates in Applied Ethics, Second Edition. Edited by Andrew I. Cohen and Christopher Heath Wellman.

and that related debates – for example, regarding entitlements of descendants of the deceased victims – take place in a liberal democratic society, broadly conceived, with no legal discrimination against such descendants. In the context of this chapter, I shall assume that information regarding the past wrong is clear and reliable. The scenario that would interest me here involves cases in which descendants of deceased victims (or persons belonging to groups targeted during the past wrong) request material redress. While non-material kinds of redress, such as apologies, are obviously important and may be requested, they will have to be examined elsewhere.

As all the wrongdoers and victims have passed away since the wrong, any policy of redress faces a "double decoupling" problem: the wrongdoers are not the (potential) payers and the victims are not the (potential) claimants (Posner and Vermeule, 2003, p. 23). Assuming the separateness between persons, this double decoupling problem will create a hurdle for the justification of material redress *once the wrong becomes a historical wrong*.

When people speak of past injustices, they typically think of horrible acts such as slavery, torture, ethnic cleansing and mass murder. Usually, such acts call for punitive measures, but these are irrelevant for historical injustices because the wrongdoers have already died. Material redress for such acts, however, may be relevant as such acts often involved stealing of property and/or land. While it is very difficult to determine the specific fiscal value of any such theft, there is certainly a material side to these wrongs. Acknowledging this does not "put a price" on such wrongs. Rather, requests for material reparations are a more modest admission that such wrongs include practical issues. Monetary compensation might be one part of a larger redress policy. It is sometimes possible to assess sums, calculate interest, and reach some quantifiable amount representing the tangible financial consequence of the wrongdoing.

One aim of compensation is that the wrongdoer will bear the burden of the loss or damage. This damage can be used to determine how much compensation is owed. In cases of historical injustices where the wrongdoers and the victims have died, it is obviously difficult to determine who should bear the cost of monetary compensation and who is eligible to receive compensation.

Any meaningful discussion regarding material redress must distinguish between compensation and restitution. I begin by following Robert Nozick's famous definition of *"compensation"*: bringing the victim to a level of well-being that she/he would have enjoyed had the wrong not occurred (Nozick, 1974, p. 57).

The advantage of Nozick's definition is that it nicely captures a strong moral intuition: if A wronged B, B will be fully compensated if compensation restores his/her well-being to the level it would have been had the wrong never occurred. In order to achieve this goal of restoring the victim's welfare to a level she/he would have enjoyed, material redress will be needed in at least some cases.

"Restitution" is one potential way to achieve "Nozickian compensation." The concept of restitution will be important to our analysis below. Requests for restitution have two typical features. First, the justification for a restitution claim is the unjust enrichment of person A following some interactions that included, or are connected to, person B. Second, the remedy amounts only to the extent of the unjust enrichment of A, which A ought to return to B. The assumption here is that person A somehow was unjustly enriched and, although no wrong was necessarily committed against B by A, A is not entitled to the given enrichment. While A is not responsible for B's loss, A is nonetheless not entitled to the enrichment. She/he should therefore return to B the extent of this enrichment. For example: Z stole B's watch and gave it to A, who did not know that the watch was stolen. If B meets A, and says "this is my watch!", A should return the watch to B. This does not imply that A is responsible for the theft or B's troubles (such as the hours she/he dedicated to trying to find the watch, emotional distress if the watch had sentimental value, etc.) (Dagan, 2004; Kull, 1995). Nozickian compensation is therefore a much wider notion than restitution. Note that, in the watch example, the watch is given to A, rather than purchased by A. The illustration also assumes that it is easy to establish proper ownership of the watch. Real-world cases, however, will often involve third parties and difficulties in identifying property. This is especially true in cases where the wrong lies in the more distant past. I shall return to these points in the closing stages of this chapter.

In some cases, restitution requests are for specific items such as a painting or a family heirloom. In other cases, requests may be for a monetary sum such as was held in a bank safe. Both are examples of restitution if the request follows the two criteria identified above: the existence of unjust enrichment and a requested remedy that does not exceed the value of unjust enrichment.

Note that restitution merely aims to undo someone's unjust enrichment, while Nozickian compensation can go further and take stock of the full impact of the loss or damage to some victim(s). Nozickian compensation, therefore, is much more comprehensive than mere restitution. The importance of these distinctions will become clearer below. Having (very briefly) clarified these concepts, we can now turn our attention to more substantive argumentation, beginning with an examination of the non-identity problem.

The Non-identity Problem

The non-identity problem presents an urgent challenge to scholars who wish to justify redressing historical injustices. Stated simply, the non-identity problem poses the following question: If living is (usually) better than non-living, and if a given historical injustice is causally connected to the existence of the descendants of deceased victims, how can a living person be wronged by something without which she/he would not exist (Morris, 1984; Parfit, 1986, pp. 351–355)? Surely the historical injustice improved her/his situation or at least did not harm him/her.

The example of wrongful life claims might help to clarify this point. Such cases involve scenarios in which the harmful act is the direct cause of the life in question, such as in cases where an MD fails to explain to parents carrying a defective gene that pregnancy will bring about a very sick child. In such scenarios, there is no counterfactually equivalent healthy child; the only options are sick child or non-existing child. Courts have struggled with the logical structure of cases in which the child, via legal guardianship, sues the MD for his/her condition. Such cases are typical instances of the non-identity problem, and the extrapolation to our subject matter of massive past wrongs and the way such wrongs are causally connected to the existence of descendants of deceased victims is obvious. Succinctly put, those descendants would not have existed if the past injustice had not taken place.

Many scholars who write about cases of historic injustice will draw on the non-identity problem and assume that it is usually better to be alive than not to exist. On this approach, unless the situation of the descendants of the victims is truly and unquestionably miserable, the historic injustice improved the descendants' situation. Some other scholars, however, maintain that it does not make sense to compare living beings to non-existent ones; at the very least, if one's existence is causally connected to a past wrong, then she/he was not harmed by it. On this approach, it is simply meaningless to compare non-existence to existence (Parfit, 1986, pp. 357–364).

Any attempt therefore by the descendants of the original victims of past wrongs to claim that they are owed compensation will be met by this objection: as their existence is causally connected to the past wrong, this wrong improved their situation (as living is better than non-living), *or*, if we accept the version that argues that it is not possible to compare living to non-living, the past wrong did not harm them, as any claim to harm will have to rest on meaningless comparisons. It is surely odd to demand compensation for an event that improved or did not harm one's situation. The non-identity argument is thus a thorny objection to redressing past wrongs, as it suggests that claims to redress are either unjustified or rest on meaningless comparisons.

Responses to the Non-identity Problem

There are two ways scholars have attempted to respond to the non-identity objection to redressing historical injustices. I call them the "identity" argument and the "timing" argument. The identity argument appeals to the significance of intergenerational group identity. The timing argument attempts to dodge non-identity worries by focusing on the rights of children of wronged parents and their descendants. I shall argue that neither approach succeeds. I start with the "identity" argument.

The identity argument

The argument from identity goes, roughly, as follows. Consider an intergenerational group whose identity is defined in terms of some shared culture, however understood. A person's identity might be strongly connected to such an intergenerational group whose past members once suffered grave wrongs because they were members of that group. Over time, new members of the group (born after the wrong) internalize new and negative norms that are the result of the wrong. These norms might include feelings of humiliation, insult, fear, and distrust of non-members of the group. Such new norms and ideas might be passed via new (or added) rites of passage, holiday customs, stories, lullabies and so on. Given how central cultural identity is to group identity, it seems that in some cases, the wrong done to the group can transform the group's identity. Of course, establishing lines of cause and effect become very complicated in such cases, but what matters here is that the group's original existence preceded the original wrong. If the past wrong harmed the intergenerational group's identity, and this harm to the group's identity persisted across generations (via holiday customs, lullabies etc.), then current, living members of the group may be harmed following a past wrong, as their current group-centered identity is harmed.

The identity-related attempt to overcome the non-identity problem, formulated more fully, is as follows: some groups have a common intergenerational identity. Call them nations or communities; the name and specific variants of the groups are relatively unimportant for the needs of this chapter as long as they meet the following criteria. Members of such intergenerational groups usually share important characteristics such as a common culture, language, religion, and history, as well as a shared aware-ness of belonging to the specified group. A characteristic crucial for the "identity" approach, however, is that such groups have existed for many years, that is, their exist-ence *precedes* the past wrong and that they remain clearly identifiable through the present day (Simmons, 2001; Weiner, 2005, ch. 1). Belonging to such a group may have formal aspects (citizenship), but in some cases a less formal criterion is sufficient (for example, being identified as belonging to such a group by members and non-members alike). While difficulties of demarcation regarding group membership always arise, let us assume for the sake of the argument that such problems can be dealt with. Membership in such groups, the argument from identity maintains, is important for the members' self-understanding.

This argument continues, maintaining that the original wrong done to the group not only harmed the members as individuals, but *also* harmed their identity as part of a group. Connecting the wrong done to the original members to the identity of the intergenerational group proves to be a crucial step as it allows the original wrong to be transformed into wrongs committed against *existing* group members. Such *current* harms may include emotional distress and feelings of insult and humiliation suffered by current members of the group following from the *earlier* substantial harms the past wrong inflicted on the (now deceased) group members. Such feelings may remain on a fairly abstract, emotional level, or they may actually impede integration and success in contemporary society even if there is no current legal discrimination against group members (Sher, 1981; Boxill, 2011, section 7).

Note that, in order to motivate such a description, the emotional attachment felt by the current members of the group towards the group's earlier incarnation should be quite strong. This pro-redress argument's view of identity exceeds a mere interest in a past wrong; we are not concerned with the sorrow of a person who simply reads books on a past wrong or visits museums, memorials and so on. The argument from identity goes much further, claiming that identification with the victims of a past wrong reduces the group's current members' abilities to function in their respective societies.

This view of the identity of group members may reflect how certain scholars under-stand identity, but it is also an essential aspect of how this argument attempts to over-come the non-identity problem. The effects considered in the non-identity problem begin immediately and "contaminate" everything that happens after the wrong; given that the existence of the descendants of the victims (born after the wrong) is causally connected to the wrong, any subsequent claim to redress for that wrong is nullified. The "identity" response must somehow avoid this "nullification" effect, and thus the justification for redressing the past wrong must be located *before* the wrong took place, otherwise, any claim for redress will be nullified by the non-identity problem.

The identity argument may offer a solution to such problems. On this argument, so strong is the emotional attachment and identification of the current members of this group with the now-deceased members of the group (who lived during the past wrong) that the past wrong can almost be said to be inflicted on the current members themselves. But there are still some problems. Since we are considering a historical wrong, all the original victims and wrongdoers have since passed away. The relevant current persons are descendants of the original victims, or current members of the group born after the wrong. As such, they are very likely affected by the non-identity problem. They would not exist had that historic wrong not taken place. So, the identity argument *seems* to fail to overcome "non-identity" objections.

At this point, the collective-identity nature of the identity approach enters the stage. Current group members identify with the pre-wrong group. Since the group's identity is intergenerational, we may have a way of overcoming the non-identity problem.

This amplified identity argument faces two serious worries, however. The first concerns the view of individual identity *as* group membership. The second concerns cross-generational group identity. I discuss each in turn.

First, the "identity" argument entails emphasizing the strength of emotional attachment and identification felt by current members towards the group that existed at the time of the original wrong. This commits the argument to a view that understands individuals as persons defined by membership in a group. Many scholars hold serious doubts regarding such accounts of individual identity, certainly in the context of liberal democratic societies. Such theorists consider such static (and collectivist) views of individual identity as highly implausible. They argue that individual identities are flexible, cosmopolitan, and hybrid, and certainly not defined by group membership (Van Den Beld, 2002).

Second, the "identity" view of cross-generational group identity seems to require that the present-day group's identity would be essentially unchanged from the pre-wrong era. Remember that the identity argument claims that as its focus is a group-identity that precedes the wrong, the non-identity problem does not apply to the current members of the group – even if they are descendants of children of the original victims who were born after the original wrong (and are therefore "contaminated" by the effects of the non-identity problem).

The key point here seems to be the identity of the group itself. If the cultural identity of the current group's individual members is strongly connected to (or "defined by") the intergenerational group, and the group's identity is sufficiently similar to the group identity that preceded the wrong, the identity argument scores an important point. If, however, the group's identity has changed since the time of the wrong, the "identity argument" fails. In particular, it would be odd indeed to argue that the problems of the group's current members are the result of the harm done to the group members at the time of the original wrong, *if these are not the same groups*.

At this point, two remarks are in place. First, the identity approach is committed to an intergenerational *static* conception of a harmed group. This sort of view is unusual. In studies of nationalism, for example, the "static" view is usually rejected as a fiction or myth, and so is no longer taken very seriously. A contrasting view sees national and cultural groups as more dynamic. These groups might (and almost surely do) change over time, and so a pre-wronged group is likely not the same as a current group. Furthermore, some such groups are simply new groups, inventing past existence for various tactical reasons. But this challenges a key premise of the identity argument. Second, it is crucial to indicate that the identity argument emphasized the magnitude of the original wrong, casting it as so substantial that it not only harmed group members at the time of the wrong, but continues to harm members in the present day. According to this argument, even the memory of this past wrong creates great difficulties for current group members. The identity argument cannot easily de-emphasize the magnitude of the original wrong, as its view – that the effects of the wrong are felt to the present day – would become less plausible. But this emphasis on the magnitude of the original wrong creates a problem from a different direction: so important an event must have had noticeable effects on the cultural identity of group members at the time of the wrong. Mass movements of people, large number of deaths, territorial changes – likely involved in a substantial historical wrong – are momentous changes that find expression in literature, music and many other cultural venues. Thus major past wrongs, which can have deep, lasting effects on the group members, affect not only the specific genetic identity of the biological descendants, but also change the cultural identity of the group and thus the identity of the group itself. In some cases, the changes might even create a new group.

The identity argument's attempt to overcome the non-identity problem dismisses the effect a historical wrong has on the genetic identity of the descendants of the original victims. It emphasizes instead the related (cultural) identity of the current members of the harmed intergenerational group as a way to avoid the non-identity problem. But if the group's (cultural) identity changed, either because group identities are not stable or because of the influence of the wrong, the non-identity problem will not have been overcome. Since it is unlikely that such a massive past wrong left the group's (cultural) identity unchanged, the non-identity problem arises in the following way: the existing group members' (cultural) identity differs from the (cultural) identity of the members at the time of the wrong – either because such group identity is not stable, or because of the influence of the wrong. If the wrong had never happened, the (cultural) identity of the current group members would be . . . well, it is impossible to know how to complete such a counterfactual. The current (cultural) identity of the group members stems directly from the wrong, and unless this group's cultural identity is absolutely miserable, the original wrong that brought about the current cultural identity improved the group members' situation. The alternative is to say that it makes no sense to compare current group members' situation with what it would have been had the historic injustice not taken place, as we simply do not know what the group identity would have been like, had the wrong not occurred. Ultimately, the harm done to the pre-original-injustice group members cannot damage any currently existing group (or their members) because *these are different groups*.

Now, this argument is likely to engender some objections, and the following two seem most pertinent. First, one can claim that the past wrong may have merely influenced the group's cultural identity, rather than changed it completely. This again implicitly accepts a static view of groups. Suppose, for the sake of the argument, we accept this implicit view. But if the past wrong was so momentous as to harm the current members of this group (and thus justify compensation), it probably changed the group's cultural identity completely. If the past wrong was only a moderate wrong, then it could be that it merely influenced the group's culture without changing it. But in such a case, it is unlikely that this modest past wrong harms the current group's members and thus justifies claims to compensation. Second, it may be the case that while a culture (the sum of norms, habits etc.) has changed, the current members still identify with the original (different) group, and thus still claim redress. This poses a difficult question of how to evaluate, or measure group membership. Is it a sum of behaviors, norms, and so on, or simply the subjective feeling of belonging? I doubt that there is a clear answer to such a question.

In the context of the non-identity problem and justifications of intergenerational redress, we can frame the problem as follows. Suppose one identifies with a group that existed before a past wrong but suppose that one's behavior clashes with the norms held by the members of that group at the time of the wrong. One's subjective feeling of membership (and therefore grievance) may nevertheless justify redress if three conditions are met:

1 Certain feelings of belonging or membership are sufficient for the ascription of group membership.
2 Felt membership must be (part of) a reliable explanation for one's (in)ability to succeed in one's society.
3 Feelings of membership must not only be an acceptable criterion for ascription of membership, but also capable of justifying holding some current people to be under a duty to bear the cost of redress.

Each one of these conditions is hardly convincing; however, put together, they are implausible.

First, purely subjective membership claims are surely odd given that groups usually require at least some actual behavior or event for admittance, such as command of language, conversion, birth to a member, and so on. Against condition (2): feelings of membership will have to be better explanations for a person's lack of success than some other mundane considerations, such as, say, an inadequate school system. Lastly, against condition (3), such subjective claims would not only have to root membership and explain lack of success, they would need to be sufficient for imposing duties on other, current non-wrongdoers to bear certain costs. The implausibility of the aggregation of these three conditions speaks for itself.

Furthermore, thinking of a policy that would attempt to implement such subjective criteria would immediately raise worries about facile manipulation – that is, such a "subjective view" would create perverse incentives to falsely portray current problems as though they were actually caused by purely subjective group membership, which might then presumably justify intergenerational compensation. This adds to the implausibility of the criteria noted.

The implausibility of this subjective approach will prove important, and perhaps even decisive, to our analysis beyond the debate regarding "how to approach or measure membership in groups." The attempt to overcome the non-identity problem through group-identity-related-claims therefore fails. I next turn to the "timing" argument.

The timing argument

The second attempt to answer the non-identity objection to redressing historical injustices concerns *timing*. If person X was alive at the time of the wrong and *already* had children, his/her interests – especially the ability to pass on resources to his/her descendants – was harmed or violated. If we assume that the resources that she/he would have passed along to his/her descendants would have been passed on to subsequent generations of descendants, then the descendants' entitlements to these resources are not undermined by the non-identity problem. Thus, the existence of original, deceased victims and their descendants is not the result of the past wrong. The timing is therefore crucial for this response: If, and only if, the harm was done to an already existing person, then the non-identity objection does not arise.

There may be some confusion regarding this last point, so further clarification is in order. The timing argument does not escape all non-identity worries about past injustices. The earlier wrong did not affect the existence of the (now deceased) original victims, but it did affect the existence of all their descendants born *after* the wrong, as their particular existence is predicated on the wrong's occurrence. The potential claim by a given descendant – "without the wrong I would have had more resources" – makes no sense if the claimant's existence is causally connected to the wrong. This is why the "timing" is so important; children of victims born before the wrong (and their descendants) do avoid the non-identity problem, but this is not so for those children and their descendants born after the wrong.

The timing argument for redress of historic injustices is therefore limited from the start. The problem is that redress claims of descendants born *after* the wrong are nullified as a result of the non-identity problem. Redress is supposed to be a response to the wrong, yet without the wrong such descendants do not exist. Thus they have no claim for redress.

Having briefly described the "timing" argument, we can now turn to a fuller examination of this attempt to justify redress. Three scholars (Boxill, 2003; Sher, 2005; Cohen, 2009) have attempted to justify redressing past wrongs and, more specifically, compensating the descendants of deceased victims of past wrongs. They argue that the duty to redress does not disappear with the original victims' passing. Rather, they insist, the original wrongdoers now have an obligation to compensate the descendants of the original victims. This is an obligation stemming from the continuing injustice of the failure to compensate the original victims, which consequently has reduced the welfare level of these victims' descendants. This view maintains that, had the original wrong not occurred, the original victims (that is parents, grandparents, etc.) would have "channeled" at least some of their additional available resources to their children. Call this "channeling" argument the "continuing injustice argument" or CIA.

Recall the non-identity objection that persons cannot claim a right to be compensated if their existence is causally linked to the past wrong. The CIA might answer: If a given victim was already a parent at the time of the original wrong, then her/his child's existence is not causally connected to the wrong, yet the child's level of welfare has been potentially lowered as a result of the original wrong and the subsequent failure to compensate. Therefore, the child is entitled to compensation sufficient to raise his/her level of welfare to what she/he would have enjoyed had the original and ensuing wrongs not occurred. In this formulation, the non-identity problem does not arise.

This sort of approach seems to rule out certain sorts of claims to redress. Consider the case of victims' children born *after* the original wrong (that is, during the *continuing* wrong of the failure to compensate the parents, grandparents, etc.). In such a case, either the original wrong or the continuing wrong may be causally connected to the birth of the child, thereby raising non-identity problems. The CIA thus avoids non-identity objections *only if* children of the victims were born before the original wrong, for only then would the wrong not be causally connected to the existence of the children (and their descendants) of the original victims. This argument, given that we are discussing children born before the historical wrong (and their descendants), avoids the non-identity problem, and it does not rely on membership in a collective. Rather, it relies on the individual right to compensation following a tangible damage that was done to a given person.

Here we must distinguish between the following two sorts of arguments: (1) the argument that the beneficiaries of the original wrongdoers ought to *return* to the descendants of the original victims any identifiable property they retain that was wrongly taken from the original victims; and (2) the argument that descendants of the original victims deserve *compensation* for any demonstrated damage caused by the ongoing failure to compensate them.

Note that the two arguments are not symmetrical; the property that the beneficiaries of the wrongdoers wrongly hold and are under a duty to restitute is distinct from the compensation to which the descendants of the victims are entitled according to the CIA. The compensation is determined by the damage done to the descendants; the duty to restitute follows the unjust enrichment of the beneficiaries of the wrongdoers. These are very different.

CIA scholars have not made this distinction, but this is a problem that complicates their attempt to justify compensation. The distinct restitution logic is clearer in this regard: if John Doe holds property that does not belong to him, he should restitute this property to the rightful owner. Problematically, however, Nozickian compensation is much wider, and includes, for example, resources that the descendants' parents, grandparents, and so on, would have accumulated (but did not) and would have later "channeled" to their descendants had the wrong never happened. This distinction is important. While the duty to restitute applies to those unjustly enriched, it is not clear who should bear the cost of Nozickian compensation, which responds to the overall damage done to the descendants of the original, now deceased, victims. The wrongdoers, let us remind the reader, have all passed away since the wrong.

Indeed, the CIA fails to establish who should bear the burden of any compensation aside from the indicated restitution. After all, the wrongdoers have all died and no living persons are guilty of the wrong. Those under a duty of restitution are unjustly enriched as a result of the wrong and, indeed, are obligated to return this property, but they *do not* inherit guilt for the past wrong or any obligation to bear the cost of Nozickian compensation. As explained earlier, one owes restitution if one enjoys unjust enrichment at the expense of identifiable victims. One owes compensation for the damage done to the victim/descendant from an earlier wrong *for which one is culpable*. But by hypothesis, the original wrong is over, all the original wrongdoers have passed away; current persons have not wronged descendants of victims of the distant original injustice. To repeat, restitution and compensation are *not* symmetrical conceptions. If we start by assuming the separateness of persons, then people cannot be guilty for the sins of their ancestors, and therefore the identity of those who will have to bear the cost of compensation (not restitution) is unclear, and CIA scholars do not provide an answer to this question. Therefore, *even if* the CIA argument can pass difficult hurdles – avoiding the non-identity problem, providing adequate counterfactuals, demonstrating continuing damage, and providing adequate information – the identity of the would-be payers of compensation remain something of a mystery. There is always, obviously, the collectivist option in which living people and dead wrongdoers are lumped together in some fashion. This view, however, clashes with the fundamental liberal commitment to the status of *each person* as an independent moral agent worthy of respect (which is why the separateness between individuals is fundamental).

A few brief comments are required regarding restitution. There are two different kinds of restitution claims that use similar logic, but differ significantly in magnitude. The first kind is large in magnitude, applying to large tracts of land, usually advanced either following colonialism or in the context of settler societies. Such claims need to somehow overcome the famous "supersession" thesis, which must be explained briefly here. Suppose that an injustice occurred at some point in the past. Imagine, for example, a certain piece of land was stolen. Suppose, further, that time has passed since the wrong took place and the circumstances that existed at the time of the wrong change. For example, the number of people living on that piece of land has grown substantially. The supersession thesis maintains that claims to restitution can be superseded by time's passage and changes in circumstance. At some point in time, returning that piece of land *solely* to the descendants of the original owners would be unjust in so far as this would harm the people now living on this piece of land (Waldron, 1992).

An important aspect of the supersession thesis is that property rights are not immune to other considerations. If a certain appropriation and the resulting ownership present a significant setback to the interests of non-owners, there is a justification for limiting such ownership. This principle – that ownership is limited by other considerations – is not, in and of itself, connected to time. In other words, if person A owns piece of land X, and B, C and D have no other place to go, A's ownership is limited by the interests of B, C and D.[3]

The novelty of the supersession argument lies in its explanation of how the consequences of a wrong that occurred at one point in time, T1 (when the circumstances did not justify limiting the ownership of a given property such as a piece of land), become at least partially just at a later point in time, T2. This change is due to a shift in circumstances that justifies limiting the property rights of the owners or their descendants, *regardless of the wrong, and regardless of the identity of the owner*. The addition of the passage of time as a variable proves, therefore, to be merely a rough (yet absolutely reasonable) way of indicating that a new situation has arisen since the wrong occurred, and that this new situation justifies a renewed evaluation of ownership claims.

The supersession argument, however, does not mean that property should simply remain where it currently resides, that is, with the current owners. It means that the interests of the past wrongdoers (or their descendants) and non-involved third parties should be taken into account when considering solutions to property rights disputes among various potential owners. But the most important consequence of the supersession idea is that it nullifies most "large" restitution claims.

The second kind of restitution claims have to do with private property on the individual level. Suppose John Doe holds property that can be traced back to a past wrong. What should happen now? In order for a successful restitution claim to go forward, and assuming that the "large" supersession issue was somehow resolved in a way that allows "small" restitution claims to go forward, several conditions must be met, including: (1) property rights need to be stable enough to withstand the passage of time; (2) the property needs to be identifiable; (3) the descendants of the deceased victim need to be identified; and (4) the current holder needs to be somehow culpably connected to a wrongdoing (which introduces difficult problems in the case of a "good faith purchaser").

Such conditions are time sensitive, and thus while legitimate claims for restitution are not impossible, restitution claims invariably fade over time. Thus, both the larger land-based claims for restitution and the more modest individual based claims are time sensitive. While it is not impossible for a successful restitution claim to pass those hurdles, it seems increasingly unlikely that many such claims will be found as time passes.

Conclusion

Material redress following past wrongs is an emotionally and politically sensitive subject. Many people are understandably sympathetic to requests for such redress; it may seem to be a fitting response to a horrific past. However, once we start our deliberations with the assumption that each individual is separated from other individuals, we move away from the antiquated idea of inherited guilt and begin to consider the challenges of figuring out precisely who owes what to whom. Overall, the case for intergenerational redress for past wrongs proves rather weak. Perhaps it is advisable to keep our focus on the needs and interests of living, contemporary people, rather than to dwell on past events.[4]

Notes

1 The interested reader may wish to consider my fuller version of this argument in Perez (2012).
2 This short phrase requires substantial elaboration, especially in the context of arguments for collective responsibility which begin from individualistic assumptions. For further discussion, see Perez (2012), chapter 4.
3 This argument owes much to the Lockean Proviso, which is outside of the domain of our inquiry here.
4 The author wishes to thank the editors of this volume for important comments on previous drafts of this chapter.

References

Boxill, B. (2003) A Lockean argument for black reparations. *The Journal of Ethics* 7 (1): 63–91.
Boxill, B. (2011) Black reparations. In *The Stanford Encyclopedia of Philosophy*. ed. Edward N. Zalta Center for the Study of Language and Information, Stanford University, Stanford, CA 94305. http://plato.stanford.edu/archives/spr2011/entries/black-reparations/ (last accessed 6/17/13).
Cohen, A.I. (2009) Compensation for historic injustice: completing the Boxill and Sher Argument. *Philosophy and Public Affairs* 37 (1): 81–102.
Dagan, H. (2004) *The Law and Ethics of Restitution*. Cambridge: Cambridge University Press.
Morris, Christopher W. (1984) Existential limits to rectification of past wrongs. *American Philosophical Quarterly*, 21 (2): 175–182.
Nozick, R. (1974) *Anarchy State and Utopia*. New York: Basic Books.

Parfit, D. (1986) *Reasons and Persons*. New York: Oxford University Press.

Perez, N. (2012) *Freedom from Past Injustices*. Edinburgh: Edinburgh University Press.

Posner, E.A. and Vermeule, A. (2003) Reparations for slavery and other historical injustices. *Columbia Law Review* 103 (3): 689–748.

Rawls, J. (1999) *A Theory of Justice*. New York: Oxford University Press.

Sher, G. (1981) Ancient wrongs and modern rights. *Philosophy and Public Affairs* 10 (1): 3–17.

Sher, G. (2005) Transgenerational compensation. *Philosophy and Public Affairs* 33 (2): 181–200.

Simmons, A.J. (2001) Historical rights and fair shares. In *Justification and Legitimacy*, pp. 222–249. Cambridge: Cambridge University Press.

Van Den Beld, T. (2002) Can collective responsibility for perpetrated evil persist over generations? *Ethical Theory and Moral Practice* 5 (2): 181–200.

Waldron, J. (1992) Superseding historic injustice. *Ethics* 103 (1): 4–28.

Weiner, B.A. (2005) *Sins of the Parents: The Politics of National Apologies in the United States*. Philadelphia, PA: Temple University Press.

Further Reading

Bauman, Z. (1996) Morality in the age of contingency. In *Detraditionalization*, ed. P. Heelas, pp. 49–58. Oxford: Blackwell.

Boxill, B. (1992) *Blacks and Social Justice*. Lanham, MD: Rowman & Littlefield.

Davis, L. (1976) Comments on Nozick's entitlement theory. *The Journal of Philosophy* 73: 836.

Giddens, A. (1991) *Modernity and Self Identity*. Stanford: Stanford University Press.

Heyd, D. (1986) Are wrongful life claims philosophically valid – a critical analysis of a recent court decision. *Israel Law Review* 21: 574.

Kull, A. (1995) Rationalizing restitution. *California Law Review* 83: 1191–1242.

Miller, D. (2008) *National Responsibility and Global Justice*. New York: Oxford University Press.

Nagel, T. (1991) *Mortal Questions*. Cambridge: Cambridge University Press.

Thompson, J. (2003) *Taking Responsibility for the Past: Reparation and Historical Injustice*. Cambridge: Polity.

Profiling

CHAPTER FOURTEEN

Bayesian Inference and Contractualist Justification on Interstate 95

Arthur Isak Applbaum[1]

In 1992, Maryland state troopers subjected Robert L. Wilkins to a search with a drug-sniffing dog after stopping him for speeding on Interstate 95. Nine out of ten drivers speed on the highway: what distinguished Wilkins was that he is black. He was also a recent Harvard Law School graduate, and now is a United States District Court judge. In settling the resulting class action suit, *Wilkins v. Maryland State Police*, Maryland troopers agreed to adopt a policy forbidding the use of race in determining whom to stop and search in their drug interdiction activities, though they denied that Wilkins's constitutional and civil rights had been violated, and denied that they routinely use race as a criterion for highway stops and searches.

In a 21-month period following the consent decree, however, a black speeder was about fifteen times more likely than a white speeder to be stopped and searched by a special drug trafficking unit on a stretch of I-95 north of Baltimore. Blacks, though only about 18 percent of all speeders, accounted for 73 percent of those searched; whites accounted for about 75 percent of all speeders, but only 20 percent of those searched.[2] The success rate of these searches was the same for both groups: about 30 percent of both blacks and whites searched were drug offenders.[3]

Despite their protestations, the Maryland troopers clearly continued to employ race as a factor in deciding whom to search, and appear to have violated the settlement reached in *Wilkins*. But it is not my purpose to enter into that dispute, or the legal dispute over whether the use of race-based generalization by police is indeed unconstitutional under a reasonable interpretation of the constitution that we currently have. Rather, I wish to ask whether the use of such generalizations is unjust, and so, the sort of practice that *ought* to be prohibited by a society that seeks to treat its members justly. Whether the practice is in fact prohibited by existing law, properly interpreted, is a matter I leave to legal scholars.

Contemporary Debates in Applied Ethics, Second Edition. Edited by Andrew I. Cohen and Christopher Heath Wellman.
© 2014 John Wiley & Sons, Inc. Published 2014 by John Wiley & Sons, Inc.

To begin, I make two very big bracketing assumptions, and then distinguish three pure cases of statistical generalization. Then, about each of the pure cases, I ask two questions.

Two Brackets

I wish, at least at the start, to set aside two important and perhaps compelling arguments against the use of racial generalizations. The first is that racial generalizations typically are the products of either faulty inference or prejudice, and so acting upon them is irrational. The second is that the persistence of racism and its consequences in America is such a grave injustice and social problem that even the use of accurate and unbiased racial generalizations is repugnant and harmful. I wish to bracket off these arguments, not because I think that they fail, but because they work too well as conversation-stoppers. I believe that there is interesting and hard conceptual work to be done about the moral permissibility of using group-based statistical generalizations in police work even when virulent prejudice is not at issue. To do so, some abstraction from the harsh realities of American racism is necessary.

Let us provisionally restrict our attention, then, to those racial generalizations that meet a minimal test of instrumental rationality, in that, use of the generalization is a means towards some given objective, such as "the efficient apprehension of violators." Generalizations that fail such a test are simply foolish, whether or not they are morally wrong. What is this minimal test of instrumental rationality? For a start, the statistical inference must be accurate. One must have reliable information, and one must ask of that information the right question. The right question is: "What is the probability that a person is a violator given that the person fits the generalization?" and not: "What is the probability that a person fits the generalization given that the person is a violator?" Suppose half of all smugglers passing through customs are Ozians. By itself, this does not tell customs officials the proportion of Ozians who are smugglers. To answer the right question, one also would need reliable information on the proportion of travellers who are Ozian and the proportion of travellers who are smugglers. If one out of every thousand travellers is an Ozian, and one out of every ten thousand travellers is a smuggler, the correct Bayesian inference is that one out of every twenty Ozians is a smuggler.[4]

With a false positive rate of 95 percent, can searching all Ozians going through customs be instrumentally rational? One cannot say without knowing more about what counts as a benefit and a cost, given the objective. Again, let us draw this with exceeding narrowness: apprehending violators is a benefit, expending scarce police resources is a cost, and nothing else counts – not fairness, not civil liberties, not the burdens that fall on the innocent false positives, etc. Under these assumptions, a generalization is instrumentally rational if the expected net benefit of finding the true positives – the Ozians who are smugglers – exceeds the cost of searching all the false positives – the innocent Ozians. Whether one in twenty is a big number or a little number, and so whether a statistical generalization about Ozians is useful to customs officials, depends on what counts as a cost and benefit. This is the lesson Bayesian decision theory holds for us.

Among the set of search strategies with positive net benefits, some are better than others. There may be a strategy more refined than "Open the luggage of all Ozians" that is more efficient: "Open the luggage of Ozian girls," or "Open the luggage of Ozians with Kansan accents." And, presumably, there are search criteria for non-Ozians as well: "Open the luggage of anyone who crosses the border frequently." If a more refined search strategy is available, not to use it is inefficient. The use of the social monitoring skills that David Wasserman discusses is a good example of a more refined search strategy (Wasserman, 1996). If it is instrumentally rational to acquire and employ the skills needed to distinguish the harmless streetplay of black teens from threatening behavior, to rely on a rougher racial generalization alone is foolish, quite apart from whether it is unfair. If there are even better strategies that make no use of group characteristics at all, so that "Open the luggage of anyone who crosses the border frequently" is more efficient than "Open the luggage of anyone who crosses the border frequently plus all Ozians," then to use the group-based generalization is foolish. But the opposite may be true, and efficiency may demand the inclusion of group-based generalizations.

I am not at all endorsing the view that law enforcement agencies should take "the efficient apprehension of violators" as their sole objective, and I certainly am not endorsing the view that the burdens that befall non-violators and the moral demands of fairness, liberty, and respect should not shape and constrain police objectives. I have elaborated an extremely narrow view of policing to make this point: since the objective of apprehending violators is at least a *part* of good policing, and since at least *some* race-based generalizations are instrumentally rational with respect to this objective, one cannot reject the use of all racial generalizations on the grounds that they are foolish. Maryland State troopers employed a more refined search strategy with whites and a less refined one with blacks, but both strategies had identical yields of 30 percent. Again, whether a false positive rate of 70 percent is a big or little number depends on what counts as a cost or benefit, but the fact that it is the *same* number for both groups shows that the use of group-based generalization here is not simply foolish. To reject at least some race-based generalizations, one must show that their use is morally wrong.

Three Cases

There are at least three conceptually distinct types of cases in which the police might use group-based selection criteria, and a number of mixes of these pure types. In the first, call it *group-based patrol*, the police, searching for as-yet undiscovered violations or seeking to deter violations, use statistical inferences from group characteristics to select those who will be subject to heightened scrutiny. Examples: stopping Ozians at customs in order to catch smugglers or shadowing young black males in department stores in order to deter shoplifting.

In the second, call it *group-based enforcement*, the police use some group-based characteristic as a criterion for selecting which known violators will be subject to law enforcement out of a larger set of known violators. Example: out of the set of all speeders, state troopers select young black males for ticketing. Though a group-based characteristic is used, there is no statistical generalization here, and the use of group characteristics has no probative value: it is already known who is and is not speeding.

Distinguish this from a mixed case, *enforce-to-patrol*, where group-based enforcement is used as way to implement group-based patrol. Here, minor violations are enforced against known minor violators who fit a group-based generalization in order to search for unknown major violations. Presumably, police engage in this search strategy when they are barred by law or policy from pure group-based patrol. Example: out of the set of all speeders, young black males are ticketed disproportionately so that the police officer can take a look inside the car for signs of drug trafficking. This is the strategy the Maryland State Police were employing on I-95.

Call the third pure case *group-based identification*: an unidentified suspect in a known violation is described as having group-based characteristics, so the police stop those who fit the group-based description. In pure group-based identification, no statistical inference is made about the likelihood of criminality among those who fit the description. Assuming that the description of the suspect is accurate, it is simply given that the violation was committed by *someone* fitting the description.[5] The inference here is about the likelihood that someone who fits a description is the particular person described. The difference is seen most clearly when the descriptive characteristics are not otherwise believed to be correlated with criminal behavior. If a thief is described as a tall redheaded woman, the instrumental rationality of stopping tall redheaded women to find *this violator* turns on the likelihood that any one tall redheaded woman in the vicinity is a particular tall redheaded woman, which – if nothing else is known about the suspect – is simply the reciprocal of the number of tall redheaded women in the vicinity. To find this particular suspect, one does not need to make any inferences about the proportion of tall redheaded women who are thieves.

When the descriptive characteristic is one that is also believed to be correlated with criminality, however, group-based identification may be mixed with group-based patrol. If a suspect is described as a young black male, the likelihood that someone who fits the description is *either* this particular suspect *or* some other violator may be high enough to pass the test of instrumental rationality, even if neither likelihood by itself does. Finally, if police are barred from engaging in pure group-based patrol, they may adopt the strategy of *identify-to-patrol*, and stop those who match the description of a particular suspect in order to implement an instrumentally rational group-based patrol.

Though the mixed cases are no doubt quite common in police practice, I will focus on the three distinct pure types. To the extent that the mixed cases are simply intermediate cases, I leave it to the reader to make the necessary interpolations. For example, as the incidence of minor violations rises in the general population, enforce-to-patrol approaches the pure patrol case. Since over 90 percent of both black and white drivers speed on I-95, *Wilkins* approaches the pure patrol case. As the incidence of major violations rises in the general population, enforce-to-patrol approaches the pure group-based enforcement case. To the extent that the mixed cases involve police lawbreaking or deception (as appears likely in *Wilkins*), the wrongs involved are not particular to the topic of statistical generalization.

Two Questions

About these three types of case, let us ask two questions. First, does an innocent false positive who is stopped by police because of the use of a group-based generalization have a reasonable complaint? Second, does a true positive, a violator, have a reasonable complaint? If, for a type of generalization, both the violators and non-violators have good grounds for objecting to their treatment, then that practice clearly lacks moral justification. If non-violators appear to have a reasonable objection, but violators do not, then perhaps there is some way to answer the objection of the non-violators. A successful answer to the non-violators will need to show why it is reasonable for them to accept the treatment to which they are subjected, and that requires showing, among other things, that being subjected to police scrutiny does not involve fundamental disrespect or indignity.

Consider first the pure group-based enforcement case: police who ticket only black speeders. Though those who are stopped are indeed violators, they have a clear objection: fairness requires that those who are alike in the relevant respects be treated alike, and those who are different in the relevant respects be treated differently. Being black is not a difference relevant to whether one should be ticketed or not, so the unequal treatment is unfair. Indeed, it is difficult to imagine an objective served by such treatment that is not straightforwardly malicious.

In contrast, a violator does not appear to have good grounds for complaint in the case of pure group-based identification. How is his treatment unfair? Like the speeder in the pure enforcement case, he indeed has committed the violation. Unlike the speeder, he is not being singled out from a larger set of violators for unequal treatment, for there is no larger set of violators in this instance – the police are responding to a report of a particular crime. Nor are the police making an inference about the propensity of members of a group to commit crimes, which may fail to treat an individual with respect, for in the pure description case the only inference is about identity, not criminality. While it is true that those violators whose identifying characteristics are less prevalent in the population are easier to catch than those with more common characteristics, why is this not simply the violator's bad luck, and our good luck, rather than unfairness? Is the tall redheaded thief who is caught treated unfairly because she is easier to catch than medium-height brown-haired thieves? I think not. Fairness does not require equal chances of success for thieves. If the tall redhead has no comparative advantage as a thief, let her choose another line of work.

Now suppose the identifying characteristic is race. Example: a theft by a tall Asian male teenager is reported in a neighborhood where Asians are few, it is instrumentally rational for police to stop all tall Asian male teenagers in the neighborhood because the chances that any one Asian teen is a particular Asian teen are high, and someone stopped by police because he fits the description is in fact that particular Asian teen described, the thief. Why is this too not simply his bad luck? Even if we suppose injustice led to the low numbers of Asians in the neighborhood, has an injustice been done to would-be thieves? Unequal opportunity to live in desirable neighborhoods is an injustice; unequal opportunity to steal from those neighborhoods is bad luck.

Can the violator claim that since, *ex ante*, the police did not know that he was the violator, his treatment should be judged as if he were not the violator? This is a puzzle we do not have to solve for our purposes, because either way, police must answer the objections of those who are stopped for identification who are found to not be the violator *ex post* – the innocent tall Asian teens. If their objections can be answered, so, *a fortiori*, can the objections of the violator.

Bayesian Inference and Contractualist Justification on Interstate 95   223

The objection of the innocent non-violator subjected to race-based instrumentally rational identification is something like this: though I have broken no law, my liberty has been infringed, my privacy violated, my dignity affronted, and my sense of security shaken. I have been selected for this treatment because of the color of my skin. If everything about me were the same, but for my skin color, I would not have been stopped and questioned. But skin color is not a relevant reason for different treatment by the police, so I am being treated unfairly.

If a reply is to succeed, it would have to show the innocent non-violator that, from some suitably constructed *ex ante* point of view, it would be reasonable for him to agree to subject himself to a general policy of properly regulated group-based identification. In part, this involves showing that all are at risk of victimization by thieves, so all should be willing to accept some burden of unwelcome police encounters in order to stop theft. Reasonable agreement is not simply a matter of an individual's calculation of self-interest, for it is reasonable for me to accept some sacrifices to prevent much greater harms to others. Nor does this involve simple utility-maximization: it is not unreasonable for me to reject great sacrifices to provide small benefits to many, many others. The reasonable burden will be lower, indeed far lower, than what an instrumentally rational search would impose. But the reasonable burden is not zero. This, roughly, is the lesson of contractualism.

How much of a burden it is reasonable to accept depends on just what the nature of the burden is, and this in turn depends on whether respect for the individual is compromised in the encounter. Whether an unwelcome search by police can ever be respectful turns on the message of the encounter, and by this I do not simply mean what literally is said by police. The reason an individual is picked out for police scrutiny itself carries a message of respect or disrespect. The wrong words can make an otherwise respectful encounter disrespectful, but the right words can go only so far in mitigating the message of an encounter that is predicated on disrespect.

Consider this message: "I'm sorry for interrupting your evening, sir, but a crime has been committed by someone who fits your general description. For the protection of all law-abiding members of society, the police have a policy of stopping and questioning those who resemble suspected criminals. Even though I have stopped you, I continue to presume that you are a law-abiding member of society, and so I hope that you can see that you and your neighbors are protected by such a policy. I trust that, upon reflection, you will find it reasonable to assume this burden. We do not suspect *you* of committing a crime, we suspect someone who resembles you. Help us confirm our presumption that you are yourself, and not the suspect you resemble, and we will trouble you no more. It is true that you were picked out for questioning in part because your skin color matches the described skin color of the suspect. But I am not supposing that every individual with your skin color is more likely to commit crimes, which I can well imagine would be insulting. I simply am supposing that individuals with your skin color are more likely to be the particular person for whom we are searching." Under *very* favorable conditions, I believe that something like this could be the message of a police identification. I am not seriously proposing that Miss Manners be hired to deliver such "Miranda apologies"; but I do think that the content of this reply involves no necessary incoherence, deception, or self-deception, and so police searches can treat persons in ways that are consistent with it. (Whether such treatment is self-defeating is another matter: perhaps the police cannot sincerely believe this about those they identify, or cannot honestly convey it, and still be effective at catching violators.)

An innocent black male teen who is stopped can accept much of this reply and still have a remaining complaint. He can grant that, looking only at the probabilities of success in a particular search, the search strategy under which the police have stopped him is both rational and reasonable, and he can grant that there is nothing inherently disrespectful about making a probabilistic inference about his identity from his appearance, which includes racial features. But even if "black male teen" when the suspect is a black male teen is as equally predictive as "tall redheaded woman" when the suspect is a tall redheaded woman, an innocent black male teen might object that he fits the description of suspects with far greater frequency than do tall redheaded women. Black male teens therefore will be overburdened by unwelcome police scrutiny, and that is unfair. Even if, implausibly, police can maintain with honesty the presumption of innocence and convey a respectful message in each instance, the cumulative message of repeated searches is degrading.

Once we allow the ugly reality of American racism to intrude, this becomes a powerful objection, but I wish to keep this reality bracketed for just a while longer so that we can isolate arguments about the use of statistical generalization. Suppose a small band of tall redheaded women committed so many thefts that police efforts at identifying them were as burdensome on the population of innocent tall redheaded women as is the burden on innocent black male teens. This would be a great misfortune for the innocent redheads, but I am not sure that it would be an injustice. The rest of us owe them something – certainly gratitude, perhaps recompense – for the troubles they undergo for our benefit. But I do not think that they can reasonably ask the police not to take their appearance into account. Still, there is an important lesson here for the treatment of innocent black males who bear a disproportionate burden of police scrutiny, even when that treatment is not tainted by biased inference and racial prejudice. They are owed gratitude, at least, for their troubles, and it is especially ungrateful to punish them with lingering suspicion.

This leads us squarely to the case of *group-based patrol*, in which police use statistical inferences about the behavior of members of groups to search for as-yet undiscovered violations or to deter violations. The examples given earlier were stopping Ozians at customs or shadowing young black males in department stores. Since we are trying to isolate the moral significance of using group-based generalizations, assume that the probabilities, benefits, and burdens are such that, if not for the fact that group characteristics are part of the search criteria, the targets of the search would have no reasonable objection. For example, suppose that if 5 percent of those who cross the border frequently are smugglers, frequent travellers would have no reasonable objection to the patrol strategy: "Open the luggage of all who cross the border frequently."

Here, there are stronger objections to overcome. One could reply to those scrutinized for purposes of identification that, in the pure case of group-based identification, no inference about the propensity of members of the group to engage in criminal activity was made. But precisely such an inference drives group-based patrol.

What is the complaint of the Ozian subjected at customs to the search strategy "Open the luggage of all Ozians" who indeed is a smuggler? Unlike the black speeders in the pure enforcement case, the Ozian smugglers in the pure patrol case cannot claim that there is no difference between them and the non-Ozians who are not searched. The difference is that the proportion of Ozians who are smugglers is much higher than the proportion of non-Ozians who are smugglers. The Ozian needs to show either that this is not a morally relevant difference, or that the relevance of the difference is counteracted by moral reasons to ignore that difference. Some of these objections appeal to the way that the group is viewed or treated, and others appeal to the way the individual is viewed or treated.

The Ozian smuggler might try to object on the grounds that the search strategy, "Open the luggage of all Ozians," amounts to an inherently disrespectful ethnic slur, for it supposes that all Ozians are criminals. But this misunderstands the connection between inference and action. The statistical inference is simply that the proportion of Ozians who are smugglers is one in twenty, while the proportion of non-Ozians who are smugglers is about one in twenty thousand. By assumption, this inference is both well-supported by evidence and true, and so cannot by itself be a slur. "Search all Ozians" is a decision rule that follows from the costs and benefits of search, not an inference about all Ozians. One may have good grounds to criticize the evaluation of costs and benefits, but not the inference itself.

The Ozian smuggler refines her objection: the inference supposes, not that every Ozian is a criminal, but that every Ozian has a higher propensity to commit a crime, and that is an ethnic slur. This is a serious objection from an innocent Ozian, and I will consider it shortly. The Ozian smuggler, however, is estopped from making this objection. She, after all, is part of the reason that the proportion of Ozians who are smugglers is higher than the proportion of Ozians who are not – her behavior helps to make true the inference that she finds insulting to her and her group.

Consider, then, the objections of the innocent Ozian. She complains that the message of group-based patrol is inherently disrespectful, both to her as an individual and to Ozians as a group. The search is disrespectful to her as an individual because it treats her as someone who has a propensity to commit a crime for no reason other than that she is a member of a group. We cannot say, as we said in the case of group-based description, that the inference is simply about whether she is a particular person. Here, the inference is about whether she is a smuggler. The search is disrespectful to Ozians as a group because it supposes that there is some trait that Ozians have that predisposes them to criminality. We cannot say, as we said in the case of group-based description, that she has been picked out simply because her appearance matches the appearance of a suspect. Here, the inference is more than skin deep: criminality is a purposeful activity, so there is some inference being made about the motivations, character, or culture of Ozians.

As before, the response to these objections will try to find an honestly respectful message in the actions of law enforcement officers. To answer the charge that group-based patrol treats the innocent individual as someone who has a propensity towards criminality, we need to show how using statistical inference to pick a search strategy is still compatible with a presumption of innocence. When Bayesian decision analysis expresses a degree of certainty about some event occurring, it makes no commitment to any one of the many underlying causal mechanisms compatible with the statistical inference made. A Bayesian may say, loosely, something like, "I believe that the probability that an Ozian is a smuggler is 5 percent," but this does not commit the Bayesian to the belief that this particular Ozian smuggles 5 percent of the time, or that she has an inclination to smuggle that has a 5 percent chance of winning out over other inclinations, or that she would turn out to be a smuggler in 5 percent of the replays if her life could be replayed, or any other formulation. To be sure, sometimes one of these is believed to be the correct causal mechanism, but it need not be. All the Bayesian needs to say is, first, that he is justified in believing that 5 percent of the Ozians passing through customs are smugglers, and second, that he has no other justified beliefs that indicate which 5 percent. Therefore, a Bayesian police officer can sincerely maintain that he is not imputing a propensity towards criminality to any particular Ozian who is searched, and so is not treating any particular Ozian with disrespect. The innocent Ozian is burdened by a loss of privacy and by the fear that law enforcement scrutiny generates, but is not burdened by disrespect or insult. The message to the innocent Ozian is, roughly, "It's nothing personal."

The innocent Ozian finds this a small comfort. The search strategy is built on the inference that Ozians as a group are far more likely to be smugglers than non-Ozians. The message of the search can be respectful to her as an individual only by supposing some distance between this "decent" Ozian and Ozians in general. If she is to accept the message, and understand herself to have been treated with respect, she needs to adopt that distance herself. But if she is connected to or identifies with other Ozians in significant ways, this is a cruel choice.

The reply to this objection is really the same as to the previous one: because Bayesianism is agnostic about underlying causal mechanisms, it does not require enforcement officials to believe anything about "Ozians in general," and it certainly does not require them to believe that Ozians have a trait that predisposes them to criminality. Ozian smugglers and Ozian innocents may be two distinct groups, one that always smuggles and the other that never does; and the causes of smuggling among the Ozian smugglers may have nothing to do with any important component of Ozian culture or identity. Perhaps only the assimilated, alienated Ozians smuggle. Obviously, if the Bayesian law enforcement officer could make these distinctions in the field, he would be foolish to not use a more refined search strategy. But just because one does not have a more refined search strategy in hand does not commit one to the view that the underlying causal mechanisms admit no further refinement. All Ozians are insulted, and the innocent Ozian needs to alienate herself to avoid insult, only if the customs official is committed to an insulting underlying mechanism. As long as the bracketing assumptions are kept in place, even group-based patrol can be given a respectful message.

To keep the bracketing assumptions in place here, however, strains credulity. In practice, it is much harder for police to believe sincerely and convey honestly the respectful message of the action when they are making inferences about criminality, rather than about identity. As the proportion of true positives picked out by a strategy rises, the cognitive discipline required to maintain respectful treatment in group-based patrol is enormous. Compare an identification case in which the search picks out two individuals who meet the description of the violator and a patrol case in which the expected rate of violation among those who fit the search criteria is one in two. A modicum of training and good will can lead police officers to recognize that at least one of the suspects in the description case is innocent, and that that should affect how both are treated. But a heroic amount of training and good will is required to get police to recognize that, in a population where half are violators, half are not, and that that should affect how all are treated.

The cognitive demand is especially great when engaged in deterrent patrol, since there is no way to confirm that one has deterred a would-be violator, and not harassed a non-violator. Pure race-based identification, if practiced with good will, has this self-regulating feature: a false positive rate that is higher than expected prompts a re-evaluation of the reasonableness and rationality of the search strategy. But in deterrent patrol, one does not know the actual false positive rate, so one never gets evidence that could weaken one's confidence in the search strategy. A department store's security guard who spends his day closely trailing young blacks is confirmed in his belief that his strategy is a good way to deter shoplifting each time an innocent kid is shamed into leaving the store.

This is a caution against all deterrent police scrutiny, on whatever criteria. But the dangers of self-confirming suspicion of racial and ethnic groups are much greater than the dangers of self-confirming suspicion of behaviors such as associating with the wrong people, hanging out in the wrong places, or wearing the wrong clothing. Though it may be unfair to be suspected because of one's clothing, one can avoid the unfairness by not wearing gang colors. One cannot avoid the unfairness of being suspected because of one's race by changing one's skin color.

Bayesian inference and contractualist justification are not in conflict. The view developed here is contractualist in that its moral evaluations appeal to the kinds of treatment it is reasonable for persons to accept. It is Bayesian in that it recognizes that the usefulness of a statistical inference for guiding action depends on what one cares about. There is no necessary conflict between contractualism and Bayesianism because, if we care to treat persons with respect, not all statistical inferences will guide action, and when inferences do guide action, they need not be disrespectful.

Race and Racism

I have asked the reader to abstract away what may, in the end, be the most important moral reasons against the use of race-based generalizations. I have done this so that the analytic structure of the problem of statistical inference in law enforcement could stand out more clearly. Not surprisingly, the case for race-based generalization is strong-est when the virulence of racism is weakest. First, I have supposed that race-based inferences are accurate, and while no doubt many are, many are not, and the injustice wrought by the use of ignorant and malevolent generalizations may be great. Second, I have supposed that the police and the rest of us can intend and convey a respectful message, when in practice the use of race-based generalizations may do the opposite: they may add insult to burden, and replace deserved gratitude with undeserved suspicion. Third, the pure cases that have been distinguished here are quite mixed in practice, and police officers face a great temptation to employ legally and morally permissible types of race-based generalization as a cover for legally and morally impermissible types. Last, and most important, I have not begun to assess the moral significance of America's long history of racism against blacks, and how that history indelibly colors the possible messages and meanings of any race-based police action.

In Chapter 15 of this volume, Deborah Hellman (Hellman, 2014, pp. 232–244) argues that we cannot fruitfully bracket the reality of racism because, in a society with our history of racial prejudice, the practice of racial profiling by police expresses a pejorative social meaning about African Americans: that blacks are naturally predisposed to be criminals. Even when attached to an otherwise accurate generalization, this expressed meaning is inescapably demeaning. When the state demeans some of its members, it fails in its obligation to treat all of its members as moral equals. In closing, I wish to comment briefly on how she might be right about what racial profiling expresses, and what would follow.

Hellman is surely right that social practices can acquire objective meanings that do not depend on the subjective understandings or intentions of any particular participants in the practice. Once green bits of paper are taken to be money, they are, objectively, money, and someone who denies that fact is making a mistake about the truth – a truth that is socially constructed, but no less objective for that.

Suppose Hellman is right about the meaning of racial profiling in our society. Consider a straightforward identification case: A crime by a tall young black male in red sneakers is reported, so a police officer stops and questions all tall young black males in red sneakers, but not whites in red sneakers or blacks in blue sneakers. Having read my article, the officer does everything she can to make the encounter respectful, but, if Hellman is correct, in our society race-based generalizations carry an inherent degrading and disrespectful meaning, so the red-sneakered blacks who are stopped have, objectively speaking, been disrespected and demeaned.

This is not far-fetched. Suppose, by analogy, the only mutually intelligible term for a dark-skinned individual in our language is "Africrim." The only way for the police to communicate with each other is to say: "Look for an Africrim in red sneakers." The stopping officer trying to be respectful can only say, "I'm terribly sorry for disturbing you, sir, and once I determine that you are not the red-sneakered Africrim we are looking for, but some other red-sneakered Africrim, you may be on your way, with the gratitude of your fellow citizens." The young man cannot but take the address as an insult, because the very language the police officer uses is insulting. Onlookers snicker and think, "There goes another Africrim, predisposed to criminality," thereby perpetuating the degrading social meaning of the use of race-based statistical inference.

If our social meanings really were like that, that would give police a very good reason to banish skin color from their identification and patrol strategies. The government indeed must treat all those under its authority as moral equals. Treatment as a moral equal does not always require equal treatment, because there are morally relevant differences that justify unequal treatment: To treat us as moral equals, a government hospital does not treat every citizen to an equal number of surgeries, whether one needs surgery or not. But if the use of racial generalizations inherently expresses unequal status in the eyes of the government, to avoid expressing unequal status the police must exclude race as a difference they take into account in their identification and patrol strategies.

Let us see what such a conclusion entails by considering a patrol case without racial generalizations. On I-95, the drug trafficking unit used a more refined search strategy with white speeders than with black speeders, so, though a black speeder was fifteen times more likely to be stopped, the troopers achieved the same true positive rate of 30 percent in both groups. What were the strategies? We are not told, but perhaps the search strategy for black speeders was "Stop young black male drivers in expensive cars," and the search strategy for white speeders was "Stop young white male drivers wearing cool sunglasses and gold chains in expensive cars." If treatment as moral equals requires excluding racial information, then the police must use the same search strategy for black and white drivers. Are they to employ the less refined strategy for all, and stop everyone at the same high rate they had been stopping blacks? Then the number of total stops will soar more than ten-fold (fifteen-fold on the three-quarters of drivers who are white) and the true positive rate – the fraction of stopped drivers who are drug violators – will plummet.[6] If race is excluded, the burden imposed on all drivers may be unreasonable, and the entire police operation may no longer satisfy a standard of probable cause. To be clear, excluding race is not simply a cost to efficiency. Because treatment as a moral equal sometimes requires unequal treatment, sometimes imposing the same treatment when differentiation is called for imposes unjust burdens. If instead, the police employ the more refined search strategy for all, and stop everyone at the same low rate they had been stopping whites, the number of drug offenders apprehended will plummet, and the I-95 operation may no longer be an instrumentally rational use of police resources.[7]

Perhaps a third search strategy somewhere in between – cool sunglasses *or* gold chains – would hit a sweet spot. Or perhaps the exclusion of racial search criteria would prompt the innovative discovery of other predictive criteria (and not simply lead to cheating by using facially color-blind proxies for race that also demean). But perhaps not. If the accurate race-based statistical generalization is sufficiently predictive, an otherwise important police practice may be rendered impossible because there is no race-neutral identification or patrol strategy that reasonably enforces the law at reasonable burdens. Since, in police work, accurate race-based statistical generalizations indeed are highly predictive, their exclusion is an unstable solution to the problem of pejorative social meanings.

Our intendedly respectful police officer has a better solution: to change what race expresses. Alas, because the meanings of words and social practices are shared, no one person can change them. If, in our fanciful analogy, the respectful police officer invented a substitute term for "Africrim," no one would understand her. But the meanings of words and the meanings expressed by social practices do change over time. No, she cannot make the change alone, but in ways that are complicated and sometimes mysterious, we change social meanings together. If one wants a hopeful example, consider recent changes in the social meaning of homosexuality. Yes, pejorative social meanings are entrenched, because we have a shameful history of racism. But our use of statistical generalization also is entrenched, because we cannot navigate the natural and social world without making inferences and predictions. Fortunately, our social meanings are less fixed than our need for instrumental rationality.

I have a brighter view of the present than Hellman: I agreed earlier that her account of what race means in our society *could* be correct, but I do not agree that it *is* correct. I think that what race expresses has already changed considerably in a generation or two, though we still have far to go. In my judgment, the efforts of reformers are more fruitfully spent trying to change what race expresses than trying to stop the use of justified, true beliefs about what race predicts. This, to be clear, is not based on any deep moral disagreement with Hellman, but on a different reading of the social facts.

Thus far, every introduction of the non-ideal conditions of the real world has made the case for racial generalizations tougher than would be the case under more favorable conditions. There is one exception to that pattern that I cannot analyze here, but it bears mentioning: one mean circumstance of the non-ideal world is that African

Americans disproportionately are the victims of crime. Though the burdens of race-based search strategies fall most heavily on law-abiding blacks, the benefits may as well. Still, no one should be forced to trade respect for safety. It is an injustice, and not simply bad luck, that many black Americans face such a bleak choice.

Notes

1 This chapter draws heavily on Applbaum (1996). Reproduced with permission of Rowman & Littlefield.
2 $(.73/.18) * (.75/.20) = 15.2$, hence blacks were about 15 times more likely to be stopped.
3 "Report of John Lamberth, Ph.D." in *Wilkins v. Maryland State Police*, 1996. Figures for other races are excluded.
4 If $p(O|S)$ is the probability of being an Ozian conditional on being a smuggler, $p(S)$ is the probability of being a smuggler, $p(O)$ is the probability of being an Ozian, and $p(S|O)$ is the probability of being a smuggler conditional on being an Ozian, then, applying Bayes's Rule: $p(O|S) * p(S) / p(O) = p(S|O)$. In the numerical example: $.5 * .0001 /.001 = .05$.
5 In actuality, descriptions are often inaccurate or accurate only by accident. Witnesses may, willfully or unwittingly, substitute their own inferences about the proportion of violators who have a characteristic, instead of describing the characteristics of a particular suspect.
6 Without knowing the true base rates of drug violators among black and white speeders, and without knowing how much the distribution of and correlation between the search criteria differ among blacks and whites, we cannot know by how much, but if we assume that the police were previously employing narrowly instrumental rational strategies, the yield on a vast increase in stops would be meager.
7 Subject to the caveats in the previous note.

References

Applbaum, A. (1996) Response: racial generalization, police discretion, and Bayesian contractualism. In *Handled with Discretion*, ed. J. Kleinig, pp. 145–158. Lanham, MD: Rowman & Littlefield.

Hellman, D. (2014) Racial profiling and the meaning of racial categories. In *Contemporary Debates in Applied Ethics*, 2nd edn, ed. A.I. Cohen and C.H. Wellman, pp. 232–244. Malden, MA: Wiley-Blackwell.

Wasserman, D. (1996) Racial generalizations and police discretion. In *Handled with Discretion*, ed. J. Kleinig. Lanham, MD: Rowman & Littlefield.

Wilkins v. Maryland State Police. (1996) Civil Action No. CCB-93–483.

Further Reading

Hill, T.E. Jr (1991) The Message of affirmative action. In *Autonomy and self respect*, pp. 189–211. Cambridge: Cambridge University Press.

Raiffa, H. (1968) *Decision Analysis*. Reading, MA: Addison-Wesley.

Scanlon, T.M. (1998) *What We Owe to Each Other*. Cambridge, MA: Harvard University Press.

Searle, J.R. (1995) *The Construction of Social Reality*. New York: The Free Press.

CHAPTER FIFTEEN

Racial Profiling and the Meaning of Racial Categories

Deborah Hellman

Introduction

What is "racial profiling" and when, if ever, is it morally wrong? I put the term in quotation marks because racial profiling can describe many different sorts of practices, including those that use race, at least in part, to determine whom to search closely at airports, highways and in high crime neighborhoods, as well as more innocuous practices such as whom advertisers should target with what sorts of advertising. If profiling is the practice of using some traits about a person to predict other traits, then the use of race by police and advertisers are both instances of racial profiling. At its root, profiling relies on statistical generalization. It is because internet users of particular races, for example, are more likely than internet users of other races to buy certain products that the sellers of those products direct advertising at these users. Some profiling seems morally troubling while other profiling does not. The aim of this chapter is to begin to unpack which is which and why.

We routinely rely on statistical generalizations about people. "Should I travel with my baby to Italy?" my friend asks. "Italians love babies," I reply. This sort of generalization, also known more pejoratively as a stereotype, underlies a profile. Profiling, understood in this way, is common and often morally permissible. Other times, it is quite the opposite. Here are some examples of profiling – some innocuous, some morally troubling, some about which people are likely to disagree:

- A police practice of watching black teens with low-slung jeans more closely than other people on the street in a high crime district
- A college admissions policy that selects students, at least in part, on the basis of SAT scores
- A government policy of searching air travelers with brown skin and Arab ethnic identities more closely than other travelers

Contemporary Debates in Applied Ethics, Second Edition. Edited by Andrew I. Cohen and Christopher Heath Wellman.
© 2014 John Wiley & Sons, Inc. Published 2014 by John Wiley & Sons, Inc.

- An advertiser's decision to target *New York Times* readers with advertising for expensive vacations.

Each of these policies, practices or decisions is a form of profiling in that each uses one trait about a person to predict other traits and relies on a statistical generalization about people with the first trait. The generalizations that underlie the practices above include the following: black teens with low-slung jeans are more likely to be criminals than the average person in the neighborhood; high school students with high SAT scores are more likely to be successful in college than students with lower scores; brown-skinned people of Arab ethnicity are more likely to be terrorists than other travelers; *New York Times* readers are more likely to spend money on vacations than the average person. When we talk about profiling, we generally have in mind only some of these sorts of examples – the first and the third from the list above. In order to understand whether police and others act wrongly in using race or ethnicity as part of a profile, we need a theory of profiling that explains why our intuitions that some profiling, like the admissions policy and advertising practice, seems morally innocuous while other profiling seems troubling. When an African-American is stopped on the highway for "driving while black," this word-play on the crime of driving while intoxicated is a clear criticism. An account of profiling should explain this difference and help us to evaluate whether these intuitions are indeed correct.

Statistical Generalization

One might be tempted to think that the problem of profiling lies in a profile's reliance on a statistical generalization. There are several forms this hypothesis might take. One might be worried that the generalization that a profile relies on is not accurate – perhaps black teens with low-slung jeans are not more likely than other people in the neighborhood to be criminals. Alternatively, one might be worried that even if the generalization is accurate, it still mischaracterizes some individuals and therefore police action based on this generalization is unjust. Not all black teens with low-slung jeans are criminals, after all. Alternatively, one might be worried about the content of the generalizations themselves. Below I consider each of these alternatives.

Accuracy

A generalization is accurate *as a generalization* even if not true of everyone. The policy of following black teens with low-slung jeans (let us use the abbreviation BTJ) more closely than others in the neighborhood is based on the generalization that BTJs are more likely to be engaged in crime than others in the neighborhood. So long as BTJs are more likely than non-BTJs to be criminals, the generalization is true. The reasons that racial profiling is an important issue to grapple with is precisely because sometimes the racial generalizations profiles rely on are in fact accurate (Monahan, 2006, pp. 417–418).

Moreover, even if the generalization is inaccurate, it is hard to see how *this* is the heart of the moral problem of racial profiling. Suppose the police use an inaccurate non-racial profile. The police closely follow women with glasses more than other people in the neighborhood based on the incorrect view that women with glasses are more likely to be criminals than others. This policy seems stupid and a waste of important police resources but not a moral wrong, much less a serious moral wrong. The intuition that racial profiling is wrong surely rests on something more serious than a concern with wasting government resources.

Generalization

Profiling consists in making decisions about people on the basis of statistical generalizations. Even where profiling relies on accurate generalizations, perhaps what makes it problematic, when it is problematic, relates to the use of a generalization to make a decision about an individual. This thought is often captured by the observation that profiling fails to treat people as individuals. It is true that in some contexts we resist making decisions about individuals based on statistical evidence alone. In evidence law, a famous example is used to exemplify this reluctance.

> While driving late at night on a dark, two-lane road, a person confronts an oncoming bus speeding down the center line of the road in the opposite direction. In the glare of the headlights, the person sees that the vehicle is a bus, but he cannot otherwise identify it. He swerves to avoid a collision, and his car hits a tree. The bus speeds past without stopping. The injured person later sues the Blue Bus Company. He proves, in addition to the facts stated above, that the Blue Bus Company owns and operates 80 percent of the buses that run on the road where the accident occurred. Can he win? (Nesson, 1985, p. 22)

While a civil finding of liability only requires that a plaintiff prove that the defendant is liable by a preponderance of the evidence – usually interpreted as proof that it is more likely than not, or 51 percent likely, that the plaintiff is responsible – still, without more, resting a decision on this statistical evidence alone is unlikely. While this conclusion is fairly entrenched in legal doctrine, it is nonetheless controversial. Some commentators point out that seemingly individualized evidence is also based on statistical generalizations. Consider eyewitness identification. A witness says he saw a bus driven by a man with a particular description leave the accident. Later the witness identifies the defendant. The view that we appropriately base a legal judgment on the witness identification is based on a statistical claim about the reliability of eyewitness identifications. Whether there is a meaningful difference between so-called statistical evidence and non-statistical evidence is an important controversy that I will wade into no further, however, as there is something more basic wrong with the supposition that the moral problem of profiling inheres in its reliance on statistical generalization.

We began this inquiry by noting that some profiling seems morally innocuous while other profiling seems morally troubling, even deeply so. In setting out to articulate a theory of when, if ever, profiling is wrong and why, I noted that any plausible account must be able to explain this fact. But seemingly innocuous profiling and seemingly problematic profiling *both* rely on statistical generalization. So, unless one thinks that racial profiling by the police is not morally different from an advertiser's use of a profile to determine whom to target with what advertising (sending promotions for expensive vacations to people who read the *New York Times* online, for example), the moral problem with profiling cannot inhere in the fact of statistical generalization by itself.

Content of the generalization

In my view, profiling can be morally problematic because of the content of the generalization relied upon. Some profiling relies on generalizations that express a complimentary or innocuous meaning and is therefore not morally troubling. An advertiser's decision to target *New York Times* readers with ads for expensive vacations says nothing insulting about either these users or others. A parent's decision to favor Italian would-be babysitters over others on the grounds that Italians love babies expresses neither disparagement of Italians nor of any other nationality, as the remainder group is too undefined to be slighted. But racial profiling by the police is different. In the next section, I explain how this practice demeans blacks and thus constitutes an important moral wrong.

What Profiling Expresses

When we talk colloquially about profiling – and use that term with its normal pejorative connotations – we are not usually thinking about advertising practices. Rather, the sort of case we generally have in mind is racial profiling by the police in the context of identifying or apprehending criminals. This paradigmatic instance of profiling is characterized by three features: it is a racial profile, the profile is used to prevent crime or apprehend criminals, and the actor doing the profiling is an agent of the state. Each of these features might make a moral difference.

Generalizing about race

Race is a broad category that includes all racial groups. But, when we talk about *racial profiling*, we generally have in mind reliance on a generalization about African-Americans or another non-majority racial group. This sort of profile relies on the generalization that blacks are more likely than non-blacks to have a particular trait (Y). As a result, targeting blacks for search or scrutiny will be more efficient in finding people with Y than would a policy that did not target blacks.

One may first be tempted to say that the generalization merely reflects a prejudice and is not accurate. While sometimes profiles rely on inaccurate generalizations about blacks, we have put this possibility aside and are considering only the moral permissibility of accurate profiles. Still, one may have a related worry. While the generalization on which the profile relies may be accurate, perhaps there are other equally good or more efficient proxies that could be used to identify people with Y, but which are missed because race is a salient category to us. It may be that clothing choice better predicts who will shoplift from the convenience store than does race but the shopkeeper follows black kids rather than kids wearing particular clothing because he notices the former correlation but not the latter. This is an important worry, as the salience of race as a category is likely to make correlations between racial group membership and other traits especially memorable. As a result, blacks are more likely to be the subjects of profiling and to the extent this is a burden to them, they are harmed.

But is it a wrong? In some sense, the use of a less efficient but accurate profile instead of a more efficient accurate profile is like the use of an inaccurate profile. It is stupid and wasteful to use a search strategy that is based on a false generalization and it is stupid and wasteful (albeit less so) to use a search strategy that does not work as well as another available better strategy. But, as I argued earlier about inaccurate profiles, the moral reaction that profiling engenders is surely not about lack of efficiency.

Still, blacks may be repeatedly unlucky in just this way. Maybe the accumulated burden makes this problem more serious. While I am sympathetic to this concern, I still do not think it taps the root of our unease about racial profiling. This is a difficult intuition to test as what we need in order to do so is a trait that is routinely used as a proxy for other traits such that it disproportionately burdens the group with the trait *but* which lacks the social significance and history of mistreatment we find with race. In order to isolate whether it is the burden itself that makes the moral difference, consider the following, perhaps fanciful, example. Let us suppose that people dispensing benefits of various sorts often work their way through possible candidates alphabetically. I have heard that law schools looking to fill curricular needs with visiting professors look alphabetically through the directory of law teachers until they find a plausible candidate. If this is so, then people whose last names begin with letters late in the alphabet are less likely to be offered this and other benefits than are people with last names beginning with letters near the front of the alphabet. Now of course this practice is not an instance of profiling. The person hiring visiting faculty does not base her decision to hire Brown on the supposition that people with names beginning early in the alphabet are likely to be better teachers than those, such as Waters, whose names begin in the latter part. Still, Waters and others like him are likely to be repeatedly harmed if this sort of practice is widespread. If this practice has social benefits, we might ask whether it is fair that Waters, and others at the end of the alphabet, bear this burden in order to generate a benefit for all. The answer to this question will depend in part on how burdensome the burden is and how beneficial and widely shared the benefit is. If the burden is not too great and the benefit significant and widely shared, it certainly seems reasonable to suppose that the burden Waters bears is reasonable.

Waters is surely unlucky that he has the trait that leads to this burden, but if the conditions above hold, the fact that he does not get various opportunities is bad luck but is not unjust. Critics of racial profiling seem to be making a stronger claim about it. They claim that it wrongs blacks; it treats them unjustly. What else might explain this intuition?

I contend that what makes racial profiling different, morally, from more generic and common forms of profiling is what it expresses. Profiling's reliance on a generalization carries a meaning. The practice expresses something about the group profiled. The practice of targeting advertising for expensive vacations to *New York Times* readers expresses the view that these readers are more likely to go on expensive vacations than non-*New York Times* readers. The practice of following black teens in rough neighborhoods expresses the view that black teens are more likely than non-black teens to commit crimes. Part of what is expressed merely recapitulates the content of the generalization itself and thus need not be insulting or demeaning. After all, if it is true that black teens are more likely than non-black teens to commit crimes, the innocent black teen need not feel impugned. But the meaning of profiling is not exhausted by the content of the generalization that underlies it.

236　　**Deborah Hellman**

Racial profiling differs from profiling on the basis of many other traits because *race* as a category has a social significance steeped in meaning that is absent, or significantly less, in the case of other traits. And this meaning is largely negative. The category "*New York Times* readers" has some additional expressive meaning but not nearly as much. We have some sense of *who those people are*. But the sense we have is amorphous, only lightly etched in our cultural understandings. Moreover, the stereotype, such as it is, is both positive and negative. To some, it signals a member of the liberal elite. To others, it connotes an educated, open-minded person. Or maybe I have these descriptions all wrong and the reader envisions something completely different or maybe no image comes to mind. This is because the category "*New York Times* reader" is not a social category with much resonance.

Contrast this with the racial category, black. The racial category calls forth a history. This history of oppression against blacks and of deep racism infuses our understanding of who "blacks" are and what being "black" entails. As the contrast between these two examples illustrates, some traits that might be used as part of profiles have little social significance (*Times* readers) or no social history at all, in which case no meaning is attached to their use. People with last names starting with letters early in the alphabet is a category of this latter sort. It has, for us, no significance. But the category "black" is dramatically different. There are many cultural understandings about what it means to be "black" that are evoked by the use of this trait.

While a racial category is not exactly a symbol, it functions in a similar way. When the Supreme Court held that the action of burning a flag was protected speech under the First Amendment (*Texas v. Johnson*, 1989), it did so because, in the Court's view, "[p]regnant with expressive content, the flag as readily signifies this Nation as does the combination of letters found in 'America'" (*Texas v. Johnson*, 1989). Burning the flag thus becomes a way of saying that America is bad, or wrong, or acting poorly. In a similar sort of way, use of a racial category has meaning. While it may be somewhat less clear what "black" means in our current social context than what the flag means, both the category and the symbol are stable enough to be meaningful. Because the category "black" does not just refer to the group of people socially considered to be black in our country but also carries with it cultural understandings of what it means to be black in America, a statistical generalization about blacks expresses more than merely the claim asserted by the generalization itself.

For example, a policy of watching black teens in convenience stores more closely than others rests on the generalization that black teens are more likely to shoplift than are non-black teens. This practice expresses both that black teens are more likely to shoplift than are non-black teens *and* something else. Because its content tracks familiar cultural stereotypes about blacks and crime, the practice also expresses something more troubling such as "blacks are naturally disposed to be criminals." While a statistical generalization about black teens and crime itself does not entail a statement about causation, the cultural baggage of the category "black" brings this meaning forward as well.

The fact that this profile expresses both that blacks are more likely than non-blacks to shoplift and that blacks are prone to crime is problematic for two reasons. First, profiling therefore has the effect of reinforcing the very racist views about blacks that color the interpretation of the statistical generalization. Second, and equally important in my view, this racial profiling also insults and possibly demeans blacks.

Racial Profiling and the Meaning of Racial Categories

Profiling and crime

The term "profiling" is most commonly used in contexts where race is used, at least in part, in order to identify criminals. Does the association of race with crime give these profiles an especially insulting or demeaning message? In order for the profile to grab on to cultural conceptions of what blacks are like, the generalization must track stereotypes already present in our culture. The black as criminal is one of these and thus using a racial profile to identify crime is likely to have the sort of symbolic meaning discussed above. But the association of blacks with crime is not the only cultural stereotype available, nor is it the only negative one. Suppose a high school used race as part of its "struggling student profile," in order to determine to whom it should provide extra help. While the students identified would benefit from the designation, use of race, especially African-American, as part of this profile would be morally troubling precisely because of cultural stereotypes about blacks as less smart. Compare these two examples with a third, in which the cultural understandings that form the background through which we understand the meaning of the profile is far thinner.

In 2005, the Federal Drug Administration (FDA) approved a drug specifically for use in African-American heart failure patients. The data on which the agency relied supported the conclusion that the drug was likely to be helpful to these patients and did not support a similar conclusion for non-African-American heart failure patients.[1] A doctor determining whom to give the drug to (rather than whom to follow or search) was thus advised by the FDA action to use race as an indicator – a profile, if you will – of who would likely benefit from the therapy. Under current law, doctors can prescribe any FDA approved drug for any patient, but the company making the drug can only promote or advertise its use by patients for whom the FDA approved its use. What do the FDA policy and the doctor's practice of using race, in part, to determine who should take the drug express?

The message expressed by the FDA policy is surely quite different than a racial profile aimed at detecting or apprehending criminals. While some critics of the FDA objected to the approval of a drug specifically for African-Americans on the grounds that it overemphasized the biological over social dimensions of racial differences in receptivity to different drugs, the difference in what is expressed by a policy of racial profiling, as compared to the FDA approval, explains and justifies the moral intuition that racial profiling is of serious moral concern while the FDA policy is not.

This account locates the wrong of racial profiling, if it is wrong, in what is expressed – the insulting or demeaning message that blacks are prone to crime – rather than in the inconvenience associated with frequent stops or being the subject of repeated surveillance. In doing so, I do not mean to minimize the hassle that these are likely to cause. Rather, I want to emphasize that what makes them both wrongs and also especially harmful – rather than simply annoying – is that they carry a pejorative meaning and thus are occasions of disrespect.

This account of what makes profiling wrong explains why the police use of race in a suspect description is generally not seen as morally troubling. If the victim of a crime identifies her attacker as a "black male" of a particular age, weight and height, for example, and the police search for a person fitting this description, the police practice expresses nothing troubling. The use of race in this context relies on a generalization about the likelihood that someone who fits the description is the person who committed the crime and thus does not insult or demean blacks as a group. The generalization does not rely on a fact about blacks as a group. Rather it relies on a generalization about the reliability of eyewitness identification. When we see that it is the content of the generalization itself that matters, because it affects what is expressed, we can easily explain why the use of race in suspect descriptions (what Applbaum in the companion piece to this chapter calls "group-based identification") differs so significantly from racial profiling.

Profiling by the Government

In the previous section, we discovered that when race is used in a profile aimed at preventing crime or apprehending criminals, its meaning is more negative and potentially insulting than when it is used in a racial profile outside the criminal context. In this section, I argue that it also makes a difference *who* employs a racial profile and, in particular, that if the government or other person or entity with significant power relies on a racial profile, it is especially morally troubling.

In order to see the difference that a powerful actor makes, compare the following two cases. First, suppose a pedestrian crosses the street as a young black man approaches. Second, suppose police officers use race, at least in part, to determine whom to watch closely in a high crime neighborhood. In both cases, race is used as part of a profile aimed at identifying a person likely to commit a crime. In the first example, the pedestrian's action is based on the statistical generalization that young black men are more likely to rob him than are other people he might pass. In the second example, the police policy is based on the statistical generalization that blacks in the neighborhood are more likely to commit crimes than are non-blacks. Both profiles use race and are aimed at identifying criminals.

In my view, the governmental action is worse, morally speaking. The governmental use of a profile differs from the private use of a profile because of the authority with which the state, and especially the police, speaks and the power it wields. The expressive content of an action or statement can vary widely depending on who acts or speaks. Consider the following comparison. Suppose a five-year-old child "orders" his parent to get him some milk. "Get me some milk," he says. Now compare this to the statement by the parent's boss to get her a report that is due. "Get me the report," the boss says. The boss has ordered the parent to get the report, but the child likely has not ordered the parent to get the milk. He has surely attempted to order his parent, but to succeed in ordering, the speaker has to have some power. I do not mean that the boss is more likely to actually manage to alter the parent's behavior than is the child; I mean rather that the child's statement just is not really an order, though it purports to be.

In other words, who a speaker or actor is changes the meaning of what is expressed. If this is right, and it seems fairly uncontroversial, then it is likely that racial profiling by the government expresses something different than racial profiling by a private person. The government actor, in most instances, will have more power and thus greater potential to subordinate or demean the person profiled than would a private citizen. While the pedestrian may well insult the nicely dressed black man whom he passes, the police officer demeans the black residents he follows. To degrade or subordinate requires that the speaker or actor have some power. Thus actions by authorities are more likely to demean or subordinate than actions by ordinary people, though sometimes private individuals have the power to do so as well.

Moreover, the government has obligations towards those over whom it exercises this power that are different from, and more demanding than, the obligations that each of us owe to other people. Each person has a moral obligation to treat another as a person with rights, interests and an independent life to lead. For this reason, the pedestrian should count the fact that the black man will be insulted as a reason not to cross the street. However, the government has additional obligations to those whom it governs. It must treat each of us in a way that not only recognizes our status as human beings – as a private person must do – but also in a way that recognizes that we are each members of equal standing. In other words, the state must treat us *as equals*. For example, it is fine for a parent to act in ways that show she loves *her* children more than other people's children, so long as she considers their interests and respects their rights. The state, however, must do more. It must act in ways that demonstrate that we matter equally to the government. When the state profiles in a way that insults or demeans those profiled, this action not only harms or wrongs the people affected, it also violates the state's obligation to treat its members *as equals*.

Racial profiling by the government is thus especially morally troubling for two reasons. First, the state has power and thus has the ability to demean those profiled, rather than merely to insult them, because of the force its expressions convey. Second, the state violates the additional moral obligation to treat those over whom it exercises this power as equals. Where government profiles convey unequal regard, they violate this additional moral obligation.

Expression and Interpretation

In the prior sections, I have argued that it is what profiling expresses that makes it morally wrong. But, one might wonder, how do we know what a particular instance of racial profiling expresses? When I say that racial profiling by the police is demeaning, I am making an interpretive claim about the best way to understand the meaning conveyed by that practice in our culture at this time. This judgment is based both on the meaning of the statistical generalization that underlies the profile and on the cultural understanding of racial classifications, especially related to crime. It also derives from the special oomph that action by the state or a state official conveys. In making this claim, I am offering an interpretive judgment about the best way to understand what racial profiling expresses. Someone else might disagree. But in order to disagree, one must get into the details; it will be a disagreement about what racial classifications signify in a particular society (such as twenty-first century America), about whether associations between race and crime are likely to convey a causal suggestion – blacks are prone to crime – as I argued. The person who disagrees must also engage in an interpretive judgment and say where I have gone wrong and what, in her view, is the better interpretation.

The fact that racial profiling relies on an interpretive judgment about the meaning of the practice does not mean that it is "just my opinion." The meaning of racial profiling is contingent in the sense that it depends on the social understanding of the meaning of racial categories in our culture. Racial categories have no meaning that is independent of human beings and the particular culture in which they are used. On the other hand, the meaning of racial profiling is objective in that I cannot simply say, "that's how it seems to me," and thereby make it the case that use of a racial category carries a particular meaning. John Searle uses the example of money to illustrate the way in which what he terms "social facts" are both dependent on human beings and thus subjective in one sense but independent of the beliefs or attitudes of particular human beings and thus objective in another sense. Pieces of paper are money only because people in a given culture believe them to be money and treat them as money. In that sense, the value of money is subjective, in that it depends on the actions and beliefs of human beings. On the other hand, the value of money is objective in that one person cannot change the value of money by his or her actions or beliefs alone. In that sense, Searle's belief that the paper in his pocket is money is objectively true (Searle, 1995, p. 32). Similarly, racial profiling expresses a particular meaning. The fact that it does so depends on the social meaning of race in our culture and is thus dependent on our beliefs and practices. At the same time, it does not depend on the beliefs and attitudes of the particular actors employing the profile. In that sense, the meaning of racial profiling is objective.

Racial profiling by the police demeans blacks because the profile relies on race, links it with crime, and because it is the government that is profiling. Where profiles use traits other than race, they may well convey something quite different, or convey nothing at all beyond the statistical generalization they rely on. The advertiser's profile of *New York Times* readers is a good example. As a result, this profile is not morally troubling. Where racial profiles use race to target something other than criminals and especially when it is used to target innocuous traits, the profiles does not denigrate either. The FDA approval of the heart failure drug BiDil provides a good example here. Finally, where private actors profile, there is less opportunity to denigrate, because most private actors speak with a weaker voice than does the government, and thus these profiles are usually less problematic. A pedestrian's use of a racial profile to judge possible attackers, while troubling, raises fewer moral concerns than use of the same profile would by the police.

Racial Profiling without Racism

Some writers evaluating the moral permissibility of profiling abstract away from the actual history of racism in our society. For example, Arthur Applbaum "brackets" the argument that "the persistence of racism and its consequences in America is such a grave injustice and social problem that even the use of accurate and unbiased racial generalizations is repugnant and harmful" (Chapter 14 in this volume, Applbaum, 2014, pp. 219–231). Thus, in order to explore the permissibility of *race-based* statistical generalization by police, he substitutes statistical generalizations about hypothetical groups (Ozians) about whom there is no social meaning or history. However, by bracketing racism, he denudes race of its actual meaning and significance in our culture so that his inquiry becomes an exploration of the significance of statistical generalization simpliciter. Race as a category is different from other categories one might generalize about because of its social meaning.

To be fair, Applbaum's question – whether race-based generalization is problematic just for being race-based – finds support in our legal doctrine about the use of racial categories. The Supreme Court treats governmental use of racial categories as inherently suspect, thus suggesting that racial categories are different in themselves. It is this suggestion that Applbaum is exploring and ultimately rejecting. However, without racism, racial categories become just like other categories and the exploration of racial profiling becomes an inquiry into the permissibility of decision-making on the basis of statistical generalization itself.

The permissibility of making policy on the basis of statistical generalization is an important question to be sure, but, as I argue above, it is not fundamentally a question of the moral permissibility of racial profiling. Decision-making on the basis of statistical generalization occurs in many contexts beyond those in which the police employ generalizations in order to identify and apprehend criminals. Some of these contexts are morally innocuous – an advertiser's decision about whom to target with what advertising, for example – while others are not. One might, for good reason, limit one's inquiry to the moral permissibility of statistical generalization by the police. But to explore this question is not the same as to explore the moral permissibility of racial profiling. The first question would focus on whether the police may initiate searches based on the "profile" that travelers who, for instance, check no baggage and travel between New York and Miami, making short trips, are more likely than the average traveler to be transporting drugs. This is, fundamentally, a question of the permissibility of police use of statistical generalization.

While racial profiling may be similar in some respects to this sort of practice, we ought not to assume it is at the start. In fact, we might begin by noting that our moral intuitions about this case and racial profiling are quite different. Why is that? One might say that in considering racial profiling without racism, one is able to isolate what part, if any, the fact that profiling relies on statistical generalization plays in the moral permissibility of profiling. But as we have seen, not all accurate statistical generalizations should be viewed the same. The problem of racial profiling does not lie in *the fact* of generalization itself. But the implication of Applbaum's argument is that the problem does not really have to do with generalization at all. This is not right. Racial profiling is morally problematic not because it relies on generalizations but because of the type of generalizations it relies on. We are able to see that the content of the generalization matters only when we keep the real facts about our society and its history of racism firmly in mind.

Mathias Risse and Richard Zeckhauser also use a bracketing approach (Risse and Zeckhauser, 2004, p. 146). They try to isolate the harm that racial profiling causes or the wrong it is from the background harm caused or wrong done by the underlying racism of society. They then go on to claim that it is society's underlying racism that is the moral problem and that the incremental difference added by racial profiling is so small as to be easily outweighed by the benefit in security it provides. In their view, the problem with racial profiling is merely "expressive," by which they mean that it works to remind blacks of racism and thereby to cause them harm. What they fail to consider is that the history of racism in society and especially the powerful images of linking blacks with crime work to change the nature of what profiling expresses such that it does not simply remind blacks of racism but itself denigrates blacks by expressing that blacks are prone to crime.

242   **Deborah Hellman**

Conclusion

My account of profiling locates its moral wrong in what it expresses. As a result, my account is able to explain why some instances of profiling are morally wrong while other instances of profiling are not. All profiles rely on statistical generalizations but the fact of generalization is not what makes profiling morally problematic, when it is. Rather, it is the content of the specific generalization that creates the problem. When a generalization tracks negative cultural stereotypes, profiling expresses something more than the claim contained in the generalization itself. When it does so, the profile may insult or demean those affected.

African-Americans are disproportionately the victims of crime, as well as being over-represented among the perpetrators. But these are not the only values on the table. We would increase security without compromising respect if each of us would accept more invasive and more frequent intrusions into our privacy. So while we ought to work to change the social meaning of racial categories, in the short run the value we should short-change is not security or equality but privacy. This approach treats each member of the community in a fair manner and has pragmatic virtues as well. By enacting a policy that errs on the side of treating everyone the same, we help to diminish the significance of racial categories and the deep associations of race and crime.

Note

1   The study on which the FDA relied was somewhat controversial. For purposes of this chapter, I will ignore that controversy.

References

Applbaum, A.I. (2014) Baysian inference and contractualist jusitification on Interstate 95. In *Contemporary Debates in Applied Ethics*, 2nd edn, ed. A.I. Cohen and C.H. Wellman, pp. 219–231. Malden, MA: Wiley-Blackwell.

Monahan, J. (2006) A jurisprudence of risk assessment: forecasting harm among prisoners, predators, and patients. *Virginia Law Review* 92: 291–435.

Nesson, C. (1985) The evidence or the event? On judicial proof and the acceptability of Vvrdicts. *Harvard Law Review* 98: 1357.

Risse, M. and Zeckhauser, R. (2004) Racial profiling. *Philosophy & Public Affairs* 32: 131–170.

Searle, J.R. (1995) *The Construction of Social Reality*. New York: The Free Press.

Further Reading

Schauer, F. (2003) *Profiles, Probabilities and Stereotypes*. Cambridge, MA: Harvard University Press.

Torture

CHAPTER SIXTEEN

Ticking Time-Bombs and Torture[1]

Fritz Allhoff

In thinking about the moral status of interrogational torture, we are commonly asked to imagine exceptional cases wherein such torture is necessary to save some significant number of lives. These cases are generally referred to as ticking time-bomb cases and invite us to think of the relationship they bear to terrorism: some terrorist has planted a bomb in a crowded metropolitan center that will kill many noncombatants unless the terrorist is tortured. But terrorism need not have anything to do with these cases nor, really, do bombs. Rather, what matters is that there is some threat to many people that can only be avoided – and which, in most formulations, certainly will be – through the torture of someone already in custody. The "ticking time-bomb" locution is therefore somewhat narrow, but not in any drastically misleading way. Furthermore, most of the contexts worth considering – by which I mean the real-world ones most closely approximating these hypothetical constructs – probably will be those involving terrorists and weapons, if not necessarily bombs. And, for purposes of engagement, there is merit in following the standard usage.

Origins of Ticking Time-Bomb Cases

Let us start by considering the origins of ticking time-bomb cases; in doing so, we will also start to understand the methodology. In terms of the philosophical literature, an early formulation owes to a seminal essay by Henry Shue:

> [S]uppose a fanatic, perfectly willing to die rather than collaborate in the thwarting of his own scheme, has set a hidden nuclear device to explode in the heart of Paris. There is no time to evacuate the innocent people or even the movable art treasures – the only hope of preventing tragedy is to torture the perpetrator, find the device, and deactivate it. (Shue, 1978, p. 141)

Contemporary Debates in Applied Ethics, Second Edition. Edited by Andrew I. Cohen and Christopher Heath Wellman.
© 2014 John Wiley & Sons, Inc. Published 2014 by John Wiley & Sons, Inc.

But not only have these cases appeared in academic journals, they have also crossed over to popular media outlets and, thereafter, into public consciousness. For example, consider the following, which comes from an essay Michael Levin wrote in *Newsweek*:

> Suppose a terrorist has hidden an atomic bomb on Manhattan Island which will detonate at noon . . . Suppose, further, that he is caught at 10 a.m . . . but preferring death to failure, won't disclose where the bomb is. What do we do? If we follow due process, wait for his lawyer, arraign him, millions of people will die. If the only way to save those lives is to subject the terrorist to the most excruciating possible pain, what grounds can there be for not doing so? I suggest that there are none. (Levin, 1982)

Long before these formulations of ticking time-bomb cases in the second half of the twentieth century, Jeremy Bentham wrote on the morality of torture almost two hundred years earlier, dating from the late 1770s (Bentham, 1973, pp. 56–62, 63–70).[2] Bentham probably had the first formulation of a case that looked anything like a ticking time-bomb case, though this came later – in 1804 – and was not otherwise attached to a systematic treatment of torture. Consider what he wrote:

> Suppose an occasion, to arise, in which a suspicion is entertained, as strong as that which would be received as a sufficient ground for arrest and commitment as for felony – a suspicion that at this very time a considerable number of individuals are actually suffering, by illegal violence inflictions equal in intensity to those which if inflicted by the hand of justice, would universally be spoken of under the name of torture. For the purpose of rescuing from torture these hundred innocents, should any scruple be made of applying equal or superior torture, to extract the requisite information from the mouth of one criminal, who having it in his power to make known the place where at this time the enormity was practicing or about to be practiced, should refuse to do so? To say nothing of wisdom, could any pretense be made so much as to the praise of blind and vulgar humanity, by the man who to save one criminal, should determine to abandon [one hundred] innocent persons to the same fate? (Twining and Twining, 1973, p. 347)

The principal difference between Bentham's case and the others previously presented is simply whether the harm that the torture aims to dispel is already active (*viz.*, the current torture of innocents) or else prospective (*viz.*, the future explosion of a bomb). Morally, there need not be any difference between these cases: what matters is whether the torture is necessary to prevent the harm. If that harm is temporally distant, then that would undermine the need to torture in so far as there might be other – and less morally offensive – ways to dispel it. But so long as the torture is necessary, then whether the harm is ongoing, imminent, or even temporally distant is irrelevant. We will return to this below but, for now, the point is merely that Bentham's case is structurally similar to the others.

Aside from this more casual presentation of a single case, Bentham also offered a more extended treatment of torture, as recorded in two manuscript fragments. It is worth considering these fragments for at least three reasons, only the first of which is historical. The second is more philosophical in that Bentham starts to elucidate some of the key logical elements of ticking time-bomb thinking, even if that discussion floats free of a particular ticking time-bomb case. And the third bears on the relationship between ticking time-bomb methodology and utilitarianism, a relationship that is more complicated than usually acknowledged. With these three reasons in mind, let us now look at some of Bentham's writings on torture.

First, Bentham asserts that torture may be applied in two cases. "The first is where the thing which a Man is required to do being a thing which the public has an interest in his doing, is a thing which for a certainty is in his power to do" (Twining and Twining, 1973, p. 312). And he continues that torture is otherwise permissible:

> where a man is required what probably though not certainly is in his power to do; and for the not doing of which it is possible that he may suffer, although he be innocent; but which the public has so great an interest in his doing that the danger of what may ensue from his not doing it is a greater danger than even that of an innocent person's suffering the greatest degree of pain that can be suffered by Torture, of the kind and in the quantity permitted to be employed. (Twining and Twining, 1973, pp. 312–313)

Then Bentham asks: "Are there in practice any cases that can be ranked under this head? If there be any, it is plain that there can be very few (Twining and Twining, 1973, p. 313). That Bentham was reserved about the extent to which torture can be justified is noteworthy: being a utilitarian hardly commits one to the promiscuous use of torture, as there are myriad utilitarian reasons to oppose it (Twining and Twining, 1973, pp. 348–350). We will discuss some of these below, but I want to get early purchase on the concept of exceptional – as opposed to normalized – torture.

After these introductory remarks, Bentham goes on to offer a series of moral rules that have to be satisfied for the legitimate application of torture. While the details of those rules need not concern us here, they are precisely the sorts of principles that undergird contemporary ticking time-bomb cases. For example, torture should not be applied without (near-) certainty that the would-be tortured has the relevant knowledge (Rule 1); that torture is only appropriate as a last resort in "cases which admit of no delay" (Rule 3); that minimal means should always be preferred to extreme ones (Rule 4); that the prospective benefits are greater than the prospective costs (Rules 5 and 7), and so on (Twining and Twining, 1973, pp. 312–315).

Importantly, many of these rules were effectively codified in Bentham's hedonic calculus, published shortly thereafter (Bentham, 2007, ch. 4). Bentham predicated his utilitarianism on seven factors, all of which are at least implicitly manifest in the ticking time-bomb cases: intensity, duration, certainty (or uncertainty), propinquity (or remoteness), fecundity, purity, and extent. While intensity and duration are rarely emphasized in the cases, they certainly *could* be, and such invocations would seemingly only make the cases more compelling: imagine that the terrorist need only be subject to a "comparatively minor and brief" form of torture to disclose the location of the bomb. (Note that this is not to suggest that torture could ever be minor – which some might argue to be incoherent – but rather that it most certainly comes in degrees and could be *comparatively* minor.)

The other features, though, are at least near-explicit in the cases. Certainty is perhaps the most conspicuous feature of ticking time-bomb cases: *everything* is certain. It is certain that the detainee is a terrorist. It is certain that he has information regarding the location of the bomb. It is certain that the torture will produce the information. It is certain that the information will lead to the timely deactiviation of the bomb. And many critics of the cases promptly seize upon all of these, which undoubtedly represents a departure from (at least almost all) actual cases (Shue, 2006, pp. 231–239). I will return to this criticism below.

Next come fecundity and purity, which are opposite sides of same coin: when we torture, we will get even more good things (fecundity) and these good things will not be offset by any bad things (purity). This fecundity is thereafter magnified by the invocation of extent, the last of Bentham's elements, which holds that, not only will a single life be saved through the torture, but rather *a lot* of lives will be saved (recall Levin's "millions"). The purity condition is fulfilled in so far as no bad consequences – aside from the pain and suffering of the tortured – are postulated. And, while it is open for the critic to say that such cases do not preclude such consequences, it is equally open to the proponent to merely issue such a stipulation, at least at this stage of the dialectic.

Critics nevertheless do complain about ticking time-bomb cases precisely on the issue of purity. They assert that torture would have to be institutionalized (Davis, 2005, pp. 161–178); including the implementation of training programs for the torturers (Wolfendale, 2006, pp. 269–287); that such institutionalization portends harms for liberal democracies; that our torturing our enemies makes it more likely that our enemies will torture us (Arrigo, 2004, pp. 4–6); that torture makes it more likely for us to perpetuate other wrongs (Fiala, 2005, pp. 127–142); and so on. We will return to these issues below but, for now, I just want to mention them.

Regardless, the proponent of ticking time-bomb methodology is still free to say: "look, that just *is not how the case goes!*"; the critic cannot load conditions into the case that are patently excluded by presupposition. To do so is simply to change the case and to ask a different question altogether, and precisely not the one that we currently care about. Rather, the question at hand is whether torture is permissible given features either stated or implied in the ticking time-bomb cases, and this is a question on which moral philosophy owes an answer. Following that inquiry, we can then think about what implications it has vis-à-vis (real world) cases that relax some of the idealizations and abstractions. But, as the dialectic goes, the aforementioned complaints are completely irrelevant. Let us now move past these methodological preliminaries and to the normative upshot.

Utilitarian Views on Torture

Of all the major moral theories, utilitarianism probably offers the most direct justification of interrogational torture. In the next section, "Deontological Views on Torture", I will argue that rights-based theories need not be opposed to torture, but making that case plausible certainly requires more work than it does in the case of utilitarianism. For example, we might think that people – terrorists or otherwise – have a right against being tortured, and being able to take rights seriously while still being able to license torture therefore presents an obvious obstacle. For the utilitarian, though, no such obstacle exists: so long as the hedonic calculus comes out right, torture is readily justified. Opponents of torture therefore can go either of two ways. First, they can reject utilitarianism, whether for more general reasons or specifically because it permits torture. In doing so, they must adopt some sort of moral theory in its place; this could either be a rights-based approach (see following section) or something else. Second, they can deny that utilitarianism would justify torture in practice, even if it could in theory. This line has the merit of being responsive to utilitarian commitments and, therefore, is able to say something to the utilitarian; the first line simply does not engage the position. For this reason, I have a lot of respect for critics who advance utilitarian arguments against torture in so far as those arguments are responsive to the dialectic: a convincing response to the utilitarian should not be that his theory is wrong – presumably he has already thought about this – but rather that his theoretical commitments are different than he might have suspected (Arrigo, 2004).

As mentioned in the previous section (Origins of Ticking Time-Bomb Cases), the basis of a utilitarian argument for torture is straightforward: it could prevent the deaths of many noncombatants. Their noncombatant status does not matter from the utilitarian perspective, so long as the aggregate value of those otherwise forfeited lives is sufficiently high to cover the costs of torture. For the utilitarian, torturing one terrorist to save one thousand other terrorists could even be justified, at least if the thousand to be saved were not going to sow too much suffering on the world. It is worth noting that the utilitarian calculation here is about preventing a utility loss rather than effecting a utility gain: it is not that torture brings more utility into the world, but rather prevents its longitudinal exit. This distinction, though, is not of much use to the utilitarian, who only cares about maximizing total aggregate happiness; whether that quantity turns out to be positive or negative is beside the point, so long as it is maximal. In other words, torture could manifest a utility loss yet still be justified so long as that loss is less than the alternatives that would be realized without torture.

Again reprising an earlier argument, there are several desiderata that the utilitarian would require torture to satisfy. First, torture should be the least harmful remedy applied, and, similarly, some insufferable form of torture should not be deployed when a lesser one would elicit the valuable information. If the information that would save lives could be gleaned through some less offensive means (e.g., simple questions), then those means should be pursued. There is little reason to torture before otherwise asking the location of the bomb. Furthermore, there should be some expectation that the torture will be efficacious, be it: against someone who we can reasonably expect to have the intelligence; against someone with the appropriate vulnerabilities; not against someone who will hopelessly deceive us with misinformation; effected on a timetable commensurable with the extant threat, and so on. There is no doubt that, in the real world, all of these requirements can get messy but, in the land of ticking time-bombs – where we currently are – they are straightforwardly stipulated.

While these stipulations are often challenged, such challenges radically misunderstand the state of play. For example, consider an ill-named paper by Vittorio Bufacchi and Jean Maria Arrigo, "Torture, Terrorism and the State: A Refutation of the Ticking Time-Bomb Argument", which is anything but (Bufacchi and Arrigo, 2006, pp. 355–373). Bufacchi and Arrigo think that the ticking time-bomb argument takes its premises to be that a terrorist is captured and that, if he is tortured, he will reveal information regarding the location of a bomb. From this, the conclusions are meant to be that torture is permissible, that information about the bomb is retrieved, and that lives are saved (Bufacchi and Arrigo, 2006, p. 360). But then they argue that this formulation has suppressed premises (e.g., that it is almost certain that the terrorist has information about the bomb) and, more to the point, that "all the premises in the argument are contentious from an empirical point of view": intelligence is never infallible, torture is not guaranteed to work, torture is not efficacious in short time periods, misinformation is revealed under torture, and so on (Bufacchi and Arrigo, 2006, pp. 361–362). Ultimately, they maintain that the ticking time-bomb argument fails by its own utilitarian lights. And this sort of strategy has been repeated elsewhere by others, be it about the institutional costs of torture (Davis, 2005, pp. 161–178) – including the costs training programs for torturers (Wolfendale, 2006, pp. 269–287) – or various other negative consequences (Brecher, 2007, chs 2–3). Assuming that all of these objections can be developed in utilitarian currency, where does the utilitarian argument for torture stand?

As discussed elsewhere, I look at ticking time-bomb cases not as arguments, but as cases (Allhoff, 2012, ch. 5). The upshot of these cases is some sort of moral judgment about the permissibility of torture, and those judgments somehow figure into our moral methodology through, for instance, a reflective equilibrium in which those judgments align with our moral principles (Rawls, 1999, pp. 42–45). We can make ticking time-bomb cases figure into some argument, but that argument looks something like this: (1) in all cases – and all else being equal – if we can choose a lesser harm to a greater one, we should; (2) in ticking time-bomb cases, torture is the lesser harm; therefore (3) we should torture in those cases. I take the first premise to be self-evident. The second is the key kernel encoded into ticking time-bomb cases. The conclusion deductively follows.

Bufacchi and Arrigo's formulation seems clunky by comparison, though maybe they aim to break out my second premise. Regardless, their strategy is just to deny some presuppositions of the cases on empirical grounds. But this move misses the point of the dialectic since the realm of discourse at this stage is non-empirical; rather, it is about the hypothetical cases. Their response is a red herring in so far as they change the question from being about ticking time-bomb cases to being about the world. Ultimately, their position has to do with torture policy, but ticking time-bomb cases are not about torture policy; they are about one-off applications of torture (Wisnewski, 2009, pp. 205–209).

What matters to the utilitarian is how the hedonic calculus plays out in the cases under consideration. For the purposes of this chapter, we are still concerned with ticking time-bomb cases, and those straightforwardly license the moral permissibility of torture under a utilitarian approach. Elsewhere, I afford empirical considerations more discussion (Allhoff, 2012, ch. 7). Note, though, that opponents of torture offer an important concession when they appeal to empirical issues. Their empirical turn presumably allows that torture *could* be justified; otherwise, there is no need – save, perhaps, dialectical expedience – to make that move. If the allowance that torture could be justified is nevertheless tightly followed by the contention that it *would* not be justified, then we just have to look at the cases and see how it comes out. By even getting to the cases, though, the opponent to torture has lost ground.

Deontological Views on Torture

Many people want little to do with utilitarianism. This is a theory that supposedly lets the sadists abuse their victims, cares more about total aggregates than fair distributions, and so on. For our purposes, though, it is this first concern that matters in so far as utilitarianism makes no provisions for deontological constraints, such as the right not to be tortured. Does everyone have such a right? Is it absolute?

To get the proper point on the first question, consider this: does the terrorist responsible for threatening many lives and unwilling to end that threat have a right against torture? I am not sure. Certainly there are things that we can do to forfeit our rights. A negligent parent forfeits his right of custodianship to his child. A drunk driver forfeits his right to operate a motor vehicle. A murderer forfeits his right to freedom when incarcerated. Perhaps, then, a terrorist can forfeit her right against torture. This suggests one way we can accept a right against torture but deny that it prohibits torture in all circumstances: we allow that people can forfeit the right. Call this the *forfeiture strategy*.

Another rights-based defense of torture denies that the right entails absolute prohibitions. Many rights prohibit people from doing things to the bearers of rights. Property rights, for instance, forbid use without permission. But in emergencies we might say these rights are permissibly infringed. Perhaps the right against torture is like that: people always have this right, but infringing it might be justifiable nevertheless. Call this the *justified infringement strategy*.

The justified infringement strategy argues against a stronger position than the forfeiture strategy. The latter simply sets aside the right against torture. But the justified infringement strategy grants both that rights against torture are universal and that they always give reasons for people not to violate them. It simply appeals to weightier reasons. This is the approach I will pursue.

The terrorist's rights are not the only rights in play. The victims of the bomb stand to have their rights disrespected as well (e.g., their rights to life). Someone's rights are going to be infringed: either the terrorist's right against torture or else the noncombatants' respective rights to life. Therefore, just trotting out the terrorist's right against torture misses the central point of ticking time-bomb cases, even from a rights-based perspective.

Imagine that A is getting ready to murder B and C and that these murders can only be prevented if D shoots A. What should D do? The deontologist can say one of two things. On the one hand, he could say that D cannot shoot A because rights are, to use some famous locutions, "trumps" (Dworkin, 1978, p. ix) or "side constraints" (Nozick, 1974, pp. 28–33). On this view, we may not disrespect some right *even if* doing so will engender fewer overall rights violations. Alternatively, the deontologist might adopt some sort of aggregative approach, which would hold that the right actions are the ones that either maximize or minimize whatever features she takes to be morally relevant (e.g., rights-preservations or rights-violations, respectively). For this latter sort of deontologist, D may shoot A: if he does, there would be one rights-infringement (*viz.*, A's) and, if he does not, there would be two (*viz.*, B's and C's). Rights-violations are minimized through the shooting, thus making it morally permissible.

Which is the more plausible view? I think that the aggregative approach makes more sense. Ronald Dworkin has argued that in cases of rights conflict, we should look to the values that suggested the right in the first place (Dworkin, 1978, p. 191). So, if individuals have a right to life, it is because life itself is something that is valuable and worth preserving. Given a conflict, for instance, where the disrespecting one person's right to life could prevent the violation of five other persons' rights to life, the values that led to the creation of the right to life would suggest infringing the one in order to prevent violating of the five. We endorse rights in the first place because we value the objects of those rights. Thus it would be permissible to act such that the underlying values and their associated objects are preserved to the highest degree possible. As I suggested above, someone who took rights seriously might think that some aggregation procedure is more attractive than viewing rights as necessarily absolute. Nozick does, however, raise a legitimate concern against aggregation. He asks: "why . . . hold that some persons have to bear some costs that benefit other persons more, for the sake of the overall social good?" (Nozick, 1974, p. 33).

One possible response is to say that the ordinary implications of rights do not apply in emergencies. Indeed, in a footnote, Nozick expresses concern with the application of his theory to "cases of catastrophic moral horror" (Nozick, 1974, p. 30). These are precisely what ticking time-bomb cases model. So, ultimately, it is unclear whether even Nozick would object to rights violations in these cases.

Let us now integrate various threads from this section. First, ticking time-bomb cases turn on conflicts of rights and not merely the rights of the terrorist. Second, torture can minimize overall rights violations in ticking time-bomb cases. Third, we can acknowledge that not all rights violations are of equal moral significance. We might even suppose for the sake of argument that the right against torture counts more than the right to life. While I suspect that this latter supposition is false, the rights of a substantial number of people who would otherwise die militates in favor of torture. To put it another way, even if the right against torture is five times more morally valuable than the right to life, the rights of the thousands that the terrorist threatens swamp his right against torture. On a straightforward rights aggregation, torture comes out to be morally permissible. And even if we were to acknowledge that there could be problems with such an approach in general – such as those suggested by Nozick – they might not apply to ticking time-bomb cases in particular. Therefore, they fail to establish the absolute moral impermissibility of torture.

Rejecting Torture: Absolutism-in-Principle and Absolutism-in-Practice

The two preceding sections evaluated the utilitarian case for torture, as well as a particular rights-based approach. Elsewhere, I have argued that a range of other moral theories (e.g., virtue theory, social contract theory, etc.) can license torture in exceptional circumstances (Allhoff, 2005, pp. 243–264). However, critics may argue that I have not grappled with absolutist moral prohibitions.

Let us say that absolutism with regards to torture can come in two different forms: absolutism-in-principle and absolutism-in-practice (henceforth let us designate these a-principle and a-practice, respectively). The difference between these two is modal: a-principle holds that torture never *could* be justified whereas a-practice holds that it could, but never *would*, be justified. In other words, a-practice allows that torture could be justified, but denies that whatever circumstances are sufficient for this justification – perhaps including those of ticking time-bomb cases – will be manifest in the real world. A-principle, by contrast, denies that such circumstances are possible at all.

First, it bears emphasizing that comparatively few people explicitly defend a-principle; even those who do defend absolutism with regards to torture overwhelmingly advance arguments defending a-practice (Mayfeld, 2008; Tindale, 2005). The limited advocacy for a-principle is not surprising given that it is a very extreme position, effectively the most extreme that one could possibly advocate. To wit, a-principle postulates some class of actions that, regardless of circumstances – whether real or imagined – is always impermissible. No view could be more extreme in so far as it is logically impossible to allow torture in any fewer cases. Regardless, some people do defend a view like this. For example, consider Kim Lane Scheppele: "I do in fact believe that torture is always and absolutely wrong, given the position we should accord to human dignity, even of terrorists" (Scheppele, 2005, p. 287). Or Ben Juratowitch: "torture is so barbaric that the right to be free from it is never defeasible. However desperate the countervailing circumstances, torture is always wrong and should never occur" (Juratowitch, 2008, p. 81). Anyone of a sufficient Kantian bent would also defend a-principle since interrogational torture treats its victim as a means to our end, operates on a maxim that generates a conflict in will, fails to respect autonomy, and so on; any act with these commitments is, to use a Kantian term, categorically impermissible.

That said, there are reasons to think that a-principle is implausible. Consider any candidate moral value with which one might construct a moral theory. Suppose there is some act, Φ, that (maximally) promotes its pre-eminent value V in some case C (hereafter, I use "promotes" to mean "maximally promotes"). Every moral theory other than a-principle treats this supposition in a straightforward way and endorses Φ. A-principle, however, needs it to be the case that Φ is ruled out a priori (i.e., without consideration of the details of C). There are two ways this exclusion could go. First, a-principle could deny that Φ promotes V (in C). So the idea would be that, necessarily, Φ is incompatible with whatever value undergirds our moral theory. Or, second, a-principle could say that even if Φ promotes V (in C), Φ is nevertheless impermissible.

This second response borders on the nonsensical in so far as some theory's commitments preclude action to promote its own pre-eminent value. It even seems contradictory in so far as the theory espouses V while, at the same time, obstructing its realization (i.e., disavowing Φ, which promotes V). This is not to object to there being *other* theoretical commitments that Φ might compromise, but that just forces us to elaborate the argument. For example, maybe Φ promotes V_1 while, at the same time, diminishing V_2; imagine that these are both values of some particular a-principle. Now there are theoretical resources to oppose Φ, namely its impact on V_2. But all this shows is that our first example was overly simplistic since it postulated only a single value for the theory. If we allow that Φ promotes $V_1, V_2, \ldots, V_n$ (i.e., all the values of a-principle), then it will still be curious that Φ is prohibited.

A-practice, unlike a-principle, makes substantive claims about the world. While a-principle holds that there are no circumstances under which torture *could* be justified, a-practice holds that there are no circumstances under which torture *would* be justified; this former claim is true or false *a priori*, whereas the latter is true or false *a posteriori*. In other words, to argue that torture is never actually justified requires some engagement with the world, or at least it does to the extent that a-practice means to be saying anything different from a-principle. To wit, a-principle also holds that torture is never actually justified, though this "actually" is redundant in so far as torture never could have been justified, whether actually (i.e., in the real world) or in imagined cases.

Various people defend a-practice, but a useful formulation is that by Daniel Statman, who argues that "[t]he moral danger of torture is so great, and the moral benefits so doubtful, that in practice torture should be considered as prohibited absolutely" (quoted in Gross, 2009, p. 146). I choose this formulation among others because it usefully sets his view apart from a-principle in so far as the conclusion pertains to the implementation of torture *in practice*. It is perfectly consistent with Statman's thesis that there are fantastic cases wherein torture could be justified, but that none of those cases would ever obtain in the real world. Henry Shue writes something similar, slightly taking himself to task for his earlier work: "I now take the most moderate position on torture, the position nearest the middle of the road, feasible *in the real world*: never again" (Shue, 2006, pp. 213–239).[3] Regarding the observation above, it is possible that Statman and Shue actually endorse a-principle, but then why do they talk about "practice" and the "real world" in their papers? Given their choice of language, a-practice is the more natural reading. Regardless, this section is not about either of their views in particular, but rather about a class of views that emphasize practice instead of theory; theirs are merely likely candidates.

I think of a-practice as making empirical claims about the world as opposed to being committed to any particular moral principles more generally. In other words, a-practice does not have any moral commitments other than the one that torture cannot be justified in the real world; such a view tells us nothing about the moral permissibility of, for example, abortion or euthanasia. This is in contrast with the views considered in the two previous sections; these were independent moral theories whose commitments we could evaluate vis-à-vis torture. Utilitarianism, for example, says that we should maximize total aggregate happiness, and then we can ask whether torture (in some particular case) does so. Not so for a-practice, which does not tell us anything other than that torture will never be justified in the real world. Aside from this sole moral commitment of a-practice, it is otherwise compatible with a wide range of moral theories. For example, there are utilitarian defenses of a-practice, and there could be myriad other defenses of it as well: Torture will never promote rights, virtues, the social contract, or whatever else takes moral priority. It is worth noting that this is not meant to be a criticism of a-practice; in fact, I take it to be a strength.

So let us ask: does the defender of a-practice think that we should torture in ticking time-bomb cases? For such a person, this question is just not interesting since it (allegedly) gains no traction with the world. What matters is not whether we should torture in ticking time-bomb cases, but rather that there are no such cases. Ethics is about action, and we can only act in the world, not in imagined hypotheticals. Elsewhere, I have defended ticking time-bomb methodology against its critics (Allhoff, 2012, ch. 5), but the criticism I now mean to engage is not concerned with the methodology per se, but rather tries to render that methodology impotent by appeal to empirical facts.

In response, there are several important points to develop. First, a-practice is only as plausible as its empirical claim that torture will *never* promote our core moral values. It is very hard to show that such a thesis is true. For example, the ivory-billed woodpecker was thought extinct since 1944, only to turn up in an Arkansas swamp in 2005. None of these woodpeckers was seen for over sixty years, and the received view was that they were gone forever. It turns out that the birds were still around and the naysayers were wrong. The point is that it only takes one case to falsify a negative existential claim, whether it be a sighting of a believed-to-be-extinct bird or some constellation of improbable features leading to justifiable torture.

The defender of a-practice is doubtless going to say that this is all hopelessly confused. His claim is not that justifiable torture is as elusive as the ivory-billed woodpecker, but rather that justifiable torture *cannot* exist (in the real world) because the benefits are too low and/or the costs too high. While it is *possible* for a rare bird to be spotted, it is not possible to justify a heinous act of torture. But, as I read this response, it just denies my premise, namely that the empirical possibility of justified torture cannot be ruled out. My position clearly has the dialectical advantage in so far as its only claim is that I allow that such and so *might* happen (in the real world) and, if so, torture could be permissible. My opponent has to say that this is false, but such empirical omniscience from a philosophical armchair is untenable.

The better way to go is something quasi-historical: to say that we have never seen a ticking time-bomb case in the real world, have no reason to think that there ever will be one, and so on. Sure, it is *possible* that such a case will arise, though we have no good reason to think that it will and, indeed, many good reasons to think that it will not. On this sort of inductive approach, we have a lot more confidence that we could find an ivory-billed woodpecker than that torture could be justified; there might have been some reason to hold out hope in the former case (e.g., ornithologists have sometimes made mistakes), but there is no such reason in the latter.

Again, these are empirical claims and ones that are rarely supported by any sort of actual empirical evidence. It is probably more fair for the defender of torture to owe us an actual ticking time-bomb case than for the critic to have to show that there have not been any, and I have discussed this elsewhere (Allhoff, 2012, ch. 7). The point at present is merely that the empirical details matter. And, in fact, a-practice rarely takes this challenge very seriously. We hear plausible-sounding claims about institutional requirements, potential for abuse, the ineffectiveness of torture, and so on, but these are almost always offered without any serious empirical engagement. Disparaging remarks about torture in Argentina and Abu Ghraib are also common, yet these are about as far from ticking time-bomb cases as one could get and still be talking about torture. That said, the empirical work is sometimes taken quite seriously, and Darius Rejali's painstaking and magisterial work is the best example, particularly its investigations into past torture regimes (Rejali, 2009). As impressive as Rejali's book is, though, it makes hardly any reference to the relevant philosophical literature. This is not necessarily a criticism of his work – it has different goals and comes from a different disciplinary orientation – the only point is that, in the end, the empirical and the philosophical need to be integrated.

Conclusion

Let me make a few final remarks about a-practice. I certainly think that this is the best way to go in opposing torture. As indicated in the two sections on deontological and utilitarian views on torture, it is just not very hard to justify torture in theory. Rather, the rub comes when moving to practice, and a-practice is unequivocal in that regard: no torture in practice, ever. Two provisional conclusions bear emphasis. First, there is strong antecedent pressure against the thesis that torture would *never* be justified for the simple reason that we do not know all the scenarios we will ultimately encounter.

It seems to me that the more plausible route is to leave open the possibility of justifiable torture, while being skeptical and protective about its application. A-practice is not content with this less-ambitious approach and, for me at least, therein lies one of its faults. Second, a-practice is highly committed to several empirical assumptions, whether about the non-existence of ticking time-bomb cases, the low benefits of torture, its high costs, and so on. All of these assumptions have to be examined and defended, and in ways that are sufficiently attuned to the appropriate philosophical issues. I submit that this has not been done by opponents of torture, quite probably because such an attempt would founder.[4]

Notes

1   This essay is adapted with permission from Fritz Allhoff (2012), ch. 5, pp. 88–92; ch. 6, pp. 114–139, especially sections 5.1, 6.1–6.5. Reproduced by permission of The University of Chicago Press.
2   He also provided one of the earlier characterizations of torture: "Torture, as I understand it, is where a person is made to suffer any violent pain of body in order to compel him to do something or to desist from doing something which done or desisted from the penal application is immediately made to cease" (Twining and Twining, 1973, p. 309).
3   In earlier work, after presentation of a ticking time-bomb case, Shue writes: "I can see no way to deny the permissibility of torture in a case *just like this*" (Shue, 1978, p. 141). Shue goes on to issue various disclaimers about whether these cases would actually obtain, but he does not rule out the possibility. The more recent paper is unequivocal in its practical prohibition on torture.
4   For more discussion on the empirical issues, however, see Allhoff (2012, ch. 7).

References

Allhoff, F. (2005) A defense of torture: separation of cases, ticking time-bombs, and moral justification. *The International Journal of Applied Philosophy* 19 (2): 243–264.
Allhoff, F. (2012) *Terrorism, Ticking Time-Bombs, and Torture: A Philosophical Analysis*. London: University Of Chicago Press.
Arrigo, J.M. (2004) A utilitarian argument against torture. *Science and Engineering Ethics* 10 (3): 543–572.
Bentham, J. (1973) Of torture. Reprinted in W.L. Twining and P.E. Twining. Bentham on torture. *The Northern Ireland Legal Quarterly* 24: 307–356.
Bentham, J. (2007 [1789]) *An Introduction to the Principles of Morals and Legislation*. Mineola, NY: Dover Publications.
Brecher, B. (2007) *Torture and the Ticking Bomb*. Malden, MA: Blackwell Publishing.
Bufacchi, V. and Arrigo, J.M. (2006) Torture, terrorism and the state: a refutation of the ticking-bomb argument. *Journal of Applied Philosophy* 23 (3): 355–373.
Davis, M. (2005) The moral justifiability of torture and other cruel, inhuman, or degrading treatment. *The International Journal of Applied Philosophy* 19 (2): 161–178.
Dworkin, R. (1978) *Taking Rights Seriously*. Cambridge, MA: Harvard University Press.
Fiala, A. (2005) A critique of exceptions: torture, terrorism, and the lesser evil argument. *The International Journal of Applied Philosophy* 20 (1): 127–142.
Gross, M. (2009) *Moral Dilemmas of Modern War: Torture, Assassination, and Blackmail in an Age of Asymmetric Conflict*. Cambridge: Cambridge University Press.

Juratowitch, B. (2008) Torture is always wrong. *Public Affairs Quarterly* 22 (2): 81–90.

Levin, M. (1982) The case for torture. *Newsweek*, February 7, 1982, p. 7.

Nozick, R. (1974) *Anarchy, State, and Utopia*. New York: Basic Books.

Rawls, J. (1999) *A Theory of Justice*. Revised edn. Cambridge, MA: Belknap Press of Harvard University Press.

Rejali, D. (2009) *Torture and Democracy*. Princeton, NJ: Princeton University Press.

Scheppele, K.L. (2005) Hypothetical torture in the War on Terrorism. *Journal of National Security Law and Policy* 1: 285.

Shue, H. (1978) Torture. *Philosophy and Public Affairs* 7: 124–143.

Shue, H. (2006) Torture in dreamland: disposing of the ticking bomb. *Case Western Reserve Journal of International Law* 37 (2 & 3): 231–239.

Tindale, C.W. (2005) Tragic choices: reaffirming absolutes in the torture debate. *The International Journal of Applied Philosophy* 19 (2): 209–222.

Twining, W.L. and Twining, P.E. (1973) Bentham on torture. *The Northern Ireland Legal Quarterly* 24: 307–356.

Wisnewski, J.J. (2009) Hearing a still-ticking bomb argument: a reply to Bufacchi and Arrigo. *Journal of Applied Philosophy* 26 (2): 205–209.

Wolfendale, J. (2006) Training torturers: a critique of the ticking-bomb argument. *Social Theory and Practice* 32 (2): 269–287.

Further Reading

Allhoff, F. (2009) The war on terror and the ethics of exceptionalism. *Journal of Military Ethics* 8 (4): 265–288.

Kershnar, S. (2005) For interrogational torture. *The International Journal of Applied Philosophy* 19 (2): 223–241.

Luban, D. (2005) Liberalism, torture, and the ticking bomb. *Virginia Law Review* 91 (6): 1425–1461.

Mayfeld, J. (2008) In defense of the absolute prohibition on torture. *Public Affairs Quarterly* 22 (2): 109–128.

Sen, A. (1988) Rights and agency. In *Consequentialism and its Critics*, ed. S. Scheffler, pp. 187–223. Oxford: Oxford University Press.

CHAPTER SEVENTEEN

Torture and its Apologists

Bob Brecher

Q: What's your definition of the word "torture"?
The President: Of what?
Q: The word "torture." What's your definition?
The President: That's defined in US law, and we don't torture.
Q: Can you give me your version of it, sir?
The President: Whatever the law says.

George Bush (2007)

Torture, Consequentialism and the "War on Terror"

For the law to say what Bush wanted it to say, the ground has – and had – to be pre-pared. This task falls no less to naive and careless lawyers and philosophers than to those politically driven to support the so-called war on terror. My target, then, is their argument that interrogational torture should be recognized as a legitimate weapon in the "war on terror" because the end it serves justifies the means in which it consists; these are the sole means available; and so torture is sometimes necessary and to that extent morally justifiable. I shall focus on the "ticking bomb" scenario: first, because necessity is claimed to arise in such situations; and second, because it is the strongest case that torture's apologists can make. So if torture is *not* justifiable even in "ticking bomb" cases, then it is never justifiable. And it is not.

I shall also restrict myself to an immanent critique: I shall not object to the moral theoretical terms in which torture's apologists make their case, namely consequential-ism. That is not because I think that consequentialism is right – I do not – but because simply to treat an argument for the moral justifiability of torture as a *reductio ad absur-dum* of consequentialism readily invites the response that an anti-consequentialist argu-ment against its justifiability constitutes just such a *reductio* of non-consequentialist

Contemporary Debates in Applied Ethics, Second Edition. Edited by Andrew I. Cohen and Christopher Heath Wellman.

conceptions of morality. Such a stand-off gets us nowhere with torture; and it is torture, not moral theory, that matters (Reader, 2011).

First, however, and contra Bush, I need at least to indicate what torture is. Christopher Tindale (1996, p. 355), adapting the 1994 United Nations General Assembly's Convention Against Torture, puts it particularly clearly: torture is

> any act by which severe pain or suffering, whether physical or mental, is intentionally inflicted on a person for such purposes as obtaining from that person or a third person information or confession, punishing that person for an act committed or suspected to have been committed, or intimidating or dehumanizing that person or other persons.

This is of course not a definition; and ex-president Bush has already shown why the demand for definition is mistaken (here as elsewhere – Brecher, 2007, pp. 3–6). Nor is it an adequate description: we shall come to that at the end. But it is enough to delineate the sort of act we are thinking about.

Second, I need to say a little about the political context in which proponents of interrogational torture make their case and about their concomitant intellectual and moral responsibility. Applied ethics, after all, consists (at least) in thinking about real moral issues; and to the extent that these concern public policy (as torture clearly does) they arise in specific political contexts. Indeed, that is why it has been said that even to engage with torture's advocates is to cede too much and, however inadvertently, to help normalize the practice: "essays . . . which do not advocate torture outright, [but] simply introduce it as a legitimate topic of debate, are even more dangerous than an explicit endorsement of torture" (Zizek, 2006). But precisely because political realities cannot be ignored, I think one has to get one's hands dirty despite that danger; just as anti-consequentialists have sometimes to engage with consequentialist arguments on their own terms. (I shall return later to the "dirty hands" argument and its misuse.)

Amnesty UK has rightly insisted that: "[G]overnments, including our own, have exploited the fear of terrorism to excuse actions that in normal circumstances would never be thought of as acceptable" (Amnesty UK, 2007). Part of that thought captures the kernel of my argument, namely that the most important consequence of recognizing interrogational torture as in certain circumstances legitimate would be to make "acceptable" a whole range of actions that are not; and that in doing so, such recognition would further the trend, especially evident since September 2001, whereby the means we are using to "defend" ourselves against the threat of "terrorism" constitute a far greater danger to what we purport to defend than do the London and Madrid bombs of 2005 or the New York and Washington planes of 2001. Of course, even the most cynical "realists" among the world's politicians rarely defend torture, at least in public, however much they order others to use it. More usually, they "redefine" it, for example as requiring "a sufficiently serious physical condition or injury such as death, organ failure, or serious impairment of body function" (Bybee, 2002); and so claim to be authorizing only "enhanced" interrogation techniques. Sometimes though, as Philippe Sands forensically shows, they go further (Sands, 2008): thus the concerted attack on the prohibition against torture mounted by the United States' government after its declaration of the "war on terror" was central to the pursuit of its ends (Project for the new American Century, 2000). To make torture an accepted and acceptable instrument of state would be to achieve public subservience unprecedented in the West since the Nazis' achievements in the 1930s. If the public can be persuaded to approve torture as a legal instrument, whether directly or via a retrospective "necessity" defense, then there is no limit on what else people can be persuaded to support, or at least to acquiesce in. Torture is the bottom line: shift that, and everything changes. That, after all, is why Israeli public opinion forced the Israeli Supreme Court's 1999 reversal of the Landau Commission's 1987 de facto acceptance of the "necessity" defense in "ticking bomb" cases, Dershowitz's objections notwithstanding (Dershowitz, 2004, p. 264).

It is in this context that we have to understand the activities of torture's apologists, the philosophers and lawyers, who, in promoting a spurious set of arguments that purport to justify torture in circumstances of extreme emergency, lend their intellectual weight to the wholesale acceptance, legitimation and normalization of torture. That is a very serious charge. So I shall pause to consider just one example, and one considerably less egregious than many. In "Torture in principle and in practice," Jeff McMahan insists that while he is opposed to torture, it is nevertheless morally justifiable *in extremis*. He opens his piece with the welcome incantation, "[T]hose of us who oppose torture . . ." (McMahan, 2008, p. 91) and compares Dershowitz: "I am against torture as a normative matter . . ." (Dershowitz, 2004, p. 266); but just a few lines later says he "will argue that the moral justifiability of torture in principle is virtually irrelevant in practice" (McMahan, 2008, p. 91). So what exactly does it mean to be "opposed to" torture, if you also think it is sometimes morally justifiable? It certainly cannot mean that he is *morally* opposed to it. So perhaps, like Dershowitz, he just does not like it – much. Even more worryingly, a serious thinker such as McMahan grants that torture is morally justifiable though "is virtually irrelevant in practice." Despite the qualification this is an extraordinary claim: both historically, as the slightest acquaintance with the history of torture makes clear (e.g., Vidal-Naquet, 1963; Langbein, 1977, 2004; Kooijmans, 1995; Peters, 1999; Elkins, 2005; Harbury, 2005), and in today's political context. It is neither McMahan nor I – nor you – who determines relevance in practice; it is the politicians, as the slightest acquaintance with, for instance, the US, Israeli or British governments' actions relating to torture over the past decade or more makes equally clear (e.g., Amnesty International, 2004; Danner, 2004; Bloche and Marks, 2005; Felner, 2005; Greenberg and Dratel, 2005; Rejali, 2009; Rendition Project, undated; Rose, 2004; Sands, 2008). Would that McMahan were an exception. But he is not.

I am in no position to speculate whether any of the apologists for torture are motivated explicitly by a desire to assist their and/or other governments' exploiting fears of "terrorism." What is clear, however, is that their arguments function politically so as to offer such assistance; and that the quality of much of the argument proffered (e.g., Dershowitz, 2002; Walzer, 2003; Elshtain, 2004; Posner, 2004; Allhoff, 2012) might reasonably lead one at least to raise the question of how they exercise their intellectual and moral responsibilities.

That is why the "ticking bomb" scenario is central; and why its consequentialist terms – though not the fantasy of its "empirical" premises – need to be accepted for the sake of argument.

First, then, I shall show (in "The 'Ticking Bomb' Scenario") that the basis of the so-called ticking bomb case is spurious. I shall then (in "The Consequences of Interrogational Torture") argue that, even if you are unconvinced that the scenario is sheer fantasy, the consequentialist argument for using torture *in extremis* nevertheless fails precisely on account of the likely social and political consequences of its use. In "A Real Case" I shall briefly discuss a genuine case where torture was considered (but not used) as a means of gaining urgent information, arguing that even in that genuine case, and even if torture might have been effective, its use remained unjustified: the end fails to justify the means, even in consequentialist terms. Finally (in "Why Torture is the Worst Thing We Can Do") I shall say a little about what torture is, suggesting why it is the worst thing that human beings can do to one another, whether on consequentialist or any other meta-ethical account; and propose that understanding why this is so sheds light on the inadequacy of any easily accounted consequentialism.

The "Ticking Bomb" Scenario

Imagine there is good reason to think that someone has planted a bomb in a city, but that no one knows where – except for one person, who is already in custody, but refuses to say where or when the bomb is going to explode. The bomb is ticking. Should the person be tortured? Many people think they should be: Dershowitz (2002), Posner (2004), Walzer (2003), and Allhoff (2012), for example. Torture, they argue, is the last and only way of preventing possibly thousands of deaths and terrible injuries and so it has to be used. The argument offered is a variant of the classic trolley problem (Thomson, 1985). Dershowitz's is particularly explicit, based on "the classic hypothetical case" involving "the train engineer whose brakes become inoperative. There is no way he can stop his speeding vehicle of death. Either he can do nothing, in which case he will plow into a busload of schoolchildren, or he can swerve onto another track, where he sees a drunk lying on the rails. (Neither decision will endanger him or his passengers.) There is no third choice. What should he do?" (Dershowitz, 2002, p. 132).

But this "ticking bomb" scenario, while perhaps at first sight plausible, makes (at least) four central and erroneous assumptions. First, it is assumed that the police, security forces or interrogators *know* their captive has the information. But that is extraordinarily unlikely (though not impossible, as the genuine *in extremis* case will later show). Determined bombers are going to be careful planners, and the authorities are unlikely to be sufficiently well informed to be certain that their suspect knows about the bomb. Dershowitz, like others, glosses over this obvious difficulty in blithely stating that "it is surely better to inflict nonlethal pain on one guilty terrorist who is illegally withholding information needed to prevent an act of terrorism than to permit a large number of innocent victims to die" (Dershowitz, 2002, p. 144). Even on the consequentialist assumption that failing to prevent something is on a par with committing it, to assume that the person in custody is a "guilty terrorist" is, or ought to be, breathtaking. As Jonathan Allen observes, having listed all the things it would be required to know, "real cases, even those that most approximate the 'ticking bomb' scenario, involve much more uncertainty" (Allen, 2005, p. 9). Even Sanford Levinson, who reluctantly supports elements of Dershowitz's argument, insists that "there is no known example of this actually occurring, in the sense of having someone in custody who knew of a bomb likely to go off within the hour" (Levinson, 2003, p. 88).

Second, how does anyone – apart from the bomber – know that time really is running out, so that more subtle interrogational techniques are useless? Again, bombers are going to do their best to leave as little time as possible between planting a bomb and its exploding (*especially* if they know they are going to be tortured if caught). At best, that can be only a suspicion, not something the authorities *know*. Furthermore, is it not likely that "most dedicated tough guys will be able to hold out for a couple of hours no matter what you do to them (bear in mind that if they know about the bomb, they'll know how long they have got to hold out . . .)" (Caola, 2005, p. 4)? And all the more so as they know also that it is only *interrogational* torture that will be used: Once the bomb explodes, the torture will of course stop – otherwise it would be punitive, not interrogational, torture.

Third, and again since the torture is interrogational only, what would captives do? Obviously they would lie to buy time; anything else would be stupid. After all, the torture – if genuinely interrogational and not punitive – would stop while the authorities checked the captives' story. In fact, they would surely lie repeatedly, so as to prolong this process until the bomb exploded. Now, you might think that this sketch is ridiculous: of course the torturers and their masters would not act in this, so to speak, gentlemanly way. Indeed not. That is why the very idea of purely interrogational torture is wholly artificial. Perhaps that is one reason why military manuals worldwide, and the US Army Field Manuals in particular, prohibit interrogational torture (see US Army, 1987, 1992). As Major Casebeer drily remarks,

> [T]he imminence of the danger requirement will probably only be met in radically under-specified thought experiments like the ticking bomb scenario (indeed, the very intelligence that will enable us to know we are facing an imminent danger will also likely serve to give us means to discover the source of the danger without having to resort to torture interrogation. (Casebeer, 2003)

Nor can the consequentialist advocate of interrogational torture take comfort from Casebeer's "probably" or "likely": concerning the future – the consequences – probability and likelihood are all there is.

Fourth, and crucially, the argument depends on the assumption that torture would in fact work. Unsurprisingly, nearly all the evidence here is again anecdotal: people have assured me that someone has assured them that torture has saved lives; others that there is no such evidence. Given that an adequate set of objective empirical studies has to remain unavailable, two points need to be made. Again, field manuals worldwide prohibit interrogational torture on grounds of its ineffectiveness (Pachecco, 1999, p. 30; Casebeer, 2003; Rose, 2004, p. 95): Are not the military more likely than academics to know what they are talking about here? Certainly Dershowitz, who is actually more thorough than any of his fellow apologists, is systematically confused about this. His central example is a 1995 case of a plot to "assassinate the pope and to crash eleven commercial airliners . . . into the Pacific" which was allegedly prevented because, "[f]or sixty-seven days, intelligence agents beat the suspect" (Dershowitz, 2002, p. 137; cf. Franklin, 2009). So much for urgency! As for effectiveness in genuine cases of real urgency (see "The Consequences of Interrogational Torture") consider waterboarding, the method most often claimed to be effective. First, if it were so effective, then would not captives say just anything to stop it (and remember that the torture would stop while their story was checked out)? Ken Loach's film, *Route Irish*, makes the point brilliantly. The fact, furthermore, that the Americans waterboarded Khalid Sheikh

Mohammed some 183 times and Abu Zubaydah 83 times (Shane, 2009) raises obvious doubt about its interrogational effectiveness.

The "ticking bomb" scenario is a fantasy, the terms of which are in serious tension, if not in contradiction, with one another. The claim that torture is necessary, therefore, is false: All that can be claimed is that it *may be* necessary and that we cannot know in advance whether it is or not. So much for "necessity *in extremis*": All that there is, is *possible* necessity. In the real world, necessity is necessarily retrospective only: Probability is all there is at the time. Concretely, therefore, there are always two risks, torturing the wrong person and resorting to torture when other methods might have succeeded. No wonder that the best evidence Dershowitz cites in favor of such post hoc necessity (albeit while not recognizing it as such) is this: "the Israeli security services claimed that, as a result of the Supreme Court's decision [to abrogate post hoc necessity as a defense against a charge of torture], at least one preventable act of terrorism had been allowed to take place, one that killed several people when a bus was bombed" (Dershowitz, 2002, p. 150). In fairness, he immediately concedes that "[W]hether this claim is true, false, or somewhere in between (*sic*) is difficult to assess" (Dershowitz, 2002, p. 150). What he fails to concede, however, is the impact that this admission should have on his argument, namely that it is based on anecdote only.

The Consequences of Interrogational Torture

Suppose, though, that these considerations fail to persuade you. Then, since all apologists for interrogational torture claim that the consequences of not torturing are worse than those of torturing, we need to attend to those consequences, and to think about them rather more carefully than they do.

First, and most obviously, interrogational torture needs professional torturers. In fact, it needs especially skilled professional torturers, just because time is claimed to be of the essence and the so-called suspect has to remain able actually to *give* the information needed. But once torture were normalized, and professional torturers treated rather like professional bomb disposal people, we would be living in a world where their services were likely to be ever more called upon, and in more and more different contexts: suspected murderers, drug-traffickers, rapists, perhaps even tax-evaders – as the (highly conservative) *Economist* (2003) long ago pointed out. In fact, the (especially) American use of torture in the first decade of the twenty-first century has already pushed us a long way down this road, as was all too predictable (e.g. Kreimer, 2003, p. 291). Even before that, this escalation had already been experienced as a result of the quasi-legality of torture in Israel from 1987 to 1999. In these twelve years, the use of torture had rapidly spread beyond interrogational settings, which is why the Israeli Supreme Court withdrew the "necessity" defense against the charge of torture in 1999 – however hypocritically (Biletzki, 2001; Parry and White, 2002, pp. 757–760).

The question of the need for professional torturers to carry out the highly specialized work of the extraordinarily urgent interrogational torture allegedly necessitated by a "ticking bomb" raises another issue, about how arguments for such torture – and others about public policy – are presented. They all too often start, and Dershowitz himself is explicit about this, with the question, "What would you do if . . .?" (Dershowitz, 2002, p. 133). For example, commenting forty years ago on the British army's "interrogation techniques" in the north of Ireland – which were in 1971 eliciting at least some disquiet in the UK – Anthony Quinton wrote:

> I do not see on what basis anyone could argue that the prohibition of torture is an absolute moral principle. . . . Consider a man caught planting a bomb in a large hospital, which no one dare touch for fear of setting it off. It was this kind of extreme situation I had in mind when I said earlier that I thought torture could be justifiable. (Quinton, 1971, p. 758)

He immediately points out two difficulties, but does not register their implications. First, he rightly notes that "any but the most sparing recourse to [torture] will nourish a guild of professional torturers, a persisting danger to society much greater, even if more long-drawn-out, than anything their employment is likely to prevent"; and second, that: "[I]f a society does not professionalise torture, then the limits of its efficiency make its application in any particular extreme situation that much more dubious" (Quinton, 1971, p. 758). But these limits *rule out* such an application, just because the scenario requires an efficiency *amateur* torturers lack. Or consider Michael Walzer, North American liberalism personified: "I would do whatever was necessary to extract information in the ticking bomb case – that is, I would make the same argument after 9/11 that I made 30 years before" (Walzer, 2003). Unless Walzer has been trained to torture, this is an extraordinarily stupid remark. The question, rather, that needs to be asked is this: "What would you require *someone else* – a professional torturer – to do *on your behalf?*"; and furthermore, not as a supererogatory or altruistic act, but as the practice of their profession.

Nor is this only a matter of the professional skill required. Even if I could do it, what *I* would do, or think I would do, in a particular situation is irrelevant to the question of what ought to be done. What you might do if someone attacked your child or parent is one thing. What it would be right or wrong to do is another: They may or may not coincide. That is why societies have legal structures rather than relying on vigilantism. The apparently innocent question, "What would you do if . . .?" is no basis for social and political policy. It is an invitation to think about what is morally right or wrong; it is not an invitation to base your judgment about that on guesses about your own likely response, or attempted response, *in extremis*.

To return to the central issue: is it not likely that torture's being a recognized weapon would lead to more, not to fewer, terrorist acts and volunteers for "suicide bombing" (Richardson, 2006)? Is this not what has actually happened since 2001? The people who planted bombs on London trains and buses in 2005, for instance, cited as their motivation the UK's role in the occupation of Iraq, with its attendant atrocities. Rightly or wrongly, sympathy for such acts would also probably increase, just as it in fact has over the past decade on account of more people coming to think of the perpetrators of atrocities as responding in kind on behalf of the victims of Western ill-treatment. In short, countries that torture can hardly claim the moral high ground (Rose, 2004, p. 72). Why should an Assad, Mubarak or Mladic, or a Sri Lankan, Chinese or Bahraini government give up their weapon of mass terror – torture – if established powers not only use it, but claim it is morally justified?

And what about the training of the torturers? They need "[S]pecial classes . . . where new torturers are shown what torture looks like, either in filmed demonstrations or even live demonstrations on actual prisoners" (Crelinsten and Schmid, 1995, p. 49) or on people – sometimes children – picked up in the favelas (Haritos-Fatouros, 1995).

(For details, see Wolfendale, 2006.) Recognizing the profession of torturer *as* a profession means recognizing also its training needs, just as with teachers, lawyers and doctors (Gray, 2003). But even that is not all. The expertise of doctors and psychologists is indispensable if torture is to be at its most effective, as the American Psychological Association knows to its cost: it took its members several years to force through a statement saying that assisting in torture was antithetical to its professional values and responsibilities (Welch, 2008). You might think such a statement superfluous; but unhappily it is not (Bloche and Marks, 2005; Medact, 2011).

Finally, among the various consequences of regarding interrogational torture as morally justifiable are two that are unavoidable because they do not depend on the contingencies of the real world. They are, if you like, logical consequences. First, it would be a *moral duty* incumbent on all able citizens to assist and/or to facilitate such assistance: For if interrogational torture is justified on account of the consequences, then anyone refusing their expert help in the service of the greatest overall happiness would be morally culpable. To have an expectation of public officials that they use torture if "necessary" is hypocritical unless you are prepared – if able – at least to assist, at most to undergo the requisite training. Principled pacifists, for example, cannot reasonably ask others to kill, or to be trained to kill, on their behalf, else the principle is hollow; and squeamishness or phobia – if genuine and thus not a contradiction of the principle – are addressed by the qualification concerning ability.

Second, if consequences determine rightness or wrongness, then surely other people besides the "captured terrorist" can be tortured in order to avoid a catastrophe. Without apparently understanding the importance of the point, Dershowitz himself recognizes this problem: "torture sometimes works. Jordan apparently broke the most notorious terrorist of the 1980s, Abu Nidal, by threatening his mother" (Dershowitz, 2002, p. 249, n.11). Quite so. However, his attempt to rescue his position from its own logical implications is ludicrous. Purporting to recognize the need to avoid arguing that "anything goes as long as the number of people tortured or killed does not exceed the number that would be saved," he then suggests that we need "other constraints on what we can properly do," which "can come from rule utilitarianisms [*sic*] or other principles of morality, such as the prohibition against deliberately punishing the innocent" (Dershowitz, 2002, p. 146). But if rule utilitarianism allows torturing suspects, then it has to allow torturing their mothers or children too. Why? Because the grounds are the same in both cases: the "best possible" outcome is all that can count. To suggest bringing in non-consequentialist principles to avoid this unpalatable conclusion is either silly or disingenuous: Consequentialism – in which terms the whole argument is made – insists that there are none. That is what makes consequentialism *consequentialism*.

The consequences of permitting interrogational torture would be – in fact already are – morally and politically disastrous.

A Real Case

As I have already suggested, the "ticking bomb" fantasy is no basis for public policy. So here is a real case. In 2002, the police in Frankfurt, Germany, really did know that it was Magnus Gäfgen who had kidnapped Jakob von Metzler, the 11-year-old son of a banker, and was holding him as ransom. They had collected conclusive evidence from

Gäfgen's apartment and had also watched him pick up the ransom money left by the father. But he refused to say where the boy was, only that he was locked up somewhere (and thus would slowly die unless he were found in time) (BBC News, 2003). The police chief concerned "ordered his men to threaten Gäfgen with violence to force a statement" (Schroeder, 2006, p. 188). The threat was not carried out; it turned out that the boy was already dead; and the ramifications of the case continue (Evans, 2011). Still, assuming that an adequately skilled torturer could have been found in time, should Gäfgen have been tortured to force him to tell the police where the boy was? No. Why not? Because the consequences in terms of society's attitudes to torture (see section "The Consequences of Interrogational Torture") would be even worse than the boy's death. It was too late: The catastrophe was no longer avoidable. And as with Jakob Metzler, so with real ticking bombs: It is already too late. That, after all, is how the real world is – and, note, for consequentialists and non-consequentialists alike. Not every catastrophe is morally avoidable.

So of course it is sometimes necessary to get one's hands dirty, as Walzer famously argues (Walzer, 1973, 2003). But here as elsewhere, Walzer, like many others, overlooks the obvious: that there is more than one way of getting your hands dirty. *Not torturing Gäfgen* – not torturing anyone, in any circumstances – is one of those ways. So the notorious and all too common insistence of some lawyers and academics that "[N]o one who doubts" that interrogational torture is justifiable "should be in a position of public responsibility" (Posner, 2004, p. 295) – since they are not prepared to get their hands dirty – remains wholly unjustified.

Why Torture is the Worst Thing We Can Do

Torture is the worst thing we do to each other. A torturous society – the sort of society increasingly legitimized since September 2001 – is the worst society we can create (Brecher, 2011). Why?

Here is an account of torture, one which makes it clear how interrogational torture is not some form of torture that, unlike others, is not depraved: "The subject of judicial or interrogational torture is 'broken' when, and only when, he has become so distraught, so unable to bear any more suffering, that he can no longer resist any request the torturer might make. The tortured then 'pours out his guts'" (Davis, 2005; 165). The capacity of the tortured person to act, that is to say, to think and then to *do* something, rather than just to *behave* in response to external stimuli, is broken. That is what makes torture *torture*, whether interrogational or, for example, intimidatory or "punitive." Consider Slavenka Drakulic's fictional but all too realistic account of rape as torture: "The person who returns to the 'women's room' the following evening is no longer A. . . . A. has left the body standing in front of them. Like her mother whose eyes lived on long after she was dead, A.'s body is still alive, but A. is dead" (Drakulic, 1999, p. 89). And because what makes us persons, rather than some other sort of being, is precisely our capacity to act, to do things, it follows that the person being tortured has, in their own eyes, ceased to be a person; that while "A.'s body is still alive . . . A. is dead". That is exactly what torturers aim for: to break a person, to make them into only a body. It is worth pausing here to emphasize that this is also why interrogational torture – as contrasted with torture used to intimidate, terrorize, punish or dehumanize – has to require extraordinary skill. The torturer has to get the person they are torturing to "pour out [their] guts," but to do so in that precise moment just before the response to "any request the torturer might make" becomes just that, an unthought response, rather than an action. Otherwise they are no longer *able* to give the torturer what they want, namely a truthful answer to the question, "Where's the bomb?"

What could be a more complete negation of a person than to break them, to make simply an object of them? Dershowitz crassly assumes, as do many others, that "[P]ain is a lesser and more remediable harm than death" (Dershowitz, 2002, p. 144). From that, he argues that, since "nonlethal torture" is less bad than death, and since the death penalty is widely accepted, torture short of torturing to death must also be justified (Dershowitz, 2002, pp. 148, 155; cf. McMahan, 2008, p. 91). But to end a person's life, terrible as it is – and Dershowitz forgets that the death penalty is not universal – is not on a par with making them into something that is not a person, whether temporarily or permanently. Perhaps it is their commitment to a consequentialist view of morality that prevents apologists for interrogational torture from understanding this simple fact. Or perhaps it is their commitment to accounting only what can be, allegedly, measured – and the termination of life puts an end precisely to measurement of pain, pleasure or whatever – that allows them to be consequentialists. (These might of course be reasons for rejecting consequentialism as an account of right and wrong.)

I shall finish by asking you, having first recalled the words of Bush that preface this chapter, to listen (Reader, 2011) to the unsurpassable words of Jean Améry, an anti-Nazi resistance fighter who eventually committed suicide some 35 years after being tortured by the Gestapo:

> Only in torture does the transformation of the person into flesh become complete. Frail in the face of violence, yelling out in pain, awaiting no help, capable of no resistance, the tortured person is only a body, and nothing else besides that. (Améry, 1980, p. 6)[1]

Note

1   My thanks to Andrew Cohen for his very helpful comments on a penultimate draft of this chapter.

References

Allen, J. (2005). Warrant to torture? A critique of Dershowitz and Levinson. ACDIS Occasional Paper, Program in Arms Control, Disarmament, and International Security, University of Illinois at Urbana-Champagne. http://acdis.illinois.edu/publications/207/publication-Warranto TortureACritiqueofDershowitzandLevinson.htmlhttp://acdis.illinois.edu/publications/ 207/publication-WarranttoTortureACritiqueofDershowitzandLevinson.html (last accessed 6/17/13).

Allhoff, F. (2012) *Terrorism, Ticking Time-Bombs, and Torture.* Chicago, IL: Chicago University Press.

Améry, J. (1980) Torture. In *At the Mind's Limit,* trans. S. and S. Rosenfeld, pp. 21–40. Bloomington, IN: Indiana University Press.

Amnesty International (2004). United States of America: Human dignity denied – torture and accountability in the "war on terror". http://www.amnesty.org/en/library/info/ AMR51/145/2004 (last accessed 6/17/13).

Amnesty UK (2007). Don't sign up to terror: unsubscribe. *Amnesty Magazine*, 145, September/October, 15.

BBC News (2003). Bank heir killer jailed for life. July 28. http://news.bbc.co.uk/1/hi/world/europe/3102313.stm (last accessed 6/17/13).

Biletzki, A. (2001). The judicial rhetoric of morality: Israel's High Court of Justice on the legality of torture. Unpublished paper. http://www.sss.ias.edu/files/papers/papernine.pdf (last accessed 6/17/13).

Bloche, G. and Marks, J. (2005) Doctors and interrogators at Guantanamo Bay. *New England Journal of Medicine* 353: 6–8.

Brecher, B. (2007) *Torture and the Ticking Bomb*. Oxford: Wiley-Blackwell.

Brecher, B. (2011) Torture: a touchstone for global social justice. In *Global social justice*, ed. H. Widdows and N. Smith, pp. 90–101. London: Routledge.

Bush, G. (2007). Conference by the President, George W. Bush, October 18. http://whitehouse.gov/news/releases/2007/10/print/20071017.html.

Bybee, J. (2002). Memorandum for Alberto R. Gonzales, counsel to the president, re: standards of conduct for interrogation under 18 U.S.C.

Caola, D. (2005). Letter to London Review of Books, June 2, p. 4.

Casebeer, W., Major (USAF) (2003). Torture interrogation of terrorists: a theory of exceptions (with notes, cautions, and warnings). http://isme.tamu.edu/JSCOPE03/Casebeer03.html (last accessed 6/17/13).

Danner, M. (2004) *Torture and Truth: America, Abu Ghraib, and the War on Terror*. New York: New York Review of Books.

Davis, M. (2005) The moral justifiability of torture and other cruel, inhuman, or degrading treatment. *International Journal of Applied Philosophy* 19: 161–178.

Dershowitz, A. (2002) *Why Terrorism Works*. New Haven and London: Yale University Press.

Dershowitz, A. (2004) Tortured reasoning. In *Torture: A Collection*, ed. S. Levinson, pp. 257–280. Oxford: Oxford University Press.

Drakulic, S. (1999) *As If I am not There*. London: Abacus.

Economist (2003). Editorial. Is torture ever justified? http://www.economist.com/node/1524784 (last accessed 6/17/13).

Elkins, C. (2005) *Britain's Gulag: The Brutal end of Empire in Kenya*. London: Pimlico.

Elshtain, J.B. (2004). Reflection on the problem of "Dirty Hands".

Evans, S. (2011). German child killer Marcus Gaefgen awarded damages. http://www.bbc.co.uk/news/world-europe-14408657 (last accessed 6/17/13).

Felner, E. (2005) Torture and terrorism: painful lessons from Israel. In *Torture*, ed. K. Roth and M. Worden, pp. 28–43. New York: New Press and Human Rights Watch.

Franklin, J. (2009) Evidence gained from torture: wishful thinking, checkability, and extreme circumstances. *Cardozo Journal of International and Comparative Law* 17: 281–290.

Gray, J. (2003). A modest proposal: for preventing torturers in liberal democracies from being abused, and for recognizing their benefit to the public. *New Statesman*, February 17, pp. 22–25.

Greenberg, K. and Dratel, J., eds (2005) *The Torture Papers*. Cambridge: Cambridge University Press.

Harbury, J. (2005) *Truth, Torture, and the American Way: The history and Consequences of U.S. Involvement in Torture*. Boston: Beacon Press.

Haritos-Fatouros, M. (1995) The official torturer. In *The Politics of Pain: Torturers and Their Masters*, ed. R. Crelinsten and A. Schmid, pp. 129–146. Boulder: Westview Press.

Kooijmans, P. (1995) Torturers and their masters. In *The Politics of Pain: Torturers and Their Masters*, ed. R. Crelinsten and A. Schmid, pp. 13–18. Boulder: Westview Press.

Kreimer, S. (2003) Too close to the rack and screw. *University of Pennsylvania Journal of Constitutional Law* 6: 278–325.

Langbein, J. (1977) *Torture and the Law of Proof: Europe and England in the Ancien Régime.* Chicago: Chicago University Press.

Langbein, J. (2004) The legal history of torture. In *Torture: A Collection,* ed. S. Levinson, pp. 93–104. Oxford: Oxford University Press.

Levinson, S. (2003). The debate on torture. *Dissent* (Summer): 79–90.

Loach, K. (Dir.) (2010). Route Irish.

McMahan, J. (2008) Torture in principle and in practice. *Public Affairs Quarterly* 22: 91–108.

Medact (2011). Preventing torture: the role of physicians and their professional organisations. http://www.medact.org/search.php (last accessed 6/17/13).

Pachecco, A. (1999). The Case Against Torture in Israel: a Compilation of Petitions, Briefs and Other Documents Submitted to the Israeli High Court of Justice. http://www.stoptorture.org.il/eng/publications.asp?menu=7&submenu=2 (last accessed 6/17/13).

Parry, J. and White, W. (2002) Interrogating suspected terrorists: should torture be an option? *University of Pittsburgh Law Review* 63: 743–766.

Peters, E. (1999) *Torture: Expanded Edition.* Philadelphia: University of Pennsylvania Press.

Posner, R. (2004) Torture, terrorism, and interrogation. In *Torture: A Collection,* ed. S. Levinson, pp. 291–298. Oxford: Oxford University Press.

Project for the New American Century (2000). Rebuilding America's defences: Strategy, forces and resources for a new century: a report of the project for the new American century. http://www.newamericancentury.org/publicationsreports.htm (last accessed 6/17/13).

Quinton, A. (1971). Views. *The Listener,* December 2, pp. 757–758.

Reader, S. (2011) Ethical necessities. *Philosophy* 86: 589–607.

Rejali, D. (2009) *Torture and Democracy.* Princeton: Princeton University Press.

Richardson, L. (2006) *What Terrorists Want.* London: John Murray.

Rose, D. (2004) *Guantanamo: America's War on Human Rights.* London: Faber & Faber.

Sands, P. (2008) *Torture Team.* London: Palgrave Macmillan.

Schroeder, D. (2006) A child's life or a "little bit of torture"? State-sanctioned violence and dignity. *Cambridge Quarterly of Healthcare Ethics* 15: 188–201.

Shane, S. (2009). Waterboarding used 266 times on 2 suspects. *New York Times.* http://www.nytimes.com/2009/04/20/world/20detain.html (last accessed 6/17/13).

The Rendition Project (2013) http://www.therenditionproject.org.uk/ (last accessed 6/17/13).

Thomson, J.J. (1985) The trolley problem. *The Yale Law Journal* 94: 1395–1415.

Tindale, C. (1996) The logic of torture. *Social Theory and Practice* 22: 349–374.

U.S. Army (1987). Intelligence interrogation. Field manual no. 34–52. http://www.loc.gov/rr/frd/Military_Law/pdf/intel_interrogation_may-1987.pdf (last accessed 6/17/13).

U.S. Army (1992). Intelligence interrogation. Field manual no. 34–52. http://www.loc.gov/rr/frd/Military_Law/pdf/intel_interrrogation_sept-1992.pdf (last accessed 6/17/13).

Vidal-Naquet, P. (1963) *Torture: Cancer of Democracy: France and Algeria 1954–62.* Harmondsworth: Penguin.

Walzer, M. (1973) Political action: the problem of dirty hands. *Philosophy and Public Affairs* 2: 160–180.

Walzer, M. (2003). Interview. *Imprints* 7 (4). http://eis.bris.ac.uk/~plcdib/imprints/michaelwalzerinterview.html (last accessed 6/17/13).

Welch, B. (2008). Torture, political manipulation and the American Psychological Association. *Counterpunch,* July 28. http://www.counterpunch.org./welch07282008.html (last accessed 6/17/13).

Wolfendale, J. (2006) Training torturers: a critique of the" "ticking bomb" argument. *Social Theory and Practice* 32: 269–287.

Zizek, S. (2006). The depraved heroes of 24 are the Himmlers of Hollywood. *The Guardian,* January 10, p. 27.

ISSUES OF PRIVACY
AND THE GOOD

Same-sex marriage

CHAPTER EIGHTEEN

Same-Sex Marriage and the Definitional Objection

John Corvino

Many marriage traditionalists believe that same-sex "marriage" is not merely bad policy, but also nonsensical on its face, because marriage is by definition male–female. Call this the *Definitional Objection* to same-sex marriage. Like Abraham Lincoln, who reportedly asked "How many legs would a dog have if we called its tail a leg?" and then insisted (correctly) that the answer is four, proponents of the Definitional Objection claim that thinking of same-sex unions as "marriages" involves a conceptual error.

In recent years the Definitional Objection has become increasingly popular. Former US Senator Rick Santorum used it on the campaign trail in his bid for the 2012 Republican presidential nomination. Waving a napkin in the air, he announced, "Marriage existed before governments existed. This is a napkin. I can call this napkin a paper towel. But it is a napkin. Why? Because it is what it is. Right? You can call it whatever you want, but it doesn't change the character of what it is" (see Towleroad, 2011). In a similar vein, National Organization for Marriage co-founder Maggie Gallagher writes, "Politicians can pass a bill saying a chicken is a duck and that doesn't make it true. Truth matters" (Gallagher, 2009). Alliance Defense Fund Attorney Jeffery Ventrella contends that "to advocate same-sex 'marriage' is logically equivalent to seeking to draw a 'square circle': One may passionately and sincerely persist in pining about square circles, but the fact of the matter is, one will never be able to actually draw one" (Ventrella, 2004/2005, p. 682).

The purpose of this chapter is to rebut the Definitional Objection, and in particular the version of it offered by New Natural Law theorists (such as Sherif Girgis, Chapter 19, in this volume). The objection is noteworthy for several reasons. First, it is favored by some of the most prominent and sophisticated opponents of same-sex marriage.[1] Second, the objection challenges the rhetoric of "marriage-equality" proponents: If marriage must be male–female *by definition*, then whatever gay men and lesbians seek when asking for equal treatment, it surely cannot be marriage. Third, it is (mostly)

Contemporary Debates in Applied Ethics, Second Edition. Edited by Andrew I. Cohen and Christopher Heath Wellman.
© 2014 John Corvino. Published 2014 by John Wiley & Sons, Inc.

immune to empirical testing, so it bypasses contentious debates about the welfare of children, the alleged promiscuity of gay men, and so on.[2]

The Definitional Objection

I want to begin by clarifying the Definitional Objection and distinguishing it from other kinds of arguments in the vicinity. Some people object to same-sex marriage on what philosophers call *consequentialist* grounds, claiming, for example, that it is bad for children, or that it would erode valuable social norms such as fidelity. These are empirical claims, testable in the world – although actually testing them is difficult, because the alleged consequences are long-term and it is hard to tease apart various possible causal factors. Such arguments contend that same-sex marriage is bad because it has *bad effects*, and the arguments stand or fall based on evidence of such effects. But that is not the sort of argument I am considering here. (I have done so at length elsewhere. See Corvino and Gallagher, 2012, esp. pp. 44–64.) Rather, I want to consider the argument that treating same-sex unions as marriage is wrong in itself, apart from any further consequences it may have. The point is not that same-sex "marriage" will have bad consequences; it is that, at some fundamental level, it is not even possible.

At first glance, it may be tempting to wonder whether the Definitional Objection merely involves a nitpicky verbal dispute. After all, committed long-term same-sex relationships – unlike square circles – are certainly possible, and in some jurisdictions they are officially recognized as legal marriages. So it might seem that the argument is less about whether something exists (same-sex marriage) than about what to call some existing thing. If members of a society call some same-sex unions "marriages," does it not follow that they are in fact marriages?

The answer to that question is "not quite" and the reason for that answer gets us to the heart of the argument. Suppose I started referring to the computer on which I am typing as a "marriage." Suppose further that I convinced others to do so, and that eventually, most people used the word "marriage" to refer to objects that they previously called "computers." It would not follow that computers are marriages. Rather, we would simply be using the word "marriage" in a different way than we once did. Computers and marriages are *different kinds* of objects, and although we can use the same word to refer to both of them, doing so would not remove their distinction in reality.

Of course, there are different sorts of realities at play here. There is what we might call the *legal* reality, and if the law recognizes something as X, then it *legally* is X, whatever X might be. For example, if the law declares a defendant guilty, then the defendant is legally guilty, even if not guilty in fact (that is, even if wrongly convicted). Same-sex legal (or civil) marriage is obviously possible. It is happening in a number of states and countries. But marriage, like guilt, is more than just a legal reality, it is a personal and social reality as well. According to proponents of the Definitional Objection, same-sex civil marriage involves a mismatch between legal reality, on the one hand, and personal and social reality, on the other. Although we can call same-sex unions "marriages" and even legally recognize them as such, doing so obscures an important natural distinction.

Sometimes, using the same word to refer to different things is unproblematic. French speakers use the word *avocat* for both lawyers and avocados, yet no one tries to make guacamole out of lawyers. When objects are more closely related, however, the potential for confusion heightens. To stick with a French-related example: most Americans use the term "champagne" to refer to any kind of sparkling wine, even though, technically speaking, *champagne* comes from the Champagne region of France. Americans who say "Let's have a glass of champagne!" are typically indifferent between a *champagne* from France, a *cava* from Spain, a *prosecco* from Italy, a sparkling wine from California and so on. By contrast, wine purists lament such verbal looseness. When they ask for "champagne" they want something specific, not *cava* or *prosecco* but *champagne*. As they see it, using the same word for all these wines obscures real (and important) differences. Moreover, this linguistic practice may have practical consequences: People who call various sparkling wines by the same name are more likely to think of them as interchangeable, and perhaps even to have less sensitive palates as a result. (The general principle is known as the *Sapir-Whorf hypothesis*: the idea that linguistic categories either determine, or at least significantly influence, conceptual categories.)

By analogy, proponents of the Definitional Objection contend that different-sex and same-sex unions are fundamentally different, and to call them by the same name is to obscure a real natural distinction. Over time, the failure to distinguish these objects in language may lead to a failure to recognize, and ultimately to appreciate, their differences in reality. In effect, it would "dull the palate" with respect to human relationships, and the stakes there are far more significant than those regarding which wine to serve. Some matters of definition are more significant than others. For example, whether we use the word "religion" in such a way as to include secular humanism may have more real-world impact than whether we use the word "planet" in such a way as to include Pluto. Given marriage's importance as a personal and social reality, how we use the word "marriage" is no small matter. But proponents of the objection are not merely worried that desensitization to "real marriage" would have bad consequences (although it might); they claim that the failure to grasp such an important moral good is bad in itself.

We still need answers to the following questions: What is "real marriage"? Why can't same-sex couples achieve it? And why is it so important to maintain a special legal category for it, rather than creating a more inclusive civil marriage institution? The most developed philosophical answer to these questions comes from a group of philosophers known as the "New Natural Law" (NNL) theorists, to whom we will shortly turn. But first, a thought experiment.

Bob and Jane and the Essence of Marriage

Imagine a young couple we will call Bob and Jane. Bob and Jane met in high school, fell in love, and started "going steady" in college. Around the time of their junior year, they began talking of marriage, and in their senior year, Bob proposed, and Jane accepted. Then tragedy struck. One night Bob was involved in a terrible car accident which paralyzed him from the waist down. As a result, Bob would never be capable of coitus: penis-in-vagina sex. Moreover, given additional injuries related to the accident, he would never be able to produce natural offspring, even with the assistance of artificial insemination. Bob offered to cancel the engagement, but Jane would have none of it. Bob persisted, "But I can't provide you with children! Or even proper lovemaking! I am scarcely a man anymore." "Nonsense," Jane replied. "You are the same man I have always loved. And the courage you have shown through this tragedy makes you even more of one. We will make this work."

After much discussion, Bob relented, and the pair wed. Although they could not engage in coitus, they pursued and enjoyed various other acts of sexual intimacy. Within a few years they adopted children: two girls and a boy. They had a happy family life, devoted to each other and their children. Bob was forever grateful for Jane's willingness to stick by him, while Jane admired Bob's courage and determination in the face of disability – a disability which soon faded into the background, as Bob pursued a successful career and family life. The two remained together for many decades, until Bob's natural death at the ripe old age of 89.

Question: Were Bob and Jane married? Obviously, they were *legally* married, but were they *really* married? We have already noted that they could not produce natural offspring together, and they could not engage in coitus. They could therefore not consummate the relationship in the traditional sense. But when we compare those facts against their love, sacrifice, and commitment; the family life they built together; and their lifelong honoring of the vow "to have and to hold, for better or for worse . . . until death do us part," only the most coldhearted would insist that their marriage was not "real." Even fewer still would insist that it is not "real enough" to be worthy of the legal recognition it received. We need not have a fully worked-out theory of what marriage is in order to recognize that, whatever it is, Bob and Jane achieved it. (In a similar way, one need not have a fully worked-out theory of what religion is in order to know that Episcopalianism counts.)

Of course, Bob and Jane were male and female, whereas our subject is same-sex marriage. But we will return to their case, so keep it in mind.

The New Natural Law (NNL) Theory

The most sophisticated attempt to argue that marriage is necessarily heterosexual comes from a group of philosophers, legal scholars, and political theorists known as the New Natural Law (NNL) theorists. Their argument rests on the idea of "basic goods," which are foundational ingredients of a flourishing human life. These goods include life and health, knowledge and aesthetic experience, friendship, integrity, and various others. Such goods are "basic" in the sense that they cannot be reduced to one another or derived from more general goods: Instead, they must be grasped through direct insight and defended dialectically, by teasing out the implications of accepting or rejecting them as goods.

NNL theorists claim that *marriage* is among the basic goods. But they do not mean marriage as a political or legal construction: They mean it as a comprehensive, "two in one flesh" union of husband and wife, something that exists prior to and independently of the state's recognizing it. Although marriage (in this special sense) may result in happiness, emotional and physical health, and so on, its purpose is not reducible to any of these. Nor is its purpose reducible to procreation – a point worth noting, since it corrects a popular misreading of NNL. According to NNL, marriage is never properly chosen merely as a means to some other thing, including children. The comprehensive union of marriage is intrinsically valuable, and therefore should be chosen for its own sake.

As a *comprehensive* union, marriage unites the partners along multiple levels, which reinforce each other. On the mental or volitional level, it requires a loving, permanent, exclusive commitment between the spouses. On the physical level, it requires that the spouses unite biologically in "reproductive-type" acts. In such acts, the male and the female become "literally, not metaphorically, one organism" (Lee and George, 1999, p. 183, n.23) – in popular parlance, the two become one. The new-natural-law theorists refer to the resulting view of marriage as the *conjugal* view.

So to sum up, the overall NNL argument against same-sex marriage looks like this: marriage is, by definition, a *comprehensive* union. A comprehensive union requires uniting on all levels, including the biological level. The only way in which human beings can truly unite biologically is in reproductive-type acts. Because same-sex partners cannot engage in reproductive-type acts, they cannot achieve the comprehensive union that real marriage requires. And it would be wrong for the state to treat their unions as marriages, because the state, which is responsible for citizens' well-being, "is justified in recognizing only real marriages as marriages," as NNL theorists Sherif Girgis, Robert P. George, and Ryan Anderson explain (Girgis *et al.*, 2010, p. 251).

The most common objection to the NNL position concerns heterosexual partners known to be sterile: the "sterile couples" or "infertile couples" objection. Consider a man and woman in their seventies who are engaged to be married. Their impending marriage cannot result in reproduction, and they both know it. Consistency seems to require either that they cannot marry or that the same-sex couple can.

The new-natural-law theorists respond that although these septuagenarians cannot reproduce, their sexual acts – unlike the same-sex couple's – can still be "of the reproductive type." Because these acts are still *coordinated toward* the common good of reproduction, they can still unite the pair in marriage. Girgis, George, and Anderson attempt to explain with an analogy:

> When Einstein and Bohr discussed a physics problem, they coordinated intellectually for an intellectual good, truth. And the intellectual union they enjoyed was real, whether or not its ultimate target (in this case, a theoretical solution) was reached – assuming, as we safely can, that both Einstein and Bohr were honestly seeking truth and not merely pretending while engaging in deception or other acts which would make their apparent intellectual union only an illusion.
>
> By extension, bodily union involves mutual coordination toward a bodily good – which is realized only through coitus. And this union occurs even when conception, the bodily good toward which sexual intercourse as a biological function is oriented, does not occur. (Girgis *et al.*, 2010, p. 254)

The problem with this explanation is that there is an important moral difference between a goal which "does not occur" even though people are "honestly seeking" it, and a goal which *cannot* occur, and which thus cannot be honestly sought by anyone aware of its impossibility. Unlike Einstein and Bohr, who are genuinely intending a solution, the heterosexual couple who know they are sterile cannot sincerely intend reproduction – and in that sense, they seem to be in the same position as the same-sex couple. (The same problem permeates Girgis's baseball analogy in this volume: It is not clear how a team's activities could be coordinated toward *winning* if they knew that there were literally no chance that they could win – say, because the rules added two points to the other team for every point the one team scored.)

Much ink has been spilled on the sterile-couples objection, and I do not want to spend excessive time on it here. (See, for instance, Koppelman, 2002.) The important thing to note – and something that all parties in the debate agree on – is this: When the NNL theorists say *reproductive-type*, they do not mean that reproduction must be possible, and they do not mean that it must be intended. So what do they mean?

As far as I can discern, what they really mean is that the sex must be *coital*: penis-in-vagina. (Some NNL theorists further narrow "reproductive type" to *uncontracepted* coitus, although there is debate on this point.) Coitus is the only act in which male and female human beings can join together to become "literally, not metaphorically one organism." And that, according to NNL, is what the comprehensive union of marriage requires.

If marriage requires "reproductive-type" acts, and "reproductive-type" means "coital," then it does indeed follow that same-sex couples cannot marry. Unfortunately for NNL, it also follows that Bob and Jane (our hypothetical couple from the last section) cannot marry.

Recall that Bob, because of his accident, was incapable of coitus. Thus Bob and Jane were unable ever to engage in "reproductive-type" acts – "acts that constitute the behavioral part of the process of reproduction, thus uniting them as a reproductive unit," to borrow language from Girgis, George, and Anderson. So the NNL account of marriage entails that Bob and Jane's so-called "marriage" was in fact a sham, and that (at least in principle) it should never have been legally recognized by the state.

If this conclusion sounds harsh, that is because it is. But it follows inescapably from the NNL view. NNL theorists need some factor by which to distinguish same-sex from different-sex couples. They cannot use ability to procreate as the relevant factor, unless they want to conclude that the septuagenarian couple (and other permanently infertile heterosexual couples) cannot marry. And they cannot simply say that the former is same-sex whereas the latter is different-sex, for that answer would be question-begging and circular. Instead, the distinguishing factor they choose is coitus.

The upshot is that the most sophisticated available argument for why marriage must be heterosexual has a bizarrely counterintuitive implication: No matter how loving and committed Bob and Jane may have been (to each other and to their adopted children), they were never in fact really married. And since the state "is justified in recognizing only real marriages as marriages," they should never have been allowed to marry legally.

Why Must Marriage Be Coital?

In Chapter 19 of this volume, Girgis concedes that Bob and Jane were never really married, at least given the "strong" version of the conjugal view, in which marriage requires the intention (and thus the expected ability) to perform coitus. He also attempts a "softer" version, on which a couple's union can be marital so long as coitus is "possible *in principle*" (Chapter 18 in this volume, Girgis, 2014, p. 300). It is not clear how this softer view gets off the ground, however. Any random male-female pair could *in principle* engage in coitus, just as they could in principle do any number of other things that Girgis's definition of marriage requires. But marriage does not consist in things that people *might* do if the world were different; it consists in what they actually do. Suppose Bob were kidnapped before the wedding and never returned to Jane. In that case, they would (sadly) never marry, even though they could marry *in principle* and even though their failure to do is involuntary. So if marriage requires comprehensive union as the NNL view defines it, and Bob and Jane are incapable of that union, then they are incapable of marriage: the strong conjugal view.

On the legal point, Girgis backpedals a bit: He responds that the state should still permit such marriages because it would be too "invasive" to inquire about such matters and because the hidden nature of Bob's incapacity means that recognition of his marriage would not undermine the proper public understanding. But there are several problems with this response. First, the question: "Do you intend to perform coitus?", written discreetly on an application form, is surely far less invasive than, say, a blood test (which several states require for a marriage license). Second, the case could be modified so that Bob's incapacity is as obvious to any court clerk as his gender: imagine, for example, that his horrible accident severed him at the waist but that he somehow survived. Third, and most important, the response does not cure the view of its counterintuitive implication: Girgis still concedes (with the strong conjugal view) that Bob and Jane were never really married; they simply "slipped by" the state's and the public's notice and are essentially engaging in an act of deception or fraud.

It is important not to misunderstand standard intuitions about this case. The common reaction is not, "Oh, isn't it unfortunate that poor Bob could never marry." The common reaction is that Bob and Jane *did* marry, not fraudulently but really, and that any theory that insists otherwise must be mistaken.

Why do we recognize Bob and Jane's union as a marriage? Salient factors include their romantic love, public lifelong commitment, mutual care and concern, sexual intimacy, and joint raising of a family. Viewed holistically, this union looks like a marriage, even though it lacks a feature (coitus) commonly associated with marriages. Yet once one concedes that marriage is possible without coitus, one removes the NNL bar to same-sex marriage.[3]

Attentive readers will notice that I have yet to offer any direct statement of my own view of marriage. The omission is deliberate, and the reason for it is twofold. First, as a complex, multifaceted, and evolving social institution, marriage does not lend itself well to pithy definitions, unless one packs those definitions with hedge-words ("typically," "primarily," "presumptively," and so on). I suspect marriage may be better understood as a family-resemblance concept – as a collection of related practices with overlapping similarities – rather than in terms of strict necessary and sufficient conditions. In order to rebut the NNL account, however, all I need to establish is that coitus is not a strictly necessary condition.

Second, in the context of same-sex-marriage debate, the demand for definitions is a recipe for question-begging: Each side typically puts forth definitions expressly with the purpose of including or excluding same-sex couples, and we get no closer to shared understanding. Girgis aims to avoid this problem by pointing to historical accounts that echo his own,[4] but revisionists can (and often do) make a similar move by pointing to cultures in history that have recognized same-sex relationships as genuine marriages.[5] They can also point to the familiar vow, which says nothing about coitus, much less children: "to have and to hold from this day forward, for better for worse, for richer for poorer, in sickness and in health, to love and to cherish, till death us do part." In any case, *what marriage has been* will not settle the question of *what marriage can, or should, be*. Whatever understanding of marriage emerges will be a result of the debate, not a premise or stipulation within it.

So having bitten the bullet on Bob and Jane's case, Girgis must pursue a different strategy: to argue that, while the conjugal view may seem counterintuitive, the so-called "revisionist" view is even more so. (As an aside, I find the term "revisionist" rather tendentious here. It is akin to my labeling Girgis's view the "discriminatory" view. So from here on I shall use "inclusivist" instead of "revisionist" to refer to any view of marriage that permits same-sex marriage.) His main objection is that the inclusivist view "cannot distinguish marriage from companionship *simpliciter*;" in a related vein, he argues that inclusivists cannot explain: (a) why marriage is inherently a sexual kind of union, as opposed to some other kind; (b) why it is permanent and exclusive, involving two and only two people; and (c) why it is connected to procreation and child-rearing (Girgis, 2014, pp. 292–295).

Before I respond to each of these specific points, let me make a general observation. Girgis seeks a "bright line" category of marital relationships, which share some unique property that distinguishes them from all other relationships and explains their distinctive features. For him, the underlying principle is *comprehensive union*, the union of both body and mind. In my view, Girgis tries to squeeze too much juice out of a single orange. While I agree that marriage is comprehensive in an important sense (though I differ with Girgis on the details), and while I also agree that, as a general matter, it is presumptively sexual, permanent, exclusive, connected to procreation and childrearing, and properly an object of state interest, I do not believe that all of these features fall neatly out of a single principle. Marriage is a human social institution with a messy history, and its internal logic evolved from multiple overlapping sources, related to children, adults, property, sex, love, religion, self-expression, and more. These sources are seldom explicit or deliberate, and they are sometimes in tension with one another. To put the point another way, marriage is not some eternal Platonic Form that various practices exemplify more or less well. Marriage's nature, such as it is, is the outcome of a complex human history. (Of course, if marriage is ordained by God, that is another matter entirely – but Girgis claims not to be invoking theological premises.)

My response to Girgis's specific criticisms of the inclusivist view will naturally reflect this more general difference between us. Take, for example, his claim that inclusivists "can *stipulate* that sex is uniquely relevant to marriage [. . .] but they cannot *explain* it" (Girgis, 2014, pp. 292–293). In one sense, there is nothing terribly difficult to explain. Humans find themselves powerfully sexually attracted to one another, and they tend to pair off in romantic couplings. Sometimes those couplings become long-term relationships, in which the partners merge their lives and their property. Sex can be an integral part of those relationships, not only because it may produce children, but also because it celebrates, replenishes, and enhances the intimacy between them. Maintaining such relationships over time is valuable, and yet also complicated and challenging; thus it is helpful to surround them with both legal recognition and social support: what we call the institution of marriage.

But is sex *essential* to marriage? There are two ways an inclusivist might answer. One is to grant that sex is an essential feature of marriage, but to argue that the sex need not be coital. Virtually everyone recognizes that there is a phenomenologically distinctive value in sexual intimacy – that the kind of pleasures it realizes are simply not interchangeable with those of deep conversation or a passion for art, as Girgis implausibly suggests (Girgis, 2014, pp. 292–293). Moreover, sexual relationships are fraught with special opportunities and risks, not only because of the possibility of children, but also because of sex's connection with deep human feelings of power and vulnerability, connection and isolation, joy and sorrow. So an inclusivist can agree with Girgis that marriage is a distinctive kind of relationship with certain essential features, including sex, but deny that the sex must be coital.

Alternatively, an inclusivist might treat marriage as a family-resemblance concept, in which sex is an important *typical* feature but not a strictly necessary one. Return to Bob and Jane, and modify their case slightly so that they never have sex of any sort, even though all other features remain the same: they fall in love, they publicly exchange marriage vows, they adopt and raise children together, and so on. I contend that it is still plausible to hold that they have a marriage, albeit a non-sexual one. But either way, the inclusivist can explain why marriage is (essentially or typically) sexual without requiring it to be coital.

Indeed, Girgis's own view does not explain the difference between sex and other human interactions as well as he thinks. For Girgis, what distinguishes sex from other activities is that only sex can unite two people as one organism. But what about non-coital forms of sex? What about romantic cuddling and kissing? Girgis is in the same boat as inclusivists when it comes to explaining why these activities are distinctive to marriage – unless he stipulates that they are all merely precursors to coitus. Such a stipulation would conflict badly with the self-understanding and lived experience of most couples.

There is also a notable oddity that arises in Girgis's discussion of comprehensive unions. He writes, "[M]arriage also includes bodily union. This is because your body is a real part of *you* . . . . [A]ny union of two people must include bodily union to be comprehensive, to avoid leaving out a basic part of each person's being" (Girgis, 2014, p. 296). But from the fact that your body is an integral part of you – a point on which Girgis and I are in full agreement – it follows that *any* personal union for human beings must in some sense be bodily: Disembodied minds do not form friendships, collaborate on scholarly projects, visit art galleries, and so on. And marriage's being *comprehensive* clearly does not require that the spouses unite in every way possible: They need not be scholarly collaborators or enjoy art together, for instance (as Girgis recognizes). So Girgis and the NNL theorists need to do a better job of explaining why comprehensive union requires *organic bodily union*, as they understand it.

Girgis complains that until the inclusivist gives a reasonably full positive account of what marriage is, it is impossible to compare the inclusivist view to the conjugal view in order to determine which fits better with our intuitions. I have two responses to this point. First, the point is true only to the extent that people's intuition that same-sex marriage is impossible outweighs their intuition that Bob and Jane were really married. Girgis believes that, given the choice between affirming that same-sex unions can qualify as marriages and denying that Bob and Jane's relationship can qualify as one, most people will choose the latter. I disagree: I suspect that most people would insist that Bob and Jane are married and then – if they still wish to oppose same-sex marriage – grant that such unions can be marriages but argue that they are nevertheless immoral. (That is, they would opt for some objection other than the Definitional

Objection.) This is, after all, the route that most opponents take with remarriage after divorce and polygamous marriage, two other situations that Girgis and his fellow NNL theorists deny to be real marriages.

As an aside, note that intuitions about divorce are quite relevant to an equality argument that inclusivists can make. When Girgis claims that same-sex marriage would "replace one view of [marriage] with another, finishing (I would say) what policies like no-fault divorce began" (Girgis, 2014, p. 291), he effectively concedes that same-sex couples are being held to a more restrictive definition of legal marriage than their heterosexual counterparts. If heterosexual couples are not required to satisfy the conjugal view (because they are allowed to divorce), is it not unfair to require it of gays and lesbians?

But there is a second response to Girgis's complaint about definitions: Unlike Girgis, I do not believe that the key question in the same-sex-marriage debate is "What is Marriage?" To see why, consider what I like to call the Marriage/Schmarriage Maneuver. Imagine a same-sex-marriage advocate who, after hearing the Definitional Objection, responds:

> You know what? You're right. This thing we're advocating isn't marriage at all. It's something else – let's call it *schmarriage*. But schmarriage is better than marriage: it's more inclusive, it helps gay people without harming straight people, etcetera. We'd all be better off if we replaced marriage with schmarriage. Now, it's unlikely that the word "schmarriage" will catch on – and besides, it's harder to say than "marriage." So from now on, let's have schmarriage – which includes both heterosexual and homosexual unions – but let's just call it by the homonym "marriage," as people currently do in New York, Iowa, Canada, South Africa and elsewhere. Okay?

Notice that in this hypothetical scenario, the question has shifted. It is no longer "What is marriage?" since our interlocutor agrees that the new institution, whatever name we use for it, is not marriage in the traditional sense of the word. And there is no hard-and-fast rule against using the same word to refer to different things: recall our earlier examples of *avocat* and "champagne." Yet the same-sex marriage debate would still rage on, because our disagreement was never really about what marriage is: It is about whether we should grant same-sex couples the same legal rights and responsibilities as different-sex couples.

Permanence, Exclusivity, and the Connection to Childrearing

I conclude that the conjugal view offers no advantage in explaining why marriage is sexual. What about the idea that marriage is exclusive and permanent? Here Girgis offers a rather peculiar argument, based on his analogy between bodily union in coitus and the union of organs in a single human body (heart, stomach, lungs, etc): "If spouses commit to form a truly organic union, it is fitting that they commit to the kind of union proper in a healthy organ: exclusive and lifelong" (Girgis, 2014, p. 299). But this conclusion does not follow. One might just as well argue that since bodily union is like the union of heart, stomach, and lungs, it should be continuous (perpetual coitus!) and also pinkish in color. The fact that coitus shares *some* features with the union of organs in a single human body does not entail that it shares *all* features, including exclusivity and permanence.

To be clear: I am not denying that exclusivity and permanence are valuable features of marriage. But the argument for them will need to come from somewhere other than the nature of "organic bodily union." While it is true that human biological reproduction requires two and only two people, one male and one female, it is equally true that individuals can form reproductive pairs with multiple others. In that sense, they are quite unlike the heart and the lungs, which (aside from transplants) are pretty much stuck where they are. But if the argument for permanence and exclusivity comes from somewhere other than organic bodily union, it will likely be just as available to inclusivists as to those who accept the conjugal view.

It is also worth noting that while permanence and exclusivity are valuable features of marriage, they should probably not be part of its *definition*. According to most people's intuitions, King Solomon really did have many wives, whether or not we approve of that choice. Ronald Reagan was really married to Nancy Davis after he divorced Jane Wyman. In so far as the conjugal view maintains otherwise, it is currently more "revisionist" than the inclusivist view. Girgis, like many proponents of the Definitional Objection, tends to vacillate between providing a descriptive account of what marriage is and a normative account of what features are desirable in marriage. Doing the latter is fine, but it moves away from the Definitional Objection, which states that same-sex unions are simply not marriages.

Girgis also argues that the inclusivist view cannot sufficiently explain marriage's connection to childrearing. Unfortunately for Girgis, the NNL understanding of "reproductive type" acts – wherein "reproductive type" acts are neither necessary nor sufficient for reproduction – leaves him vulnerable to the same charge. Recall that, given the NNL understanding, "reproductive type" acts are not necessary for reproduction because reproduction may occur via *in vitro fertilization* and other technologies; they are not sufficient because coitus frequently does not result in reproduction. The upshot is that children are only *indirectly* related to marriage in Girgis's account.

Again, to be clear: no sane person in this debate denies that children are a crucial rationale for the institution of marriage. But it is important to ask why this is so. The answer is not – or at least, not merely – that their presence "fulfills and extends a marriage, by fulfilling and extending the act that embodies (consummates) the *commitment* of marriage: sexual intercourse, or the generative act," as Girgis puts it (Girgis, 2014, p. 297). The answer is far simpler, and it does not require accepting anything like the NNL conjugal view: Children need stability and permanence. That is as true for Bob and Jane's (adopted) children as it is for the natural offspring of any couple. Indeed, it is true for the many children currently being raised by same-sex couples. When children arrive in a family, whether by natural birth, artificial insemination, surrogacy or adoption, they need to know that their parents are going to stick around for them and for each other, despite the vicissitudes of romantic passion. Marriage helps to ensure that.

Conclusion

The NNL conjugal view appears at first glance to have some real advantages in the marriage debate. It neatly captures familiar wedding rhetoric about how "the two become one," and it provides a simple, bright-line distinction between marital relationships and all other kinds. But that simplicity comes at too great a cost. It ignores marriage's complex history: sometimes monogamous, sometimes polygamous; sometimes egalitarian, sometimes hierarchical; sometimes with shared domestic life, sometimes without; sometimes child-focused, sometimes (unfortunately) not. Moreover, it conflicts with strong intuitions about contemporary cases, such as that of Bob and Jane.

The correct view is, necessarily, more complex. Marriage is defined by a list of typical features, including a public lifelong commitment, mutual care and concern, sexual intimacy, and joint raising of offspring. But "typical" does not mean "strictly necessary," and for any of these features (and many more) it is not difficult to find examples of genuine marriages which lack the feature. As anthropologists know well, there will be some "gray-area" cases, where it is unclear whether or not to classify the arrangements as marriages. One does not settle such cases simply by throwing down a definition, but rather by engaging in an ongoing process of reconciling competing intuitions, both within oneself and across individuals and cultures.

In the midst of this process, it is easy to forget that the salient debate here is a moral and political debate, not a conceptual or epistemological one. In other words, our ultimate question is not "What is marriage?" but rather "How should we treat gay and lesbian individuals and couples as members of a larger society?" Focusing on *what marriage is* may elucidate that debate, as Girgis hopes. But it may, I fear, instead distract from it.

Notes

1 In addition to those mentioned, see Girgis *et al.*, 2010, p. 247: "[T]he nature of marriage (that is, its essential features, what it fundamentally is) should settle this debate."
2 I say "mostly" because, in an argument of this sort, there are relevant empirical questions about how competent speakers use the term "marriage."
3 Some have told me that Bob and Jane's case is "exceptional," and we do not define marriage by the exceptions. But this response, aside from being vague, essentially gives up the argument: once we allow room for "exceptions" to the rule that marriage must include coitus, we can simply include same-sex marriages among them.
4 Although this does not quite avoid the problem in the way he thinks. He writes, "Nor can animus against a particular group have given rise to the view, which was developed and implemented long before the 19th century medicalization of homosexuality and the subsequent rise of gay cultural identity, let alone the Stonewall riots" (Girgis, 2014, p. 291). One does not need to recognize "gay cultural identity" to display explicit or implicit animus toward persons in same-sex relationships, as the Inquisition (among other oppressive moments in history) made abundantly clear.
5 For a useful history of marriage, including cultures that recognize same-sex marriage, see Coontz, 2005.

References

Coontz, S. (2005) *Marriage: A History*. New York: Viking.
Corvino, J. and Gallagher, M. (2012) *Debating Same-Sex Marriage*. New York: Oxford University Press.

Gallagher, M. (2009). The Maine vote for marriage. Posted to Real Clear Politics, November 5, 2009. http://www.realclearpolitics.com/articles/2009/11/05/the_maine_vote_for_marriage_99020.html (last accessed 6/17/13).

Girgis, S. (2014) Making sense of marriage. In *Contemporary Debates in Applied Ethics*, 2nd edn, ed. A.I. Cohen and C.H. Wellman, pp. 290–303. Malden, MA: Wiley-Blackwell.

Girgis, S., George, R.P., and Anderson, R. (2010) What is marriage? *Harvard Journal of Law and Public Policy* 34: 245–287.

Koppelman, A. (2002) *The Gay-Rights Question in Contemporary American Law*. Chicago, IL: University of Chicago Press.

Lee, P. and George, R.P. (1999) What sex can be: self-alienation, illusion, or one-flesh union. In *In Defense of Natural Law*, ed. R.P. George, pp. 161–183. Oxford: Oxford University Press.

Towleroad (2011). Rick Santorum compares marriage to a napkin [video]. Posted August 9, 2011. http://www.towleroad.com/2011/08/santorumnapkin.html#ixzz1XqWnk2fR (last accessed 6/17/13).

Ventrella, J. (2004/2005) Square circles?!! Restoring rationality to the same-sex "marriage" debate. *Hastings Constitutional Law Quarterly* 32: 681–724.

Further Reading

Corvino, J. (2005) Homosexuality and the PIB argument. *Ethics* 115: 501–534.

Graff, E.J. (1999) *What is Marriage For? The Strange Social History of Our Most Intimate Institution*. Boston: Beacon Press Books.

Macedo, S. (1995) Homosexuality and the conservative mind. *Georgetown Law Review* 84: 261–300; and Reply to Critics, in the same issue, 329–337.

Mohr, R.D. (2005) *The Long Arc of Justice: Lesbian and Gay Marriage, Equality, and Rights*. New York: Columbia University Press.

Nussbaum, M., ed. (1998) *Sex, Preference, and Family: Essays on Law and Nature*. New York: Oxford University Press.

Rauch, J. (2004) *Gay Marriage: Why it is Good for Gays, Good for Straights, and Good for America*. New York: Henry Holt and Co.

Sullivan, A., ed. (1997) *Same-Sex Marriage: Pro and Con: A Reader*. New York: Vintage Books.

Wardle, L.D., Strasser, M., Duncan, W.C., and Coolidge, D.O., eds (2003) *Marriage and Same-Sex Unions: A Debate*. Westport, CT: Praeger.

Wolfson, E. (2004) *Why Marriage Matters: America, Equality, and Gay People's Right to Marry*. New York: Simon & Schuster.

CHAPTER NINETEEN

Making Sense of Marriage

Sherif Girgis

Introduction

Our national debate over gay marriage often seems like a chaos of loud, disconnected slogans, but in its essence it is a conflict between two visions of marriage. On what I will call the *conjugal view*, marriage is essentially a *comprehensive* union. Joining spouses in body as well as mind, it is begun by commitment and sealed by sexual intercourse. So completed in the acts of bodily union by which new life is made, marriage is itself deepened by procreation, and calls for that broad sharing of domestic life uniquely fit for family life. Uniting spouses in these all-encompassing ways, it also calls for all-encompassing commitment: permanent and exclusive. Comprehensive union is valuable in itself, but its link to children's welfare makes marriage a public good that the state should recognize and support.[1]

The second view, which I call the *revisionist view*, rejects the criteria for comprehensiveness proposed in the conjugal view. What sets marriage apart from other close bonds, on this view, is an affective, emotional union of special intensity, usually lived out in home life and enhanced by mutually agreeable sexual activity. While emotion lasts, these unions are distinctively valuable in themselves, and the state has an interest in recognizing them to keep them stable.

It should be clear that, on the conjugal view, only sexually complementary couples (one man and one woman, as it is said) can enjoy the bodily union required for comprehensive, marital union. On this view, two men or two women cannot form a true marriage. It is equally clear that any two people can enjoy the emotional union at the heart of the revisionist view. Marriage, on this view, is possible between two men or two women. And no doubt some opposite-sex couples think of their bond in terms the revisionist view would commend. But these two different stances toward same-sex

Contemporary Debates in Applied Ethics, Second Edition. Edited by Andrew I. Cohen and Christopher Heath Wellman.

relationships are merely logical applications of the two views of marriage, not judgments on anyone's personal worthiness to receive a public good. It cannot be said too clearly or too often: This debate is not about whether to expand marriage, but whether to replace one view of it with another, finishing (I would say) what major policy changes, such as no-fault divorce, began.

Here I will only touch on the possible harms of that legal redefinition. I will argue mainly for the conjugal account of what *marriage is*. In doing so, I take no position on whether same-sex erotic desire can be changed, or on the moral status of same-sex sexual acts. (I also rest nothing on the fallacious "perverted faculty" argument, which considers it wrong to use organs against their natural purposes.) I infer nothing about how marriage *must* be from how it always *has* been. And I require no theological premises. Indeed, ancient thinkers who had no contact with religions such as Judaism or Christianity – including Aristotle, Plato, Socrates, Musonius Rufus, Xenophanes, and Plutarch – reached views of marriage in line with the one I defend.

Nor can animus against a particular group have produced this view, which was implemented long before the nineteenth century medicalization of homosexuality and the subsequent rise of gay cultural identity, let alone the Stonewall riots. Indeed, some cultures, such as certain ancient Greek communities, clearly took the conjugal view for granted while celebrating homoeroticism. (Yes, same-sex sexual acts were long treated as immoral – but so were analogous acts between a man and woman.)

Finally, my argument cannot be refuted by pure appeals to equality or neutrality. Theories (or laws) that distinguish marriage at all will *always* leave some bonds out; that does not mean they all violate equality. To know when it is a true marriage that a theory (or the state) refuses to recognize, arbitrarily, and when what is excluded is something else entirely, one must first have some idea of what marriage is, and why we legally recognize it. The conjugal and revisionist views are two answers to this central question: both morally charged, both accepted by some citizens and religions and rejected by others. Neither answer is "neutral." Which, then, is right? What is marriage?

Marriage as a Basic Human Good

Same-sex *civil marriage* exists in many jurisdictions. Yet I am arguing that same-sex marriage is, in fact, impossible. What does this mean? And how can it amount to anything more than a quibble about dictionary entries? To answer that I offer a bit of background.

There are, according to a large family of ethical theories (objective-list theories, in Derek Parfit's terminology), certain conditions or activities that make us better off, whether or not they bring further goods. I call these conditions or activities (health, knowledge, skillful performance, and critical aesthetic appreciation, for example) *basic human goods*.

These goods can be realized in many ways, but they have some requirements. Friendship, for all its cultural variety, has an objective core, fixed by our social nature: mutually acknowledged good will and cooperation. Without that, two people's connection simply lacks the distinctive value (and duties) of friendship. To overlook this is to err about a human good, not just a word.

Marriage, I submit, is a basic human good, and partakes of the objectivity and basic determinacy common to all such goods. For all its cultural varieties, shaped by shifting historical demands, marriage as a human good has a core, fixed by the demands of our nature as sexual-reproductive beings. And to deviate from it is to miss a crucial part of a basic good, not just a definition.

To agree that goods have some objective features, one need not believe in God, just in some constants of human nature – at least across some time span. If our species evolved into one that reproduced asexually, then that (new) species would be one for which nothing like marriage existed. But this, far from undermining the conjugal view, reinforces it.

What is clear, at any rate, is that most people on both sides of this debate believe marriage is not just close friendship, but has a distinctive benefit and distinctive *requirements*. Most agree that certain relationships simply cannot count as marriage, and that some other kinds simply must. (Indeed, most revisionists implicitly deny that marriage is just a construct: Their arguments often presuppose a *natural right* to marriage that laws might violate by being too narrow.)

Consider the content of our common intuitions about marriage: Marriage is inherently sexual, it is uniquely enriched by family life, and it uniquely requires permanent and exclusive commitment to begin at all.[2] What best explains this mix of characteristics? I contend that these are characteristics of a basic human good that only a couple of one man and one woman can realize. The movement to redefine marriage in the law, should it succeed, will legitimate and entrench in our public institutions a profound error about this human good. Because the law is a teacher, this error will spread and many people will internalize it and live it out, to the harm of individuals and indeed their societies.

Problems with the Revisionist View

My main objection to the revisionist view is that it has no cogent way of distinguishing marriage from companionship *simpliciter*. This objection is alone decisive, if it is sound. But I have never seen a good reply to it – in my counterpart Corvino's piece in Chapter 18, or elsewhere.

Consider Oscar and Alfred. They share a home and domestic duties. Their unparalleled mutual trust makes each want the other to be the one to visit and manage his care if he is ill, and inherit his assets if he dies. Each offers the other ready counsel in distress, security amid hardship, company in defeat and from every personal victory, a two-way tie. They face the world together.

On the revisionist view, it seems obvious that Oscar and Alfred have every right to march up to the courthouse and demand recognition as spouses. But consider the additional detail that Oscar and Alfred are bachelor brothers who have never considered a sexual relationship. Here, I believe, the clear consequences of the revisionist theory conflict with near-universal judgments and intuitions. Oscar and Alfred's relationship may be worthy of great respect, but their bond should not be recognized as a marriage, because marriage somehow involves sex.

At this point, most revisionists would agree, and hurriedly stipulate that sex is uniquely relevant to marriage. They certainly can *stipulate* as much, but they cannot *explain* it. They may say sex generally fosters and expresses the emotional intimacy that *really* makes a marriage, but they cannot maintain that sex is in that respect unique. Two celibate monks can share deep conversation, cooperation amid hardship, custody of an orphaned child, or a passion for art, and feel themselves to be twin souls. That does not give their bond the distinctive value and norms of a marriage.

But why? *Why* is sex inherently and categorically more marital than other pleasing activities that build attachment? What about sex, *apart* from its emotional effects, makes it critical to marriage?

Corvino's *causal* account of the connection – that marriage often includes sex because the couplings we call "marriages" often result from sexual attraction – is unresponsive here. It does not explain how sex might accomplish something so different from other activities that it can set a whole class of bonds apart (in terms of the basic good they realize and the commitment they require) from the spectrum of friendships sealed by non-sexual activities and compatible with a variety of commitments. On examination, I argue in the next section, only *coital* sex does any such thing; only it constitutes a *bodily* union as non-sexual acts cannot.

There is more. Suppose that Oscar, Alfred, and a third man – Herman – are romantically involved, as a triad (like a "throuple" recently profiled in *New York Magazine*). If one dies, the other two are co-heirs. If one is ill, either can visit. They advise and console each other, and share major experiences. If Oscar and Alfred could form a marriage, why not Oscar, Alfred, and Herman?

Again, the revisionist could *stipulate* that the emotional union of marriage should include only two people. But why? Indeed, *how*? It seems difficult to restrict the range of emotion. We cherish sincere and legitimate feelings of tenderness and confidence of varying intensity toward friends and relatives in connection with a range of activities. The kind of romantic union revisionists want to single out may indeed have a particular emotional hue, but does this shade of difference suffice here? Is it really enough to mark off a distinct human good?

The revisionist theory looks even shakier when we remember that marriage involves a free *commitment* to exclusivity and so presupposes that some future behavior or state of affairs is under our direct control. Is it reasonable to think that the main content of our solemn promises is to restrict certain vaguely defined feelings to one person on earth?

It gets us no farther to invoke empirical data about preferences. Yes, most people in love want monogamy, but most want the opposite sex, too. Neither preference is universal, and each seems to be losing normative ground. Rank-and-file revisionists too hastily dismiss the claims of polyamorists, who may find in their group sexual unions a superior package of emotional benefits, including variety and freedom from the suspicion and deceit that often mar officially monogamous relationships. The result, some say, is greater fulfillment and stability overall. What is the revisionist's reply?

Another problem arises when we consider marriage's distinctive relation to children. Traditional jokes and modern sociology alike show that child-rearing is no foolproof aphrodisiac: Its pressures may take an emotional toll on a relationship. The conjugal view can acknowledge this yet explain (as I sketch below) how family life enriches marriage as such, even as compared with other stable bonds – say, our monks raising an orphan. Revisionists cannot explain this or any other systematic difference between marriage and deep friendship.

Corvino replies that the revisionist need not give a complete theory of marriage. I agree, not least because there are no "complete" theories: Any plausible account will draw imperfectly sharp borders, where close cases always lurk. But we do have to sharpen our accounts enough to decide the issue at hand. In a debate about whether same-sex partnerships can be distinguished in principle from the spectrum of affective bonds that all agree are (despite surface similarities) not marriages, the revisionist will have to offer *enough* of a theory to set committed same-sex partnerships apart from companionship generally. This Corvino fails to do.

Nor am I setting us up to beg the question by just tailoring our views to reach our conclusions, as Corvino fears. I am suggesting that we each offer *enough* of a positive account of marriage to: (a) support our conclusions about same-sex unions, and (b) render our own view more plausible overall. We can do the latter by showing that our account has better overall fit with shared intuitions, or better explains the unity of the features it ascribes to marriage.

What, after all, is the alternative? Corvino does point to similarities between same-sex relationships and undisputedly marital bonds, but *any* two things are similar in some ways, different in others. What Corvino must show is that same-sex unions are similar enough to clear cases of marriage. But this presupposes a view of what is "enough" to make a marriage: a theory of marriage. I suggest only that Corvino make his explicit, so that we can assess its implications.

Consider Corvino's argument that: (1) "Bob and Jane" in his example are clearly married, but (2) my view cannot distinguish their union from gay partnerships (Chapter 18 in this volume, Corvino, 2014, pp. 279–282). This is no *positive* case for the revisionist view, which it treats as a default – vindicated if my objections fail. But the nearly universal consensus on marriage until now surely lays the burden of proof on revisionists. And even as against my view, Corvino's example succeeds only if (a) our judgment that Bob and Jane are married is firmer than our judgment that two men cannot be, and (b) the revisionist view is more plausible than what I call below the "softer" conjugal view, on which Bob and Jane *are* married. Neither can be shown, especially when we consider the revisionist view's broader implications.

At this point, some people reply by changing the subject: by asking if anyone is *really* clamoring for other bonds to be recognized as marriages. Well, first, yes. In *Beyond Same-Sex Marriage*, over 300 prominent activists and academics such as Gloria Steinem and Kenji Yoshino demanded legal recognition of open and multiple-partner relationships, even certain non-sexual ones. Nor are such relationships unheard of: *Newsweek* reports that there are more than 500,000 polyamorous households in the United States alone.

Second, my chief task here is not to predict empirical effects, but to assess arguments. Revisionists (at least the most cogent) argue, effectively, that two men can marry because marriage is essentially about emotional union and domestic life. But that argument cannot explain other, much less controversial features of what makes a marriage: for example, limitation to two people, sexuality, and sexual exclusivity. Indeed, it renders these other features just as optional (just as arbitrary and discriminatory to require) as complementarity. So the best revisionist argument gets marriage wrong.

And this is related to the harms of implementing the revisionist view. This is no *mere* conceptual dispute, for the concept we dispute is of a basic human good. The more the law teaches us to internalize a revisionist view of this good, marriage – as it would if same-sex unions were recognized – the more monogamy, sexual exclusivity, and other marital norms would seem merely optional. That would make them harder to respect in practice, which would be bad in itself, since marriage is good in itself. But it would also harm children's development, and thereby the common good in all its dimensions, as my co-authors and I argue at length in *What Is Marriage? Man and Woman: A Defense*. If the policy status quo (including no-fault divorce) already erodes key marital norms, then resisting the recognition of same-sex unions is not an end, but just one step toward rebuilding a sound marriage culture.

Sketching an Account of Marriage

What, then, *is* marriage? Corvino and I agree that marriage is special, but perhaps we can best explain its peculiarities if we get a vocabulary and a set of concepts that apply to all voluntary relationships.

Two people (or more) enter a voluntary relationship by committing to do certain things – to engage in certain characteristic activities – that aim at certain shared goods. The parties also commit to protect and facilitate their pursuit of common goods. This commitment is specified by certain norms that govern the parties' behavior as long as the relationship lasts. When people commit to pursue given goods through embodying or characterizing activities under the restraint of certain norms, the result is what we call a relationship, or a union, or a community.

Let me give an example. Alvin, Simon and Theodore want to acquire knowledge, and for that purpose form a scholarly community. They commit to cooperate in research, publication, and other activities ordered to gaining knowledge. These activities characterize and distinctly build up their kind of bond: they make it most present and real. And for all these reasons, their bond demands a commitment shaped by norms that specially serve the search for truth – for example, high standards of accuracy even when this means tedium or embarrassment or economic loss.

We have identified three generic features by which we might classify kinds of union, for they are plausibly what *make* a union: unifying activities, unifying goods, unifying commitments. The union we know as marriage is special because it is in these respects *comprehensive*. It is comprehensive:

1 in the most basic dimensions in which it unites two people, because it unites spouses in their minds and bodies;
2 in the goods with respect to which it unites them, because in uniting them with respect to procreation, it directs them to family life, and its broad domestic sharing;
3 in the kind of commitment that it calls for, namely a permanent and exclusive commitment.

Comprehensive unifying acts: mind and body

Marriage clearly requires a unity of minds and wills. But marriage also includes bodily union. This is because your body is a real part of *you*, not a vehicle driven by the "real" you, your mind. This point is of fundamental and pervasive importance in ethics. If a man ruins your car, he vandalizes your property, but if he slices your leg, he injures *you*. More positively (and to inch closer to our subject), spouses find it uniquely fitting when their legal children are *also* a mixture of their two bodies. These and many other points highlight a fact of great moral significance: *we* include our bodies. Thus, any union of two people must include bodily union to be comprehensive, to avoid leaving out a basic part of each person's being.[3]

Most will agree with this last point, and say that this is obviously where sex comes in. It is sexual activity that satisfies the criterion of bodily union, which makes two people one flesh. The only contested question is why.

My answer[4] begins, again, with a more general philosophical reflection. Why do we ever say disparate parts form a real unity? Why, for example, do your many organs form truly one body? If the key were spatial proximity, a house of cards would be a unity in just the same sense, and your coffee cup would be part of you for as long as you held it. Nor is the key just a common genome: permanently separated parts of you might share that with you, as might your identical twin; and a transplant patient's heart can really become a part of him.

So, what makes for unity is rather activity toward common ends. Two things are parts of a greater whole – *are one* – if they *act as one*; and they act as one if they coordinate *toward one end that encompasses them both*. Your organs form one body because they are coordinated for the single biological purpose of sustaining your biological life.

Now here is the critical point: Even separate organisms, such as a human male and female, can achieve, in a limited but real sense, the very kind of union enjoyed by parts of a single organism, to the extent their bodies can coordinate toward a common biological end. This coordination is what happens in exactly one act: sexual intercourse, coitus.

Consider this: Humans, like other animals, have certain bodily, biological functions, such as digestion, toward which various parts of them coordinate. For all of these functions, individuals are naturally sufficient. We can walk, see, and digest as individuals. But there is one function that no individual can perform without another: reproduction. The coordinated action of coitus is, biologically, the first step of the reproductive process. Here the whole is the couple; the good, their reproduction; the unifying activity, their coordination toward that good in coitus. Achieving the good would crown the union (and deepens its basis by making them jointly responsible for offspring), but the coordination is enough to create it.

Traditional expressions reflect the view that coitus is specially unitive and hence specially marital. The "marital act" involves the most distinctively marital behavior (bodily union in coitus), chosen for distinctively marital reasons: to make spousal love concrete, to unite as spouses do, to extend their union of hearts and minds on to the bodily plane. And like other interpersonal unions that are valuable in themselves, a husband and wife's loving marital union is valuable even when conception is neither sought nor achieved.

Some mockingly ask what could be so special about "penis-in-vagina" sex. One might as well ask revisionists what could be so special about orgasms. With the right (unfair) description, any view can be scorned. The question is not whether there is a description that obscures the special value of conjugal acts, but whether there is a true description that highlights it. *Organic bodily union* and *life-giving act* are both related to the concept of *comprehensive union*. They make the special value of marriage luminous. But they apply only to husband and wife.

The centrality of this act explains why marriage, unlike many other enriching bonds, is only available to a biologically complementary couple – to husband and wife. If we want marriage to require comprehensiveness, we need bodily union, and if we are looking for bodily union – coordination toward a single bodily end of the whole pair – there is no substitute for coitus.[5]

Comprehensive unifying goods: procreation and domestic life

We have seen that marriage unites spouses in mind and in body, and is in that sense uniquely encompassing or comprehensive. Because it is oriented to children and family life, marriage also uniquely requires spouses to be open to the whole range of human goods.

The connection between marital commitment and parenting is intuitive, but it is easily misstated or misunderstood. Of course, children are not sufficient to create marriage, but neither are they necessary. Marriage is not a *means* to procreation, but it is *oriented* to procreation – inherently enriched or fulfilled by it and hence shaped by its demands. And procreation fulfills and extends a marriage by fulfilling and extending the act that embodies (consummates) the *commitment* of marriage: sexual intercourse, or the generative act.

To make this clearer, let me return to the notion of embodying activities developed in my general discussion of voluntary relationships. In an ordinary friendship, each friend unites with the other in mind and will in order to know and pursue the other's good. So friendships are sealed – friends are most obviously and truly *befriending* – in conversations and common pursuits. Scholarly relationships, oriented to knowledge, are embodied in joint inquiry and publication. In short, there is a parallel between the type of commitment that forms a bond, and the type of activity that most embodies that commitment and bond. And this holds for marriage. The embodying act of coitus, by its nature (even apart from partners' expectations or wishes), has procreation as its biological end. Marriage is ordered to family life because the same act that makes marital love also makes new life.

This orientation is unique to sexually complementary couples. Partners in other kinds of bonds may regard sex as a unique seal of their commitment, but neither the bonds nor their sealing activity will have an inherent orientation to procreation. It is no surprise that Corvino argues that marriage and children are only connected because children need stability (Corvino, 2014, p. 287). It follows from this, I think, that childrearing has no more significance or fittingness for married couples than for two sisters committed to cohabiting who believe childrearing would enhance their bond.

Some worry that the conjugal view here excludes infertile couples. It does not. Consider an analogy to, say, baseball. Teammates embody their union mainly by engaging in activity aimed at winning games. But they get the chief benefit of their union, and their commitment is rational and fulfilling, whether they win or lose on a given day, indeed, whether or not they *ever* win.

The same is true of marriage. Because marriage is ordered to childrearing, spouses arrange their common life as devoted and responsible parents would. They help each other to develop a comprehensive range of virtues, and to pursue a comprehensive range of goods – as is proper for those charged with guiding the all-around development of new human beings. They do this in the context of a permanent and exclusive union that includes generous sharing – which cultivates the stability and harmony children need. But such development and sharing, including the bodily union of the generative act, are possible and inherently valuable for spouses even when they do not conceive.

I have said "inherently valuable," and I say it again. The baseball analogy does not imply that infertile couples are "losers"; infertility reflects nothing of the couple's effort or character. But infertility is on this view a regrettable *loss* for spouses, precisely as spouses. It is a missed opportunity for their union to be, in a new and literal sense, *embodied*. What is more, infertile couples themselves are often the most emphatic on this point.

On the other hand, procreation is not the sole point, or even necessarily the most important part, of a marriage. Indeed, fixating on procreation can harm a marriage and obscure its nature. To treat a good in itself as merely a means to an end is to disrespect that good and to enjoy it imperfectly: A baseball team that focuses exclusively on winning will not enjoy much camaraderie or love of the game.

And yet, in baseball as in marriage, the end plays a crucial role even when beyond reach. Desired or not, achieved or not, procreation and winning each distinguish a practice by shaping some of its activities, especially activities that give the practice some of its distinctive value. If a group of people commits only to running laps around the field, and if they do not commit to doing anything ordered to winning a baseball game, they are not a baseball team. In the same way, a couple is not married unless they commit to bodily union in coitus, to an act and way of life *oriented* to procreation and family life.[6]

Finally, the life-sharing that most forms of community call for is limited, because the common values that define them are limited. (Sports may call for regular weekly or monthly cooperation, but there is no loss in not *living* with your bowling partner.) But marriage unites spouses in mind and body, and is oriented to producing not just one or another human value but whole new persons, new centers of value. So it calls for the broad sharing of life that would be needed for helping new human beings develop their capacities for pursuing *every* basic kind of value. Spouses benefit as spouses from *some* cooperation intellectually, in recreation, and so on. Thus, again, the conjugal view makes sense of marriage's links to family and domestic sharing.

Comprehensive commitment: norms of permanence and exclusivity

We have seen that marriage is comprehensive in so far as it: (1) unites spouses comprehensively (i.e., in mind and body), and (2) unites them in pursuit of a comprehensive range of goods, the range of goods proper to childrearing and family life. It remains only to show that marriage (3) inherently calls for comprehensive commitment, whatever the partners' temperament or taste. I submit that it is the first two facts that explain the third.

The unity of organs into a healthy whole is total and lasting for the life of the parts (i.e., till death). If spouses commit to form a truly organic union, it is fitting that they commit to the kind of union proper in a healthy organ: exclusive and lifelong.[7] Again, the marital commitment also involves spouses in the open-ended task of family life, an adventure that requires coordination of the whole of their lives and unconditioned commitment.

In short, a union comprehensive in these two senses must also be temporally comprehensive: through time (vowed permanence) and at each time (vowed exclusivity). These requirements do not merely fit a conceptual definition of marriage-as-comprehensive. They also happen to serve childrearing in a more functional sense, by excluding the destabilizing practices of infidelity and divorce. This harmony between conceptual elegance and practical wisdom is fitting, and should bolster the plausibility of the conjugal view.

The conjugal view has a further advantage: It shows not only that exclusivity is necessary, but how it is *possible*. Remember that revisionists believe marriage is, at its core, the most intense of emotional unions. Let us grant for argument's sake that they can give adequate reason to think this union should be exclusive; they cannot show why it should be exclusive with respect to sexual relations. Any given couple may feel that exclusivity on balance enhances their emotional bond, but what if they, like many in open relationships, say that it does not? The revisionist, who sets marriage apart by its degree of emotional intensity, has no answer. But on the conjugal view, the answer is immediate. Marriage is distinguished not by degree but by the *type* of cooperation it involves: bodily union and its natural fulfillment, children and family life. It is therefore not at all arbitrary to isolate sexual activity (and anything that fosters romantic desire, which *seeks* such union) as central to exclusivity.

In non-marital relationships, of course, various kinds of exclusivity can be useful, as can permanence. Stability is a value for many of our bonds. But its costs can sometimes outweigh its benefits. The conjugal view explains why marriage is an exception to this logic, but anyone who takes the revisionist view must conclude that marriage is no exception at all. Corvino *asserts* that permanence and exclusivity are "valuable" for marriage (Corvino, 2014, p. 287); he says nothing *at all* to show even that these features of the relationship are systematically preferable, for spouses as such, to alternative commitments.

Accept it or reject it, my judgment about nature and properties of marriage is nothing new, obscure, or discredited. The three great philosophers of antiquity – Socrates, Plato and Aristotle – as well as Xenophanes and Stoics such as Musonius Rufus, defended something quite like the conjugal view, sometimes amid homoerotic cultures. Especially clear is Plutarch's statement in *Erotikos* that marriage as a class of friendship is uniquely embodied in coitus (which he calls a *renewal* of marriage). Plutarch also says, in his *Life of Solon*, that intercourse with an infertile spouse realizes the good of marriage – something that these other ancient thinkers took for granted, even as they (like Plutarch) denied that other sexual acts could do the same.[8]

Then there is the history of the common law, an important guide to enduring moral understandings shared by large portions of the community. For hundreds of years (a) infertility was no ground for declaring a civil marriage void, and (b) only coitus was recognized as completing a marriage. Let us draw reasonable inferences: If the point of marriage is to keep parents together for the good of the children or potential children, why not let the clearly infertile dissolve their marriages? If the legislation aimed at stigmatizing or devaluing people attracted to their own sex, why not permit all heterosexual acts to consummate a marriage? There is no puzzle at all here if we assume that the law reflected this rational judgment: The uniquely comprehensive unions consummated by coitus are valuable in themselves, and different in kind from other bonds – that is, the conjugal view.

Closing Thoughts

Hopefully readers find the conjugal view, as I have explained it, in harmony with their deepest intuitions and best-grounded judgments about marriage. But there may still be some tension: The conjugal view may seem harsh in some cases. Here Corvino's argument that my theory excludes from marriage a paraplegic and his partner may be troubling (see Corvino, 2014, pp. 279–280). Let us think through the example.

Certainly, on the conjugal view, a marriage is *incomplete* without consummation, but this does not necessarily settle the ethical question of whether a paraplegic can form a true marriage at all, much less the distinct policy question of which bonds to recognize. I believe there are two possible responses, both of which comport with the theory laid out above.

The first response, what I would call *the strong view*, is to say that you cannot commit to marriage unless you intend coitus, and you cannot intend coitus unless you reasonably expect it to be possible in practice. This would certainly mean that a paraplegic cannot form a true marriage. Still, a good marriage policy would go on recognizing marriage in such cases. For one thing, inclusion is not harmful: Recognizing the marriage would not undermine the public understanding of marriage as conjugal union. For another, exclusion is harmful: Serious enforcement requires (at a minimum) asking highly invasive questions. This kind of interference with personal privacy requires more justification than the desire to bring legal reality into perfect conformity with moral reality, quite apart from the social effects of such conformity.

The second response, which I would call *the softer view*, is to say marriage only requires the intent to perform coitus when reasonably feasible, and this intention in turn requires only that coitus be possible *in principle*. Maybe the paraplegic's partnership is just on a spectrum with other opposite-sex unions: Each could consummate given normal conditions (good health, time to reach arousal, etc.), even though the former would remain unconsummated.

I admit that the strong view, as applied to these facts, sits ill with many people's intuitions. But even if reflection and debate eventually commit all supporters of the conjugal view to the strong view (and at present we are a house divided on the question), the position as a whole will remain much stronger than the revisionist view. On this philosophical question, consider three points.

First, if you have an intelligible, interesting, and plausible view of marriage it will almost certainly exclude some relationships whose participants want their bond included. But no plausible view would equate "non-marital" with "trivial." Everyone can form worthy, loving and sustaining relationships of various kinds. Indeed, it would be surprising if the conjugal view – which posits both volitional and bodily criteria for the comprehensive union of marriage – left out only same-sex bonds.

Second, what we want as moral philosophers is not a theory that justifies the exact constellation of dominant intuitions in our society here and now. These intuitions are shaped by many factors, only some of which are reliable. They are where we start our reasoning, but rarely where we end. What we (should) want is a line of *best* fit with our practices, experience, and judgments about how human beings are constituted (as mind–body unities) and how clearly distinct goods like friendship are structured. After all, again, the revisionist view has no basis in principle – none at all – for distinguishing marriage from companionship.

Third, Corvino has not raised a new *kind* of difficulty for the conjugal view, or indeed any view on which marriage is a determinate type of relationship. If a Purple Heart paraplegic Marine cannot form a true marriage, it is not any less bad than that a man without desire for women cannot marry, or that a woman who cannot find a mate cannot marry, or that an only child who is too tied up caring for her ailing mother cannot marry. These are all people of equal dignity for whom marriage would be a real fulfillment but is practically impossible through no fault of their own. So I do not think Corvino's example rebuts the conjugal view – either immediately, or on reflection and further comparison with the revisionist.

We might still regret some of its implications, but we must be careful not to think we can cure regrettable facts by redefinition. It would be bad if paraplegics could not really marry, but in that case reality, not anybody's theory, would be to blame. It is clear, to take a pointed example, that some people do not have the psychological ability to form marriages in the revisionist's sense. That highlights a regrettable loss – some people's inability to form a certain human good – but is no argument against the revisionist view.

Even more to the point, the revisionist's philosophical mistakes about marriage also exaggerate the badness of this loss. For revisionists marriage is simply and categorically the deepest and hence most valuable bond. To have something else is always to have something less. On the conjugal view, to have something else is just to have *something else* (a certain form of friendship), excellent in its incommensurable way, with its own characteristic scale and forms of depth and mutual presence and care. So the unmarried are not denied the pinnacle of social fulfillment.

Of course people who cannot marry, or who cannot marry the one they want, may seek some kind of relationship with romance and domestic life. The point of this argument is not to say that these relationships are bad or impossible. What I argue is that comprehensive union is crucial to making sense of the desirability and fittingness of the *combination* of these and other characteristic features of marriage, and thus of the distinctive value of whatever basic good (besides friendship) is on offer in their vicinity. And that if we treat fragments of this good as basically like the whole, we will not see or reliably live out their unity. Corvino seems discreetly to agree with me here: If sexual complementarity is optional we cannot be quite sure about any of the other distinctively marital norms. Perhaps there is just a "family" of configurations whose "resemblance" we intuit and so class together as marriages.

To close, let me reiterate why we should care about this detailed analysis of bodily union. For sometimes in investigating even the most straightforward topic, we find ourselves, as it were, in a dark forest, cutting our way through vines, and losing heart. But that is no discredit to the topic. Thus consent is clearly critical to the value of sex, but if we spent hours considering what it required (there are very hard cases), we might get lost in all the distinctions. It is then that we must pull back to remember our destination, the value of our inquiry. I have belabored the importance of bodily union here not because it is all that makes a marriage, but because I think it is essential to marriage and has lately been neglected.

Keeping this all in mind, one can indeed – as for centuries, almost every culture did – see something morally distinctive, even awe-inspiring, and crucial for marriage, in the sort of act that unites generation to generation as one blood, and man to woman as one flesh.

Notes

1   For a more complete defense of the view I sketch here, see Girgis, *et al.*, 2012, on which I draw at various points here.

2   The existence of polygamous cultures does not prove that monogamy is not part of the good of marriage. No significant moral truth has enjoyed universal assent. And yet it is natural to think that the most basic ethical principles would be most widely held; while *derived* ones would have patchier assent, since reaching them requires applying more principles. In that case, the historical record is no embarrassment for the conjugal view. Features that the latter considers most basic to marriage – such as bodily union and connection to family life – are nearly universal in practice. And what the conjugal view treats as *grounded in* these basics – permanent, exclusive commitment – is less represented. Hence widespread polygamy (historically), contrasted with the nearly perfect consensus on spouses' sexual complementarity. Moreover, even if polygamy partially realizes the good of marriage because it still includes this good's most basic features, the same cannot be said of same-sex unions.

3   Here I take for granted that, though we are essentially body–mind composites, all our unions do not extend equally on both planes. Silently consenting to an agreement unites people; a marital act unites people. And plainly, the latter involves bodily union as the former does not, even though neither unites *mere* minds or *mere* bodies.

4   The following discussion owes much to the work of Germain Grisez (1993) and Alexander Pruss (2012).

5   Some might be tempted to say the pursuit or enjoyment of sexual pleasure unites spouses, but this cannot be all. Pleasure adds value only when taken in some independent good. Even if it were inherently valuable, it would benefit the partners as individuals, not as a whole. Pleasure, after all, is private like other mental states. This point suggests the last reason – pleasure is a feature of experience, and is thus not really bodily in the relevant sense.

6   Of course, it is our scoring conventions that make hitting the ball ordered to winning games, but coitus is ordered to reproduction by nature. So coitus remains coordination toward reproduction, *whatever* the spouses' beliefs about conception, even if (as in Corvino's example [Corvino, 2014, pp. 281–282]) a team is no longer playing baseball if new scoring rules make a win impossible. Recall, too, that mating (*behavior*) is necessary, not sufficient for a marital *act*. For that, spouses must be choosing this behavior to embody their comprehensive union – and thus, for example, unwilling to seek it with others.

7   I am not inferring that *x* is a property of marriage from the fact that something like *x* is a property of bodily union. I am pointing to harmonies among the three ways in which marriage is comprehensive (in its distinctive acts, goods, and commitment), to highlight the unity of the conjugal view, thus bolstering it. Why comprehensive in *these* three respects? I suspect that relationship differences orthogonal to these dimensions will not differ much in terms of their value, while relationships that do differ in important ways, will differ along these dimensions. So, no, comprehensive cannot mean "comprehensive in *every* dimension."

But the same holds of most revisionists' master principle: A spouse cannot be your "number one partner" or "soul-mate" in *every* activity and domain.

8   See the essays on sex and marriage in Finnis (2011).

References

Corvino, J. (2014) Same-sex marriage and the definitional objection. In *Contemporary Debates in Applied Ethics*, 2nd edn, ed. A.I. Cohen and C.H. Wellman, pp. 277–289. Malden, MA: Wiley-Blackwell.

Finnis, J. (2011) *Collected Essays of John Finnis*, vol. III. Oxford & New York: Oxford University Press.

Girgis, S., Anderson, R.T., and George, R.P. (2012) *What is Marriage? Man and Woman: A Defense.* New York: Encounter Books.

Grisez, G. (1993) Why is every marriage a permanent and exclusive union? In *Way of the Lord Jesus*, vol. 2. San Jose, CA: Franciscan Press.

Pruss, A. (2012) *One Body*. South Bend, IN: University of Notre Dame Press.

Further Reading

Bennett, J. (2009) Only You. And You. And You: Polyamory – relationships with multiple, mutually consenting partners – has a coming-out party. *Newsweek*, July 29, 2009. http://www.newsweek.com/2009/07/28/only-you-and-you-and-you.html (last accessed 6/17/13).

BeyondMarriage.org (2006) Beyond Same-Sex Marriage: A New Strategic Vision For All Our Families & Relationships. July 26, 2006. http://beyondmarriage.org/full_statement.html (last accessed 6/17/13).

Cherlin, A. (2009) *The Marriage-Go-Round*. New York: Knopf.

Gallagher, M. and Corvino, J. (2012) *Debating Same-Sex Marriage*. New York: Oxford University Press.

George, R.P. and Elshtain, J.B. (2010) *Meaning of Marriage: Family, State, Market, & Morals*. New York: Scepter Publisher.

Lee, P. and George, R.P. (2008) *Body-Self Dualism in Contemporary Ethics and Politics*. New York: Cambridge University Press.

Oppenheimer, M. (2011). Married, with infidelities. *New York Times*, June 30, 2011. http://www.nytimes.com/2011/07/03/magazine/infidelity-will-keep-us-together.html?pagewanted=all (last accessed 6/17/13).

Pornography

CHAPTER TWENTY

The Right to Get Turned On: Pornography, Autonomy, Equality

Andrew Altman

Introduction

Debates over whether adults have a right to produce, distribute, and view pornographic materials have typically proceeded on the premise that freedom of speech is the central liberty at stake. Those who argue that there is a moral "right to pornography" contend that it is part of a person's freedom of speech. Those who argue that there is no such right contend that pornographic material is "low value" speech or more like conduct than speech. They proceed to claim that some other value such as sexual equality between men and women overrides an individual's claim to have access to pornography.

I believe that the premise behind this debate is mistaken. While there are certain respects in which freedom of speech is at stake in the matter of pornography, such freedom is not the central liberty relevant to the issue. Rather, the right to pornography should be understood primarily as an element of another form of freedom: sexual autonomy. Individuals ought to have a broad liberty to define and enact their own sexuality. Persons who view pornography are exercising their sexual autonomy, and the debate over pornography should be seen from the standpoint of that liberty.

When seen from such a standpoint, the claim that there is a right to pornography is analogous to claims that there is a right to use contraceptives, to engage in sexual relations outside marriage, and to engage in homosexual activity. Freedoms that protect sexuality-defining decisions get closer to the heart of the pornography issue than freedoms that protect speech and other activities whose primary intent is to communicate ideas or attitudes.

The principle of sexual autonomy has its limits. The moral right to have sex without being married does not include the moral right to have sex with children or with a

Contemporary Debates in Applied Ethics, Second Edition. Edited by Andrew I. Cohen and Christopher Heath Wellman.

non-consenting adult. A moral right to pornography does not include the moral right to buy or possess photographs of children having sex, or of people who are actually being raped or sexually assaulted. However, I will argue that sexual autonomy does entail a moral right to buy and possess a wide range of pornographic materials, including those that depict sexual violence.

What is Pornography?

It is not realistic to think that there is a succinct definition of pornography that would prove acceptable to the different sides in the debate and capture all of the material that might reasonably be thought pornographic. This does not mean that we should remain content with Justice Potter Stewart's attitude: "I know it when I see it" (*Jacobellis* v. *Ohio*, 1964, p. 197). Rather, we can formulate a concise description of a class of materials that includes many, if not all, of the materials which the different sides in the debate could agree are reasonably described as pornographic. The description would be a kind of starting point that could be qualified and expanded in various ways as the debate proceeded. The point is that we need some reasonable starting point that can be accepted without unfairly tilting the debate over the existence of a moral right to pornography.

My suggestion for such a starting point is this: pornography is sexually explicit material, in words or images, which is intended by its creators to excite sexually those who are willing viewers of the material. By a "willing viewer," I mean a person who voluntarily pays something – in time, effort, or money – to view the material and who is willing to pay because he expects to become sexually aroused by viewing it. Thus, pornography is a commodity which represents a kind of sexual meeting of the minds between producer and consumer: the producer intends that the consumer be sexually aroused by the product and the consumer pays for the product in the expectation of becoming aroused by it.

The intention to cause sexual arousal is clearly not the only one for which a producer of pornography may be acting. Commercial producers intend to make money. However, the intent to cause sexual arousal is central, even in the commercial case. The producers intend to make money by creating a product which causes sexual arousal and the buyer expects to be aroused by viewing the product.

In contrast, consider the authors or publishers of a medical textbook which contains photographs of sexual organs and their various diseases. Such persons intend to make money. However, it is not their intention to make it by causing sexual arousal but rather by communicating medical information. Moreover, buyers of medical textbooks do not generally purchase them in order to stimulate themselves sexually: there is no sexual meeting of the minds between the authors or publishers and the consumers.

It is an important fact about human sexuality that different people are sexually excited by very different kinds of sexually explicit material. The makers of pornography know this fact well. Much hardcore pornography is explicitly addressed to the viewer's preference for particular types of sexual content: oral, anal, sadomasochist, gay, lesbian, and so on.

It seems clear that the vast majority of pornography in contemporary society is directed at males. Among all of the hours spent watching pornography, the vast majority of those hours belong to men. However, even within the group of heterosexual men,

308　　**Andrew Altman**

there are differences in the pornographic content which they willingly seek out. In addition, empirical studies show that a significant percentage of willing viewers of pornography are women (Slade, 2001, p. 967).

Sexual Autonomy

Individuals have a right to a substantial degree of control over their own lives. This right does not mean that any individual has the liberty to do whatever she or he chooses: one person's liberty is limited by the duties that she has toward others. Moreover, individual control is invariably exercised within a social context created by the choices and actions of other people who are exercising control over their own lives. Yet it would be mistaken to think that individual control is rendered factually impossible by the unchosen character of our social context or morally meaningless by the existence of duties we owe to others. Persons are not puppets of their social circumstances, nor are they smothered by moral duties owed to others. Rather, they are agents who have the broad right to decide for themselves how to live their lives. Other individuals and the government have a duty to respect those decisions.

Under the rules of traditional sexual morality, a person's sexual life was, to a large extent, not his or her own: the rules imposed a highly confining set of duties on sexual choices and actions. In particular, sexual activity was condemned as "unnatural" if it was outside heterosexual marriage or if the activities were undertaken for purposes other than procreation. Traditional sexual morality looked askance on pornography because such materials excite passions that do not stay neatly confined within the narrow channels of sexual activity that traditional morality deemed the only natural and acceptable way of expressing human sexuality. Accordingly, pornography was seen as corrupting individual character and subverting the proper order of society.

The sexual revolution of the 1960s replaced the traditional sexual morality with a liberal one. This liberal morality located a person's sexual life much more within his or her own dominion than did traditional morality. One way of characterizing the liberal rules is to say that they left adults morally free to engage in the sexual activities of their choice, so long as the activities had no direct unwilling victims. This characterization will require some qualification, but it does help to highlight the difference between traditional and liberal sexual morality.

From the liberal viewpoint, traditional sexual morality violated the rights of the individual by treating a person's sexual choices as if they belonged to society. Where the traditional morality reigned, sexuality was conscripted by society to promote its interest in procreation and in preserving a certain model of the family. Individuals were expected to follow the "appropriate" social scripts, which were defined by gender and restricted a person to marital (heterosexual) intercourse without the use of contraceptives. Liberal morality does not deny the importance of procreation or family, but it does assert that adult individuals have the right to decide for themselves when and whether to have children and when and whether to engage in sexual activity for purposes other than procreation. And the liberal view is that this right of sexual autonomy is possessed equally by each adult. David Richards, a leading proponent of a liberal sexual morality, puts the central point plainly: "Legal enforcement of a particular sexual ideal fails . . . to accord due respect to individual autonomy" (Richards, 1982, 99).

The Right to Get Turned On: Pornography, Autonomy, Equality 309

The new liberal principles cast a very different light on pornography than did the traditional morality. There is nothing inherent to the activities of producing or consuming pornography which raises a presumption that there is some direct unwilling victim of the activities. Pornography does not necessarily involve children or any unwilling adult. The sole participants in the production and use of pornographic materials may be consenting adults, and, in such a situation, the strong liberal presumption is that those adults have a moral right to do what they are doing. The basis of this presumption is the idea that the sexuality-defining decisions of adults are up to them, and those decisions include ones that involve voluntary association for purposes of sexual pleasure or for profit from the manufacture of materials that help produce sexual arousal.

Accordingly, on the liberal sexual morality, a right to pornography is akin to the right to use contraceptives: adults must be free to manufacture and use pornographic materials, just as they must be free to make and use contraceptive devices, and others must not interfere with those choices. Other sexuality-defining activities, such as the right to engage in homosexual activity, are also central to the liberal sexual morality.[1] Some people may be revolted by homosexuality and regard it as depraved, just as some are revolted by pornography and regard it as depraved. But such attitudes are not adequate grounds, on the liberal view, for restricting a person's sexual activities.

At the same time, it is important to understand that any reasonable version of liberal sexual morality must go beyond the idea that there is an absolute right to choose one's sexual activities as long as there is no direct unwilling victim. Some room must be left for the possibility that, in some circumstances, such choices are outside the boundaries of the person's right to sexual autonomy. In the next two sections, we will examine some possible circumstances that mark the limits of an individual's right. For the present, the key point is that, for a reasonable version of liberal morality, any restriction on the right of sexual autonomy must rest on considerations that possess considerable weight and are supported by clear and convincing evidence.

It is also important to note that the liberal claim that individuals have a broad right to define their own sexual identity is compatible with the idea that some of the activities which individuals have a right to engage in are, nonetheless, morally deficient. For example, one may agree that an adult has the right to view violent pornography but still contend that any adult who does seek sexual arousal by viewing violent sexual images has a morally deficient character. Put another way, it is consistent for a liberal to assert that a person who has an impeccable character would refrain from certain activities, even though people have a right to do those activities.[2]

Liberal sexual morality has become the dominant morality of contemporary society, although the traditional morality still survives and exerts some influence. Defenders of traditional morality claim that liberal "permissiveness" leads to social disintegration. Thus, Robert George, a contemporary proponent of the traditional view, asserts: "it is plain that moral decay has profoundly damaged the morally valuable institutions of marriage and the family" (George, 1993, p. 36).

It is true that divorce rates are much higher than in past generations, and family life has taken on a very different shape. However, one cannot infer that profound moral damage has been done without making many unproven assumptions about how much better family was in "the good old days," when marriages were often forcibly held together by the economic dependence of the wife and the powerful social stigma of divorce. While it would be wrong for liberals to presume that liberal society is, in every aspect, better than traditional society was, there are two important respects in which liberals should insist that people are better off under the liberal morality. First, men and women are freer to define a central aspect of their existence, their sexuality, in ways that fit their individual character, and, secondly, women are freer and more equal participants in society. Without attempting any full-scale assessment of the traditional morality, in the sections on "Sexual Inequality" and "Sexual Identity," I will elaborate on these two considerations in favor of liberal sexual morality. However, the principal task of the remainder of this chapter is to examine critically several feminist arguments which, if sound, would show that any liberal right to pornography must be far more limited than I have suggested.

Sexual Violence

Suppose that the viewing of certain types of pornography has very harmful indirect effects on unwilling victims. For example, consider pornographic movies which depict the gang rape of a woman. Even assuming that all of the participants in such movies are consenting adults – so that the rapes are staged and not real – it is possible that the movies could lead some male viewers to "imitate" what they see and commit real rapes. Similar possibilities could obtain for other kinds of violent pornography.

Moreover, in contemporary society, there are many willing viewers of violent pornography: the material is commercially produced and widely distributed. Even if most viewers do not directly violate anyone's rights, some of them may be prompted to commit sexual violence as a result of their exposure to violent pornography. Accordingly, Helen Longino expresses the view of many feminist thinkers when she claims: "Pornography, especially violent pornography, is implicated in the committing of crimes of violence against women" (Longino, 1995, p. 41). Longino proceeds to argue on the basis of her claim that the access of adults to pornography made by adults should be legally restricted. In the light of such an argument, it is important to address the question of whether the right to view pornography reaches its limit when sexual violence is depicted.[3]

It is true that a willing viewer of violent pornography who becomes sexually aroused does not necessarily harm any unwilling victim. Under liberal principles, this means that there is a presumption that the viewer is simply exercising his right of sexual autonomy. But we should not ignore the societal consequences of the availability of violent pornography in deciding whether that presumption is overridden by countervailing considerations.

If the availability of violent pornography led to substantial increases in sexual violence, then the victims of this increased violence would be paying the price for the availability of violent pornography to all adults. And it seems wrong to make those victims pay such a steep price so that some can have ready access to violent sexual materials for purposes of sexually arousing themselves. In such a situation, it would appear that any presumptive right to violent pornography would be overridden by countervailing considerations.

Notice that the considerations here consist precisely of rights-based concerns to which a liberal sexual morality must give considerable weight. The victims of the criminals who commit pornography-inspired sex crimes have their basic liberal right to sexual autonomy violated egregiously by the perpetrators. However, there are obstacles that need to be surmounted before one can reasonably conclude that, in contemporary society, any right to pornography must stop short of including a right to pornographic materials depicting sexual violence.

First, there must be clear evidence of a causal connection between the production of violent pornography and sexual violence. In the absence of such evidence, there are insufficient grounds for limiting the right of sexual autonomy so as to leave out a right to make and view violent pornography. Yet, the evidence for the existence of a causal connection is, at best, mixed.

Experimental studies suggest that when males repeatedly view violence against women in films, they tend to undergo attitudinal changes that make them desensitized to such violence and more accepting of it.[4] However, the films used in the studies were R-rated "slasher movies," such as *Texas Chainsaw Massacre*, which lacked the sort of graphic depictions of sexual activity characteristic of paradigm cases of pornography. Moreover, the extrapolation from the experimental studies to conclusions about sexual crimes is rather tenuous: no one knows how long the attitudinal changes measured by the studies persist or whether they produce behavioral changes leading to the perpetration of sex crimes.

Since the 1960s, violent pornography has become much more readily accessible in many countries, including the United States. The incidence of sexual crime has also increased in those countries. However, data collected over many decades in the United States show that the number of rapes rises in virtual lockstep with the rate of nonsexual assaults (Kutchinsky, 1991, p. 55). It is not plausible to think that violent pornography causes a rise in nonsexual violence.[5] Indeed, much more reasonable is the hypothesis that sociological variables such as poverty rates and the extent of alcohol consumption explain the equal increases in both sexual and nonsexual violence.

On the other hand, there are studies that provide some evidence for the conclusion that sexual crimes increase as a result of an increase in the availability of pornography. One such study found that the rise in rape rates around the world was traceable to pornography. However, other studies have found no correlation and some have even concluded that rape drops as a result of the availability of pornography (Slade, 2001, pp. 997–998).

The existing state of the evidence, then, is quite far from clearly establishing any causal connection between violent pornography and sexual violence, and appears to weigh against any such connection. Yet, even if a causal connection between violent pornography and sexual violence were clearly established, it would still be insufficient to conclude that, in contemporary society, the production, distribution, and viewing of violent pornography lay beyond the limits of an adult's right of sexual autonomy. Additionally, one would need to justify selecting out such pornography and distinguishing it from the myriad of other forms of media violence that have the potential to cause violence.

Consider the "slasher films" mentioned earlier. It is reasonable to suspect that such films and much else in the mass media cause at least some amount of violence against women, sexual and otherwise. However, it is unreasonable to deny that adults have a right to produce, distribute, and view such movies, even if we were to assume the existence of an established causal relation between the films and sexual violence. Adults who find the films entertaining are subject to criticism for getting enjoyment from watching depictions of terrified women inhumanely attacked. However, these adults do not violate anyone's rights by getting their enjoyment in that way. The situation with respect to viewing violent pornography is different only in the respect that watching such pornography is typically an exercise of sexual autonomy. To the extent that viewing "slasher films" is seen as nonsexual entertainment, the right to see them would actually be *less* strong than the right to view violent pornography.

Accordingly, it is unclear how one could justify selecting out violent pornography as setting a limit to the individual's right of sexual autonomy, while at the same time conceding that there is a right to view forms of media which, as far as we know, could contribute just as much to sexual violence as does violent pornography. It might be argued that violent pornography is a more powerful stimulus to sexual violence. However, we have seen that the evidence of any causal connection between pornography and violence is mixed. And there is simply no evidence indicating the relative contribution which different factors make to the overall level of sexual violence in society.

It may seem that liberal sexual morality is indifferent to the actual violence that may be caused by the production and viewing of the depictions of sexual violence found in films and other media. However, we must be careful in our understanding of what the liberal right of sexual autonomy involves. I have argued that it does include the right to produce and view violent pornography. But liberal sexual morality also holds that each adult has an equal right to sexual autonomy. If sexual violence is widespread in society, as it is in ours, then liberal morality cannot simply brush off that fact. Widespread sexual violence means widespread violation of the equal right of sexual autonomy. Liberal morality demands that something be done about it. But there are ways of reducing levels of sexual violence without placing the production and viewing of violent pornography – or any other media depictions of violence against women – beyond the bounds of the right of autonomy.

The most straightforward ways involve more vigorous prosecution of, and more serious punishments for, crimes of sexual violence. In a similar vein, laws regarding rape and sexual assault can and should be changed, so that the women who are the victims of such crimes are treated in a respectful manner by the criminal justice system. Additionally, efforts at educating individuals – especially young men – about sexual violence should be more seriously pursued.[6] In sum, then, subscribing to a liberal sexual morality does not require that one ignore or exhibit indifference to the level of sexual violence in society and its harmful impact on women.

Sexual Inequality

Even if we set aside the issue of whether violent pornography causes sexual violence, the question remains as to whether pornography in general helps to maintain many of the important social and economic inequalities that disadvantage females. Many feminists assert that pornography plays a pivotal role in maintaining such sexual inequalities, and they cast the issue of pornography as one that is "not a moral issue," but rather is a matter of the civil rights of women (MacKinnon, 1988, pp. 146–162).

For example, Catherine Itzin claims that "women are oppressed in every aspect of their public and private lives," and she sees pornography as playing a central role in maintaining the system of oppression. Itzin proceeds to defend "civil sex discrimination legislation against pornography [that] would enable women to take action on grounds of harm done to them by pornography" (Itzin, 1992, p. 424). The legislation is seen as a kind of civil rights law for women.[7]

There is little doubt that the vast bulk of pornography willingly viewed by heterosexual men – whether violent or not – involves women in positions of sexual servility or subordination: the women are there to serve the sexual pleasure of the men. And serve it they do, not only to the men who are their "co-stars" in the movie or photograph, but also to the men who masturbate to the scene or who have sex with their partners while using the scene to help arouse them. These facts are what lead some feminists to argue that pornography is unique in its power to create a psychological nexus between the social subordination of women and the sexual pleasure of men, and so is unique in its power to create and sustain patterns of sex inequality that severely disadvantage females. Catharine MacKinnon puts the matter plainly: "Pornography is masturbation material. . . . With pornography, men masturbate to women being exposed, humiliated, violated, degraded, mutilated, dismembered, bound, gagged, tortured and killed. . . . Men come doing this" (MacKinnon, 1993, p. 17).

MacKinnon is right to take the focus off pornography as a form of speech and to look instead at its role in sexual behavior. However, there is a crucial consideration which renders her line of thinking problematic as a viable basis for rejecting a right to pornography. The evidence does not support the idea of any robust correlation, much less a causal relation, between the level of sex inequality in a society and the availability of pornography in it. Quite the opposite: the most repressive countries in the world for women are ones where pornography is least available. Compared to Saudi Arabia, the United States is awash in pornography. Indeed, MacKinnon herself insists that the United States is "a society saturated with pornography" (1993: 7) – a description which might be arguably applied to the United States but clearly does not apply to Saudi Arabia. Nonetheless, on the indices of sex inequality developed by the United Nations Development Program, the United States and other Western countries where pornography circulates widely are the nations with the highest levels of *equality*, while Saudi Arabia and other sexually repressive regimes have among the highest levels of inequality (United Nations Development Program, 2002, pp. 222–242). Thus, it is hard to credit the notion that pornography is a kind of causal linchpin in the creation and maintenance of large inequalities between males and females.

There is certainly an analogy between the ways in which much pornography depicts women in relation to men and the ways in which social practices actually treat women in relation to men. In much pornography, there is a sexual hierarchy dominated by men; in much of society, there is a social hierarchy dominated by men. Moreover, it is plausible to think that pornography plays some causal role in the perpetuation of sexual hierarchy. But as with the matter of sexual violence, any limitation of the right of adults to sexual autonomy requires more than a plausible belief that some indeterminate degree of connection exists between pornography and sexual hierarchy.

Making Pornography

Much pornography depicts the subordination of women. Even though the symbolic representation of inequality is not the same as the inequality that is represented, it may

314   **Andrew Altman**

be argued that in making pornography, women humiliate and subordinate themselves. They get on their knees and suck on men's cocks. They let men ejaculate into their mouth and on their face and breasts. They have several men simultaneously penetrating their anus, vagina, and mouth. They are tied up and gagged. The humiliation seems all the more acute because it is done before cameras that will circulate the images to untold numbers of men to view. One might claim that this means that making pornography is making female inequality, and not simply depicting it.

However, context counts in deciding whether a person's sexual conduct is a form of humiliation and subordination. It is difficult to see why fellatio is any more inherently degrading than cunnilingus, or why either form of oral sex has that feature. If the parties are adults and consent, the assessment of the activity as humiliating is highly contestable. Multiple penetration also seems inherently innocuous.

Nonetheless, the key point is this: even if we grant that much pornography does involve women performing humiliating or degrading sexual acts, it does not follow that the actors have no right to participate in making such material or that viewers have no right to see it. A willingness to sexually degrade oneself before a camera for commercial purposes may constitute a serious deficiency in one's character. A willingness to view such pornography may also reflect a character flaw. But the men and women who perform in such pornography have a right to make their choices, and consumers have a right to view the commercial product.

If women are intimidated by violent threats into performing in pornography, then their rights have been violated and their victimizers ought to be prosecuted and punished. But it is simply an ideological prejudice to assume a priori that any woman who performs humiliating or degrading sexual acts in pornography has been threatened or coerced in some way. Especially in matters of sex, the line between humiliation, on one side, and breaking the procrustean bed of traditional morality, on the other, is a very tricky one to draw.

Some feminist advocates of laws against pornography claim that physical threats, violence, and economic coercion against women pervade the actual operation of the pornography industry (Dworkin, 2000, pp. 27–29). It may be said that the only way to stop such threats is by closing down the industry. But even if that were true, it would not justify closing down the industry. It does not make sense to think that the only industries that should be allowed are those that can operate without anyone abusing them by threatening violence. Such abuse can be found in any industry. Criminal prosecution of the perpetrators should be the main remedy for physical abuse and coercion in the pornography industry.

Moreover, there are less draconian ways of diminishing violence in the industry than shutting the industry down. For example, some feminists have argued for the unionization of women who work in pornography and other sex-related industries (Cornell, 2000, p. 552). While unionization efforts may not have good prospects at present, especially in the United States, the prospects for banning pornography under a civil rights approach are no better. And the unionization strategy has the decided advantage of treating women in the pornography industry as agents who are capable of exercising their own right of sexual autonomy.

Some of the females who get caught up in the pornography industry are legal minors. The industry executives who intentionally, or negligently, hire minors ought to be prosecuted and punished. Legal minors may have some aspects of the right of sexual autonomy (for example, a 17-year-old girl has the right to purchase and use contraceptives), but the law should rest on the premise that minors are too easily manipulated by industry executives and other adults with vested interests to have a right to decide for themselves to perform in commercial pornographic films or pose for pornographic pictures.

Some feminists contend that women accede to make pornography only because they have no other economic options (except perhaps prostitution, a close cousin of pornography). This contention may have some truth in countries of the underdeveloped world, where educational opportunities for women are highly restricted, rampant sexism operates in all quarters of life, and economic opportunities even for many men are bleak at best. However, in the economically advanced liberal democracies, the situation of almost all women is drastically better, and claims of economic coercion are considerably less plausible as a result.

The clear conclusion seems to be that uncoerced adults have a right to be legally free to make, market, and view pornography. However, it might be objected that if some women voluntarily choose to make pornography in which they are engaged in humiliating or degrading conduct, then their actions affect all women in a detrimental way. The idea here is that the manufacture and circulation of such pornography shapes the sex-role expectations of men and women in society at large, and it does so by showing women as the sexual servants of men. The result is that individuals are not free to control their sexual identities: just as much as in a society ruled by traditional sexual morality, sexual identities are controlled by social forces which are beyond their control and which are hostile to their basic interests.

Sexual Identity

It must be admitted that, even in a society governed by a liberal sexual morality, the sexual autonomy of a person is significantly circumscribed. There is a built-in tension between living in a society and possessing the autonomy to define oneself sexually or in any other way. Without connections to other people in an organized and ongoing system of relations, the life-options of the individual would be radically limited. But those connections also mean that a person's life-defining choices are not entirely her own. The patterns of behavior and attitude that other people adopt not only establish pathways through life which would not otherwise exist, but also create barriers and limits on the individual's exercise of her autonomy. The ability of the individual to shape her own identity is both enabled by, and held hostage to, the actions and attitudes of other people.

There is no solution to this problem. The conditions of meaningful autonomy are also conditions that can inhibit the exercise of such autonomy. Nonetheless, even though this conflict cannot be eliminated, it can be mitigated. And some kinds of society do a much better job of mitigating it than others. Societies with a liberal sexual morality are much better in this respect than those with a traditional sexual morality, and that is the decisive consideration in favor of the liberal morality. Individuals have many more meaningful options in living out the sexual aspects of their lives: their sexuality is not held hostage to what other people do and think to nearly the extent that is found in traditional societies. The grip of pre-existing social scripts that define a sexual identity for each person is dramatically weaker in liberal societies and the power of individuals to shape a centrally important aspect of their lives is correspondingly greater.

However, even in a liberal society, there is no escaping the fact that how other women act and think affects the opportunities and obstacles for any given woman's efforts to define her own sexual identity. The same is true, of course, for men, but the problem of concern here is the willingness of some women to participate in the creation of pornography in which they engage in conduct that is humiliating and servile. Such conduct may be voluntary on the part of the woman, but – the claim goes – it also makes it more difficult for other women to define their own sexual identities as the equals of men.

I think that it is reasonable to hold that the existence of such pornography makes it more difficult for women to live their lives as the sexual equals of men – that is, more difficult relative to a society which was ruled by a liberal sexual morality and had fewer women, or none at all, who were willing to engage in humiliating conduct as part of the production of pornographic materials. However, women are far better off in societies where a liberal sexual morality dominates than they are in traditional societies, even when the liberal ones contain much pornography degrading to women. Although the freedom of women to humiliate or degrade themselves in making pornography creates costs that all women in a liberal society bear, the gains for women that have resulted from society moving to a liberal sexual morality from a traditional morality far outweighs the costs.

It might be argued that the costs are still too great, and I would not dissent. However, there are ways to lessen those costs without incursions on the right to sexual autonomy. Those ways are likely to be far more effective in promoting sexual equality than restricting the freedom of willing adults to view pornography made by willing adults.

Conclusion

The recognition of a right to sexual autonomy is critical in adequately addressing the issue of pornography. There are other important dimensions of the issue, including the levels of sexual violence perpetrated against women and the social inequalities that systematically disadvantage women. Also relevant is the question of whether there is some character defect in those who make and enjoy pornographic materials.

However, liberal sexual morality correctly places the right of sexual autonomy at the center of the pornography issue. In doing so, the liberal morality places a substantial burden on those who argue for legal restrictions on the access of adults to pornography made by consenting adults. Those who argue for such restrictions tacitly concede that the burden is theirs, as they make claims aimed at meeting it, for example, that pornography causes sexual violence, reinforces sexual hierarchy, and involves non-consenting women who are forced to perform.

When examined carefully, though, we find that the burden has not been met. The empirical claims are insufficiently verified, and some of the empirical assertions, even if substantiated, would be inadequate to justify restricting an adult's right of sexual autonomy. We are left, then, with the claim that the producers and viewers of pornography exhibit a defect of moral character. Such a claim is consistent with a liberal sexual morality. However, it is also inadequate to justify restrictions on adults who willingly create and view pornography.

Notes

1 Cf. Richards, 1982, pp. 29 and 39.
2 Cf. Driver, 1992; Waldron, 1993, ch. 3.
3 Longino also contends that pornography defames women by communicating falsehoods about them and reinforces the societal oppression of women. The oppression argument is considered in the section "Sexual Inequality". The defamation argument would license sweeping restrictions on communication, including political expression.
4 See, for example, Linz *et al.* (1984).
5 Kutchinsky (1991) also found that in West Germany, Denmark, and Sweden, rape increased less than nonsexual assault, despite the greatly increased availability of violent pornography in those countries as well.
6 Many thinkers assert that pornography fosters the myth that women enjoy being forced to have sex (the rape-myth) and some studies support the assertion. However, other studies show that better educating young men can counteract their acceptance of the rape-myth. Moreover, mainstream movies in which rapes take place also appear to foster the rape-myth (see Slade, 2001, pp. 992–993).
7 Catharine MacKinnon and Andrea Dworkin helped draft anti-pornography, civil rights laws in the United States, but the courts have found them to be unconstitutional on free-speech grounds (see *American Booksellers* v. *Hudnut*, 1985).

References

American Booksellers v. *Hudnut* (1985) 771 F.2d 323 (7th Cir.).
Cornell, D. (2000) Pornography's temptation. In *Feminism and Pornography*, ed. D. Cornell, pp. 552–568. New York: Oxford University Press.
Driver, J. (1992) The suberogatory. *Australasian Journal of Philosophy* 70: 286–295.
Dworkin, A. (2000) Against the male flood. In *Feminism and Pornography*, ed. D. Cornell, pp. 19–44. New York: Oxford University Press.
George, R.P. (1993) *Making Men Moral*. Oxford: Oxford University Press.
Itzin, C. (1992) Legislating against pornography without censorship. In *Pornography: Women, Violence, and Civil Liberties*, ed. C. Itzin, pp. 401–434. New York: Oxford University Press.
Jacobellis v. *Ohio* (1964) 378 US 184.
Kutchinsky, B. (1991) Pornography and rape: theory and practice. *International Journal of Law and Psychiatry* 14: 47–64.
Linz, D., Donnerstein, E., and Penrod, S. (1984) The effects of multiple exposures to filmed violence against women. *Journal of Communication* 34: 130–147.
Longino, H. (1995) Pornography, oppression, and freedom: a closer look. In *The Problem of Pornography*, ed. S. Dwyer, pp. 34–47. Belmont, CA: Wadsworth Publishing.
MacKinnon, C. (1988) *Feminism Unmodified*. Cambridge, MA: Harvard University Press.
MacKinnon, C. (1993) *Only Words*. Cambridge, MA: Harvard University Press.
Richards, D.A.J. (1982) *Sex, Drugs, Death, and the Law*. Totowa, NJ: Rowman & Littlefield.
Slade, J.W. (2001) *Pornography and Sexual Representation*, Vol. III. Westport, CT: Greenwood Press.
United Nations Development Program (2002) *Human Development Report 2002*. New York: Oxford University Press.
Waldron, J. (1993) *Liberal Rights*. New York: Cambridge University Press.

CHAPTER TWENTY-ONE

"The Price We Pay"?
Pornography and Harm

Susan J. Brison

Defenders of civil liberties have typically held, with J.S. Mill, that governments may justifiably exercise power over individuals, against their will, only to prevent harm to others (Mill, 1978, ch. 1).[1] Until the 1970s, liberals and libertarians assumed that since producers and consumers of pornography clearly did not harm anyone else, the only reasons their opponents had for regulating pornography were that they considered it harmful to the producers or consumers, that they thought it an offensive nuisance, and that they objected, on moral or religious grounds, to certain private sexual pleasures of others. None of these reasons was taken to provide grounds for regulating pornography, however, since individuals are considered to be the best judges of what is in their own interest (and, in any case, they cannot be harmed by something to which they consent), what is merely offensive may be avoided (with the help of plain brown wrappers and zoning restrictions), and the private sexual activities, of consenting adults anyway, are no one else's – certainly not the state's – business.

In the 1970s, however, the nature of the pornography debate changed as an emerging group of feminists argued that what is wrong with pornography is not that it morally defiles its producers and consumers, nor that it is offensive or sinful, but, rather, that it is a species of hate literature as well as a particularly insidious method of sexist socialization. Susan Brownmiller was one of the first to take this stance in proclaiming that "[p]ornography is the undiluted essence of anti-female propaganda" (1975, p. 443). On this view, pornography (of the violent degrading variety) harms women by sexualizing misogynistic violence. According to Catharine MacKinnon, "[p]ornography sexualizes rape, battery, sexual harassment, prostitution, and child sexual abuse; it thereby celebrates, promotes, authorizes, and legitimizes them" (1987: 171).

The claim that women are harmed by pornography has changed the nature of the pornography debate, which is, for the most part, no longer a debate between liberals who subscribe to Mill's harm principle and legal moralists who hold that the state can

Contemporary Debates in Applied Ethics, Second Edition. Edited by Andrew I. Cohen and Christopher Heath Wellman.

legitimately legislate against so-called "morals offenses" that do not harm any non-consenting adults. Rather, the main academic debates now take place among those who subscribe to Mill's harm principle, but disagree about what its implications are for the legal regulation of pornography. Some theorists hold that violent degrading pornography does not harm anyone and, thus, cannot justifiably be legally regulated, socially stigmatized, or morally condemned. Others maintain that, although it is harmful to women, it cannot justifiably be regulated by either the civil or the criminal law, since that would cause even greater harms and/or violate the legal rights of pornography producers and consumers, but that, nevertheless, private individuals should do what they can (through social pressure, educational campaigns, boycotts, etc.) to put an end to it. Still others claim that such pornography harms women by violating their civil right to be free from sex discrimination and should, for that reason, be addressed by the law (as well as by other means), just as other forms of sex discrimination are. But others argue that restricting such pornography violates the moral rights of pornography producers and consumers and, thus, restrictions are morally impermissible. Later in this chapter I will argue that there is no moral right to such pornography.

What is Pornography?

First, however, I need to articulate what is at issue, but this is hard to do, given various obstacles to describing the material in question accurately. (I have encountered the same problem in writing about sexual violence.) There is too much at stake to be put off writing about issues of urgent import to women because of squeamishness or fear of academic impropriety – but how can one write about this particular issue without reproducing the violent degrading pornography itself? (Recall the labeling of Anita Hill as "a little nutty and a little slutty" because she repeated, in public, the sexually demeaning language that Clarence Thomas had uttered to her in private.) However, if one does not write graphically about the content of violent degrading pornography, one risks being viewed as either crazy ("she must be imagining things!") or too prudish to talk frankly about sex. And what tone should one adopt – one of scholarly detachment or of outrage? There is a double bind here, similar to that faced by rape victims on the witness stand. If they appear calm and rational enough for their testimony to be credible, that may be taken as evidence that they cannot have been raped. But if they are emotional and out of control enough to appear traumatized, then their testimony is not considered reliable.

Any critic of violent degrading pornography risks being viewed not only as prudish (especially if the critic is a woman), but also as meddling in others' "private" business, since we tend not to see the harm in pornography – harm which is often made invisible and considered unspeakable. But "we" used not to see the harm in depriving women and minorities of their civil rights. And "we" used not to see the harm in distributing postcards depicting and celebrating lynchings. More recently, "we" did not see the harm in marital or "date" rape, spousal battering, or sexual harassment. Even now, as Richard Delgado and Jean Stefancic point out:

> [M]embers of the empowered group may simply announce to the disaffected that they do not see their problem, that they have looked for evidence of harm but cannot find it. Later

generations may well marvel, "how could they have been so blind?" But paradigms change slowly. In the meantime, one may describe oneself as a cautious and principled social scientist interested only in the truth. And one's opponent, by a neat reversal, becomes an intolerant zealot willing to trample on the liberties of others without good cause. (Delgado and Stefancic, 1997, p. 37)

A further problem arises in critically analyzing violent degrading pornography, deriving from precisely those harmful aspects of it being critiqued, which is that descriptions of it and quotations from it can themselves be degrading, or even retraumatizing, especially for women who have been victimized by sexual violence. But one thing that is clear is that feminist critics of such pornography are *not* criticizing it on the grounds that it is erotic, or sexually arousing, or that it constitutes "obscenity," defined by the Court as "works which, taken as a whole, appeal to the prurient interest in sex, which portray sexual conduct in a patently offensive way, and which, taken as a whole, do not have serious literary, artistic, political or scientific value" (*Miller v. California*, 1973, p. 24). Those who work on this issue – and have familiarized themselves with the real world of the pornography industry – know all too well that pornography is not merely offensive. In contrast, here is how some of them define "pornography":

[T]he graphic sexually explicit subordination of women through pictures or words that also includes women dehumanized as sexual objects, things, or commodities; enjoying pain or humiliation or rape; being tied up, cut up, mutilated, bruised, or physically hurt; in postures of sexual submission, servility or display; reduced to body parts, penetrated by objects or animals, or presented in scenarios of degradation, injury, torture; shown as filthy or inferior; bleeding, bruised, or hurt in a context that makes these conditions sexual. (MacKinnon, 1987, p. 176)[2]

I define "pornography," for the purposes of this chapter, as violent degrading misogynistic hate speech (where "speech" includes words, pictures, films, etc.). I will argue that, if pornography unjustly harms women (as there is reason to suppose it does), then there is no moral right to produce, sell, or consume it. (I will not here be arguing for or against its legal restriction and no position on that issue is dictated by my arguments against the alleged moral right.)

Pornography and Harm

I cannot hope to portray adequately the harms inflicted on girls and women in the production of pornography (for the reasons given above), but there is plenty of research documenting them. One of the most powerful forms of evidence for such harms is the first-person testimony of "participants" in pornography. (Those who are interested in reading more about this are referred to the Lederer, 1980; Attorney-General's Commission on Pornography, 1986; MacKinnon, 1987; Itzen, 1992; MacKinnon, 1993; Russell, 1993; Lederer and Delgado, 1995; MacKinnon and Dworkin, 1997.) A not uncommon scenario in which a girl becomes trapped in the pornography industry is described by Evelina Giobbe in her testimony to the US Attorney-General's Commission on Pornography. After running away from home at age 13 and being raped her during her first night on the streets, Giobbe was befriended by a man who seemed initially kind and concerned, but who, after taking nude photographs of her, sold her to a pimp who raped and battered her, threatening her life and those of her family until she "agreed" to work as a prostitute for him. Her "customers" knew she was an adolescent and sexually inexperienced. "So," she testified, "they showed me pornography to teach me and ignored my tears and they positioned my body like the women in the pictures, and used me." She tried on many occasions to escape, but, as a teenager with no resources, cut off from friends and family, who believed she was a criminal, she was an easy mark for her pimp: "He would drag me down streets, out of restaurants, even into taxis, all the while beating me while I protested, crying and begging passers-by for help. No one wanted to get involved" (quoted in Russell, 1993, p. 38). She was later sold to another pimp who "was a pornographer and the most brutal of all." According to her testimony, he recruited other girls and women into pornography by advertising for models:

> When a woman answered his ad, he'd offer to put her portfolio together for free, be her agent, and make her a "star." He'd then use magazines like *Playboy* to convince her to pose for "soft-core" porn. He'd then engage her in a love affair and smooth talk her into prostitution. "Just long enough," he would say, "to get enough money to finance your career as a model." If sweet talk didn't work, violence and blackmail did. She became one of us. (Quoted in Russell, 1993, p. 39)

Giobbe escaped the pornography industry by chance, after "destroy[ing] herself with heroin" and becoming "no longer usable." She considers herself one of the lucky ones – "a rare survivor" (quoted in Russell, 1993, pp. 39–40). And this was *before* the AIDS epidemic.

More recently, according to an article in the *Sunday New York Times Magazine*, pornography – of an increasingly violent sort – has played an important role in the global sex trafficking of girls and women who, lured by promises of employment (for example, as nannies or waitresses), end up trapped in foreign countries, with no money, no (legal) papers, no family or friends, and no ability to speak the local language. Immigrations and Customs Enforcement (ICE) agents at the Cyber Crimes Center in Fairfax, Virginia are "tracking a clear spike in the demand for harder-core pornography on the Internet. 'We've become desensitized by the soft stuff; now we need a harder and harder hit', says ICE Special Agent Perry Woo." With ICE agents, the author of the article looked up a website purporting to offer sex slaves for sale: "There were streams of Web pages of thumbnail images of young girls of every ethnicity in obvious distress, bound, gagged, contorted. The agents in the room pointed out probable injuries from torture" (Landesman, 2004, p. 72). " 'With new Internet technology', Woo said, 'pornography is becoming more pervasive. With Web cams we're seeing more live molestation of children' " (Landesman, 2004, p. 74).

It is not enough to say that the participants in pornography consent, *even* in the case of adult women who apparently do, given the road many have been led (or dragged) down, since childhood in some cases, to get to that point. Genuine autonomous consent requires the ability to evaluate critically and to choose from a range of significant and worthwhile options. Even if all the participants genuinely consented to their use in the pornography industry, however, we would need to consider how pornography influences how *other* non-consenting women are viewed and treated. Compare the (thankfully imaginary) scenario in which some blacks consented to act servile or even to play the part of slaves – who are humiliated, beaten, and whipped for the pleasure of their masters. Suppose a *lot* of whites got off on this and some people got a lot of money from tapping into (and pumping up) the desire for such films. And suppose the widespread consumption of such entertainment – a multibillion-dollar industry, in fact – influenced how whites generally viewed and treated blacks, making it harder than it would otherwise be for blacks to overcome a brutal and ongoing legacy of hate and oppression. It is unimaginable that we would tolerate such "entertainment" simply because some people got off on it.

To give another analogy, the fact that scabs will work for less money (in worse conditions) than strikers harms the strikers. It makes it harder for the strikers to work under fair conditions. Sure, the scabs benefit; however, that is not the point. The point is that the strikers suffer. Suppose there were "slave auction" clubs where some blacks allowed themselves to be brutalized and degraded for the pleasure of their white customers. Suppose the black "performers" determined that, given the options, it was in their best interest to make money in this way. Their financial gain – imagine that they are highly paid – more than compensates for the social harm to them as individuals of being subjected to a slightly increased risk (resulting from the prevalence of such clubs) of being degraded and brutalized outside their workplace. Some of them even enjoy the work, having a level of ironic detachment that enables them to view their customers as pathetic or contemptible. Some, who do not actually enjoy their work, do not suffer distress, since they manage to dissociate during it. Others are distressed by it, but they have determined that the financial benefit outweighs the psychic and physical pain. For those blacks who did not work in the clubs, however, there would be nothing that compensated for their slightly increased risk of being degraded and brutalized as a result of it. They would be better off if the clubs did not exist. The work done by the blacks in the clubs would make it harder for other blacks to live their lives free of fear.

The harms caused by pornography to non-participants in its production – often called "indirect" or "diffuse" harms, which makes them sound less real and less serious than they actually are – include: (1) harms to those who have pornography forced on them, (2) increased or reinforced discrimination against – and sexual abuse of – girls and women, (3) harms to boys and men whose attitudes toward women and whose sexual desires are influenced by pornography, and (4) harms to those who have already been victimized by sexual violence. The first three categories of harm have been amply documented (Lederer, 1980; Attorney-General's Commission on Pornography, 1986; MacKinnon, 1987; Itzen, 1992; MacKinnon, 1993; Russell, 1993; Lederer and Delgado, 1995; MacKinnon and Dworkin, 1997). That the proliferation of pornography leads to attitudinal changes in men, which, in turn, lead to harmful behavior, should not be surprising, especially given the high rates of exposure to pornography of pre-teen and teenage boys. On the contrary, as Frederick Schauer, Frank Stanton Professor of the First Amendment at the John F. Kennedy School of Government at Harvard University, testified at the Pornography Civil Rights Hearing in Boston, Massachusetts on March 16, 1992:

> I find it a constant source of astonishment that a society that so easily and correctly accepts the possibility that a cute drawing of a camel can have such an effect on the number of people who take up smoking, has such difficulty accepting the proposition that endorsing

images of rape or other forms of sexual violence can have an effect on the number of people who take up rape. (Cited in MacKinnon and Dworkin, 1997, p. 396)

One might object, though, that pornography is merely a symptom (of a misogynistic, patriarchal society), not a cause. Even if this were the case, however, that would not mean that we should not be concerned about it. The fact that there are so few female legislators in the United States at the federal level (and that it is still inconceivable that a woman could be elected president) is a symptom, not a cause, of patriarchy. But this does not mean that we should not do anything about the political status quo. In any case, pornography is more than a mere symptom: it fosters and perpetuates the sexist attitudes that are essential for its enjoyment, even if it does not create them.

It should be noted here that the fact that the *point* of pornography (from the standpoint of the producers) is to make money by giving pleasure does not mean that it cannot also be harmfully degrading. On the contrary, it is pleasurable (and profitable) *precisely because* it is degrading to others. And it is reasonable to expect a spill-over effect in the public domain, since its enjoyment requires the adoption of certain attitudes. Compare the case of pornography with that of sexist humor. Until quite recently, it used to be maintained that women who were offended by sexist jokes were simply humorless. After all, it was held, one can laugh at a sexist joke (because it is funny) and not *be* a sexist. Now it is widely acknowledged that such jokes are funny only if one holds certain sexist beliefs: in other words, the humor is contingent upon the beliefs.[3] With regard to pornographic depictions, it would be difficult to argue that the degradation and subordination of women they involve are merely incidental to their ability to arouse. The arousal is dependent on the depiction of degradation, just as, in sexist humor, the humor is dependent on the sexism. I stress this in order to deflect the objection that the *point* of pornography is to give pleasure, not to defame or degrade women.

It might be argued that one could laugh at sexist jokes and enjoy sexist pornography *in private* without this having any effect on one's ability to view women as equals *in public* and to treat them accordingly. But are we really so good at keeping our private and public attitudes distinct? Suppose it became known that a white public official – say, a judge – privately relished racist humor, collected racist paraphernalia, and showed old racist films at home for the entertainment of his close friends and family. Although one might not want there to be laws against such reprehensible behavior (for their enforcement would require gross invasions of privacy), one would presumably consider such *private* behavior to compromise the integrity of the judge's public position. (Were this judge's pastime to be made public during his confirmation hearings for a seat on the Supreme Court, for example, it would presumably defeat his nomination.)

It is easier for us, now, to see the harm in the dehumanization of blacks and Jews in racist and anti-Semitic propaganda. We are well aware that the Nazis' campaign to exterminate the Jews utilized anti-Semitic propaganda which portrayed Jews as disgusting, disease-ridden vermin. In addition, "Nazis made Jews do things that would further associate them with the disgusting," making them scrub latrines to which they were then denied access (Nussbaum, 2001, p. 348). This in turn made them appear less than human. As Primo Levi observed in *The Drowned and the Saved*:

> The SS escorts did not hide their amusement at the sight of men and women squatting wherever they could, on the platforms and in the middle of the tracks, and the German

passengers openly expressed their disgust: people like this deserve their fate, just look how they behave. These are not *Menschen*, human beings, but animals, it's as clear as day. (Quoted in Nussbaum, 2001, p. 348)

It is harder for us to see the same process of dehumanization at work when girls and women are routinely portrayed as being worthy of degradation, torture, and even death. But empirical studies have shown that exposure to such portrayals increases the likelihood that people will take actual sexual violence less seriously – and even consider it to be justified in some cases (see MacKinnon 1993, pp. 46–60; Russell, 1993, pp. 113–213; Lederer and Delgado, 1995, pp. 61–112).

There is another connection between the dehumanization of girls and women in pornography and their brutalization in rape, battering, forced prostitution, and sexual murder, which is that, in a society where women are victimized in these ways at an alarming rate, it shows a callous disregard for the actual victims to have depictions of sexual violence bought and sold as entertainment. For a short while, after 9/11, we empathized so much with the victims of the terrorist attacks that films of similarly horrifying attacks were withdrawn because they were no longer considered entertaining. But victims of sexual violence are given so little respect that many of us see nothing wrong with being entertained by depictions of what they have had to endure.

If we take seriously the harm of pornography, then we want to know what to do about it. Should the government intervene by regulating it? The standard debate over pornography has framed it as a free speech issue. The drafters of an anti-pornography ordinance adopted by the city of Indianapolis argued that pornography constitutes a violation of the civil rights of women. In response to those who asserted that the First Amendment protected pornography, they argued that pornography violated the First Amendment rights of women (by "silencing" them – depriving them of credibility and making "no" appear to mean "yes" in rape scenarios) as well as their Fourteenth Amendment rights to equal protection. In his opinion in *American Booksellers Association* v. *Hudnut*, which ruled unconstitutional the Indianapolis anti-pornography ordinance, Judge Frank Easterbrook acknowledged that pornography harms women in very significant and concrete ways:

Depictions of subordination tend to perpetuate subordination. The subordinate status of women in turn leads to affront and lower pay at work, insult and injury at home, battery and rape on the streets. In the language of the legislature, "[p]ornography is central in creating and maintaining sex as a basis of discrimination. Pornography is a systematic practice of exploitation and subordination based on sex which differentially harms women. The bigotry and contempt it produces, with the acts of aggression it fosters, harm women's opportunities for equality and rights [of all kinds]." Indianapolis Code §16-1(a) (2). Yet this simply demonstrates the power of pornography as speech" (*American Booksellers Association, Inc.* v. *Hudnut*, 1985, p. 329).[4]

Easterbrook seems to take the harms of pornography seriously, but he then goes on to talk about its "unhappy effects" which he considers to be the result of "mental intermediation." He assumes that speech has no (or merely negligible) effects that are not under the conscious control of the audience, although this assumption is undermined not only by the widely acknowledged power of advertising, but also by recent work in cognitive neuroscience on the prevalence of unconscious imitation in human beings.[5] It might be argued, though, that, if we consider the producers of pornography to be even partially responsible for the violence perpetrated by some of its consumers, then we must consider the perpetrators *not* to be responsible or to be less than fully responsible for their crimes. But this does not follow. Even if the perpetrators are considered to be 100 percent responsible, some responsibility can still be attributed to the pornographers. (In fact, two or more people can each be 100 percent responsible for the same crime, as in the case of multiple snipers who simultaneously fire many shots, fatally wounding their victim.)

The courts have, for now, decided that even if serious harm to women results from it, pornography is, qua speech, protected (except for that material which also meets the legal definition of obscenity). That is, there is, currently, a *legal* right to it, falling under the right to free speech. But *should* there be?

A Moral Right to Pornography?

Of course we value freedom of speech. But how should we value it? What should we do when speech is genuinely harmful? Traditionally, in the United States, the right to free speech is held to be of such high importance that it trumps just about everything else. For example, in the *Hudnut* case, discussed above, it was acknowledged that the pornography producers' and consumers' right to free speech was in conflict with women's right to equal protection, but it was asserted (without argumentation) that the free speech right had priority. Acceptance of this claim without requiring a defense of it, however, amounts to adopting a kind of free speech fundamentalism. To see how untenable such a view is, suppose that uttering the words "you're dead" caused everyone within earshot (but the speaker) to fall down dead. Would anyone seriously say that such speech deserved protection? Granted, the harms of pornography are less obvious and less severe, but there is sufficient evidence for them for it to be reasonable to require an argument for why the legal right to it should take priority over others' legal rights not to be subjected to such harms.

If we reject free speech fundamentalism, the question of whether pornography should be legally restricted becomes much more complicated. My aim here is not to articulate or defend a position on this question, but I do want to stress that whatever view we take on it should be informed by an understanding of the harms of pornography – the price some people pay so that other people may get off on it.

In Chapter 20 in this volume, "The Right to Get Turned On: Pornography, Autonomy, Equality," Andrew Altman (Altman, 2014, pp. 308–318) shifts the debate over pornography in a promising way by arguing that there is a *moral* right to (even violent misogynistic) pornography, falling not under a right to free speech, but, rather, under a right to sexual autonomy (which also covers the right to use contraceptives and the right to homosexual sex).[6] On this view, which Altman dubs "liberal sexual morality," whatever harm results from pornography is just the price we pay for the right to sexual autonomy. Sexuality is an important, arguably central, aspect of a flourishing human life. Sexual expression is one of the primary ways we define ourselves and our relations to others, and a healthy society should value and celebrate it. But what does it add to these claims to say that we have a moral *right* to sexual autonomy? And, if we do have such a right, does it include a right to produce, distribute, and consume pornography (defined, as above, as violent degrading misogynistic hate speech)?

Although philosophers disagree about the nature of rights (and, indeed, even about whether such things exist at all), most hold that to say that someone, X, has a moral right to do something, y, means that others are under a duty not to interfere with X's doing y. (Of course, X's right is limited by others' rights, as expressed by the saying "your right to swing your arm ends at my face.") But beyond this, there is little agreement. Some hold that rights are natural, inalienable, and God-given. Others hold that rights-talk is just shorthand for talk about those interests that are especially important to us (e.g., because protecting them tends to increase our welfare). Some hold that we have positive rights, just by virtue of being human, such that other people are under an obligation to provide us with whatever we need to exercise those rights. (If there is a positive right to education, for example, then society has an obligation to provide free public education for all.) Others hold that we have only negative rights (unless individuals *grant* us positive rights by, for example, making promises to assist us), which require only that other people do not interfere with our exercising those rights. (The right to privacy, if taken to be simply a right to be left alone, is an example of a negative right.)

On any account, the concept of a right is diffuse. To say that X has a moral right to do y does not, by itself, say very much, unless we specify what others are required to do (or to refrain from doing) in order not to violate that right. There is a wide range of different responses to X's doing y, given that X has a right to do y – from complete acceptance (or perhaps even positive support) to something just short of physical restraint or intervention. Where is the alleged right to pornography located on this spectrum of moral assessment?

Altman considers the right to pornography and the right to sexual orientation to have the same foundation in a right to sexual autonomy. What should our (society's) attitude be toward the exercising of that right? Should we tolerate it, that is, have no laws against it, while allowing private individuals to lobby against it or to try to dissuade people from it? Or should we actively embrace it? Assimilating the right to pornography to the right to sexual orientation muddies the waters here. Presumably, according to liberal sexual morality, the right to sexual orientation requires more than mere tolerance. It requires society's complete acceptance (and, I would argue, positive support, given that prejudice and violence against gays and lesbians persist in our society). It is wrong to hold that gays and lesbians have "bad characters" or to try to get them to "reform."

The right to pornography, however, does not lie on the same end of the spectrum, since Altman claims that getting off on pornography is a sign of a bad character. Some feminists and liberals who defend a legal right to pornography hold at the same time that all sorts of private pressure – protests, boycotts, educational campaigns – should be brought to bear on the pornographers. Altman's position is that there is not just a legal right, but also a *moral* right to pornography, even if there is something bad about exercising it. There are persuasive reasons for holding that we have legal rights to do some things that are morally wrong, in cases in which enforcement would be impossible or would involve gross violations of privacy. But Altman seems to hold that we have a *moral* right to do some things that are morally wrong. What does this mean? It cannot mean that people have a right to do things that are wrong in that they harm others. It might mean that people have the right to do things that other people consider wrong (but that are not harmful to others) – that is, people have the right to do harmless things that other people morally disapprove of. However, if the behavior (e.g., engaging in homosexual sex), is not unjustly harming others, then liberals who subscribe to Mill's harm principle have no grounds for considering it to be wrong.

So where should the right to pornography be located on the spectrum of moral assessment? There is no one answer to this question. We need to look at particular cases. Suppose I have a 21-year-old son – leaving aside the question of whether minors have a right to pornography – who is a heavy consumer of pornography (of the kind I have been talking about). What does his (alleged) right to pornography entail? Given my opposition to pornography, presumably I would not be under an obligation positively to support his pornography habit by buying it for him. But would I have to pretend that I am not aware of it? Would I be under a duty not to try to dissuade him from viewing pornography? Would his sister be under a duty not to throw the magazines out when she saw them in common areas of the house? Would it be wrong for his buddies to try to talk him out of it? Would his teachers have a duty to refrain from arguing against it? Would it be wrong for his neighbors to boycott the local convenience store that sold it? Would his girlfriend (or boyfriend) who became convinced it was ruining their relationship be under a moral duty not to rip it out of his hands? If the answer to each of the above questions is "no," which I think it is, then it is not clear what, if anything, his right entitles him to.[7] What is clear is that, if a right to pornography exists, it is quite unlike a right to engage in homosexual sex or to use contraceptives, and is located at the opposite end of the spectrum of moral assessment.

Perhaps there is, nevertheless, something special about sexual arousal ("getting turned on") that gives it special moral status. But Altman has not said what makes sexual arousal different (in a morally significant way) from other forms of arousal – for example, that of racial animus. It makes sense to say that there is a right to be turned on – not a special right, but, rather, one falling under a general right to liberty, but this general right to liberty is delimited by the harm principle. There is no general right to have pleasurable feelings (of any sort, sexual or otherwise) that override others' rights not to be harmed. There is no moral right to achieve a feeling of comfort by unjustly discriminating against homosexuals on the grounds that associating with them makes you uncomfortable. Likewise, there is no moral right to achieve a feeling of superiority (no matter how pleasurable such a feeling might be) by discriminating against those of a different race. And it does not matter how central to one's self-definition the feeling in question might be. For parents, the satisfaction of ensuring the good upbringing and education of their children is of paramount importance, and yet this degree of importance does not give racist parents the right to racially segregated housing or schools.

It might be argued that sexual arousal is special in that it is a bodily pleasure and, thus, more natural, possibly even immutable. Even if this were so, it would not follow that one has a right to achieve it by any means necessary. To take an example of another kind of "bodily" pleasure, suppose that there are gustatory pleasures that can be achieved only in immoral ways – for example, by eating live monkey brains (which some people used to do), or organs or flesh "donated" by (or purchased from) living human beings, or food that has been stolen from the people on the verge of starvation. That there is a (general) right to enjoy eating what one chooses to eat – it would be (in general) wrong, for example, for me to force you to eat, or not to eat, something – does not mean that one has a right to eat whatever gives one pleasure.

But it is not the case that what people find sexually arousing is a simple biological fact about them, a given, something immutable. People can be conditioned to be aroused by any number of things. In one study, for example, men were conditioned to be aroused by a picture of a woman's boot (Russell, 1993, p. 129). Emotions, especially ones with strong physiological components, such as sexual arousal, *feel* natural. They do not seem to be socially constructed, because we do not (at the time) consciously choose them: they just *are*. But emotions are, at least to some extent, learned reactions to things. There are gender differences in emotional reactions; for example, men tend to get angry in some situations in which women tend to feel not angry, but hurt. But this does not mean that such differences are *natural*.

Given the wide variety of sexual fantasies and fetishes we know about, it is conceivable that just about *anything* could be a turn on for someone – looking at photos of dead, naked bodies piled in mass graves in Nazi death camps, for example, or looking at photos of lynched black men. According to liberal sexual morality, the only reason for supposing that there might not be a moral right to make a profit from and get off on such "pornography" would be that the photographed people are posthumously harmed by it (given that they did not consent to their images being used in this way). But suppose they had consented. Or suppose, more plausibly, that the images were computer-generated – completely realistic-looking, but not images *of* actual individuals. Liberal sexual morality would have to allow (some) people to make money by others' getting turned on by these images. Not only that, but, given that sexual desires are malleable, the pornographer also has a right to make money by acculturating others to be turned on by such images. (In other words, the pornographer has a right to turn the world into a place where people get turned on by such images.) And, if our attitude toward this is grounded in the right to sexual autonomy, it should be similar to our attitude toward homosexuality: we should not merely tolerate it, we should come to accept and support it.

While conceding that there are limits to the right to sexual autonomy – it is constrained by the harm principle – Altman assumes (as most liberals do) that one cannot be harmed by something to which one consents. I argued earlier that the way many models get lured into the pornography industry should make us at least question the extent to which they are consenting to what is being done to them. But suppose they do consent. Does that mean that we must tolerate the production and use of whatever pornography results? Unfortunately, one does not have to construct a thought experiment to test our intuitions about this. According to *The New York Times*, Armin Meiwes, "[a] German computer technician who killed and ate a willing victim he found through the Internet" was recently convicted of manslaughter. His "victim," Bernd-Jürgen Brandes, had "responded to an Internet posting by Mr Meiwes seeking someone willing to be 'slaughtered'." " 'Both were looking for the ultimate kick'," the judge said. It was "an evening of sexual role-playing and violence, much of it videotaped by Mr Meiwes," enough to convince the court that the "victim" had consented (Landler, 2004, p. A3). Does the right to sexual autonomy include the rights to produce, sell, and get turned on by the videotape of this "slaughter" – a real-life instance of a snuff film? If we cannot *prove* that there is a causal connection between the film and harm to others, the answer, according to liberal sexual morality, is "yes."

Altman claims that "even if a causal connection between violent pornography and sexual violence were clearly established, it would still be insufficient to conclude that, in contemporary society, the production, distribution and viewing of violent pornography lay beyond the limits of an adult's right to sexual autonomy" because *other* media – he cites "slasher films" – arguably "cause at least some amount of violence against women, sexual and otherwise. However, it is unreasonable to deny that adults have a right to produce, distribute, and view such movies" (Altman, 2014, p. 312). Why, if one has established that, say, "slasher films" are harmful, we must hold that adults have a right to them is not explained. But even if we agree that adults have the right to produce/consume non-pornographic media even if it is as harmful as pornography, it does not follow that adults have the right to produce/consume pornography. To assume that it does would be like arguing against prohibiting driving while talking on cell phones on the grounds that this is not the *only* thing (or even the main thing) contributing to automobile accidents.

Altman accepts that "it is reasonable to hold that the existence of . . . pornography makes it more difficult for women to live their lives as the sexual equals of men – i.e., more difficult relative to a society which was ruled by a liberal sexual morality and had fewer women, or none at all, who were willing to engage in humiliating conduct as part of the production of pornographic materials" (Altman, 2014, p. 317), but he notes that women are better off in a society with liberal sexual morality than in a society with traditional sexual morality (for example, Saudi Arabia). I agree, but surely these are not the only two possibilities. I would advocate the alternative of a progressive sexual morality. What might that look like? We do not even know. Even our most deep-seated assumptions about sexuality may turn out to be mistaken. We used to view rape as being motivated purely by lust and battering as a way of showing spousal love. Some of us still do. Gradually, however, we are breaking the link between sexuality and violence. Perhaps some day we will have reached the point where sexual violence is no longer arousing, where it makes no sense to talk of killing and being killed as the "ultimate" sexual "kick."

According to liberal sexual morality, the harms of pornography are the price we pay for having the right to sexual autonomy in other areas – for example, the right to have sex (including homosexual sex) outside of marriage and the right to use contraceptives. But this view (of the right to sexual autonomy as an all-or-nothing package) is formed in response to legal moralism, and makes sense only if one considers all these rights to be rights to do harmless things that some people nevertheless morally condemn. In such cases, proponents of liberal sexual morality say: "If you don't like it, don't look at it (or hear about it or think about it)." This is a satisfactory response only if the behavior in question is not harming anyone. But as our views about what constitutes harm have changed, our views of what is our business have also changed. Just as we no longer look the other way in response to marital or "date" rape, domestic violence, and sexual harassment, we should no longer accept pornography's harms as the price we pay for sexual autonomy.

Notes

I would like to thank Ann Bumpus, Christopher Wellman, and Thomas Trezise for helpful discussions of many issues in this article. My deepest thanks go to Margaret Little who gave me invaluable comments on several drafts.

1 Mill considered his harm principle to apply equally to governmental regulation and to "the moral coercion of public opinion." The harm principle states that, "the only purpose for which power can be rightfully exercised over any member of a civilized community, against his will, is to prevent harm to others" (1978: 9). Mill does not specify what counts as harm. Following Joel Feinberg (1984), I consider it to be a wrongful setback to one's significant interests.

2 This is the definition used in the anti-pornography ordinance drafted by Andrea Dworkin and Catharine MacKinnon, passed by the city of Indianapolis, but ruled unconstitutional by the courts.

3 For a persuasive argument to that effect, see de Sousa (1987). In comparing sexist fantasies with sexist and racist humor, one might reply, however, that we have less control over, and thus are less responsible for, our fantasies than our jokes. This seems right, to the extent that we can refrain from laughing at or telling certain jokes (even though we might not be able to resist finding them funny). But the same distinction applies to fantasies. We do not always choose the fantasies that occur to us, but we can choose whether or not to cultivate them (voluntarily return to them repeatedly, make or view films about them, etc.). Even in the case of dreams, over which we, at the time, anyway, have no control, a white male liberal would be alarmed if he often had pleasurable dreams of watching blacks getting lynched. This would presumably prompt some probing of his unconscious attitudes about blacks.

4 This view cannot consistently be held, however, by liberals and feminists who support laws against sex or race discrimination and segregation in schools, workplaces, and even private clubs. One does not hear the argument that if segregation harms minorities' opportunities for equal rights this simply demonstrates the power of freedom of association, which is also protected by the First Amendment.

5 The recent research discussed in Hurley (2004) suggests that the imitation of others' behavior, including others' violent acts, is not a consciously mediated process, under the autonomous control of the viewers/imitators.

6 Since some theorists ground the right to free speech in a right to autonomy, however, there may not be such a sharp distinction between these two approaches. See Brison (1998).

7 I also mean for the above thought experiment to illustrate the fact that the nature of the duty one has with respect to the holder of the alleged moral right to pornography depends on one's relationship to the right-holder. Presumably a neighbor would be under a duty not to snatch pornography out of the right-holder's hands. But if someone *else*, the right-holder's lover, say, is under no such duty, then it is not clear what the right amounts to.

References

Altman, A. (2014) The right to get turned on: pornography, autonomy, equality. In *Contemporary Debates in Applied Ethics*, 2nd edn, ed. A.I. Cohen and C.H. Wellman, pp. 308–318. Malden, MA: Wiley-Blackwell.

American Booksellers Association, Inc. v. Hudnut (1985) 771 F.2d 323.

Attorney General's Commission on Pornography (1986) *Final Report*. Washington, DC: US Department of Justice.

Brison, S.J. (1998) The autonomy defense of free speech. *Ethics* 108: 312–339.

Brownmiller, S. (1975) *Against Our Will: Men, Women and Rape*. New York: Bantam Books.

Delgado, R. and Stefancic, J. (1997) *Must We Defend Nazis? Hate Speech, Pornography, and the New First Amendment*. New York: New York University Press.

de Sousa, R. (1987) When is it wrong to laugh? In *The Rationality of Emotion*, p. 275. Cambridge, MA: MIT Press.

Feinberg, J. (1984) *The Moral Limits of the Criminal Law*, Vol. 1, *Harm to Others*. New York: Oxford University Press.

Hurley, S.L. (2004) Imitation, media violence, and freedom of speech. *Philosophical Studies* 117 (1–2) (January): 165–218.

Itzen, C., ed. (1992) *Pornography: Women, Violence and Civil Liberties*. New York: Oxford University Press.

Landesman, P. (2004) The girls next door. *Sunday New York Times Magazine*, January 25, pp. 30–39, 66–74.

Landler, M. (2004) German court convicts Internet cannibal of manslaughter. *New York Times*, January 31, p. A3.

Lederer, L., ed. (1980) *Take Back the Night: Women on Pornography*. New York: William Morrow and Co., Inc.

Lederer, L.J. and Delgado, R., eds (1995) *The Price We Pay: The Case Against Racist Speech, Hate Propaganda, and Pornography*. New York: Hill and Wang.

MacKinnon, C.A. (1987) *Feminism Unmodified: Discourses on Life and Law*. Cambridge, MA: Harvard University Press.

MacKinnon, C.A. (1993) *Only Words*. Cambridge, MA: Harvard University Press.

MacKinnon, C.A. and Dworkin, A., eds (1997) *In Harm's Way: The Pornography Civil Rights Hearings*. Cambridge, MA: Harvard University Press.

Mill, J.S. (1978) *On Liberty*. Indianapolis, IN: Hackett Publishing Co. (Originally published 1859).

Miller v. California (1973) 413 US 15.

Nussbaum, M. (2001) *Upheavals of Thought*. Cambridge: Cambridge University Press.

Russell, D.E.H., ed. (1993) *Making Violence Sexy: Feminist Views on Pornography*. Buckingham: Open University Press.

Further Reading

Dwyer, S., ed. (1995) *The Problem of Pornography*. New York: Wadsworth.

Kappeler, S. (1986) *The Pornography of Representation*. Minneapolis, MN: University of Minnesota Press.

Drugs

CHAPTER TWENTY-TWO

In Favor of Drug Decriminalization

Douglas Husak

I propose to divide my argument in favor of drug decriminalization into three sections. Each section defends the following conclusions. First, a great deal of uncertainty about whether or not drugs should be decriminalized stems from confusion about the question itself. I will try to clarify the issue that needs to be resolved. Second, at least three different kinds of approaches might be taken to answer the question of whether or not drugs should be decriminalized. I will briefly discuss each of these approaches and identify the strategy I believe a legal philosopher should prefer. Finally, the debate about drug decriminalization is somewhat unlike controversies about whether other kinds of conduct should be decriminalized. I will comment briefly on the most important difference between debates about drug decriminalization and other such controversies.

The Meaning of Decriminalization

Philosophers are at their best in clarifying issues. We cannot hope to answer a question unless we understand exactly what it asks. First, one might think we need to identify what *drugs* are. In fact, no accepted definition allows us to categorize whether or not a given substance is a drug. Without an adequate account, how can we be expected to decide whether drugs should be decriminalized? Surely the debate would be improved if a satisfactory definition were available. Despite the problems caused by the absence of a definition, however, I believe we can make progress on questions of criminalization or decriminalization. We can neglect borderline cases and move directly to a discussion of paradigm cases, that is, substances that no reasonable person would fail to categorize as a drug. Should we decriminalize the use of any (currently illicit) drug such as marijuana? Cocaine? The opiates? And should we criminalize the use of any (currently licit) drug such as alcohol? Tobacco? Caffeine? Indeed, sensible discussions

Contemporary Debates in Applied Ethics, Second Edition. Edited by Andrew I. Cohen and Christopher Heath Wellman.
© 2014 John Wiley & Sons, Inc. Published 2014 by John Wiley & Sons, Inc.

of whether criminal punishments are justified must proceed drug by drug; the case in favor of criminalization is far more plausible for some substances than for others. Despite the fact that a detailed analysis would examine one drug at a time, in what follows I will discuss those considerations I believe are applicable to *all* substances we agree to be drugs.

Whether or not we understand exactly what drugs are, it may be surprising that there also is no consensus about the meaning of such terms as *legalization* or *decriminalization*. This lack of consensus is more serious and is bound to compound our confusion. I make little effort to canvass what other commentators take these terms to mean. What *I* mean by proposals to "decriminalize" the use of a given drug is simple – deceptively so. I mean that the *use* of that drug would not be a criminal offense. I take it to be a conceptual truth that criminal laws are those laws that subject persons to state punishment. Thus anyone who thinks that the use of a given drug should be decriminalized believes that persons should not be punished merely for using that drug. Public opinion polls reveal tremendous confusion about this topic. Many respondents report that they do not want to see a given drug decriminalized, but do not favor punishing anyone who merely uses it. If my definition is accepted, this response is incoherent. If persons no longer are eligible for punishment for using a drug, the use of that drug has been decriminalized.

I now want to mention five reasons why I contend that this definition of decriminalization is deceptively simple. First, there is little punishment for mere *use* today. Of course, use is and ought to be punished in situations in which it is especially risky – as when driving or hunting, for example. Still, in most (but not all) jurisdictions throughout the United States, what is punished is *possession* rather than use. Technically, then, drug use per se is not criminalized in most places. But I take the fact that laws punish possession rather than use to be relatively unimportant. Possession is punished rather than use mostly because it is so much easier to prove. Thus I will ignore this technicality in what follows, and will continue to suppose that drug decriminalization pertains to use.

Second, I have indicated that the question to be debated is whether or not drugs should be decriminalized. But expressing the issue in this general way is misleading. Arguably, we cannot decide whether to criminalize or decriminalize drug use without specifying the *purpose* for which that drug is used. If so, our topic is not really about drug use per se, but rather about a particular *kind* of drug use – that is, a particular reason for using drugs. We might well decide that one and the same drug – cocaine, morphine, or Prozac, for example – should be criminalized for some purposes, but decriminalized for others. Presumably, the use of these drugs should not be a criminal offense when people have a medical reason to take them. But public attitudes and state policy often are entirely different when that same drug is used for a non-medical purpose. Although drugs can be used for several different kinds of non-medical purposes, one such purpose is especially significant. This use is *recreational*. It is hard to be precise in characterizing when use is recreational. Roughly, people engage in recreational activities – whether or not these activities involve a drug – in order to seek pleasure, euphoria, satisfaction, or some other positive psychological state. When a drug is used in order to attain a positive psychological state – a drug *high* – I will call that use recreational. Under our present policy, a recreational user of a given drug may face severe punishment, but none at all if his use of that same drug is medical.

336   **Douglas Husak**

As I have indicated, drugs are used for a variety of objectives apart from the expectation of attaining a "high." Jurisdictions must decide whether to allow an exception to their prohibitory regime when drugs are taken for religious reasons, for example. One issue about whether a purpose is legitimate has recently come to dominate all others. The question whether marijuana should be permitted for medical purposes has received an extraordinary amount of attention from academics as well as from the public at large. As of 2012, eighteen states and the District of Columbia allow some system of medical marijuana, and legislative battles can be anticipated in most of the jurisdictions that continue to resist this trend. Even in states that have not authorized it, a necessity or choice-of-evils defense from criminal liability might be available to patients who use medical marijuana. Should states really punish sick persons who use marijuana to relieve their symptoms? Of course, it is controversial how much drug use should be categorized as medical. Some commentators are persuaded that many users of marijuana and other drugs *self-medicate*. That is, they consume drugs in order to return to a condition of normalcy from which they deviate when abstinent. If this hypothesis is correct, it casts the permissibility of drug use in a more favorable light. Nonetheless, I have little to say about this hypothesis because I think that the determination of whether a given substance has medical benefits should be made almost solely by health professionals. Moreover, I do not share the foremost concern that has led many citizens to oppose medical marijuana. Some critics predict that the use of marijuana for medical purposes is the first step down the road to decriminalization for non-medical reasons. I suspect this prediction is accurate, even though I depart from these critics about whether that result is sufficiently worrisome to justify punishing those persons whose use *is* medical.

Third, decriminalization of a drug means that persons are not subject to punishment simply for using it, but there is no agreement about what kinds of state response amount to *punishments*. Many reformers argue that drug users should be fined rather than imprisoned. Others argue that drug users should be made to undergo treatment. Still others believe the state should be allowed to seize any illicit drugs persons possess but no additional action should be taken. Each of these ideas has been described as a form of decriminalization. Whether these proposals are compatible with what I mean by decriminalization depends on whether fines, coerced treatment, or seizure of drugs are modes of punishment, rather than alternatives to punishment. I tend to think that these state responses *are* modes of punishment. Even though these modes of punishment almost certainly are preferable to what we now do to drug users, these responses are ruled out by decriminalization as I construe it. But that is a relatively minor quibble I hope will not distract us. Whatever one takes punishment to be, *that* is what decriminalization forbids the state from doing to people who merely use drugs. The conceptual lesson is clear. Unless one specifies exactly what sanctions the state is authorized to inflict on drug users, the claim that some jurisdictions have decriminalized that drug is bound to create uncertainty and confusion.

Fourth, decriminalization as I define it has no implications for what should be done to persons who *produce* or *sell* drugs. Therefore, it is not really a comprehensive drug policy that can rival the status quo. As Peter de Marneffe (2013) argues, the considerations that count in favor of decriminalizing use are different from those that pertain to the decriminalization of manufacture and exchange. Thus I recommend that we put production and sale aside. This recommendation is bound to disappoint some reformers. Many thinkers are attracted to decriminalization, or are reluctantly driven to support it, because they hope to end the violence, black market, and involvement of organized crime in drug transactions today. These goals are worthwhile, but drug decriminalization per se does not achieve them. I think we should begin by getting clear on what the law should do to drug users, and *then* move to the issue of how production and sale should be regulated. Again, I am aware that many thoughtful people believe that these topics should all be tackled simultaneously. But I think it is easier to take one step at a time.

Fifth and finally, I admit that there is something odd about my understanding of decriminalization. What I call decriminalization in the context of drugs is comparable to what was called *prohibition* in the context of alcohol from 1920 to 1933. In that notorious era, production and sale were banned, but not use or possession. If we replicated the approach of alcohol prohibition in our drug policy, I would call it decriminalization. The oddity of this description underscores the fact that our response to illicit drug users today is far more punitive than anything our country ever did to drinkers of alcohol.

Three Approaches to Decriminalization

At least three different approaches are available to anyone who argues that drug use should be decriminalized. Some of these strategies are more persuasive than others, and each has its strengths and weaknesses. I will briefly summarize two of these approaches before identifying the particular strategy I believe should be especially attractive to a philosopher. Of course, these approaches can be combined to form a stronger case for drug decriminalization than any single rationale.

Any good reason to criminalize a kind of behavior invokes a theory of criminalization. We cannot decide whether we have a good reason to punish persons who use drugs unless we know what would count as a good reason to punish anyone for anything. We do not really *have* a good theory of criminalization in the real world, unless "more is better" qualifies as such a theory. Still, I hope we can make progress in debating the issue of drug decriminalization even without a comprehensive theory of criminalization.

The first argument in favor of drug decriminalization might be called *utilitarian* or *consequentialist*. The basic idea is that no law should be enacted unless it works. A law works, this argument continues, when it reduces the prevalence of the type of conduct it prohibits without causing unintended effects that are worse than the proscribed conduct itself. More specifically, a criminal law against the use of drugs does not work if it is ineffective or counterproductive.

Critics of criminalization have argued for decades that our drug laws do not work. In the first place, they have done little to reduce the incidence of drug use itself. Despite billions of dollars invested, and millions of persons arrested and convicted, rates of drug use remain high. Even when the use of a given drug falls dramatically in the span of a generation – as with cocaine and LSD – it is hard to believe that criminal punishments account for these patterns. Drug trends are difficult to explain, but punitive state action is only a small part of the story. To some extent, new drugs have replaced the old. Non-medical use of prescription drugs currently exceeds the use of cocaine, opiates, and

LSD combined. The threat of punishment is not an especially effective deterrent because illegal drug use is so difficult to detect. Individual users are not foolish to believe they are unlikely to get caught and punished.

An even more worrisome allegation made by utilitarians is that drug laws are counterproductive, causing more harm than good. Drug prohibitions have several pernicious side effects; I will mention only five, but the list could be expanded. First, the enforcement of drug laws has eroded privacy and civil liberties. Second, the health of users is unnecessarily damaged because buyers do not know the strength or purity of the substances they consume. Third, efforts to slow production where drugs are produced have caused unbelievable violence and corruption, especially south of the US border. Fourth, arrests and punishments cost billions of dollars of tax resources that could be put to better purposes.

A fifth counterproductive effect of drug prohibitions is especially worrisome to philosophers who care about social justice. When the state cannot possibly punish *all* of the people who commit a crime, it can only punish *some*. Inevitably, those who get arrested, prosecuted and sentenced are the least powerful. Drug criminalization would have vanished long ago if whites had been sent to prison for drug offenses at the same rate as blacks and Hispanics. Although minorities are no more likely than whites to use illicit drugs, they are far more likely to be arrested, prosecuted, and punished when they do.[1] The enforcement of drug prohibitions has been especially devastating to minority communities. And enforcement will remain selective, since not every offender can possibly be punished.

But decriminalization arguments need not be utilitarian or consequentialist. A second argument for decriminalization maintains that drug prohibitions violate moral and legal rights. The enforcement of these laws has always raised concerns about privacy. According to some philosophers – whom we might call *libertarians* – drug prohibitions themselves also violate the right to do whatever you want with your own body unless you harm others. Of course, the existence of this right is controversial. Arguably, whether a person has a right to do something to his own body depends on what happens when he does it. But additional rights might be violated by drug prohibitions. These rights become easier to identify if we return to the issue of exactly what it is that drug laws are designed to prevent. As I have indicated, use per se is rarely prohibited. Instead, the use of most drugs is prohibited only for a given purpose – to produce a drug "high." In case there is any doubt, let me cite the California criminal statute governing nitrous oxide. This statute makes it a crime for "any person [to possess] nitrous oxide . . . with the intent to breathe [or] inhale . . . for purposes of causing a condition of intoxication, elation, euphoria, dizziness, stupefaction, or dulling of the senses or for the purpose of, in any manner, changing . . . mental processes" (California Penal Code, 2012, §381(b)). The ultimate objective of this statute is to prevent a person from breathing something in order to change her mental state. It is hard to see how this objective is legitimate in a country committed to freedom of thought.

Although utilitarian and rights-based arguments for decriminalization are powerful, I prefer a third strategy. Again, we must begin by asking the right question. In my judgment, the most fundamental issue is not whether to *decriminalize* the use of any or all drugs, but whether to *criminalize* the use of any or all drugs. We should not presuppose that the status quo is just; it must be defended. If I have posed the right question to ask, I believe its answer is as follows. The best reason *not* to criminalize drug use is that no argument *in favor* of criminalizing drug use is any good – not nearly good enough to justify the punishment of drug users. Admittedly, this argument is necessarily inconclusive, placing those who favor decriminalization in the unenviable position of trying to prove a negative. How can anyone hope to show that no argument *in favor* of criminalizing drug use is very good? Until an argument has been put on the table, there is nothing to which one can respond. All that can be done is to respond to the best arguments that have been given.

Three arguments are heard most frequently in favor of punishing drug users. First, drugs are bad for the development and maturation of children and adolescents. Any debate about decriminalization is certain to turn to the effects of drugs on our nation's youth. Second, drugs are unhealthy both physically and psychologically. Advocates of decriminalization can expect to be reminded of a study purporting to show that drugs kill brain cells, destroy memory, or drain motivation. Third, drugs are correlated with violent behavior and criminal activity. Many persons arrested for serious crimes test positive for drugs. A vast literature explores these three arguments, each of which must be examined carefully. Although I do not argue the point here, none of these arguments can withstand close scrutiny in my judgment. Alone or in combination, they do not show the state is justified in prohibiting drugs and punishing persons who use them.

A fourth argument is sometimes produced. Many people oppose decriminalization because they predict it would cause a huge increase in drug consumption. The specter of greater drug use is the trump card my opponents play when supporting our existing policy. I am very skeptical of the accuracy of these predictions; they should be taken with a grain of salt. Recall the meaning of decriminalization. The only change that this policy requires is that the state would not *punish* anyone simply for using a drug. The state may adopt any number of devices to discourage drug use, as long as these devices are not punitive. Even more importantly, institutions other than the state can and do play a significant role in discouraging drug use. After decriminalization, some of these institutions might exert even more influence. Private businesses, schools, insurance companies, and universities, to cite just a few examples, might adopt policies to deter drug use. Suppose that employers fired or denied promotions to workers who use cocaine. Suppose that schools barred students who drink alcohol from participating in extracurricular activities. Suppose that insurance companies charged higher premiums to policy holders who smoke tobacco. Suppose that colleges denied loans and grants to undergraduates who use marijuana. I do not endorse any of these ideas; many seem unwise and destined to backfire. I simply point out that these institutions could continue to do a great deal to decrease drug consumption even if use were decriminalized.

But an even more important point is that these predictions may not be relevant to the topic at hand. Philosophers should search for a respectable reason to criminalize drug use. Predictions about how decriminalization will cause an increase in drug consumption simply do not provide such a reason. Indeed, this reason could be given against repealing virtually *any* law, however unjustified it may be. Let me illustrate this point by providing an example of an imaginary crime that I assume everyone would agree to be unjustified. Imagine the state sought to curb obesity by enacting a statute criminalizing the eating of pizza. Suppose a group of philosophers were convened to debate whether we should *change* this law and *de*criminalize the consumption of pizza. Someone would be bound to protest that repealing this offense would cause persons to eat greater amounts of pizza. They would probably be correct. But surely this prediction would not serve to justify retaining this imaginary crime. If we lack a good reason to attack the problem of obesity by punishing pizza eaters in the first place, the effects of repeal on pizza consumption would not provide such a reason. And so with drugs. This prediction does not provide a good reason to continue to impose punishments unless we already *have* such a reason. Of course, whether we have such a reason is precisely the point at issue.

One reason I prefer my third strategy for decriminalization is that it makes minimal assumptions about controversial empirical and normative questions. First, it does not require that we take a stand on whether our regime of drug prohibitions can be tweaked to become more effective and less counterproductive. Moreover, it makes minimal assumptions about justice. Philosophers famously disagree about the content of principles of justice. It seems safe to assume, however, that no one should be punished unless there are excellent reasons for doing so. Punishment, after all, is the worst thing our state is allowed to do to us. The imposition of punishment must satisfy a very demanding standard of justification. Even though principles of justice are enormously controversial, it is hard to believe that anyone rejects the minimal assumption I make here.

We have excellent reasons to punish people who commit theft or rape. These offenses *harm* others by violating their rights. But this rationale cannot explain why drug users should be punished. I do not think there is any sense of harm or any theory of rights that can be invoked to show that I harm someone or violate his rights when I inject heroin or smoke crack. At most, I *risk* harm to myself and/or to others when I use a drug. Should users be punished for creating these risks? Notice the enormous burden any such argument for criminalization would have to bear. Well over 120 million living Americans have used an illicit drug at some point in their lives – about half of our population aged 12 and over. Over 22 million Americans use an illicit drug each month – on literally billions of occasions.[2] Very few of these occasions produced any harm. Longitudinal studies do not indicate that the population of persons who ever have used illicit drugs is very different from the population of lifetime abstainers in any ways that seem relevant to criminalization.[3] Thus an argument for criminalization would have to justify punishing many people whose behavior is innocuous in order to try to prevent a risk that materializes in a very tiny percentage of cases.

My case for drug decriminalization depends on the claim that no good argument in favor of criminalization has yet been defended. It is utterly astounding, I think, that no sound rationale for drug prohibitions has ever been produced. When asked to recommend the best book or article that makes a philosophically respectable case for punishing drug users, I am embarrassed to say that I have little to suggest. Let me cut directly to my own conclusions. Although some drugs are worse than others, I have yet to hear a persuasive case for punishing people merely for using *any* existing drug. But no single argument for decriminalization can hope to respond to each argument for criminalization. We must respond argument by argument and, I think, drug by drug. I do not know anyone who proposes to punish persons who use caffeine, for example. Surely this consensus exists because of empirical facts about caffeine – about how it affects those who use it as well as how it affects society generally. Despite the problems caused by tobacco and alcohol, no prominent figure advocates that we punish persons simply for smoking or drinking. These facts about licit substances are important in the present context.

If neither tobacco nor alcohol should be punished, we must ask how the drugs we think *should* be punished differ. If no morally relevant differences between these substances can be identified, we should treat them similarly for legal purposes.

If there is a good reason to criminalize the use of any drug that actually exists, we have yet to find it. We need a better reason to criminalize something other than predictions about how its frequency would increase if punishments were not imposed. These predictions are dubious both normatively and (in this case) empirically. Despite my uncertainty about the future, there is *one* prediction about which we can be absolutely confident. After decriminalization, those who use illicit drugs will not face arrest and prosecution. The lives of drug users would not be devastated by a state that subjects them to punishment. The single prediction we can safely make about decriminalization is that it will not devastate the lives of the hundreds of thousands of people who otherwise would be punished for the crime of using drugs.

How Debates about Drug Decriminalization Differ from other Decriminalization Debates

As I have indicated, any debate about whether to punish anyone for anything implicitly invokes a theory of criminalization. I have also said that we lack such a theory. But at least *one* principle in a theory of criminalization seems secure: No state should punish persons for engaging in conduct that is morally permissible. Imagine that someone asks *why* he is punished, what he is punished *for*, or what he has done wrong to *deserve* his punishment. No one who punishes him would respond that he has not done anything wrong, but his punishment is justified nonetheless. This response would be nearly unintelligible. If I am correct about the close connection between wrongfulness and justified punishment, it is crucial to determine the moral status of anything we propose to criminalize. Despite the obvious importance of this issue, advocates of drug decriminalization have treated it in a peculiar way. In other contexts, live controversies about criminalization begin with direct efforts to assess the morality of the conduct in question. Consider the recent history of the debate about the criminalization of homosexual relations, for example. No philosopher opposed the criminalization of this conduct on the ground that the prohibition was ineffective and/or counterproductive. Instead, they tended to argue that criminalization is misguided because homosexual relations are morally permissible. Of course, they were obligated to respond to the (weak) arguments in favor of the contrary position. Still, criminalization is objectionable in principle if the underlying conduct is morally permissible. Thus they had little need to assess the contingent and uncertain empirical considerations involving the enforcement of prohibitions of gay sex.

It is noteworthy that the debate about drug decriminalization has proceeded differently. Commentators rarely purport to address the moral status of drug use itself. In fact, philosophically sophisticated arguments about the morality of drug use are difficult to find anywhere. Opponents of decriminalization frequently assert that drug use is wrongful, but make almost no effort to substantiate their charge. Typically, they do nothing more than cite public opinion polls that support their position. Still, many of the most vocal champions of decriminalization offer no more detail in their rebuttal. Parents in particular are enormously conflicted about what to say to their children about drugs. They often are quick to announce that they do not personally condone drug use and are embarrassed about their former history of experimentation: a handicap or a youthful indiscretion, as they frequently describe it. Libertarians do not provide an exception to this generalization. Many who emphasize personal responsibility and defend a limited role for government insist that sane adults have a right to do whatever they want with their own bodies. This anti-paternalist position, however, is hardly a ringing endorsement of drug use. On the libertarian view, drug use falls under the same principle as the decision to commit suicide or to mutilate oneself. In fact, this position is compatible with the claim that drug use is wrongful but is protected by a right to do wrong. As far as I am aware, almost no one has expressed comparable sentiments in the history of debates about the proscription of homosexual relations.

What *should* philosophers think about the moral status of drug use? To underscore the novelty of this question, I place it in a specific context. Suppose you are a professor of moral philosophy and a student enters your office and announces his intention to use a given drug: crack or heroin, let us suppose. His friends report this drug to be exhilarating, and he wants to experience the sensation himself. He claims to be motivated by the same curiosity that has led him to hang-glide and scuba-dive. The student is not at all interested in his legal obligations or worried about the risks of being caught and arrested. Suppose further that you set aside the possible consequences for your job or professional reputation should your advice become public and brought to the attention of the dean or department chair. The student seeks advice solely from a moral perspective, and that is the perspective from which you purport to advise him. To the best of my knowledge, this basic issue is almost completely untouched in the relevant commentary produced by moral philosophers who debate drug decriminalization. Most professors would have an easier time counseling students who seek advice about obtaining an abortion.

It would be impossible to offer sensible advice to this student without identifying the possible benefits of drug use. Even minimal risks are foolish in the absence of good reasons to take them. It is easy to recommend literature that recites the harms of drug experimentation, but credible commentary on the benefits of drug use is much more difficult to find. As long as drug use is thought to produce all harms and no benefits, its moral status is easy to specify. Presumably, however, most people consume drugs largely because they regard their effects as pleasurable and euphoric. Is pleasure itself a good? If so, is the pleasure caused by drug use any different? Is it good to satisfy curiosity about how something feels? And how are these goods (if they exist at all) to be balanced against the risks drug use imposes? Should this calculus include other advantages frequently alleged on behalf of drug use: spiritual enlightenment, artistic creativity, consciousness expansion, and the like? Even those academic philosophers who regard psychological hedonic states as intrinsic goods – or even as the *only* intrinsic good – have precious little to say about drugs. But even if we would refuse to be connected to a "pleasure machine" that guaranteed we would have more positive psychological sensations than we experience in our existing reality, the possible use of drugs to supplement a good life must be taken seriously.

These largely untouched questions become more vexing as we move from real to imaginary drugs – even though intuitions about thought experiments are a staple of moral and legal philosophy on a wide range of other topics. Consider, by way of contrast, how philosophical reflection on the permissibility of torture is driven by intuitions about "ticking bomb" scenarios – whether or not such scenarios have ever been actualized in the real world. A comparable use of thought-experiments is conspicuously absent in questions about drugs and drug policy. Advocates of decriminalization should be asked whether any drugs can be imagined that it would be *im*permissible to use. What are the properties of these drugs that render their use impermissible, and how confident should we be that these properties are absent in the drugs we propose to allow? Prohibitionists should be asked a parallel question. Can they imagine drugs that it would be *permissible* to use? What are the properties of these drugs that render their use permissible, and how confident should we be that the drugs to be proscribed lack these properties? These sorts of questions are familiar to moral and legal theorists in other contexts. By focusing on imaginary cases, we have some chance of making progress on moral questions about drug use without the need to take a stand on contested empirical issues.

I think it is easy to imagine a drug that people should be punished for using. Some such drugs are vividly portrayed in great works of fiction. Consider the substance that transformed Dr Jekyll into Mr Hyde. If a drug literally turns users into homicidal monsters, we would have excellent reasons to prohibit its consumption. Fortunately, no such drug actually exists. Are there good reasons to criminalize the use of any real drug? Those who answer affirmatively have the burden of showing why the state is *justified* in doing so – in punishing persons who merely use that drug.

Conclusion

We need to be clear about the meaning of terms such as criminalization and decriminalization in order to have any reasonable chance of deciding whether users of any or all drugs should be punished. I have contended that a drug is decriminalized when persons who use it are no longer subject to state punishment. If I am correct, the most basic question in the entire debate is not whether or why given drugs should be *decriminalized*, but whether or why given drugs should be *criminalized*. Finally, I have suggested that a sensible position on this debate should not evade the issue of whether and under what circumstances the use of a given drug is morally permissible.

Notes

1   For some useful information on minorities, drugs, and criminal justice, see Erik Cain (2011).
2   An excellent source of drug use data can be found at DrugWarFacts.org.
3   Many longitudinal studies of drug users have been conducted. The most famous is Shedler and Block (1990).

References

Cain, E. (2011) The war on drugs is a war on minorities and the poor. *Forbes*, June 28. http://www.forbes.com/sites/erikkain/2011/06/28/the-war-on-drugs-is-a-war-on-minorities-and-the-poor/ (last accessed 6/17/13).

California Penal Code (2012) §381(b). http://codes.lp.findlaw.com/cacode/PEN/3/1/10/s381b (last accessed 6/17/13).

de Marneffe, P. (2013) Vice laws and self-sovereignty. *Criminal Law and Philosophy* 7(1): 29–41.

Shedler, J. and Block, J. (1990) Adolescent drug use and psychological health. *American Psychologist* 612. http://druglibrary.org/schaffer/kids/Adolescent_Drug_Use_ALL.htm (last accessed 6/17/13).

Further Reading

Heyman, G.M. (2009) *Addiction: A Disorder of Choice.* Cambridge: Harvard University Press.

Husak, D. (2002) *Legalize This! The Case for Decriminalizing Drugs.* London: Verso.

MacCoun, R.J. and Reuter, P. (2001) *Drug War Heresies: Learning from other Vices, Times, & Places.* Cambridge: Cambridge University Press.

CHAPTER TWENTY-THREE

Against the Legalization of Drugs

Peter de Marneffe

Introduction

By the *legalization of drugs* I mean the removal of criminal penalties for the manufacture, sale, and possession of large quantities of recreational drugs, such as marijuana, cocaine, heroin, and methamphetamine. In this chapter, I present an argument against drug legalization in this sense. But I do not argue against *drug decriminalization*, by which I mean the removal of criminal penalties for recreational drug use and the possession of small quantities of recreational drugs. Although I am against drug legalization, I am for drug decriminalization. So one of my goals here is to explain why this position makes sense as a matter of principle.

The argument against drug legalization is simple. If drugs are legalized, they will be less expensive and more available. If drugs are less expensive and more available, drug use will increase, and with it, proportionately, drug abuse. So if drugs are legalized, there will be more drug abuse. By *drug abuse* I mean drug use that is likely to cause harm.

Ineffectiveness Objection

A common objection is that drug laws do not work. The imagined proof is that people still use drugs even though they are illegal. But this is a bad argument. People are still murdered even though murder is illegal, and we do not conclude that murder laws do not work or that they ought to be repealed. This is because we think these laws work well enough in reducing murder rates to justify the various costs of enforcing them. So even if drug laws do not eliminate drug abuse, they might likewise reduce it by enough to justify their costs.

Why should we think that drug laws reduce drug abuse? For one thing, our general knowledge of human psychology and economic behavior provides a good basis for

Contemporary Debates in Applied Ethics, Second Edition. Edited by Andrew I. Cohen and Christopher Heath Wellman.
© 2014 John Wiley & Sons, Inc. Published 2014 by John Wiley & Sons, Inc.

predicting that drug use will increase if drugs are legalized. People use drugs because they enjoy them. If it is easier and less expensive to do something enjoyable, more people will do it and those who do it already will do it more often. Laws against the manufacture and sale of drugs make drugs less available, because they prohibit their sale in convenient locations, such as the local drug or liquor or grocery store, and more expensive, because the retail price of illegal drugs reflects the risk to manufacturers and sellers of being arrested and having their goods confiscated. So if drugs are legalized, the price will fall and they will be easier to get. "Hey honey, feel like some heroin tonight?" "Sure, why not stop at Walgreens on the way home from picking up the kids?"

The claim that drug laws reduce drug abuse is also supported by the available empirical evidence. During Prohibition it was illegal to manufacture, sell, and transport "intoxicating liquors" (but not illegal to drink alcoholic beverages or to make them at home for one's own use). During this same period, deaths from cirrhosis of the liver and admissions to state hospitals for alcoholic psychosis declined dramatically compared to the previous decade (Warburton, 1932, pp. 86, 89). Because cirrhosis and alcoholic psychosis are highly correlated with heavy drinking, this is good evidence that Prohibition reduced heavy drinking substantially (Miron and Zwiebel, 1991). Recent studies of alcohol consumption also conclude that heavy drinking declines with increases in price and decreases in availability (Edwards et al., 1994; Cook, 2007). Further evidence that drug use is correlated with availability is that the use of controlled psychoactive drugs is significantly higher among physicians and other healthcare professionals (who have much greater access to these drugs) than it is among the general population (Goode, 2012, pp. 454–455), and that veterans who reported using heroin in Vietnam, where it was legal, reported not using it on returning to the USA, where it was illegal (Robins et al., 1974).

For all these reasons it is a safe bet that drug abuse would increase if drugs were legalized, and it is hard to find an expert on drug policy who denies this. This alone, however, does not settle whether laws against drugs are a good policy because we do not know by how much drug abuse would increase if drugs were legalized and we do not know how much harm would result from this increase in drug abuse. It is important to recognize, too, that drug laws also cause harm by creating a black market, which fosters violence and government corruption, and by sending people to prison. It is possible that the harms created by drug laws outweigh their benefits in reducing drug abuse. I will say more about this possibility below, but first I address some philosophical objections to drug laws.

Paternalism Objection

One objection is that drug laws are paternalistic: they limit people's liberty for their own good. A related objection is that drug laws are moralistic: they impose the view that drug use is wrong on everyone, including those who think it is good. It is true that drug use can be harmful, but most people who use drugs do not use them in a way that harms someone or that creates a significant risk of harm. This is true even of so-called "hard drugs" such as heroin and cocaine. Is it not wrong for the government to prohibit us from doing something we enjoy if it causes no harm?

To oppose drug legalization, however, is to oppose the removal of penalties for the *commercial manufacture* and *sale* and *possession of large quantities* of drugs; it is not to support criminal penalties for the use or possession of small quantities of drugs. To oppose drug legalization is therefore not to hold that anyone should be prohibited from doing something they enjoy for their own good, or that the government should impose the controversial view that drug use is wrong on everyone.

Violation of Rights Objection

A more fundamental objection to drug laws is that they violate our rights. I believe there is some truth to this. So I want to explain why it makes sense to oppose drug legalization even though some drug laws do violate our rights.

Each of us has a right of self-sovereignty: a moral right to control our own minds and bodies. Laws that prohibit people from using drugs or from possessing small quantities of them violate this right because the choice to use drugs involves an important form of control over our minds and bodies, and recreational drug use does not usually harm anyone or pose a serious risk of harm. The choice to use drugs involves an important form of control over our minds partly because recreational drug use is a form of mood control, which is an important aspect of controlling our minds. There are also perceptual experiences that we can have only as the result of using certain drugs, such as LSD, and certain kinds of euphoria that we can experience only as the result of using certain drugs, such as heroin. The choice to put a drug into one's body – to snort it, smoke it, inject it, or ingest it – is also an important form of control over one's body. Because we have a right to control our own minds and bodies, the government is justified in prohibiting us from using a drug only if the choice to use this drug is likely to harm someone, which is not true of most recreational drug use. Laws that prohibit us from using recreational drugs therefore violate our right of self-sovereignty and for this reason should be repealed.

The choice to manufacture or sell drugs, in contrast, does not involve an important form of control over one's own mind or body – no more than the choice to manufacture or sell any commercial product does. These are choices to engage in a commercial enterprise for profit, and may therefore be regulated or restricted for reasons of public welfare, just as any other commercial enterprise may be. One might think that there is something "hypocritical" or "inconsistent" about prohibiting the manufacture and sale of drugs and not prohibiting their possession and use, but this is confused. If one opposes drug legalization on the ground that the government should do whatever it can to reduce drug abuse, regardless of whether it violates anyone's rights, then it would be inconsistent to oppose drug criminalization. But it is not inconsistent to oppose drug criminalization if one opposes drug legalization on the ground that the government should do whatever it can to reduce drug abuse consistent with respect for individual rights. This is because it makes sense to hold that whereas drug criminalization violates the right of self-sovereignty, non-legalization does not (de Marneffe, 2013).

Some might argue that non-legalization violates the right of self-sovereignty too, because it is not possible to use drugs if no one is legally permitted to sell them. But this is obviously false because people still use drugs even though selling them is illegal. Although this fact is sometimes cited to demonstrate the futility of drug control, ironically it makes drug control easier to justify. If drug non-legalization really did make it impossible to use drugs, and so to have the unique experiences they provide, this policy would arguably violate the right of self-sovereignty on this ground. But drug control laws do not make drug use impossible; they only increase the price and reduce the availability of drugs. This is no more a violation of self-sovereignty than a decision by the local supermarket not to carry a certain food or to double its price.

High Costs of the Drug War Objection

Laws against the manufacture and sale of drugs might of course still be a bad policy even if they do not violate the right of self-sovereignty. This is because these laws have costs, and these costs might outweigh the benefits of these laws in reducing drug abuse. Laws against the manufacture and sale of drugs create a black market, which fosters violence, because when disputes arise in an illegal trade the disputants cannot go to the legal system for resolution. The black market also fosters government corruption, because those in an illegal trade must pay government officials for protection from arrest and confiscation. Drug laws also cost money to enforce, which might be better spent in other ways. Finally, drug laws result in some people being arrested and imprisoned and being left with criminal records. It is certainly possible that these costs outweigh the benefits of drug control in reducing drug abuse.

It is important to understand, though, that drug control policy need not be as costly as the so-called War on Drugs, which is current US policy. So even if the War on Drugs is too costly, as critics maintain, it does not follow that drugs should be legalized. The case against drug legalization rests on the assumption that the benefits of drug control in reducing drug abuse are sufficient to justify the costs of drug control *once these costs are reduced as much as possible consistent with effective drug control*. By *effective* drug control, I mean a policy that reduces drug abuse substantially compared to the amount of drug abuse that would exist if drugs were legalized. I do not mean a policy that eliminates drug abuse altogether. It is no more possible to eliminate drug abuse than it is to eliminate crime. But just as effective crime control is still possible, effective drug control is possible too. And if it is possible to have effective drug control without the high costs of the War on Drugs, then the benefits of prohibiting the manufacture and sale of drugs are more likely to justify the costs.

One compelling objection to the War on Drugs is to the sentencing rules for drug law violations, which require judges to impose long prison terms for drug trafficking offenses. Critics rightly argue that mandatory sentences and long prison terms for selling drugs are morally indefensible. These are not, however, necessary features of effective drug control policy. They are not features of European drug control policy, for example. So it makes sense to oppose harsh mandatory penalties while also opposing drug legalization.

Drug control works primarily by increasing price and reducing availability, which can be accomplished by reliably enforcing laws against the manufacture and sale of drugs with moderate penalties. Where it is illegal to manufacture and sell drugs, most business persons avoid the drug trade because they do not want to be arrested and have their goods confiscated. This reduces supply, which increases price. Where it is illegal to sell drugs, stores that aim to retain their licenses also do not sell them, which reduces availability. Heavy penalties no doubt drive the price up even higher and decrease availability even more by increasing the risks of drug trafficking – but the biggest increases in price and the biggest reductions in availability come simply from the illegality of the trade itself together with reliable enforcement of laws against manufacture and sale (Kleiman *et al.*, 2011, pp. 48–50). If effective drug control does not require harsh mandatory penalties, then the fact that such penalties are unjustifiable is not a good argument for drug legalization.

Effect on Imprisoned Youths Objection

Another objection to US drug control policy is that it results in many young people being arrested, imprisoned, and left with criminal records, who would otherwise not suffer these misfortunes. Some might retort that if a person chooses to deal drugs illegally, he cannot legitimately complain about the foreseeable consequences of his choice. But this response is inadequate because by making drugs illegal the government creates a hazard that otherwise would not exist. By making the manufacture and sale of drugs illegal, the government creates a lucrative illegal market, and the money-making opportunities that this market creates are attractive, especially to young people who lack a college education or special training, because they can make much more money by dealing drugs than by doing anything else. When the government creates a system of penalties for manufacturing and selling drugs it therefore creates a hazard; it creates a tempting opportunity to make money and then imposes penalties for making money in this way.

In general, the government has an obligation to reduce the risk to individuals of being harmed by the hazards it creates. When the government tests weapons, for example, it must take care that people do not wander into the testing areas. Bright signs are not enough; it must also build fences and monitor against trespass. The government also has an obligation to help young people avoid the worst consequences of their willingness to take unwise risks. It has an obligation to require teenagers to wear helmets when they ride a motorcycle, for example. So when the government creates the hazard of imprisonment by making the manufacture and sale of drugs illegal, it must guard against the likelihood of imprisonment, and it must take special care to reduce this likelihood for young people who commonly lack a proper appreciation of the negative impact that conviction and imprisonment will have on their lives. For all these reasons, the government must structure drug laws so that young people have an adequate opportunity to avoid being imprisoned for drug offenses, and to avoid acquiring a criminal record. This means, among other things, that no one should be arrested for a drug offense prior to receiving an official warning; no penalty for a first conviction should involve prison time; initial jail or prison sentences should be short and subject to judicial discretion; and imprisonment for subsequent convictions should increase in length only gradually and also be subject to judicial discretion.

Racial Discrimination Objection

A related objection to the War on Drugs is that those imprisoned for drug offenses in the USA are disproportionately black inner city males (Alexander, 2012). This objection would be addressed to some degree by the changes in sentencing policy just proposed, but one might predict that any effective drug control policy would result in the same sort of disproportionality, which some might see as an argument for drug legalization. However, it also is important to consider the potential negative impact of drug legalization on inner city communities. Drug legalization will result in a substantial increase in drug abuse. Drug abuse commonly leads parents to neglect their children, and to neglect their own health and jobs, which harms their children indirectly. Drug abuse also distracts teenagers from their schoolwork, interferes with the development of a sense of responsibility, and makes young people less likely to develop the skills necessary for acquiring good jobs as adults. If drugs are legalized, there will therefore be more child neglect as a result and more truancy by teenagers. This is likely to have an even more devastating impact on the life prospects of young people in non-affluent inner city communities than it has on the life prospects of young people in affluent suburbs. I suspect this is the primary reason why many inner city community leaders oppose drug legalization.

It is true that incarcerating large numbers of inner city youths for drug offenses also has a negative impact on inner city communities. A man who is in jail cannot be present as a parent or make money to support his children, and a person with a criminal record has a harder time finding a decent job. These consequences alone would warrant drug legalization if there were no downside. If we assume, however, that drug legalization would result in a substantial increase in child neglect and adolescent truancy, then legalization does not seem like a good way to improve the life prospects of inner city youth overall. It seems better to maintain laws against the manufacture and sale of drugs, and reduce the number of those who are convicted and imprisoned for drug offenses. This would be consistent with effective drug control because the number of dealers in prison could be reduced dramatically without making drugs noticeably cheaper or easier to get (Kleiman *et al.*, 2011, p. 203).

Increase in Violence Objection

Another objection to US drug control policy is that it has increased violence in other countries, particularly Mexico. Americans enjoy using drugs and are willing to pay for them. Because it is illegal to manufacture and sell drugs in the USA, American drug control policy creates opportunities for people south of the border to get rich by making drugs and selling them wholesale to retailers north of the border. Because those in the drug trade use violence to control market share and to intimidate law enforcement, US drug laws result in violence. If drugs were legalized in the USA, the recreational drug market would presumably be taken over by large US drug, liquor, and food companies and it would not be possible for anyone in Mexico to get rich by selling illegal drugs to Americans, which would eliminate the associated violence there.

Drug legalization, however, is not the only way to reduce drug-related violence abroad. Here are some alternative strategies:

- The USA might legalize the private production of marijuana for personal use (the way it was legal during Prohibition to make alcoholic beverages at home). Because much of the Mexican drug trade is in marijuana, this would reduce its profitability, and so presumably the associated violence.

- The USA might also concentrate its drug enforcement efforts in Mexico on the most violent drug trafficking organizations, as opposed to concentrating on the biggest and most profitable organizations, which would create incentives for those in the Mexican drug trade to be less violent.
- The USA might also ease border control at entry points not on the US-Mexico border. The violence in Mexico is created partly by the fact that it is the primary conduit of cocaine from South and Central America to North America. If the USA were to loosen border control in Florida, fewer drugs would travel through Mexico. Because the USA imports so many goods, it is not possible to stop drugs from coming into this country. Some would cite this as proof that drug control is futile, but this conclusion is unwarranted because border controls still raise the retail price of drugs substantially, which results in less drug abuse (Kleiman *et al.*, 2011, pp. 162–163). The suggestion here is that a general policy of border control is consistent with US law enforcement experimenting with different border control policies with an eye to reducing violence abroad (Kleiman *et al.*, 2011, p. 170).

None of these proposals would eliminate drug-related violence in Mexico, but it is unrealistic to think that criminal violence in Mexico would be eliminated by drug legalization in the USA. After all, what will career criminals in Mexico do once they cannot make money via the drug trade? Presumably they will turn to other criminal activities, such as kidnapping, extortion, and human trafficking, which also involve violence.

Corruption of Foreign Governments Objection

Another objection to US drug control policy is that it fosters the corruption of foreign governments. Because those in the foreign drug trade need protection from arrest, prosecution, and confiscation of assets, because they are willing to pay government officials to look the other way, and because some government officials are willing to accept this payment, the drug trade increases government corruption. If drugs were legalized in the USA, this would destroy the illegal market abroad, which would remove an important contributing factor in government corruption.

It is naive, though, to think that US drug control policy is the primary cause of government corruption abroad. Although we associate police corruption with drug trafficking, the latter tends to flourish where government officials are already corrupt (Kleiman *et al.*, 2011, p. 177). Although it provides a good plot line for movies and television shows, drug trafficking has not in fact resulted in the widespread financial corruption of US police. One explanation for this is the existence of multiple US enforcement agencies – federal, state, municipal – which have overlapping jurisdictions (Kleiman *et al.*, 2011, p. 65). This arrangement reduces bribery, because in paying protection money to one government agency, a criminal organization does not gain protection from the others, which makes bribery less cost-effective. Overlapping jurisdictions also increases interagency monitoring, which functions as a check on corruption within any one agency. A skeptic might observe that there were also overlapping jurisdictions during Prohibition when US police were notoriously corrupt. But US police departments are now far more professionalized than they were in the 1920s, when positions on municipal police forces were commonly doled out as political spoils by political bosses. A police officer must now receive specialized training and pass the kinds of tests that are required of all civil service employees, and the internal affairs divisions of police departments are now much more effective at monitoring the illegal activities of their members. In any case, the fact that there has been relatively little drug-related financial corruption of US law enforcement shows that a fully profession-alized police force with overlapping jurisdictions and strong internal affairs divisions can effectively resist financial corruption even where drugs are illegal. In contrast, a foreign police force that is not fully professionalized will be susceptible to financial cor-ruption regardless of whether the USA legalizes drugs.

The Inconsistency Objection

Another argument against drug laws is that it is hypocritical or inconsistent for our government to prohibit the manufacture and sale of heroin, cocaine, and metham-phetamine while permitting the manufacture and sale of alcohol and cigarettes. Drinking and smoking cause far more harm than other kinds of recreational drug use. This is partly because there is so much more drinking and smoking, which is partly because the manufacture and sale of alcohol and cigarettes are legal. But drinking and smoking are also inherently more harmful than other forms of drug use. Drinking alcohol is correlated much more highly with violence, property crime, and accidental injury than the use of heroin is, and a regular user of heroin who uses it safely – in moder-ate doses with clean equipment – does not face any significant health risk as a result, whereas cigarette smoking is known to cause heart and lung disease. So it can seem that if the government is justified in prohibiting the manufacture and sale of heroin, it must also be justified in prohibiting the manufacture and sale of alcohol and cigarettes.

This would be a good objection to drug laws if laws against the manufacture and sale of alcoholic beverages and cigarettes were wrong in principle, but it is hard to see why they would be. After all, drinking and smoking cause a lot of harm and neither policy would violate the right of self-sovereignty discussed above, because a law that prohibits only the manufacture and sale of a drug does not prohibit its possession or make its use impossible. Of course, the suggestion that alcohol prohibition might be justified is commonly dismissed with the incantation that Prohibition was a disastrous failure, but historians agree that Prohibition succeeded in substantially reducing heavy drinking, and it would have been even more effective had its enforcement been ade-quately funded and had it been administered from the outset by law enforcement pro-fessionals instead of by political appointees (Okrent, 2010, pp. 134–145, 254–261). Prohibition did fail politically, but so did Reconstruction and the Equal Rights Amend-ment. The fact that a policy is rejected or abandoned does not show that it was wrong in principle. Finally, it is worth noting that alcohol prohibition still exists in some parts of this country, on Indian reservations, for example, and that these polices make sense as part of an effort to reduce alcoholism and the harms associated with it.

It is not necessary, though, to advocate alcohol prohibition in order to defend other drug laws, because there are relevant differences between them. For one thing, the institution of alcohol prohibition now is likely not to reduce heavy drinking by as much as drug non-legalization reduces drug abuse. Drinking is widely accepted and a part of normal social rituals, in a way that heroin, cocaine, and methamphetamine use is not. This means that alcohol prohibition now would not work in tandem with a strong social stigma, which would presumably reduce its effectiveness in reducing alcohol abuse. It is possible, too, that in an environment of social acceptance, sharply increasing the excise taxes on alcoholic beverages would achieve almost as much as prohibition in reducing the harms caused by heavy drinking with none of the costs of prohibition (though it is worth noting here that liquor industry lobbying has been more effective in preventing excise tax increases than it was in preventing Prohibition). There are also important ways in which instituting alcohol prohibition now would be more burdensome than continuing with drug non-legalization. Many people have built their lives around the alcoholic beverage industry. If alcohol were now prohibited, many of these people would lose their jobs, and many companies, restaurants and bars would go out of business, which would be a serious hardship for owners and employees. In contrast, people who go into the drug trade do so knowing that it is illegal. So the burden on them of maintaining drug laws is not as great as the burden that alcohol prohibition would impose on those who have built their lives around the liquor trade on the assumption that the manufacture and sale of alcohol will remain legal. Ironically, it is drug *legalization* that would burden those in the illegal drug trade, in much the same way as Prohibition burdened those in the legal liquor trade: by depriving them of their livelihood.

There are also important differences between illicit drugs and cigarettes. Drug legalization, I assume, would result in a substantial increase in drug abuse, which, I assume, would also result in a substantial increase in child neglect and adolescent truancy, which would have a substantial negative impact on the life prospects of many young people. Cigarette smoking, in contrast, does not make someone a worse parent or a worse student or employee. Furthermore, because heavy smoking typically has a negative impact on a person's life only toward the end when he or she is older, smoking as a young person is less likely than adolescent drug abuse to have a negative impact on the *kind* of life a person has. Finally, although psychologically challenging, it is quite possible to quit smoking as an adult and so to reduce the long-term health consequences of starting to smoke as a teenager – much easier than it is to reverse the long-term negative consequences of having had inadequate parenting or having failed out of high school as the result of drug abuse. Given these differences between the consequences of smoking and drug abuse, one can consistently oppose the legalization of drugs for the reasons I have given here without advocating prohibiting the manufacture and sale of cigarettes.

In explaining above how one might consistently oppose the legalization of drugs without advocating alcohol prohibition, I observed that drinking is so widespread and socially accepted that alcohol prohibition is likely to reduce heavy drinking by less than drug abuse is reduced by laws against the manufacture and sale of illicit drugs. This same point might now be given as an argument for legalizing marijuana: marijuana use is so widespread and socially accepted that laws against the manufacture and sale of marijuana do not do very much to reduce it. It might also be argued that legalizing marijuana would not result in a dramatic increase in drug abuse because marijuana is less subject to abuse than other drugs (including alcohol). Finally, legalizing marijuana in the USA would dramatically reduce the drug trade in Mexico, which would result in a corresponding reduction in violence and government corruption there. Should not marijuana be legalized, then, even if other drugs should not be?

In this chapter I am arguing against the view that the manufacture and sale of *all* drugs should be legalized; I am not arguing that there is *no* drug that should be legalized. Suppose that marijuana legalization would not result in a substantial increase in drug abuse. Suppose that most of those who would use marijuana if it were legalized are already using it and using it almost as much as they want to. Or suppose that marijuana use itself is harmless and does not lead to the use of more harmful drugs. If either of these things is true, then marijuana should be legalized. It is also possible, though, that, as a result of legalization, many more young people would use marijuana than do now, and that a sizable fraction of them would use it in ways that interfere with their education or employment, and that a sizable fraction of them would go on to abuse more harmful drugs who would otherwise never have tried them. Because I am not sure that these things would not happen, I do not support legalizing marijuana. With more information, though, I might change my mind. So it is important to make clear that whether a drug should be legalized depends on the consequences of legalizing it, and not on whether any *other* drug should be legalized. Hence, even if marijuana should be legalized, it would not follow that heroin, cocaine, and methamphetamine should be legalized too.

Unhealthy Foods Objection

Another argument against drug laws is that if the government is justified in prohibiting us from putting a drug into our bodies for our own good, then it is also justified in prohibiting us from putting unhealthy foods into our bodies for our own good. The suggestion that the government is entitled to control what we eat strikes many of us as outrageous. Why is it not likewise outrageous for the government to prohibit us from using recreational drugs?

For the reasons given above, I think it is. Laws that prohibit us from using drugs – or drinking alcohol or smoking cigarettes – violate our right of self-sovereignty in the same way that laws that prohibit us from eating high fat or high sugar foods would. However, just as laws that prohibit the manufacture and sale of drugs do not violate our self-sovereignty, laws that regulate the sale of fatty or sugary foods do not either. So if the government prohibits fast food restaurants from selling humongous hamburgers, or prohibits convenience stores from selling sugary soda in giant cups, or prohibits vending machines in schools from stocking items with high fat or sugar content, no one's right of self-sovereignty is violated. Whether these policies are a good idea is a separate question, but if they are a bad idea, it is not because they violate anyone's rights.

No Scientific Proof Objection

In arguing against drug legalization, I assume that drug abuse would increase substantially if drugs were legalized. Some might now object that there is no proof of this, and this is true, but there is also no proof that murder rates will rise if murder is decriminalized. That is, this assumption is not warranted by any set of controlled laboratory experiments or randomized field trials. Should murder therefore be decriminalized? Obviously not. Some might say that the freedom to murder is not a very important liberty, so the standard of proof need not be so high. But most of us also support on the basis of assumptions for which there is no scientific proof policies that do impinge on important liberties. For example, many of us support restrictions on campaign contributions on the assumption that unrestricted contributions would result in more political corruption. But there is no scientific proof of this, and restrictions on campaign contributions impinge on the important freedom of political speech. Many of us also support immigration laws on the assumption that unrestricted immigration would lower our quality of life. But there is also no scientific proof of this, and freedom of movement is also an important liberty. Should we withdraw our support for these policies just because we support them on the basis of scientifically unproven assumptions? I think not. In general we are justified in supporting a legal restriction for a reason if two conditions are met: (a) this reason would justify this restriction if it was based on true assumptions, and (b) we are warranted by the available evidence in believing that the relevant assumptions are true. So if we are warranted by the available evidence in believing that drug abuse will increase if drugs are legalized, then we are justified in making this assumption for the purpose of evaluating drug control policy. And we are warranted in making this assumption – by what we know about patterns of alcohol and drug consumption and more generally about human psychology and economic behavior.

Conclusion

If drug abuse would increase substantially if drugs were legalized, and laws that prohibit the manufacture and sale of drugs do not violate our right of self-sovereignty, and effective drug control requires only moderate penalties reliably and conscientiously enforced, then it makes sense to oppose drug legalization. This, in essence, is the argument I have made here. In evaluating drug policy, it is important, too, to consider how public policy would be shaped if drugs were legalized. Beer, liquor, and cigarette companies already do as much as they can to prevent the government from adopting policies that would reduce drinking and smoking and so their associated harms. They do as much as they can to prevent increases in excise taxes, which increase the price of alcohol and cigarettes, and so reduce their sales, and so smoking and drinking. They do as much as they can to prevent restrictions on the hours and locations of the sale of alcohol and cigarettes. They do as much as they can to prevent licensing and rationing policies, which would reduce the amount of alcohol consumed by problem drinkers. And they do as much as they can to make their products attractive through advertising, particularly to young people. We should expect that if drugs are legalized, drug companies will behave in the same way: that they will do everything they can to prevent the enactment of laws that restrict the marketing and sale of heroin, cocaine, and methamphetamine, and that they will do everything they can to market these drugs successfully, particularly to young people, who will be their most profitable market. Because drug use is currently stigmatized, drug companies are unlikely to be as successful as liquor companies in preventing sound public policy, at least initially. But if we envision a world in which legal drug companies are legally trying to persuade consumers to buy recreational drugs from legal vendors and legally trying to prevent any socially responsible legislation that reduces their legal sales, it is hard to envision a world that does not have much more drug abuse.

References

Alexander, M. (2012) *The New Jim Crow: Mass Incarceration in the Age of Colorblindness.* New York: New Press.

Cook, P.J. (2007) *Paying the Tab: The Economics of Alcohol Policy.* Princeton, NJ: Princeton University Press.

de Marneffe, P. (2013) Vice laws and self-sovereignty. *Criminal Law and Philosophy* 7: 29–41.

Edwards, G., Anderson, P., Babor, T.F. *et al.* (1994) *Alcohol Policy and the Public Good.* New York: Oxford University Press.

Goode, E. (2012) *Drugs in American Society,* 8th edn. New York: McGraw-Hill.

Kleiman, M.A.R. *et al.* (2011) *Drugs and Drug Policy: What Everyone Needs to Know.* New York: Oxford University Press.

Miron, J.A. and Zwiebel, J. (1991) Alcohol consumption during prohibition. *American Economic Review* 81: 242–247.

Okrent, D. (2010) *Last Call: The Rise and Fall of Prohibition.* New York: Scribner.

Robins, L.N. *et al.* (1974) Drug use by U.S. army in Vietnam: a follow-up on their return home. *American Journal of Epidemiology* 99: 235–249.

Warburton, C. (1932) *The Economic Results of Prohibition.* New York: Columbia University Press.

ISSUES OF COSMOPOLITANISM AND COMMUNITY

Immigration

CHAPTER TWENTY-FOUR

Immigration: The Case for Limits

David Miller

It is not easy to write about immigration from a philosophical perspective – not easy at least if you are writing in a society (and this now includes most societies in the western world) in which immigration has become a highly charged political issue. Those who speak freely and openly about the issue tend to come from the far Right: they are fascists or racists who believe that it is wrong in principle for their political community to admit immigrants who do not conform to the approved cultural or racial stereotype. Most liberal, conservative, and social democratic politicians support quite strict immigration controls in practice, but they generally refrain from spelling out the justification for such controls, preferring instead to highlight the practical difficulties involved in resettling immigrants, and raising the spectre of a right-wing backlash if too many immigrants are admitted. Why are they so reticent? One reason is that it is not easy to set out the arguments for limiting immigration without at the same time projecting a negative image of those immigrants who have already been admitted, thereby playing directly into the hands of the far Right ideologues who would like to see such immigrants deprived of their full rights of citizenship and/or repatriated to their countries of origin. Is it possible *both* to argue that every member of the political community, native or immigrant, must be treated as a full citizen, enjoying equal status and the equal respect of his or her fellows, *and* to argue that there are good grounds for setting upper bounds both to the rate and the overall numbers of immigrants who are admitted? Yes, it is, but it requires dexterity, and always carries with it the risk of being misunderstood.

In this chapter, I shall explain why nation-states may be justified in imposing restrictive immigration policies if they so choose. The argument is laid out in three stages. First, I canvass three arguments that purport to justify an unlimited right of migration between states and show why each of them fails. Second, I give two reasons, one having to do with culture, the other with population, that can justify states in limiting

Contemporary Debates in Applied Ethics, Second Edition. Edited by Andrew I. Cohen and Christopher Heath Wellman.
© 2014 John Wiley & Sons, Inc. Published 2014 by John Wiley & Sons, Inc.

immigration. Third, I consider whether states nonetheless have a duty to admit a special class of potential immigrants – namely refugees – and also how far they are allowed to pick and choose among the immigrants they do admit. The third section, in other words, lays down some conditions that an ethical immigration policy must meet. But I begin by showing why there is no general right to choose one's country of residence or citizenship.

Can There be an Unlimited Right of Migration Between States?

Liberal political philosophers who write about migration usually begin from the premise that people should be allowed to choose where in the world to locate themselves unless it can be shown that allowing an unlimited right of migration would have harmful consequences that outweigh the value of freedom of choice (see, for instance, Carens, 1987; Hampton, 1995). In other words, the central value appealed to is simply freedom itself. Just as I should be free to decide who to marry, what job to take, what religion (if any) to profess, so I should be free to decide whether to live in Nigeria, or France, or the United States. Now these philosophers usually concede that in practice some limits may have to be placed on this freedom – for instance, if high rates of migration would result in social chaos or the breakdown of liberal states that could not accommodate so many migrants without losing their liberal character. In these instances, the exercise of free choice would become self-defeating. But the presumption is that people should be free to choose where to live unless there are strong reasons for restricting their choice.

I want to challenge this presumption. Of course there is always *some* value in people having more options to choose between, in this case options as to where to live, but we usually draw a line between *basic* freedoms that people should have as a matter of right and what we might call *bare* freedoms that do not warrant that kind of protection. It would be good from my point of view if I were free to purchase an Aston Martin tomorrow, but that is not going to count as a morally significant freedom – my desire is not one that imposes any kind of obligation on others to meet it. In order to argue against immigration restrictions, therefore, liberal philosophers must do more than show that there is some value to people in being able to migrate, or that they often *want* to migrate (as indeed they do, in increasing numbers). It needs to be demonstrated that this freedom has the kind of weight or significance that could turn it into a right, and that should therefore prohibit states from pursuing immigration policies that limit freedom of movement.

I shall examine three arguments that have been offered to defend a right to migrate. The first starts with the general right to freedom of movement, and claims that this must include the freedom to move into, and take up residence in, states other than one's own. The second begins with a person's right to *exit* from her current state – a right that is widely recognized in international law – and claims that a right of exit is pointless unless it is matched by a right of entry into other states. The third appeals to international distributive justice. Given the huge inequalities in living standards that currently exist between rich and poor states, it is said, people who live in poor states have a claim of justice that can only be met by allowing them to migrate and take advantage of the opportunities that rich states provide.

The idea of a right to freedom of movement is not in itself objectionable. We are talking here about what are usually called basic rights or human rights, and I shall assume (since there is no space to defend the point) that such rights are justified by pointing to the vital interests that they protect (Shue, 1980; Nickel, 1987; Griffin, 2001). They correspond to conditions in whose absence human beings cannot live decent lives, no matter what particular values and plans of life they choose to pursue. Being able to move freely in physical space is just such a condition, as we can see by thinking about people whose legs are shackled or who are confined in small spaces. A wider freedom of movement can also be justified by thinking about the interests that it serves instrumentally: if I cannot move about over a fairly wide area, it may be impossible for me to find a job, to practice my religion, or to find a suitable marriage partner. Since these all qualify as vital interests, it is fairly clear that freedom of movement qualifies as a basic human right.

What is less clear, however, is the physical extent of that right, in the sense of how much of the earth's surface I must be able to move to in order to say that I enjoy it. Even in liberal societies that make no attempt to confine people within particular geographical areas, freedom of movement is severely restricted in a number of ways. I cannot, in general, move to places that other people's bodies now occupy (I cannot just push them aside). I cannot move on to private property without the consent of its owner, except perhaps in emergencies or where a special right of access exists – and since most land is privately owned, this means that a large proportion of physical space does not fall within the ambit of a *right* to free movement. Even access to public space is heavily regulated: there are traffic laws that tell me where and at what speed I may drive my car, parks have opening and closing hours, the police can control my movements up and down the streets, and so forth. These are very familiar observations, but they are worth making simply to highlight how hedged about with qualifications the existing right of free movement in liberal societies actually is. Yet few would argue that because of these limitations, people in these societies are deprived of one of their human rights. Some liberals might argue in favor of expanding the right – for instance, Britain saw a protracted campaign to establish a legal right to roam on uncultivated privately owned land such as moors and fells, a right that finally became effective in 2005. But even the advocates of such a right would be hard-pressed to show that some vital interest was being injured by the more restrictive property laws that have existed up to now.

The point here is that liberal societies in general offer their members *sufficient* freedom of movement to protect the interests that the human right to free movement is intended to protect, even though the extent of free movement is very far from absolute. So how could one attempt to show that the right in question must include the right to move to some other country and settle there? What vital interest requires the right to be interpreted in such an extensive way? Contingently, of course, it may be true that moving to another country is the only way for an individual to escape persecution, to find work, to obtain necessary medical care, and so forth. In these circumstances the person concerned may have the right to move, not to any state that she chooses, but to *some* state where these interests can be protected. But here the right to move serves only as a remedial right: its existence depends on the fact that the person's vital interests cannot be secured in the country where she currently resides. In a world of decent states – states that were able to secure their citizens' basic rights to security, food, work, medical care, and so forth – the right to move across borders could not be justified in this way.

Our present world is not, of course, a world of decent states, and this gives rise to the issue of refugees, which I shall discuss in the final section of this chapter. But if we leave aside for the moment cases where the right to move freely across borders depends upon the right to avoid persecution, starvation, or other threats to basic interests, how might we try to give it a more general rationale? One reason a person may want to migrate is in order to participate in a culture that does not exist in his native land – for instance he wants to work at an occupation for which there is no demand at home, or to join a religious community which, again, is not represented in the country from which he comes. These might be central components in his plan of life, so he will find it very frustrating if he is not able to move. But does this ground a right to free movement across borders? It seems to me that it does not. What a person can legitimately demand access to is an *adequate* range of options to choose between – a reasonable choice of occupation, religion, cultural activities, marriage partners, and so forth. Adequacy here is defined in terms of generic human interests rather than in terms of the interests of any one person in particular – so, for example, a would-be opera singer living in a society which provides for various forms of musical expression, but not for opera, can have an adequate range of options in this area even though the option she most prefers is not available. So long as they adhere to the standards of decency sketched above, all contemporary states are able to provide such an adequate range internally. So although people certainly have an *interest* in being able to migrate internationally, they do not have a basic interest of the kind that would be required to ground a human right. It is more like my interest in having an Aston Martin than my interest in having access to *some* means of physical mobility.

I turn next to the argument that because people have a right to leave the society they currently belong to, they must also have a right to enter other societies, since the first right is practically meaningless unless the second exists – there is no unoccupied space in the world to exit *to*, so unless the right to leave society A is accompanied by the right to enter societies B, C, D, and so on, it has no real force (Dummett, 1992; Cole, 2000).

The right of exit is certainly an important human right, but once again it is worth examining why it has the significance that it does. Its importance is partly instrumental: knowing that their subjects have the right to leave inhibits states from mistreating them in various ways, so it helps to preserve the conditions of what I earlier called "decency." However, even in the case of decent states the right of exit remains important, and that is because by being deprived of exit rights individuals are forced to remain in association with others whom they may find deeply uncongenial – think of the militant atheist in a society where almost everyone devoutly practices the same religion, or the religious puritan in a society where most people behave like libertines. On the other hand, the right of exit from state A does not appear to entail an unrestricted right to enter any society of the immigrant's choice – indeed, it seems that it can be exercised provided that at least one other society, society B say, is willing to take him in. It might seem that we can generate a general right to migrate by iteration: the person who leaves A for B then has the right to exit from B, which entails that C, at least, must grant him the right to enter, and so forth. But this move fails, because our person's right of exit from A depended on the claim that he might find continued association with the other citizens of A intolerable, and he cannot plausibly continue making the same claim in the case of each society that is willing to take him in. Given the political and cultural diversity of societies in the real world, it is simply unconvincing to argue that only an unlimited choice of which one to join will prevent people being forced into associations that are repugnant to them.

It is also important to stress that there are many rights whose exercise is contingent on finding partners who are willing to cooperate in the exercise, and it may be that the right of exit falls into this category. Take the right to marry as an example. This is a right held against the state to allow people to marry the partners of their choice (and perhaps to provide the legal framework within which marriages can be contracted). It is obviously not a right to have a marriage partner provided – whether any given person can exercise the right depends entirely on whether he is able to find someone willing to marry him, and many people are not so lucky. The right of exit is a right held against a person's current state of residence not to prevent her from leaving the state (and perhaps aiding her in that endeavor by, say, providing a passport). But it does not entail an obligation on any other state to let that person in. Obviously, if no state were ever to grant entry rights to people who were not already its citizens, the right of exit would have no value. But suppose states are generally willing to consider entry applications from people who want to migrate, and that most people would get offers from at least one such state: then the position as far as the right of exit goes is pretty much the same as with the right to marry, whereby no means everyone is able to wed the partner they would ideally like to have, but most have the opportunity to marry *someone*.

So once the right of exit is properly understood, it does not entail an unlimited right to migrate to the society of one's choice. But now, finally, in this part of the chapter, I want to consider an argument for migration rights that appeals to distributive justice. It begins from the assumption of the fundamental moral equality of human beings. It then points out that, in the world in which we live, people's life prospects depend heavily on the society into which they happens to be born, so that the only way to achieve equal opportunities is to allow people to move to the places where they can develop and exercise their talents, through employment and in other ways. In other words, there is something fundamentally unfair about a world in which people are condemned to relative poverty through no fault of their own when others have much greater opportunities, whereas if people were free to live and work wherever they wished, then each person could choose whether to stay in the community that had raised him or to look for a better life elsewhere.

The question we must ask here is whether justice demands equality of opportunity at the global level, as the argument I have just sketched assumes, or whether this principle only applies *inside* societies, among those who are already citizens of the same political community (see, for instance, Caney, 2001). Note to begin with that embracing the moral equality of all human beings – accepting that every human being is equally an object of moral concern – does not yet tell us what we are required to do for them as a result of that equality. One answer *might* be that we should attempt to provide everyone with equal opportunities to pursue their goals in life. But another, equally plausible, answer is that we should play our part in ensuring that their basic rights are respected, where these are understood as rights to a certain minimum level of security, freedom, resources, and so forth – a level adequate to protect their basic interests, as suggested earlier in this chapter. These basic rights can be universally protected and yet some people have greater opportunities than others to pursue certain aims, as a result of living in more affluent or culturally richer societies.

Is it nonetheless unfair if opportunities are unequal in this way? That depends upon what we believe about the scope of distributive justice, the kind of justice that involves comparing how well different people are faring by some standard. According to Michael Walzer, "the idea of distributive justice presupposes a bounded world within which distributions take place: a group of people committed to dividing, exchanging, and sharing social goods, first of all among themselves" (Walzer, 1983, p. 31). The main reason that Walzer gives for this view is that the very goods whose distribution is a matter of justice gain their meaning and value within particular political communities. Another relevant consideration is that the stock of goods that is available at any time to be divided up will depend on the past history of the community in question, including decisions about, for example, the economic system under which production will take place. These considerations tell against the view that justice at global level should be understood in terms of the equal distribution, at any moment, of a single good, whether this good is understood as "resources" or "opportunity" or "welfare" (Miller, 1999). The basic rights view avoids these difficulties, because it is plausible to think that whatever the cultural values of a particular society, and whatever its historical record, no human being should be allowed to fall below the minimum level of provision that protects his or her basic interests.

But what if somebody does fall below this threshold? Does this not give him the right to migrate to a place where the minimum level is guaranteed? Perhaps, but it depends on whether the minimum could be provided in the political community he belongs to now, or whether that community is so oppressive, or so dysfunctional, that escape is the only option. So here we encounter again the issue of refugees, to be discussed in my final section. Meanwhile, the lesson for other states, confronted with people whose lives are less than decent, is that they have a choice: they must either ensure that the basic rights of such people are protected in the places where they live – by aid, by intervention, or by some other means – or they must help them to move to other communities where their lives will be better. Simply shutting one's borders and doing nothing else is not a morally defensible option here. People everywhere have a right to a decent life. But before jumping to the conclusion that the way to respond to global injustice is to encourage people whose lives are less than decent to migrate elsewhere, we should consider the fact that this policy will do little to help the very poor, who are unlikely to have the resources to move to a richer country. Indeed, a policy of open migration may make such people worse off still, if it allows doctors, engineers, and other professionals to move from economically undeveloped to economically developed societies in search of higher incomes, thereby depriving their countries of origin of vital skills. Equalizing opportunity for the few may diminish opportunities for the many. Persisting global injustice does impose on rich states the obligation to make a serious contribution to the relief of global poverty, but in most instances they should contribute to improving conditions of life on the ground, as it were, rather than bypassing the problem by allowing (inevitably selective) inward migration.

Justifications for Limiting Immigration

I have shown that there is no general right to migrate to the country of one's choice. Does it follow that states have a free hand in choosing who, if anyone, to admit to membership? One might think that it does, using the analogy of a private club. Suppose that the members of a tennis club decide that once the membership roster has reached 100, no new members will be taken in. They do not have to justify this decision to would-be members who are excluded: if they decide that 100 members is enough, that is entirely their prerogative. But notice what makes this argument convincing. First, the benefit that is being denied to new applicants is the (relatively superficial) benefit of being able to play tennis. Second, it is a reasonable assumption that the rejected applicants can join another club, or start one of their own. It would be different if the tennis club occupied the only site within a 50-mile radius that is suitable for laying tennis courts: we might then think that they had some obligation to admit new members up to a reasonable total. In the case of states, the advantages that they deny to would-be immigrants who are refused entry are very substantial; and because states monopolize stretches of territory, and in other ways provide benefits that cannot be replicated elsewhere, the "go and start your own club" response to immigrants is not very plausible.

So in order to show that states are entitled to close their borders to immigrants, we have to do more than show that the latter lack the human right to migrate. Potential immigrants have a *claim* to be let in – if nothing else they usually have a strong *desire* to enter – and so any state that wants to control immigration must have good reasons for doing so. In this section, I shall outline two good reasons that states may have for restricting immigration. One has to do with preserving culture, the other with controlling population. I do not claim that these reasons will apply to every state, but they do apply to many liberal democracies that are currently having to decide how to respond to potentially very large flows of immigrants from less economically developed societies (other states may face larger flows still, but the political issues will be different).

The first reason assumes that the states in question require a common public culture that in part constitutes the political identity of their members, and that serves valuable functions in supporting democracy and other social goals. There is no space here to justify this assumption in any detail, so I must refer the reader to other writings where I have tried to do so (Miller, 1995, 2000). What I want to do here is to consider how the need to protect the public culture bears upon the issue of immigration. In general terms we can say: (a) that immigrants will enter with cultural values, including *political* values, that are more or less different from the public culture of the community they enter; (b) that as a result of living in that community, they will absorb some part of the existing public culture, modifying their own values in the process; and (c) that their presence will also change the public culture in various ways – for instance, a society in which an established religion had formed an important part of national identity will typically exhibit greater religious diversity after accepting immigrants, and as a consequence religion will play a less significant part in defining that identity.

Immigration, in other words, is likely to change a society's public culture rather than destroy it. And since public cultures always change over time, as a result of social factors that are quite independent of immigration (participation in the established religion might have been declining in any case), it does not, on the face of it, seem that states have any good reason to restrict immigration on that basis. They might have reason to limit the *flow* of immigrants, on the grounds that the process of acculturation outlined above may break down if too many come in too quickly. But so long as a viable public culture is maintained, it should not matter that its character changes as a result of taking in people with different cultural values (Perry, 1995).

What this overlooks, however, is that the public culture of their country is something that people have an interest in controlling: they want to be able to shape the way that their nation develops, including the values that are contained in the public culture. They may not of course succeed: valued cultural features can be eroded by economic and other forces that evade political control. But they may certainly have good reason to try, and in particular to try to maintain cultural continuity over time, so that they can see themselves as the bearers of an identifiable cultural tradition that stretches backward historically. Cultural continuity, it should be stressed, is not the same as cultural rigidity: the most valuable cultures are those that can develop and adapt to new circumstances, including the presence of new subcultures associated with immigrants.

Consider the example of language. In many states today the national language is under pressure from the spread of international languages, especially English. People have an incentive to learn and use one of the international languages for economic and other purposes, and so there is a danger that the national language will wither away over the course of two or three generations. If this were to happen, one of the community's most important distinguishing characteristics would have disappeared, its literature would become inaccessible except in translation, and so forth. So the states in question adopt policies to ensure, for instance, that the national language is used in schools and in the media, and that exposure to foreign languages through imports is restricted. What effect would a significant influx of immigrants who did not already speak the national language have in these circumstances? It is likely that their choice of second language would be English, or one of the other international languages. So their presence would increase the incentive among natives to defect from use of the national language in everyday transactions, and make the project of language-preservation harder to carry through. The state has good reason to limit immigration, or at least to differentiate sharply among prospective immigrants between those who speak the national language and those who do not, as the government of Quebec has done.

Language is not the only feature to which the argument for cultural continuity applies. There is an internal relationship between a nation's culture and its physical shape – its public and religious buildings, the way its towns and villages are laid out, the pattern of the landscape, and so forth. People feel at home in a place, in part, because they can see that their surroundings bear the imprint of past generations whose values were recognizably their own. This does not rule out cultural change, but again it gives a reason for wanting to stay in control of the process – for teaching children to value their cultural heritage and to regard themselves as having a responsibility to preserve the parts of it that are worth preserving, for example. The "any public culture will do" position ignores this internal connection between the cultural and physical features of the community.

How restrictive an immigration policy this dictates depends on the empirical question of how easy or difficult it is to create a symbiosis between the existing public culture and the new cultural values of the immigrants, and this will vary hugely from case to case (in particular the experience of immigration itself is quite central to the public cultures of some states, but not to others). Most liberal democracies are now multicultural, and this is widely regarded as a source of cultural richness. But the more culturally diverse a society becomes, the greater need it has for a unifying public culture to bind its members together, and this culture has to connect to the history and physical shape of the society in question – it cannot be invented from scratch (Miller, 1995, ch. 4; Kymlicka, 2001, esp. part IV). So a political judgment needs to be made about the scale and type of immigration that will enrich rather than dislocate the existing public culture.

The second reason for states to limit immigration that I want to consider concerns population size.[1] This is a huge, and hugely controversial, topic, and all I can do here is to sketch an argument that links together the issues of immigration and population control. The latter issue really arises at two different levels: global and national. At the global level, there is a concern that the carrying capacity of the Earth may be stretched to breaking point if the total number of human beings continues to rise as it has over the last half century or so. At national level, there is a concern about the effect of population growth on quality of life and the natural environment. Let me look at each level in turn.

Although there is disagreement about just how many people the Earth can sustain before resource depletion – the availability of water, for example – becomes acute, it would be hard to maintain that there is *no* upper limit. Although projections of population growth over the century ahead indicate a leveling off in the rate of increase, we must also expect – indeed should welcome – increases in the standard of living in the developing world that will mean that resource consumption per capita will also rise significantly. In such a world it is in all our interests that states whose populations are growing rapidly should adopt birth control measures and other policies to restrict the rate of growth, as both China and India have done in past decades. But such states have little or no incentive to adopt such policies if they can "export" their surplus population through international migration, and since the policies in question are usually unpopular, they have a positive incentive not to pursue them. A viable population policy at global level requires each state to be responsible for stabilizing, or even possibly reducing, its population over time, and this is going to be impossible to achieve if there are no restrictions on the movement of people between states.

At national level, the effects of population growth may be less catastrophic, but can still be detrimental to important cultural values. What we think about this issue may be conditioned to some extent by the population density of the state in which we live. Those of us who live in relatively small and crowded states experience daily the way in which the sheer number of our fellow citizens, with their needs for housing, mobility, recreation, and so forth, impacts on the physical environment, so that it becomes harder to enjoy access to open space, to move from place to place without encountering congestion, to preserve important wildlife habitats, and so on. It is true, of course, that the problems arise not simply from population size, but also from a population that wants to live in a certain way – to move around a lot, to have high levels of consumption, and so on – so we could deal with them by collectively changing the way that we live, rather than by restricting or reducing population size (De-Shalit, 2000). Perhaps we should. But this, it seems to me, is a matter for political decision: members of a territorial community have the right to decide whether to restrict their numbers, or to live in a more ecologically and humanly sound way, or to do neither and bear the costs of a high-consumption, high-mobility lifestyle in a crowded territory. If restricting numbers is part of the solution, then controlling immigration is a natural corollary.

What I have tried to do in this section is to suggest why states may have good reason to limit immigration. I concede that would-be immigrants may have a strong interest in being admitted – a strong economic interest, for example – but in general they have no obligation-conferring *right* to be admitted, for reasons given in the previous section. On the other side, nation-states have a strong and legitimate interest in determining who comes in and who does not. Without the right to exclude, they could not be what Michael Walzer has called "communities of character": "historically stable, ongoing associations of men and women with some special commitment to one another and some special sense of their common life" (1983: 62). It remains now to see what conditions an admissions policy must meet if it is to be ethically justified.

Conditions for an Ethical Immigration Policy

I shall consider two issues. The first is the issue of refugees, usually defined as people who have fled their home country as a result of a well-founded fear of persecution or violence. What obligations do states have to admit persons in that category? The second is the issue of discrimination in admissions policy. If a state decides to admit some immigrants (who are not refugees) but refuses entry to others, what criteria can it legitimately use in making its selection?

As I indicated in the first section of this chapter, people whose basic rights are being threatened or violated in their current place of residence clearly do have the right to move to somewhere that offers them greater security. Prima facie, then, states have an obligation to admit refugees, indeed "refugees" defined more broadly than is often the case to include people who are being deprived of rights to subsistence, basic healthcare, and so on (Shacknove, 1985; Gibney, 1999). But this need not involve treating them as long-term immigrants. They may be offered temporary sanctuary in states that are able to protect them, and then be asked to return to their original country of citizenship when the threat has passed (Hathaway and Neve, 1997). Moreover, rather than encouraging long-distance migration, it may be preferable to establish safety zones for refugees close to their homes and then deal with the cause of the rights violations directly – whether this means sending in food and medical aid or intervening to remove a genocidal regime from power. There is obviously a danger that the temporary solution becomes semi-permanent, and this is unacceptable because refugees are owed more than the immediate protection of their basic rights – they are owed something like the chance to make a proper life for themselves. But liberals who rightly give a high moral priority to protecting the human rights of vulnerable people are regrettably often unwilling to countenance intervention in states that are plainly violating these rights.

If protection on the ground is not possible, the question then arises *which* state should take in the refugees. It is natural to see the obligation as shared among all those states that are able to provide refuge, and in an ideal world one might envisage some formal mechanism for distributing refugees among them. However, the difficulties in devising such a scheme are formidable (see Hathaway and Neve, 1997; Schuck, 1997). To obtain agreement from different states about what each state's refugee quota should be, one would presumably need to start with simple and relatively uncontroversial criteria such as population or per capita GNP. But this leaves out of the picture many other factors, such as population density, the overall rate of immigration into each state, cultural factors that make absorption of particular groups of refugees particularly easy or difficult, and so forth – that would differentially affect the willingness of political communities to accept refugees and make agreement on a scheme very unlikely. Furthermore, the proposed quota system pays no attention to the choices of the refugees themselves as to where to apply for sanctuary, unless it is accompanied by a compensatory scheme that allows states that take in more refugees than their quota prescribes to receive financial transfers from states that take in less.

Realistically, therefore, states have to be given considerable autonomy to decide how to respond to particular asylum applications: besides the refugee's own choice, they are entitled to consider the overall number of applications they face, the demands that temporary or longer-term accommodation of refugees will place on existing citizens, and whether there exists any special link between the refugee and the host community – for instance, similarities of language or culture, or a sense of historical responsibility on the part of the receiving state (which might see itself as somehow implicated among the causes of the crisis that has produced the refugees). If states are given this autonomy, there can be no guarantee that every bona fide refugee will find a state willing to take him or her in. Here we simply face a clash between two moral intuitions: on the one hand, every refugee is a person with basic human rights that deserve protection; on the other, the responsibility for ensuring this is diffused among states in such a way that we cannot say that any particular state S has an obligation to admit refugee R. Each state is at some point entitled to say that it has done enough to cope with the refugee crisis. So the best we can hope for is that informal mechanisms will continue to evolve which make all refugees the *special* responsibility of one state or another (Miller, 2001).

The second issue is discrimination among migrants who are not refugees. Currently, states do discriminate on a variety of different grounds, effectively selecting the migrants they want to take in. Can this be justified? Well, given that states are entitled to put a ceiling on the numbers of people they take in, for reasons canvassed in the previous section, they need to select somehow, if only by lottery (as the United States began to do in 1995 for certain categories of immigrant). So what grounds can they legitimately use? It seems to me that receiving states are entitled to consider the benefit they would receive from admitting a would-be migrant as well as the strength of the migrant's own claim to move. So it is acceptable to give precedence to people whose cultural values are closer to those of the existing population – for instance, to those who already speak the native language. This is a direct corollary of the argument in the previous section about cultural self-determination. Next in order of priority come those who possess skills and talents that are needed by the receiving community.[2] Their claim is weakened, as suggested earlier, by the likelihood that in taking them in, the receiving state is also depriving their country of origin of a valuable resource (medical expertise, for example). In such cases, the greater the interest the potential host country has in admitting the would-be migrant, the more likely it is that admitting her will make life worse for those she leaves behind. So although it is reasonable for the receiving state to make decisions based on how much the immigrant can be expected to contribute economically if admitted, this criterion should be used with caution. What cannot be defended in any circumstances is discrimination on grounds of race, sex, or, in most instances, religion – religion could be a relevant criterion only where it continues to form an essential part of the public culture, as in the case of the state of Israel.

If nation-states are allowed to decide how many immigrants to admit in the first place, why can't they pick and choose among potential immigrants on whatever grounds they like – admitting only red-haired women if that is what their current membership prefers? I have tried to hold a balance between the interest that migrants have in entering the country they want to live in, and the interest that political communities have in determining their own character. Although the first of these interests is not strong enough to justify a right of migration, it is still substantial, and so the immigrants who are refused entry are owed an explanation. To be told that they belong to the wrong race, or sex (or have hair of the wrong color) is insulting, given that these features do not connect to anything of real significance to the society they want to join. Even tennis clubs are not entitled to discriminate among applicants on grounds such as these.

Let me conclude by underlining the importance of admitting all long-term immigrants to full and equal citizenship in the receiving society (this does not apply to refugees who are admitted temporarily until it is safe to return to their country of origin, but it does apply to refugees as soon as it becomes clear that return is not a realistic option for them). Controls on immigration must be coupled with active policies to ensure that immigrants are brought into the political life of the community, and acquire the linguistic and other skills that they require to function as active citizens (Kymlicka, 2001, ch. 8). In several states immigrants are now encouraged to take citizenship classes leading up to a formal admissions ceremony, and this is a welcome development in so far as it recognizes that becoming a citizen is not something that just happens spontaneously. Precisely because they aim to be "communities of character," with distinct public cultures to which new immigrants can contribute, democratic states must bring immigrants into political dialogue with natives. What is unacceptable is the emergence of a permanent class of non-citizens, whether these are guest workers, illegal immigrants, or asylum seekers waiting to have their applications adjudicated. The underlying political philosophy which informs this chapter sees democratic states as political communities formed on the basis of equality among their members, and just as this gives such states the right to exclude, it also imposes the obligation to protect the equal status of all those who live within their borders.

Notes

Earlier versions of this chapter were presented to the Nuffield Political Theory Workshop; the Politics, Law and Society Colloquium at University College London; and the Department of Government, University of Essex. I am very grateful to these audiences for their criticisms and suggestions, and especially to Clare Chambers, Matthew Gibney, Cecile Laborde, and Tiziana Torresi for their written comments on previous drafts.

1   For some reason this issue is rarely considered in philosophical discussions of immigration. An exception, albeit a brief one, is Barry (1992).
2   Another criterion that is often used in practice is having family ties to people who already have citizenship in the state in question, and this seems perfectly justifiable, but I am considering claims that have to do with features of the immigrants themselves.

References

Barry, B. (1992) The quest for consistency: a sceptical view. In *Free Movement: Ethical Issues in the Transnational Migration of People and Money*, ed. B. Barry and R.E. Goodin, pp. 279–287. Hemel Hempstead: Harvester Wheatsheaf.
Caney, S. (2001) Cosmopolitan justice and equalizing opportunities. *Metaphilosophy* 32: 113–134.

Carens, J. (1987) Aliens and citizens: the case for open borders. *Review of Politics* 49: 251–273.

Cole, P. (2000) *Philosophies of Exclusion: Liberal Political Theory and Immigration*. Edinburgh: Edinburgh University Press.

De-Shalit, A. (2000) Sustainability and population policies: myths, truths and half-baked ideas. In *Global Sustainable Development in the 21st Century*, ed. K. Lee, A. Holland, and D. McNeill, pp. 188–200. Edinburgh: Edinburgh University Press.

Dummett, A. (1992) The transnational migration of people seen from within a natural law tradition. In *Free Movement: Ethical Issues in the Transnational Migration of People and Money*, ed. B. Barry and R.E. Goodin, pp. 169–180. Hemel Hempstead: Harvester Wheatsheaf.

Gibney, M.J. (1999) Liberal democratic states and responsibilities to refugees. *American Political Science Review* 93: 169–181.

Griffin, J. (2001) First steps in an account of human rights. *European Journal of Philosophy* 9: 306–327.

Hampton, J. (1995) Immigration, identity, and justice. In *Justice in Immigration*, ed. W.F. Schwartz, pp. 67–93. Cambridge: Cambridge University Press.

Hathaway, J.C. and Neve, R.A. (1997) Making international refugee law relevant again: a proposal for collectivized and solution-oriented protection. *Harvard Human Rights Journal* 10: 115–211.

Kymlicka, W. (2001) *Politics in the Vernacular: Nationalism, Multiculturalism and Citizenship*. Oxford: Oxford University Press.

Miller, D. (1995) *On Nationality*. Oxford: Clarendon Press.

Miller, D. (1999) Justice and global inequality. In *Inequality, Globalization, and World Politics*, ed. A. Hurrell and N. Woods, pp. 187–210. Oxford: Oxford University Press.

Miller, D. (2000) *Citizenship and National Identity*. Cambridge: Polity.

Miller, D. (2001) Distributing responsibilities. *Journal of Political Philosophy* 9: 453–471.

Nickel, J. (1987) *Making Sense of Human Rights*. Berkeley, CA: University of California Press.

Perry, S.R. (1995) Immigration, justice, and culture. In *Justice in Immigration*, ed. W.F. Schwartz, pp. 94–135. Cambridge: Cambridge University Press.

Schuck, P.A. (1997) Refugee burden-sharing: a modest proposal. *Yale Journal of International Law* 22: 243–297.

Shacknove, A. (1985) Who is a refugee? *Ethics* 95: 274–284.

Shue, H. (1980) *Basic Rights: Subsistence, Affluence, and US Foreign Policy*. Princeton, NJ: Princeton University Press.

Walzer, M. (1983) *Spheres of Justice: A Defence of Pluralism and Equality*. Oxford: Martin Robertson.

Further Reading

Barry, B. and Goodin, R.E., eds (1992) *Free Movement: Ethical Issues in the Transnational Migration of People and Money*. Hemel Hempstead: Harvester Wheatsheaf.

Dummett, M. (2001) *On Immigration and Refugees*. London: Routledge.

Gibney, M., ed. (1988) *Open Borders? Closed Societies? The Ethical and Political Issues*. New York: Greenwood Press.

Joppke, C. (1999) *Immigration and the Nation-state*. Oxford: Oxford University Press.

Schwartz, W.F., ed. (1995) *Justice in Immigration*. Cambridge: Cambridge University Press.

Whelan, F.G. (1981) Citizenship and the right to leave. *American Political Science Review* 75: 636–653.

CHAPTER TWENTY-FIVE

The Case for Open Immigration

Chandran Kukathas

People favor or are opposed to immigration for a variety of reasons. It is therefore difficult to tie views about immigration to ideological positions. While it seems obvious that political conservatives are the most unlikely to defend freedom of movement, and that socialists and liberals (classical and modern) are very likely to favor more open borders, in reality wariness (if not outright hostility) to immigration can be found among all groups. Even libertarian anarchists have advanced reasons to restrict the movement of peoples.

The purpose of this chapter is to make a case for greater freedom of movement or, simply, freedom of immigration. Its aim is to defend immigration against critics of all stripes, and also to defend immigration against some of its less enthusiastic friends.

To put a case for free immigration is not easy. Though it may be simple enough to enunciate political principles and stand doggedly by them, in questions of public policy coherence and consistency are merely necessary, but not sufficient, virtues. The feasibility of any policy proposal is also important, and political theory needs to be alive to this. "How open can borders be?" is an obvious question that it may not be possible to evade. The defense of free immigration offered here is, I hope, sensitive to this requirement. Nonetheless, it is an important part of its purpose to suggest that, in the end, political theory needs also to be suspicious of feasibility considerations, particularly when they lead us to morally troubling conclusions.

Before proceeding to the defense of free immigration, however, it will be important to understand what precisely immigration amounts to, and to recognize the nature of the *problem* of immigration as it exists in the world today. This is the task of the first section of this chapter. The second section defines and offers a short defense of free immigration. The three sections that follow then consider various challenges to the principle of free immigration coming from economic, national, and security perspectives, and argue that each challenge can be met. The final section offers some general

Contemporary Debates in Applied Ethics, Second Edition. Edited by Andrew I. Cohen and Christopher Heath Wellman.

reflections on the dilemmas of contemporary immigration policy, before restating more forcefully the case for the free movement of peoples.

The Problem of Immigration in the Modern World

Toward the end of the twentieth century, more than 100 million people lived outside of the states of which they are citizens (Trebilcock, 1995, p. 219). But this figure does not come close to identifying the numbers of people who are moving about from country to country across the globe. Many people move between countries as tourists, businessmen, sportswomen or performers without ever stopping to "live" in a country – let alone with any intention to settle in a foreign land. Global human movement is a fact of life, as it has been for centuries, if not for all of human history. This has always had its own difficulties. But the problem of immigration is a problem of a particular kind, for immigrants are people who aim to stop rather than simply to pass through – though, as we shall see, the definition of "stopping" is not an easy one to establish. The migration of people is a problem in the modern world because that world is a world of states, and states guard (sometimes jealously) the right to determine who may settle within their borders. Immigration may be defined as the movement of a person or persons from one state into another for the purpose of temporary or permanent settlement (Kukathas, 2002).

Modern states are reluctant to allow people to enter and settle within their borders at will for a variety of reasons. Security is one important consideration, though different states have different security concerns. The United States at present fears terrorist attacks and has tightened its immigration laws in part because of concerns for the safety of its citizens. China, on the other hand, has different security concerns since its political system does not permit much internal freedom of movement and could not tolerate an uncontrolled influx of foreigners into a population that harbors dissidents who would challenge the authority of the government. For states such as Israel, security is a prominent concern, but perhaps one no more important than the desire to preserve a certain cultural integrity. A state founded as a Jewish homeland cannot allow immigration to transform it into a multicultural polity.

For modern liberal democratic states, however, there are a number of important reasons why immigration is problematic. These states, including Canada, the United States, Australia, Britain, and several countries in Western Europe, are particularly popular destinations for immigrants, whether because they are refugees seeking safe havens, or simply people looking to improve their prospects of a better life. One important reason why immigration is a problem in these cases is that immigrants impose costs on society even as they bring benefits. While economists tend to agree that the consequences of free movement are generally positive, since competitive labor markets make for a more efficient use of resources (Simon, 1990; Sykes, 1995, pp. 159–160), not all nations may benefit immediately from an influx of immigrants. Nor do the burdens of accommodating or adjusting to immigrants fall equally on all within a society – much will depend on who the immigrants are, where they settle, and with whom they end up competing for jobs, real estate, and public facilities. Even if the benefits of immigration outweigh the costs to the nation, those who are adversely affected by an influx of settlers will object; and in liberal democratic states this will translate into electorally significant opposition.

Another important reason why immigration is a problem in liberal democratic states is that these states are, to varying degrees, welfare states. The state in such societies provides a range of benefits, including education, unemployment relief, retirement income, medical care, as well as numerous programs to serve particular interests. Immigrants are potential recipients of these services and benefits, and any state considering the level of immigration it will accept will have to consider how likely immigrants are to consume these benefits, how much they might consume, whether or not they are going to be able to finance the extra costs from the lifetime tax contributions of these immigrants, and what are the short-term implications of accepting immigrants who begin by consuming more in benefits than they pay in taxes. Consequently, such states are reluctant to accept immigrants who are infirm, or too old to contribute enough in taxes in their remaining working lives to cover the costs of medical care and retirement subsidies.

Under these circumstances liberal democratic governments will go to great lengths to limit immigration, though they will face pressures both to admit and refuse entry to applicants seeking to enter their countries. The pressures to admit will come from businesses looking for cheaper labor, from humanitarian groups calling for the admission of refugees, and from families and ethnic communities pressing to have relatives join them from their countries of origin. The pressures to refuse entry will generally come from labor unions, from "nativist" groups, and from conservatives concerned about the cultural and economic impact of settlers, particularly if the settlers are predominantly from ethnically different countries. The lengths to which liberal democratic states might go to discourage immigration is well illustrated by the reaction of the Australian government in August 2001 to the appearance near its coastal waters of a Norwegian merchant vessel, *The Tampa*, bearing refugees rescued at sea. The vessel was denied permission to enter Australian waters and to offload its human cargo, which was shipped to the island of Nauru to prevent the refugees from appealing for asylum in Australia (Marr and Wilkinson, 2003). More recently, the United States responded to the crisis in Haiti in February 2004 by intervening to encourage the departure of President Aristide, and to restore some degree of order, because it feared an exodus of Haitian refugees making their way to Florida.

Immigration is a problem largely because of the nature of the modern state. Most states, and certainly all liberal democratic states, regard their people as "citizens" or "members" of the state. Membership is not standard, and the nature of membership has a substantial bearing on the rights that individuals have within a state. Full membership might amount to citizenship and include the right to vote and stand for public office. (Though it is worth noting that in the United States, for example, even full citizenship does not entitle a member to stand for the office of President if he or she was not born in the country.) "Permanent resident" status might give one the right to work and to change employer at will, and also to draw on health, education, and welfare services, but not provide security against deportation. Status as a "guest-worker" or temporary resident might provide fewer rights still. Modern states restrict immigration because they must manage access to the goods for which immigrants and natives would compete. Modern states are like clubs that are reluctant to accept new members unless they can be assured that they have more to gain by admitting people than they have by keeping them out.

In Defense of Free Immigration

Given that immigrants will compete for goods and resources with natives, why should states open their borders when it is their task to manage affairs within their domains? Does the idea of open immigration not go against the principles of good husbandry?

There are many reasons why borders should be open and the movement of people should be free. But before considering these reasons more closely, it should be admitted that the prospect of states opening their borders completely is a remote one. Even as the European Union expands its membership and facilitates freer movement among its denizens, to take one possible counter-example to this claim, it continues to control entry into Europe – and is feeling the pressure from member states to tighten restrictions on entry from refugees and displaced people. "Open borders" is not a policy option currently being considered by any state. Nonetheless, the case for open borders should be considered, though in the end, as we shall see, it cannot be defended without rethinking the idea of the state.

There are two major reasons for favoring open borders. The first is a principle of freedom, and the second a principle of humanity.

Open borders are consistent with – and on occasion, protect – freedom in a number of ways. The first, and most obvious, is that closed borders restrict freedom of movement. Borders prevent people from moving into territories whose governments forbid them to enter; and to the extent that they cannot enter any other territory, borders confine them within their designated boundaries. This fact is not sufficient to establish that so confining people is indefensible; but if freedom is held to be an important value, then there is at least a case for saying that very weighty reasons are necessary to restrict it.

Several other considerations suggest that such reasons would have to be weighty indeed. First, to keep borders closed would mean to keep out people who would, as a consequence, lose not only the freedom to move but also the freedom they might be seeking in an attempt to flee unjust or tyrannical regimes. The effect of this is to deny people the freedom they would gain by leaving their societies and to diminish the incentive of tyrannical regimes to reform the conditions endured by their captive peoples. Second, closing borders means denying people the freedom to sell their labor, and denying others the freedom to buy it. Good reasons are needed to justify abridging this particular freedom, since to deny someone the liberty to exchange his labor is to deny him a very significant liberty. Third, and more generally, keeping borders closed would mean restricting people's freedom to associate. It would require keeping apart people who wish to come together whether for love, or friendship, or for the sake of fulfilling important duties, such as caring for children or parents.

Now, to be sure, defenders of restricted immigration do not generally argue that borders should be completely sealed, or that no one should be admitted. Many concede that exceptions should be made for refugees, that some people should be allowed to come into a country to work, and that some provision should be made for admitting people who wish to rejoin their families. Those who want restricted or controlled immigration are not indifferent to freedom. Nonetheless, even those who argue for generous levels of immigration by implication maintain that people should be turned away at the border. This in itself is a limitation of liberty, for which good reasons must be given. In the end, or so I will argue, the reasons that have been offered are not weighty enough to justify restricting freedom even to a limited degree.

The second reason for favoring open borders is a principle of humanity. The great majority of the people of the world live in poverty, and for a significant number of them the most promising way of improving their condition is to move. This would remain true even if efforts to reduce trade barriers were successful, rich countries agreed to invest more in poorer ones, and much greater amounts of aid were made available to the developing world. For even if the general condition of a society were good, the situation of particular individuals would often be poor, and for some of them immigration would offer the best prospect of improving their condition. To say to such people that they are forbidden to cross a border in order to improve their condition is to say to them that it is justified that they be denied the opportunity to get out of poverty, or even destitution. And clearly there are many people who share this plight, for numerous illegal immigrants take substantial risks to move from one country to another – courting not only discomfort and even death by traveling under cover in dangerous conditions, but also punishment at the hands of the authorities if caught.

A principle of humanity suggests that very good reasons must be offered to justify turning the disadvantaged away. It would be bad enough to meet such people with indifference and to deny them positive assistance. It would be even worse to deny them the opportunity to help themselves. To go to the length of denying one's fellow citizens the right to help those who are badly off, whether by employing them or by simply taking them in, seems even more difficult to justify – if, indeed, it is not entirely perverse.

Not all people who look to move are poor or disadvantaged. Nor do all of them care about freedom. But if freedom and humanity are important and weighty values, the prima facie case for open borders is a strong one, since very substantial considerations will have to be adduced to warrant ignoring or repudiating them. I suggest that no such considerations are to be found. But to show this, it is necessary to look more closely at arguments that restrictions of immigration are defensible, and indeed desirable.

Economic Arguments Against Open Borders

It is sometimes argued that there are strong economic arguments for limiting immigration. There are two kinds of concern here. The first is about the impact of migrants on the local market economy: large numbers of people entering a society can change the balance of an economy, driving down wages or pushing up the prices of some goods such as real estate – to the disadvantage of many people in the native population. The second is about the impact of migrants on the cost and availability of goods and services supplied through the state: education, healthcare, welfare, and the publicly funded infrastructure of roads, parks, and other non-excludable goods. Do these concerns warrant closing borders to immigrants?

In the end, the answer must be that they do not. But the reasons why are not as straightforward as might be anticipated. If our concern is the impact of migrants on the local market economy, one argument often advanced by economists is that, on balance, the net impact of immigrants is mildly positive. While immigrants do take jobs that might have gone to locals and drive down wages, while driving up some prices, they also have a positive impact on the economy. Migrants expand the size of the work-force and extend the division of labor, so society gains from the benefits this brings. As new consumers, they expand the size of the domestic market and help to lower prices for many goods. Measuring the precise impact of any cohort of immigrants is difficult; but the overall impact is, at best, positive and, at worst, only mildly negative – even with respect to employment. Moreover, the global effect of migration is positive, as it involves a movement of people from places where they are less productive and often unable to make a living to places where they are both more productive and better off – and in many cases no longer a burden on their societies.

The problem, however, is that whatever the overall impact of migration, particular persons will do badly out of it. An influx of cheap labor may be good for society overall, but bad for those who are put out of work or forced to accept lower wages. It is to these people that the critic of open borders will point to illustrate the economic costs of immigration. Why should *they* bear the costs? Equally, why should other societies be happy about the brain-drain that is also an aspect of immigration, as skilled people leave their native countries for better opportunities abroad?

While it is true that the burdens and benefits of immigration do not fall evenly or equitably on all members of a host society, open borders are defensible nonetheless for a number of reasons. First, it has to be asked why it must be assumed that locals are entitled to the benefits they enjoy as people who have immediate access to particular markets. As residents or citizens, these people enjoy the rents they secure by virtue of an arrangement that excludes others from entering a particular market.[1] Such arrange-ments are commonplace in every society, and indeed in the world as a whole. Often those who find a resource to exploit, or a demand which they are particularly able to fulfill, are unable to resist the temptation to ensure that they enjoy the gains to be had in exploiting that resource or fulfilling that demand by preventing others from doing the same. Yet it is unclear that there is any principle that can justify granting to some persons privileged access to such rents. To be sure, many of the most egregious examples of rent-seeking (and rent-protecting) behavior are to be found in the activities of capitalist firms and industries. But this does not make such activity defensible, since it serves simply to protect the well-off from having to share the wealth into which they have tapped with those who would like to secure a little of that same wealth for themselves.

If we are considering labor markets, there is no good reason to exclude outsiders from offering their labor in competition with locals. While it may disadvantage locals to have to compete, it is equally true that outsiders will be disadvantaged if they are forbidden to do so. Also, locals who would benefit from the greater availability of labor would also be disadvantaged by the exclusion of outsiders. To prevent, say, firms from hiring outside labor would be no more justifiable on economic grounds than preventing firms from moving their operations abroad to take advantage of cheaper or more pro-ductive labor in other countries.

The same arguments hold if we are considering the case of people who wish to move to a different country to sell not their labor but their wares – perhaps by setting up a business. There is no more a justification for preventing them from doing this than there is for preventing them from trading their goods from abroad. Restricting access to markets certainly benefits some people, but at the expense of others, and generally to the disadvantage of all. If particular privileges should be accorded to some because of their state membership, the justification cannot be economic in the first of the two senses distinguished.

In the second sense of economic, however, the argument for restricting immigration is not that access to particular markets should be limited, but that the economic benefits dispensed by the state must be limited if economic resources and indeed the social system more generally are to be properly managed. Immigration dulls the edge of good husbandry. For some libertarians, the concern here is that open borders – or even increased immigration – will impose a greater tax burden on existing members of society as the poor and disabled move to states with more generous welfare provisions, as well as subsidized education and healthcare. Indeed, a number of libertarians have argued that until the welfare state is abolished, immigration will have to be tightly controlled in countries like the United States (Hoppe, 1998).

Here it would not be enough to point out that, to the extent that immigrants join the workforce, they would also contribute to the revenues of the state through taxes, even as they consume resources dispensed by the state. Open immigration might well encourage people to move with the intention of taking advantage of benefits that exceed their tax contributions. People on low incomes and with children or elderly or infirm dependents would find it advantageous to move to countries with generous public education and healthcare. This could impose a significant additional burden on taxpaying individuals and firms, or pressure a state with fiscal problems to reduce the quality of its services. Immigration is a problem for welfare states – understanding welfare in its broadest sense to include health and education services as well as unemployment relief and disability benefits.

The problem here is a significant one. But it should be noted that it is not a problem that results from the movement of the rich or able, only one that results from the movement of the poor. The independently wealthy, and the well-off moving into well-paying jobs, will contribute to the state's coffers through direct and indirect taxes, and may well pay for more than they consume. The poor will in all likelihood be net consumers of tax dollars – at least at the outset. An important purpose of closed borders is to keep out the poor.

If the concern is to preserve the integrity of the welfare state, however, the most that could be justified is restricting membership of the welfare system. The movement of people into a country could then be free. Such restricted forms of immigration would still impose serious disadvantages upon poorer people, for whom the attraction of immigration would diminish if they were obliged to fund their own healthcare and pay for the education of their children. Yet for many it would be better than no opportunity to move at all. Certainly, immigration with limited entitlements would be attractive to young and able people with dependents, since the opportunity to work abroad and remit money home might significantly improve all their lives.

Nonetheless, it would not do to be too sanguine about the possibility of such an arrangement. Most states would baulk at the suggestion of such arrangements, and even advocates of open immigration may reject the idea of different classes of membership. Moreover, immigrants paying taxes may feel disgruntled if their taxes do not buy them equal entitlements. In the end, it may be that the existence of the welfare state makes open borders, or even extensive immigration, very difficult – if not impossible. From the perspective of a principle of freedom, or a principle of humanity, I suggest, the standard of open borders should prevail. To defend closed borders a principle of nationality would have to take precedence. We should turn then to look more closely at the argument from nationality.

Nationality and Immigration

Implicit in most arguments for closed borders or restricted immigration is an assumption that the good or well-being of the members of a polity should take precedence – to a significant degree, even if not absolutely – over the good of outsiders. From this perspective, that one of my fellow countrymen is harmed or made worse off is a weighty consideration when assessing any policy, in a way that the impact of that policy on foreigners is not. Defenders of this perspective may disagree about the extent to which the interests of outsiders should be discounted; and indeed some may hold that rich nations owe substantial obligations of justice to the world's poor. But they are agreed that something more is owed to one's own country and its people. And this justifies protecting one's nation from the impact of open or substantial immigration. (For contrasting views see Goodin, 1988; Miller, 1988.)

Immigration, on this view, may be damaging for a number of different reasons. We have already considered some of the economic consequences of immigration; but there are other problems as well. First, immigration in substantial numbers, even if it takes place over a long period of time, "has the effect of changing the recipient area" (Barry, 1992, p. 281). The influx of Indian workers in the nineteenth century changed Fiji from an island of Polynesian people to one that is bicultural, just as the movement of Indians and Chinese to Malaya turned that society into a multicultural one. The fear of many people is that immigration will change a society's character, and perhaps undermine or displace an ancient identity (Casey, 1982). The cultural character of Britain or France cannot remain the same if substantial numbers of people move there from Africa or Asia.

Second, immigration from culturally different people may be damaging to wealthy countries to the extent that their wealth is dependent upon the existence of a political culture, and economic and social institutions, that are especially conducive to wealth-creation. Immigration from people who do not share the same values, and who would not help to sustain the same institutions, may ultimately undermine those institutions (Buchanan, 1995). If so, this may be good reason to restrict immigration not only by number but also by culture.

Third, immigration may make it very difficult for a society to develop or sustain a level of social solidarity that is necessary for a state to work well, and particularly for it to uphold principles of social justice. This argument has been developed especially forcefully by David Miller, who suggests that if immigration exceeds the absorptive capacities of a society, the bonds of social solidarity may break down. The nation is a natural reference group when people ask whether or not they are getting a fair share of society's resources. If people have different understandings of what their rights and obligations are and disagree about what they may legitimately claim, it may become impossible to establish and operate appropriate standards of social justice (Miller, 1995, 1999a). For all of these reasons, then, open borders cannot be justified. Or so it is argued.

While all of these considerations are weighty, they do not suffice to warrant limitations on freedom of movement. First, while it is true that immigrants do change the character of a place – sometimes dramatically – it is not evident that this is necessarily a bad thing. More to the point, it is difficult to know how much change is desirable, partly because the results will not be known for some time and partly because different people – even in relatively homogeneous societies – want different things. It is perfectly understandable that some people want things to remain the way they have been during their lifetimes. Yet it is no less understandable that others want changes they regard as improvements. The Know-Nothings of nineteenth-century America were completely hostile to Catholic, and especially Irish, immigration; though Irish Americans were all too ready to welcome to the United States even more settlers from Ireland. In the end, our capacity to shape society or preserve its character may be as limited as our capacity to know how much (or how little) change is really desirable – even if we could agree on what sort of character we would like our societies to have.

It is also worth bearing in mind that many societies have experienced significant cultural or social transformations and not only survived but prospered. The United States in the nineteenth century welcomed immigrants from all over the world, incorporated large parts of what was once Mexico into its territory, overturned a three-century-old tradition of slavery and yet began the twentieth century a prosperous and vibrant democracy. Canada and Australia have seen their societies transformed by postwar immigration into multicultural polities, while continuing to enjoy economic growth and social stability. And the European Union continues to expand its membership by admitting states from Eastern Europe – and perhaps, eventually, Turkey – in a way that makes it possible for peoples from diverse ethnic, religious, and political traditions to move freely from one end of the continent to the other, without fearing a loss in prosperity; though there can be no doubt that this development will bring with it significant cultural changes to many of Europe's communities.

Social and cultural change can be effected by large-scale immigration, and its significance should not be discounted. But neither should it be overestimated. Nor should too much weight be given to the possibility that immigration from poor nations to rich ones will undermine the institutions of wealth-creation – though it is surely a possibility. If anything, it is perhaps more likely that immigrants who move to wealthy countries will do so because they want to take advantage of the opportunities it offers, and that they will assimilate by adopting the practices that bring success to the natives. In any case, if our interest is in wealth-creation, it is more likely that this skill will be taught to those who enter a rich country than that it will be exported successfully to some countries that are poor.

The most challenging argument against open immigration, however, is that institutions of social justice can only be built if social solidarity is preserved – and that immigration may undermine that solidarity if it is not appropriately restricted. If we accept that social justice is an important concern, then Miller's analysis and argument are powerful and convincing. The only way to resist them is to question the very idea that the nation-state is the appropriate site for the settlement of questions of distributive justice. And indeed that is what we need to do.

There are a number of reasons why we should be suspicious of the idea that the nation-state is the site of distributive justice, but the most powerful have been advanced by Miller in his own critique of the idea of global social justice. Miller maintains that principles of social justice are always, "as a matter of psychological fact, applied within bounded communities" (1999a, p. 18). It is easier for us to make judgments of justice in small communities such as workplaces, but difficult in units larger than nation-states. We make such judgments by comparing ourselves with others. But it is difficult for us to compare ourselves with people who are remote from our own circumstances, such as people in other countries. We can more readily make judgments based on comparisons with people who belong to our own reference group – people with whom we are likely to share some common conceptions of value. When conceptions of the value of a resource differ, it becomes very difficult to establish common standards of distributive justice, since the very question of what counts as a resource to be distributed may be impossible to settle. And when we consider that different communities have conflicting views about how trade-offs should be made, for example, between the consumption of what the earth will produce and the preservation of the natural environment, it would be difficult for one community to demand a share of another's resources on the basis of its own determination of the "true value" of those resources (Miller, 1999b, pp. 193–196). Global social justice is difficult to defend.

Yet all the things that make global justice problematic also go to make problematic social justice *within* the nation-state. Certainly, some nation-states are so large that it is difficult to see how they could really share a single conception of social justice. China and India between them hold more than a third of the world's population, and harbor different languages, religions, and customs. Even the United States, though much smaller, is sufficiently diverse that there are noticeable differences among significant groups about morality and justice – from California, to Utah, to Louisiana. Britain and France are smaller still, but are home to a diversity of religions and ethnicities. If the preservation of a shared ethos or sense of social justice is an important reason to restrict immigration, then, it might be defensible if we are considering small, homogeneous nations such as Iceland or Tahiti. It might also be defensible for a state such as Israel, though it might be more difficult to make this case the more it is a multicultural (or bicultural) state. But in larger states, which are diverse and already have a long history of immigration, the idea of a shared conception of social justice might be too much to hope for. Certainly, the vigorous debates among philosophers about social justice suggest that there is no substantial agreement on this question even among a group as homogeneous as the academy. Miller's point about the nature of social justice is a telling one; but it also tells against his own defense of restricted immigration. (For a fuller critique of Miller's view, see also Kukathas, 2002.)

Even if states were plausible sites of social justice, however, there is another issue that has to be raised. Is it right that the preservation of local institutions of social justice take precedence over the humanitarian concerns that make open immigration desirable? As was noted earlier, immigration barriers operate largely to limit the movement of the world's poor. It seems odd to suggest that this can be defended by appeal to the importance of social justice. If the price of social justice is exclusion of the worst-off from the lands that offer the greatest opportunity, this may be a mark against the ideal of social justice.

To be fair, however, it should be acknowledged that defenders of social justice or the primacy of membership (Walzer, 1983) generally acknowledge the need to make special provision for the world's poor. In this regard, they suggest that refugees may have a special claim to be allowed to immigrate and resettle to escape persecution. But here a number of problems arise. First, the line distinguishing a refugee and what we might term an "economic migrant" is a very fine one. As it stands, the 1951 United Nations

Convention relating to the Status of Refugees adopts a very narrow definition of refugee to include only persons with a well-founded fear of persecution for reasons of race, religion, nationality, membership of a particular social group or political opinion. Those people fleeing war, natural disaster, or famine are, on this definition, not refugees. Second, even on this narrow definition, at the start of the twenty-first century, there are more than 20 million people in the world who count as refugees who have yet to be resettled. The problem these two points pose is that making an exception for refugees requires a very significant increase in immigration – even if the narrow definition of refugee is used. If a more humane definition were adopted – one that recognized as refugees people fleeing war zones, for example – an even greater number of immigrants would have to be accepted. Yet then, if the standard of humanity is the appropriate standard, it is difficult to see why any sharp distinction should be made between the desperate fleeing war and the destitute struggling to make a living.

It would perhaps be too much to hope or expect that states – especially wealthy ones – will readily lower the barriers to the free movement of peoples. As it stands, the world of states has struggled to relocate the refugees for whom it has acknowledged responsibility. Indeed, it is sobering to remember that immigration controls were tightened with the invention of the passport during the First World War precisely to control refugee flows. Nonetheless, on this much at least, both the defenders of open borders and the advocates of restrictions can agree: that at present the borders are too securely sealed.

Immigration and Security

One reason for greater restriction of immigration, which clearly has assumed enormous significance in recent times, is the need for security. Can immigration be free in an age of terror?

Security from terrorist attack, it should be noted, is only one kind of security. Even before terror became a serious concern, modern states have been anxious about the security of political systems from foreign threats, and the security of society against international criminal organizations. Smugglers, traffickers in illegal goods (from drugs to rare wildlife to historical artifacts), and slave-traders of various kinds operate across boundaries to violate the laws of host states. Nonetheless, the threat of terror has added significantly to the security concerns of a number of western states. Does this give us greater reason to restrict immigration, or show that the idea of open borders is simply untenable?

In the end, I suggest that security concerns do not do much to diminish the case for open borders. This is not to say that security concerns are unfounded or should not be addressed. But it is to say that immigration controls are not they key. There are a number of reasons why. First, while it is easy to restrict legal immigration, it is another matter to control illegal immigration. Limiting legal immigration is unlikely to deter either criminals or subversive agents from moving between states. Borders are porous even when they are closed. Second, limiting immigration seldom means limiting the movement of people more generally, since many more people move from one country to another as tourists, or students, or businessmen, or government officials than they do as immigrants intending to settle in a new land. If security is a concern, tourism should be more severely limited in many countries than it presently is. If a person is likely to pose a threat to a country's security, it would be odd to think it acceptable for him to be granted a tourist visa for one, three, or six months. Equally, if a person is considered safe to be awarded a three-month tourist visa, it is hard to see why he should be denied the right to permanent residence *on security grounds*. It might well be that in times of insecurity greater vigilance is necessary: greater scrutiny of many aspects of the behavior of people – including travelers – may be warranted, just as one would expect the police to establish road blocks and search cars when there is an escaped criminal in the vicinity. It is not evident, however, that this would justify further restrictions on immigration rather than simply greater effort to discover who poses a threat to society, to try avert the threat, and to apprehend the particular persons who are menaces.

There are, however, reasons not to place too much weight on the importance of security, for like all things, the search for security comes with costs of its own. In the case of the search for security through immigration controls, the cost is borne not only in the financial expense that is incurred but also in the impact that controls on immigrants and immigration have on society more generally. Immigration control requires the surveillance of people moving in and out of the country, and to some degree of people moving about within the country. But it is not possible to do this with immigrants or outsiders generally without also placing one's own citizens under surveillance. In dangerous times this may be unavoidable, at least to some degree. But the risks it brings are substantial. Even if the burdens imposed upon citizens and residents are trivial, they may be burdens all the same – and for some more than others. Furthermore, there is always a risk that impositions designed to meet a particular danger will remain in place long after the danger has passed. (Malaysia's Internal Security Act, which, among other things, sanctions arrest and detention without trial, was passed at the height of the communist insurgency in the 1960s, but remains in place 25 years after the emergency ended.) Liberal democracies, in particular, should be wary of state controls advocated in the name of national security – particularly since the trade-off is a loss of liberty.

Concluding Reflections

Whatever the merits of the case for open borders, it is highly unlikely that we will see an end to immigration controls at any time soon – for reasons that were canvassed at the beginning of this chapter. In one important respect, free migration is entirely unfeasible: it is politically untenable.

One reason why it is politically untenable is that most voters in wealthy countries do not favor immigration, particularly by the poor. Another is that states themselves do not favor uncontrolled population movements. In a world order shaped by the Westphalian model of states operating within strict geographical boundaries, and dominated by the imperative to secure the welfare of members, the free movement of peoples is not a strong possibility. The inclination of most people to hold on to the advantages they possess also makes it unlikely that nations will open up their borders to allow others to come and take a greater share of what they control.

Yet if the free movement of peoples is not politically feasible, how can there be a case for open borders? Surely, political theory, in considering issues of public policy, should keep its focus on the world of the possible rather than on impossible ideals.

There is a good deal of truth to this. But there is, nonetheless, good reason for putting the case for open immigration. One important consideration is that many feasibility problems have their roots not in the nature of things but in our way of thinking about them. Many of the reasons open immigration is not possible right now have less to do with the disadvantages it might bring than with an unwarranted concern about its dangers. Even to the extent that the source of the problem for open immigration lies in the nature of things, however, it is worth considering the case for open borders because it forces us to confront the inconsistency between moral ideals and our existing social and political arrangements. One of the reasons why open immigration is not possible is that it is not compatible with the modern welfare state. While one obvious response to this is to say, "so much the worse for open immigration," it is no less possible to ask whether the welfare state is what needs rethinking.

Note

1   "Rent" is money someone pays to have access to some capital asset (such as land, a dwelling, or a means of transport) that he or she does not, or cannot, own outright. Persons who engage in "rent-seeking" seek money from rents instead of from profits or wage income.

References

Barry, B. (1992) The quest for consistency: a sceptical view. In *Free Movement: Ethical Issues in the Transnational Migration of People and Money*, ed. B. Barry and R.E. Goodin, pp. 279–287. University Park, PA: Pennsylvania State University Press.

Buchanan, J. (1995) A two-country parable. In *Justice in Immigration*, ed. W.F. Schwartz, pp. 63–66. Cambridge: Cambridge University Press.

Casey, J. (1982) One nation: the politics of race. *Salisbury Review* 1: 23–28.

Goodin, R.E. (1988) What's so special about our fellow countrymen? *Ethics* 98 (4): 663–686.

Hoppe, H.H. (1998) The case for free trade and restricted immigration. *Journal of Libertarian Studies* 13 (2): 221–233.

Kukathas, C. (2002) Immigration. In *The Oxford Handbook of Practical Ethics*, ed. H. LaFollette, pp. 567–590. New York: Oxford University Press.

Marr, D. and Wilkinson, M. (2003) *Dark Victory*. Crow's Nest, NSW, Australia: Allen & Unwin.

Miller, D. (1988) The ethical significance of nationality. *Ethics* 98 (4): 647–662.

Miller, D. (1995) *On Nationality*. Oxford: Clarendon Press.

Miller, D. (1999a) *Principles of Social Justice*. Cambridge, MA: Harvard University Press.

Miller, D. (1999b) Justice and global inequality. In *Inequality, Globalization and World Politics*, ed. A. Hurrell and N. Woods, pp. 187–210. Oxford: Oxford University Press.

Simon, J. (1990) *The Economic Consequences of Immigration*. Oxford: Blackwell.

Sykes, A.O. (1995) The welfare economics of immigration law: a theoretical survey with an analysis of US policy. In *Justice in Immigration*, ed. W.F. Schwartz, pp. 158–200. Cambridge: Cambridge University Press.

Trebilcock, M.J. (1995) The case for a liberal immigration policy. In *Justice in Immigration*, ed. W.F. Schwartz, pp. 219–246. Cambridge: Cambridge University Press.

Walzer, M. (1983) *Spheres of Justice: A Defence of Pluralism and Equality*. Oxford: Martin Robertson.

Humanitarian intervention

CHAPTER TWENTY-SIX

The Moral Structure of Humanitarian Intervention

Fernando R. Tesón

When can states intervene in other states to protect persons against their own govern-ments? Globalization has certainly brought about a reevaluation of state sovereignty, and while it may be too optimistic to say that liberal political principles have prevailed globally, governments have lost a considerable amount of the political space in which they could treat their citizens as they pleased. Almost everyone agrees that truculent rights violations (summary executions, arbitrary arrest, torture, violent suppressions of dissent) are morally abhorrent. The question remains, however: what may *outsiders* permissibly do? When and why does tyranny authorize foreign intervention?

In this chapter I confine the discussion to the moral permissibility of *military* intervention to protect persons against their own governments – humanitarian inter-vention, for short. Important as they are, for reasons of space I will not discuss here the morality of non-military diplomatic measures that governments may adopt to respond to tyranny, although the same basic principles I propose here apply, *mutatis mutandi*, to non-military diplomacy as well.

Just Cause: Humanitarian Intervention as Defense of Persons

A humanitarian intervention is a species of war. Let us start, then, with the legitimate *causes* of war. Any use of violence maims, kills, and destroys, and for that reason it is presumptively prohibited. Because war is the most terrifying and destructive form of violence, the presumption against it is particularly strong. War is in principle prohib-ited; justified wars are the exception. Humanitarian intervention, if it is to be justified at all, must be a response to serious wrongs.

Let us start with a generally accepted *casus belli*: international aggression. The war to repel an aggressor is a justified war in national self-defense. While the contours of

Contemporary Debates in Applied Ethics, Second Edition. Edited by Andrew I. Cohen and Christopher Heath Wellman.
© 2014 John Wiley & Sons, Inc. Published 2014 by John Wiley & Sons, Inc.

self-defense are not easy to draw, international lawyers and philosophers agree that defensive wars are justified. But why? The natural reply is that citizens who have been attacked are defending *themselves and their compatriots* against the aggressor. Now consider humanitarian intervention. Just as national self-defense is justified as a defense of persons (myself and my compatriots) against an aggressor, so humanitarian intervention is justified as a defense of persons (foreigners) against their own government. Self-defense and humanitarian intervention have the same rationale. While there are important differences between them, it is misleading to consider self-defense as a defensive war and humanitarian intervention as an offensive war. Both wars in self-defense and wars in defense of others (humanitarian interventions) are wars in defense of persons. The main difference between the two is that when the country is attacked the government has a *duty* to fight, grounded in a fiduciary obligation to its citizens – after all, one of the *raisons d'être* of government is to defend us. A humanitarian intervention, on the contrary, generates in principle only a *permission* to act, because the intervener is not bound to foreigners in the same way and thus cannot lightly impose costs on its citizens to save others. (However, if the cost of intervention is very low and the rights violations very severe, then the permission may morph into a moral *obligation* to intervene. For example, some observers thought that the failure of Western powers to intervene in the Rwandan genocide of 1994 was a moral failure.)

Critics of humanitarian intervention think that national borders are morally significant barriers against intervention. But it is hard to see why this should be so. Consider the following hypothetical examples.

Genocide in Rodelia   The provincial government of a federal state, Rodelia, has unleashed its troops against an ethnic group within that province. Casualties mount rapidly. The Rodelian federal government sends federal troops to the province to stop the genocide. The provincial army resists, however, and a civil war follows. After several months of fighting with significant military and civilian casualties, the Rodelian federal troops subdue the rebels and save the victims.

Revolution in Andinia   A military junta in Andinia has overthrown the democratic government. It has indefinitely suspended constitutional liberties and started an aggressive persecution of dissenters that includes torture and summary executions. Democratic forces regroup and take to the streets to fight the junta. After several months of fierce fighting, with significant military and civilian casualties, the revolutionaries succeed in deposing the régime and restoring the liberal constitution.

Most people would endorse the use of force in these two cases. Yet, in the Rodelia case many will object if *foreign* armies invaded to stop the atrocities in the Rodelian province. (Imagine, for instance, that the Rodelian federal government is unable to act.) And in the Andinia case they will likewise object to *foreign* troops aiding the liberal revolutionaries against the junta. In both cases, non-interventionists will oppose foreign involvement even if everything else remains the same, that is, even if the number of casualties and other destruction caused by the violence would remain constant. In fact, non-interventionists would likely oppose intervention even if foreign interventions would predictably cause more good than harm (reduce casualties and so forth). Yet there is no moral difference between the internal political violence in these imaginary cases and the foreign interventions having identical purpose and effect. The rationale for opposing intervention in these cases is hard to see. If opposition to humanitarian intervention is not grounded in state sovereignty but on other factors such as the impermissibility of killing innocents, then the non-interventionist cannot justify her endorsement of the use of internal force to stop genocide, or liberal revolutions to depose tyrants, because in these instances innocent bystanders also die. There is no relevant difference between the national army rescuing victims of genocide and a foreign army doing the same thing. Similarly, if a revolution is justified, foreign aid to them is too, *if* everything else remains equal. Indeed, foreign aid might be the revolutionaries' only hope for liberation.

The Question of Legitimacy

Some have suggested that the concept of legitimacy plays a crucial role in justifying or condemning intervention (Tesón, 2005; Altman and Wellman, 2010, p. 78). On this view, humanitarian wars are impermissible against *legitimate* regimes precisely because those regimes are legitimate, and permissible (sometimes) against *illegitimate* regimes precisely because those regimes are illegitimate. It is true that only intervention against illegitimate regimes is (sometimes) permissible. But the statement identifies the wrong reason for permissible intervention. Permissible intervention does not aim at restoring political legitimacy, but at ending or preventing impermissible violence against persons. All states perform acts of coercion, acts of violence, against their citizens in their territories. Some of those acts, let us assume, are morally justified and some are not. We may say, perhaps, that coercive acts of the state that are consistent with the moral rights of subjects are justified and coercive acts of the state that are inconsistent with those rights are unjustified. Any intervention to frustrate justified acts of coercion will be impermissible (even non-military interferences against such acts will be impermissible, let alone military ones).

For a humanitarian intervention to even begin to be justified, then, it must be aimed at unjustified acts of state coercion. However, not all unjustified acts of coercion constitute *casus belli*. Governments perpetrate mild wrongs and serious wrongs. A humanitarian intervention is, in principle, justified only to end or prevent the most seriously wrong acts of coercion perpetrated by governments. This is so, not because states have a "right" to perform mildly wrong acts of coercion, but because war to redress those mild wrongs will often be disproportionate. For this reason, whether or not governments are legitimate (see, e.g., Chapter 27 in this volume, Van der Vossen, 2014, pp. 408–411]) is irrelevant for purposes of justifying humanitarian intervention. Intervention is impermissible against any act that is not seriously wrong *if*, as is almost always the case, the military intervention will impose significant costs. For the same reason, that a state is internally illegitimate (on whatever standard of legitimacy one chooses) is insufficient reason to intervene, as the prospective intervener must comply with the strictures of the proportionality principle as well.

Let us consider first states that are illegitimate, judged under some standard of substantive justice. These states perform unjustified acts of coercion. But here again, it will often be the case that a military intervention would be so costly as to be disproportionate or counterproductive. When the military invasion is likely to be costly in terms of blood and treasure, it will be permitted only to end or prevent serious assaults on persons – seriously wrongful acts of state coercion – because only in those cases will the intervention be proportionate; only then will the costs of war be justified. The interesting case is one where the government has rendered itself guilty of less extreme rights violations. Consider the case of a military junta that has taken power by undemocratic means and has suspended constitutional guarantees – perhaps to prevent revolutionary activity. Let us assume that this undemocratic seizure of power violates the moral rights of the subjects, and (to avoid complications) that this junta has little support in the population. Now suppose that the democratic forces can secure the help of a powerful neighbor, and suppose further that the mere border-crossing by the foreign army will predictably cause the junta to surrender. It seems to me that this military intervention is morally justified, because the military invasion will *not* impose unacceptable costs (since the junta will surrender without resistance), *and* the violations of rights are relatively serious (although not egregious). The intervention, in other words, is proportionate. So the general proposition that only severe tyranny justifies humanitarian intervention (Tesón, 2005) will be true in most cases, but it is not strictly correct. The proposition has a hidden premise: that in less serious cases the intervention will impose unacceptable costs, as wars are prone to do. But if the military intervention, as in the example, is not costly, then the threshold of justification, in terms of the gravity of the cause, is lower. So the answer to the question, "Is it justified to intervene by force to restore democracy?" is that it depends. If the intervention will kill many innocents, destroy vast amounts of property, or will likely make things worse, the answer is no. If, on the contrary, the intervention will not have those dire consequences, then the answer may be yes. I hasten to add that the epistemic barriers faced by prospective interveners in calculating the likely costs of the war are such that perhaps the international norm should confine the permissibility of intervention to the really serious cases of tyranny.

Now consider legitimate states. Suppose a state is considered legitimate, again, on some unspecified standard of substantive justice, such as general compliance with human rights. These states also perpetrate unjustified acts of coercion – usually on a minor scale, and usually less serious. These illegitimate acts of coercion, these violations of the moral rights of persons (say, incarceration of morally innocent people or systematic governmental acts of theft) are *not* deserving of protection. If someone could press a magic button and stop the rights violations without any collateral cost, then that "intervention" would be morally permissible. But because *military* humanitarian intervention is extremely costly, it will never be allowed against the impermissible acts of coercion of otherwise (presumed) legitimate states. For one thing, these states have reasonable avenues of redress against rights violations. More generally, these states have valuable institutions that, on the whole, allow persons to pursue their personal projects, and, for that reason, violence against them, even if animated by a just cause such as ending illegitimate acts of coercion, is banned.

We can see, then, that the concept of state legitimacy does not do any work in the justification of humanitarian intervention. Contrary to some suggestions (Copp, 1999; Rawls, 1999; Van der Vossen, 2014, pp. 405–408 [Chapter 27 in this volume]), states do not have any *rights-based* shield against foreign intervention directed at ending their wrongful acts of coercion. States do not have a general right to rule, that is, a right to rule *beyond* what the rights of the subjects (one could say, the social contract) allow. All the work is done by the requirement of proportionality in war. War is justified *only* to end serious and systematic rights violations, that is, to protect persons against attacks by their own governments, because otherwise war in most cases would be disproportionate. If the rights violations are less severe, the principle of proportionality indicates less severe remedies. Because often war is the most severe remedy in terms of blood and treasure, it must be reserved to redress the most urgent situations. To see this, imagine yet another even less severe situation. Imagine a state that harasses political dissidents; for example, it closes newspapers that are critical of government, and it persecutes dissidents with phony charges of tax evasion and the like. Outsiders (say, liberal democracies) are perfectly authorized to put diplomatic pressure, even coercive pressure such as economic sanctions, on this regime to stop these rights violations, but are not authorized to invade – unless, as already indicated, the invasion itself will be costless (which is unlikely in such a scenario). The reason for the prohibition on invasion is not (as conventional wisdom would have it) that sovereignty shields non-egregious governmental misdeeds. The reason is that a military intervention would be patently disproportionate.

Sovereignty and Culture

Someone may object that in their path toward deposing bad regimes, interventions destroy political cultures. The principle of state sovereignty not only protects governments: it protects cultures as well. The problem with humanitarian intervention is that it violates sovereignty in this other sense. The idea is that there is something valuable in confining political processes to the citizens of the state, and that the principle of sovereignty protects precisely this collective autonomy (Walzer, 1980). Foreigners who use force to alter these processes are disrespecting the citizens of the target state.

Let us start with the idea that within the state there is something worth protecting against foreign intervention. However, if the government has turned against its own citizens, that is, perpetrates seriously wrong acts of coercion against them, what justification could there be for prohibiting the victims from securing foreign rescue? Some authors have advanced what I will call the *Popular Will argument*. Intervention is banned because a substantial number or a majority of the population oppose the intervention. This opposition will likely translate into resistance, and it is this expectation of resistance that makes intervention wrong (Walzer, 1980).

The Popular Will argument, however, is fatally flawed. To see why, let us first disaggregate the state where the violations occur. We have (roughly) three parties: the tyrant (joined by his henchmen and collaborators), his victims, and the bystanders. I concede at once that the intervener should not try to rescue victims from tyranny if they do not want to be rescued. But this principle identifies the victims, and the victims alone, as those whose consent matters. Others (collaborators and bystanders) are not entitled to veto the intervention on behalf of those who are victimized by the regime. Collaborators are accomplices in a crime, and bystanders, while they are not accomplices, do not have standing to resist attempts to rescue the victims. None of those groups has a valid communal interest in the tyrannical governance of their community. If they resist the intervention, they will be fighting an unjust war: a war on behalf of tyranny. It is simply unacceptable to vindicate these persons' desires to oppress their fellow citizens. Under any plausible theory of democracy, tyranny is not one of the things a majority can impose on a minority. While it is possible that, all things considered, the intervention predictably will have unacceptable costs, the reason to criticize the intervention will not be that those who resist the invader are protecting something valuable, but rather that the overall consequences of intervening are unacceptable.

Of course, it is always a difficult question whether any one intervention can pass this proportionality test. However, the objection that foreign intervention violates state sovereignty has some force when the well-intentioned intervener (often unwittingly) significantly changes *legitimate* social arrangements. When citizens of the state have cemented their social relations through cumulative processes over time, the resulting institutions are prima facie worthy of respect by foreigners, provided the institutions are morally justified. One can think of sovereignty as protecting non-oppressive social structures from foreign interference. To this extent, and to this extent only, the sovereignty argument against intervention is plausible. Foreigners do not have the right to change non-oppressive political structures.

By the same token, cumulative historical processes do not deserve respect if they are oppressive. I am thinking here, not of oppressive regimes, but of social institutions and practices that originated from the bottom up, as it were, but are unacceptable from the standpoint of justice. For example, the historical subjugation of women in some cultures is not acceptable from the standpoint of justice, so the objection that a foreign intervention will alter *that* practice is unavailable to those who oppose the intervention. It does not follow, of course, that military intervention is justified *just* to reverse those oppressive social practices. But the critic of intervention cannot point to those oppressive practices as valuable elements of society that (he thinks) the intervention will upset. Sovereignty does not protect those.

Legitimate states, then, are morally protected from intervention, not because they are legitimate (whatever the standard of legitimacy may be), but because a war against those states will always be disproportionate. However, the illegitimate acts of coercion committed by otherwise legitimate states are *not* protected against non-military forms of diplomacy. Legitimate social structures are themselves protected even in illegitimate states. Illegitimate states – here meaning *tyrannical* states – are not protected, but again, this does not mean that intervention against any one of those states is necessarily permissible. Also, if unaided revolution against a tyrant is justified, then aided revolution is also justified, provided that the other factors remain constant. If those factors do not remain constant (for example, if the foreign intervention predictably will greatly increase civilian casualties), then it may be unjustified under the doctrine of proportionality and the revolutionaries would have to fight alone.

Proportionality and Double Effect

But when are humanitarian interventions permissible all things considered? Assuming just cause in the sense described, the main reason to oppose a particular humanitarian war is that it is likely to be disproportionate. The cost of the intervention in terms of deaths (especially of innocents) may be unacceptable, or the intervention may be counterproductive in other ways – for example, it may give rise to evils (a new tyranny or anarchy) that are worse than the evil the intervention suppressed. For example, people have condemned the 2003 Iraq war, not because it was wrong to depose Saddam

Hussein, but because the invasion started a complex and violent causal chain with terrible consequences (deaths of many innocents, widespread destruction). But suppose the Coalition would have been able to depose Saddam with no adverse consequences – surgically, as it were. Surely the war would not have been so widely condemned. This raises the question of proportionality. The general idea is that even when just cause is present, a military intervention may be impermissible because of its bad consequences. Just cause is a necessary but not sufficient justification for war. This is why we should be reluctant to encourage any war, even when it has a just cause and the warriors harbor genuinely noble purposes.

The idea of proportionality is highly complex, because it cannot be measured simply in terms of costs and benefits. Thus, an intervener who wants to overthrow a tyrant cannot permissibly achieve that justified aim by rounding up and murdering innocents, even if that action would cause fewer deaths than allowing the tyrant to continue in power. Something more sophisticated is needed, and the just war literature has offered the Doctrine of Double Effect. Space prevents me from discussing in detail this complex idea, but here is a proposed summary.

A humanitarian intervention will be justified if, and only if:

1 The intervener uses permissible means of fighting (e.g., permissible weapons).
2 The commanders of the intervener do not directly intend the deaths of innocent persons. If they foresee those deaths, they must try to minimize them, even at some cost to themselves. The intervening forces must transfer some risk to themselves.
3 It has a just cause. Only the defense of persons is a just cause. Put differently: only the act of stopping or preventing serious violence against persons is a just cause. (This clause is strictly circumscribed in sections 5 and 6 below.)
4 The intervention is materially conducive to the realization of the just cause. (This rule condemns cases where the intensity of the intervention is unnecessary to the realization of the just cause.)
5 The degree to which the intervention is materially conducive to the realization of the just cause is great enough to compensate for the costs of the war, in particular collateral deaths. (This rule establishes a requirement to weigh harms and benefits.)
6 The just cause mentioned in (3), ending tyranny, must be compelling enough, in a moral sense, to compensate for the costs of the war, in particular collateral harm. This condition recognizes that there are degrees of moral urgency, so that not all just causes will justify collateral harm. The more compelling the cause, the lower the threshold for collateral harm. (For example, stopping genocide is more compelling than restoring a deposed democratic government; therefore, the threshold for intervention will be higher in the latter case.)

The Question of Intent

Many believe that a humanitarian intervener must have an altruistic intent: saving the victims. If he does not, if the intervener acts for selfish reasons (such as economic or political gain), then the intervention will not be justified. It is unclear that humanitarian intent should be a requirement of legitimacy; it is possible to argue that only outcomes matter. But let us assume that humanitarian intent is required. When does a government have humanitarian intent? The proof of intent is hard to come by. Lawyers think that intent is revealed by what the intervener says: if the intervener announces that it intends to liberate victims of genocide, then that is its intent. On close examination, however, this performative theory of intention, that what matters is what governments say they are doing, is untenable. Simply put, governments may lie about why they are doing what they are doing, or they may be mistaken about why they are doing what they are doing and about which rule, if any, is available to justify their behavior. Whether a military action is justifiable as humanitarian cannot depend on the government saying so.

So we are left with real, not expressed, intent. Following John Stuart Mill (Mill, 1998, p. 65, n.2) I propose to distinguish between *intent* and *motive*. If a government invades to stop genocide and stops it, then it had the requisite intent, even if it had a further (non-altruistic) motive. In other words: even if right intent should be a requirement for the legitimacy of humanitarian intervention, the presence of non-altruistic motives does not invalidate the goodness of the act of rescue. A selfish motive allows us to criticize the *intervener*, the person who undertakes the intervention, but it does not affect the goodness of the act, the act of rescue. It is perfectly possible to say that India's intervention in Bangladesh in 1971 was good because it saved the Bengalis from genocide, but that the Indian government itself was open to criticism for harboring self-interested motives (such as, perhaps, the desire for regional hegemony). The intervener, then, may intend the rescue as an *end* in itself (in which case intent and motive coincide) or as a *means* for a further end. In either case, the legitimacy of the intervention will be judged by the principles of proportionality discussed above.

Who May Intervene?

Here I examine a different family of reasons not to intervene: those relating, not to the target state, but to the intervener. There are two issues. The first is whether a liberal government may validly use the state's resources (troops and taxes) to liberate victims of tyranny in other societies. The second is whether or not it is desirable to require an international authorization for any humanitarian intervention. I will discuss each in turn.

The internal legitimacy of humanitarian intervention

A government, we saw, has a fiduciary duty toward its subjects. This includes the obligation to respect the moral rights of persons at home. But because morality is universal and all persons have rights, governments also have an obligation to respect those rights abroad. Governments may not treat foreigners as things, nor can they collude with other governments in the oppression of their citizens. Thus governments have an obligation not to cooperate with tyranny. This purely negative obligation can be reinforced by adding a softer obligation to promote good institutions (that is, rights-respecting institutions) globally, provided of course that this can be done at reasonable cost (Buchanan, 1999).

These principles for a liberal foreign policy yield the following consequence for the ethics of intervention: if promoting human rights globally includes sometimes (rarely, to be sure) saving foreigners from severe tyranny, then the government can *tax* citizens for that purpose. If the government in a liberal state can tax citizens to contribute, say, to the establishment of a human rights court, then it can tax citizens to support a voluntary army to free foreigners from tyranny. In both cases the government is using the citizen's economic resources (taxes) to improve the lives of others. In neither case is the government literally forcing people to defend others. So both cases are morally indistinguishable from the standpoint of the taxpayer. It may well be that neither taxation is justified, of course. The point here is that taxing to finance (say) a human rights court and taxing to finance a humanitarian intervention entail the same demands on the taxpayers. The degree of intrusion into the citizens' freedom is the same. Humanitarian intervention is not a more egregious case of governmental demand on the citizens' resources. (I set aside further complications here, such as the problem of taxing persons to finance activities, such as wars, that they consider immoral. This difficulty can be overcome only if one accepts that a liberal government may implement sufficiently compelling principles of global justice, just as it can overrule citizens' moral objections internally in the implementation of domestic justice. The argument here is agnostic on that question.)

The more pressing question is whether the government can validly send people to fight for the freedoms of foreigners. Assuming an intervention is otherwise justified, *who* may the government send to fight? There are four theoretical possibilities: spontaneous private brigades, forcibly conscripted army, voluntary army, and mercenaries.

Volunteers may form private brigades to invade the country and end tyranny there. This possibility raises a host of questions about the permissibility of private wars. However, in theory a private brigade is a satisfactory solution to this problem (i.e., whether the government can legitimately use the citizens' resources to liberate foreigners) because the government does not force anyone to do anything: the privateers voluntarily undertake the intervention. The government does not even tax anybody. As a matter of principle, it is arbitrary to require that their rescue mission should be approved or endorsed by their government. It is true that international law places liability on the state of the privateers, but perhaps this is yet another state-centered feature of international law that requires revision. Nonetheless, for all their advantages, relying on privateers is likely to be highly inefficient for obvious reasons. The coordinating function of the government plays a decisive role in military issues, at least as long as the world is populated by the current kind of nation-states.

Conscription is a serious violation of personal freedom, and it is especially objectionable to forcibly enlist citizens to conduct a humanitarian intervention. To see why, let us start with self-defense. The government forcibly enlists persons and orders them to fight under threat of grave harm. Such intrusion can be justified, if at all, only under the most extreme of circumstances. Let us assume that the country is invaded by an evil and ruthless enemy. Our democracy and everything we value will collapse unless we can succeed in our defensive war. Someone may argue that the government may never coerce persons into fighting, not even in these extreme cases. If defensive force is merely permitted, not obligatory, the victim of an attack is free to decide whether he will fight for his life, property, or freedom. He may choose not to fight and submit instead to the aggressor. Others (the government especially) cannot coerce him into combat.

If the government cannot coerce a person to fight in his own or his fellow citizens' defense, a fortiori the government cannot force people to defend foreigners. This libertarian argument, then, contends that the only legitimate collective force occurs when citizens rise spontaneously against the aggressor. Those who choose not to fight are within their rights and should be left alone.

Perhaps this argument is right, and no conscription is ever justified, even in these extreme cases. But even conceding, for the sake of argument, that conscription can be justified for national self-defense, it cannot be justified for humanitarian intervention. Let us consider how conscription for national self-defense may be justified. The argument rests on the fact that national defense is a public good. If people are allowed to choose individually whether they should contribute to national defense against the kind of serious threats we imagine, they will be tempted to free-ride on the defense efforts of others. National defense is vulnerable to market failure: everyone wants to repel the aggressor, but each one hopes that others will risk their lives to do so. Because everyone reasons similarly, the public good (defense) is under-produced and the state succumbs to the aggressor. One can, of course, recast this argument in normative terms: we have sometimes, in extreme circumstances, a moral duty to defend our fellow citizens and the social contract. This argument, then, reluctantly accepts that the government may in extraordinary circumstances use the military draft to defend the state against attack.

However, this argument does not support the legitimacy of conscription for humanitarian intervention. The public-goods argument that served as a justification for military conscription for self-defense does not work as forcefully in the case of a war to defend distant others. The argument depends on the assumption that the good in question is demanded by a sufficient number of people. Because the demand for national defense is likely to be strong, conscription is arguably needed to eliminate free riders. Those who object to national self-defense, where they face their own destruction, are hardly credible: they are likely to be free riders. But, while humanitarian intervention may also be a public good in the sense that allows for opportunistic moves *ex post* (people who would agree *ex ante* to intervene would refuse to fight once the dangers are known), the demand for humanitarian intervention by the public will surely be much weaker than the demand for national defense. There will be genuine objectors who are not, by definition, opportunistic agents. Therefore, a liberal argument must balance respect for these genuine dissenters against the moral urgency to liberate victims of foreign tyranny. The duty to promote global human rights is weaker than the duty to support one's nation's own legitimate self-defense: it must cohere with other moral-political considerations, such as the need to respect non-opportunistic exercises of individual autonomy. It follows that persons have a moral veto to the attempt by the state to force them to fight and risk their lives to save foreigners.

If this analysis is correct and conscription is excluded for humanitarian interventions, the two alternatives left are voluntary armies and mercenaries. Voluntary soldiers have contractually consented to fight in wars, the justice of which is decided case by case by their employer, the liberal government. Voluntary soldiers have agreed to fight in cases where the government believes there is a sufficiently grave reason to fight; this includes humanitarian interventions. The advantages of a voluntary army are well known. For one thing, the market is likely to select the better fighters. Also, a voluntary army mitigates the problem of conscientious objection to particular wars, because the voluntarily-enlisted soldier has contractually authorized the government to decide on the justice of particular wars.

Another solution is to hire mercenaries. Mercenaries are private entrepreneurs who offer military services for a price. As long as they have existed (since Antiquity), mercenaries have been the target of scorn and contempt. However, an unprejudiced look at the issue reveals that most of the reasons for this hostility are questionable. It is argued, first, that killing for money is morally wrong (Pattison, 2009). But surely enlisted soldiers kill for money too: it is their profession. It is not altogether clear what the mix of self-interested and altruistic motives is in either group. Arguably, enlisted soldiers risk their lives not so much (or not only) for love of country as for solidarity with comrades. If this is true, then it seems equally true of mercenaries, who presumably take pride in what they do and feel solidarity for their comrades also. The point here is that monetary compensation is one element in a richer range of motives, many of which are self-interested. Enlisted soldiers receive salary, benefits (including generous educational benefits), prestige, and social esteem. All of these motives are self-interested, yet a romantic tradition has emphasized instead the honor and glory of fighting for one's country. Because this altruistic motive seems to be lacking in mercenaries, they are supposed to be bad people. But once we disaggregate motivation in both cases and we see the patriotic prejudice for what it is, the two cases do not seem that different.

Another reason offered against mercenaries is that using them will weaken the communal bonds in society (Pattison, 2009). But communal bonds are not very useful if the country loses the war. Imagine a country that is fighting a just war against a powerful enemy. At one point the voluntary army becomes insufficient. Should then the government conscript soldiers or hire mercenaries? The answer is not clear at all. Conscription, we saw, is highly objectionable because it intrudes massively into people's liberty. Hiring mercenaries solves that problem. Moreover, mercenaries are professional soldiers who will presumably increase the chances of victory, so they are preferable for efficiency reasons as well. The hostility to mercenaries is the result of the patriotic prejudice: that it is honorable to fight and die for your homeland, and that this noble motive should not be tainted by the profit motive. But as we said, the enlisted soldiers are also fighting for a host of motives, many of which are self-interested, and there is no reason to think that they are nobler persons just because they become a part of the state's bureaucracy, as opposed to being mere contractors.

The problem of authority

Assuming humanitarian intervention is sometimes permissible along the suggested lines, should authorization by an international organization (such as the United Nations Security Council) be required? Or are unilateral (that is, unauthorized) humanitarian interventions acceptable? Any procedure must avoid over-intervention and abuse, on the one hand, and under-intervention and inertia, on the other. That is a daunting challenge in institutional design. Recent experiences show that both dangers are real, and, alas, the victims of such failures will often be the most vulnerable persons. Authors have proposed various alternatives, in addition to authorization by the UN Security Council: authorization by an alliance of democracies (Buchanan and Keohane, 2004), or by an independent committee of experts (Tesón, 2006).

Contrary to convention (International Commission on Intervention and State Sovereignty, 2001), I suggest that if the intervention is substantively justified a state may permissibly intervene without anyone's authorization. The general reason is that, unfortunately, war cannot be completely eradicated as a form of self-help in a semi-anarchical world. Each sovereign state establishes the monopoly of force within that state. States have police, courts, and armies, and individual violence is narrowly confined. The modern state outlaws interpersonal violence by monopolizing force. It prohibits both opportunistic plundering and private retribution, while providing means of redress to those wronged by the aggressive behavior of others. The international society, however, is not a state. Its central feature is precisely that it does not in fact establish a global monopoly of force on any higher sovereign. At its inception, the United Nations Charter was meant to create a system of collective security that would replace unilateral wars, but its weaknesses are well known. Suffice it to say that international law does not provide a satisfactory remedy for wronged nations, groups, or persons. To be sure, the international society has made great strides in developing norms to regulate war. The United Nations Charter reads as if it empowers the Security Council to respond to humanitarian crises. Yet the mechanisms available to prevent or end those crises are woefully weak. Before the United Nations Security Council authorizes the use of force against a criminal regime and a government is willing and able to act on that authorization, many improbable things have to happen. For one thing, the five permanent members of the Security Council have the right to veto. This means that the Security Council will not authorize force against permanent members or their allies. (As I write these lines, the tyrannical Syrian regime is killing people with impunity, as the Security Council is paralyzed by the veto of Russia and China.) As important, while domestic law *obligates* the government to act in response to private violence, international law (in the best scenario, when the United Nations Security Council functions properly) merely *authorizes* governments to respond to major humanitarian crises. Assuming, against the facts, that the Security Council acts expeditiously in every case of tyranny and authorizes the use of force to address them, governments are still not bound to react. Note that only a powerful state can act. But the powerful state will not act unless its government believes the national interest to be at stake. (This, again, is a rosy scenario: politicians will go to war if *their* own interests are thereby advanced – especially their electoral interests.) States in a position to respond to tyranny can simply decline the invitation. In short, there is no effective global mechanism to prevent war, and as a result self-help is omnipresent in international relations. Given these institutional failures, and the sobering fact that states and others will continue to murder their own citizens, humanitarian intervention cannot be excluded as the only way, sometimes, to save people.

References

Altman, A. and Wellman, C.H. (2010) *A Liberal Theory of International Justice*. Oxford: Oxford University Press.

Buchanan, A. (1999) The internal legitimacy of humanitarian intervention. *The Journal of Political Philosophy* 7: 71–87.

Buchanan, A. and Keohane, R. (2004) Governing the preventive use of force. *Ethics and International Affairs* 18: 1–22.

Copp, D. (1999) The idea of a legitimate state. *Philosophy and Public Affairs* 28: 3–45.

International Commission on Intervention and State Sovereignty (2001) *The Responsibility to Protect*. Ottawa, Canada: International Development Research Centre.

Mill, J.S. (1998) *Utilitarianism*, ed. R. Crisp. Oxford: Oxford University Press.

Pattison, J. (2009) Deeper objections to the privatisation of military force. *The Journal of Political Philosophy* 18: 425–447.

Rawls, J. (1999) *The Law of Peoples*. Cambridge: Harvard University Press.

Tesón, F.R. (2005) *Humanitarian Intervention: An Inquiry into Law and Morality*, 3rd edn. Ardsley-on-Hudson, NY: Transnational Publishers.

Tesón, F.R. (2006) The vexing problem of authority in humanitarian intervention: a proposal. *Wisconsin International Law Journal* 24: 761–772.

Van der Vossen, B. (2014) The morality of humanitarian intervention. In *Contemporary Debates in Applied Ethics*, 2nd edn, ed. A.I. Cohen and C.H. Wellman, pp. 404–416. Malden, MA: Wiley-Blackwell.

Walzer, M. (1980) The moral standing of states: a response to four critics. *Philosophy and Public Affairs* 9: 209–229.

Further Reading

Dobos, N. (2012) *Intervention and Revolution*. Cambridge: Cambridge University Press.

Holzgrefe, J.L. and Keohane, R., eds (2003) *Humanitarian Intervention: Ethical, Legal, and Political Dilemmas*. Cambridge: Cambridge University Press.

Van der Vossen, B. (2012) The asymmetry of legitimacy. *Law and Philosophy* 31: 565–592.

Walzer, M. (2006) *Just and Unjust Wars*, 2nd edn. New York: Basic Books.

CHAPTER TWENTY-SEVEN

The Morality of Humanitarian Intervention

Bas van der Vossen

State sovereignty has long been taken to pose a strong barrier against humanitarian intervention. It can be overcome, it is said, only by the need to stop a supreme emergency. Given the dominion of states over their internal affairs, and given the significant costs of military action, only such exceptional situations are thought to enable morally permissible interventions.

Over the past decades most philosophers have come to reject this view. According to them, not all states enjoy the rights of sovereignty, but only those states that satisfy the moral requirements of *legitimacy* enjoy the rights of sovereignty. On this view, states enjoy their rights (when they do) only because they stand in a morally important relation with their citizens, and this relation does not allow for human rights violations. This implies a significantly more permissive stance to the morality of humanitarian intervention. After all, states might be illegitimate even if there is no supreme humanitarian emergency happening and thus be without rights-based protections against intervention. That is not to say that intervention is morally permissible all things considered, but it does defuse a very important objection: that intervention would violate sovereignty.

No doubt, this view is a great improvement. It recognizes that human rights are the most important moral values around. And it recognizes that states are morally valuable only in so far as they serve the people. However, more recently philosophers have also started to question whether this revised view may still not take human rights sufficiently seriously. They suggest that states simply have no rights against interventions aimed at ending human rights violations at all. Instead, whether or not intervention is morally permissible depends only on whether it can be done effectively, without excessive risks, and so on.

How should we think about these issues? Are legitimate states morally protected against intervention only if they violate no human rights? If so, what would that mean

Contemporary Debates in Applied Ethics, Second Edition. Edited by Andrew I. Cohen and Christopher Heath Wellman.

for the morality of humanitarian intervention more broadly? Most importantly: how might we know? In what follows I hope to make some progress towards answering these questions. The first section, "Interstate Morality", outlines a framework for thinking about them. Then, the second section, "Legitimacy and Non-intervention", applies this framework to the question of the rights of legitimate states and the third section, "Intervention in illegitimate states", to intervention more broadly.

Interstate Morality

International morality at its core forms a body of regulatory norms for interstate conduct. The rules of morality are such that, when observed by states, they render international affairs just, peaceful, and prosperous for all. Candidate moral rules can be evaluated with an eye on this goal. That is, we can ask whether this or that moral rule, if accepted, would steer the actions of conscientious (though not infallible) agents in the right direction. Vice versa, moral rules that would, if generally followed by conscientious agents, predictably lead to suboptimal states of affairs ought to be rejected in favor of ones that work better. Moral rules, we might say, have a point; and unless they serve their point, they are no good.

To see the how moral rules have a point, consider a homely example: property rights. Suppose we think that the point of property rights is to further people's vitally important interests. What would be the extent of those rights? Should they prohibit or allow trespass? One suggestion might be that, given the point of property rights, all and only those actions by others that harm the interests of the possessor should be ruled out by the rights in question. So, trespass on a piece of land is prohibited unless it can be done without setting back any of the interests of its owner.

Would that be a compelling understanding of property rights? The answer, of course, is no. For one, such an understanding would leave individual agents without clear guidance about what to do. To determine whether they ought to respect the property rights of others, they would need to thoroughly investigate the likely effects of their trespass, the interests of those owners, and so forth. But given the difficulty of obtaining such information, and well-known psychological biases, such a rule would likely lead even conscientious agents to trespass even when the interests of the owner might be set back. It would fail to give owners the kind of genuine control over their property that they need, and thus fail to serve its point: protecting the interests of property owners.

The correct moral rules of property, then, should reflect the ways actual people are likely to act in real circumstances. And this means that they will have to be complex. Part of this complexity will be that property rights prohibit trespass even in cases where it might not end up setting back anyone's interests.

We can ask the same questions about humanitarian intervention and the rights of legitimate states. And just as asking these questions about property rights revealed that such rights are complex – and can prohibit even victimless trespass – so too, I will argue, the morality of intervention is complex – and can prohibit even victimless interventions.

To understand the morality of intervention, then, we must ask what its point is: what rules and rights would lead to a morally desirable international scene? One thing we can say is that these moral rules and rights must take into account structural features of the world to which they apply. Just as individual property rights should take into account the difficulty of obtaining information about the interests of owners, the biases of would-be trespassers, and other facts, so too the rules for intervention ought to take into account structural facts about states and state officials.

The content of international morality, in other words, must be sensitive to certain empirical facts. Among these are the nature of the agents that are to follow them, the decision-making processes such agents employ, and the contexts in which they find themselves. Unless international morality is sensitive to such facts, it runs the risk of leading even conscientious agents to act in ways that are counterproductive or even self-defeating, thus bringing about morally sub-optimal states of affairs.

This is still very abstract. Let us contrast what I am proposing here with a different approach, one in which the morality of interstate conduct is treated, by and large, as an extension of the morality of interpersonal conduct. Here, state actions are permissible if they avoid violating people's rights, do not bring about morally undesirable consequences, and so on. And we might conclude from the claim that a particular act would be permissible for an individual to perform that it must also be permissible for a state to perform.

Consider Christopher Wellman's proposal that "even a legitimate state has no principled objection to outsiders' intervening in its internal affairs if this interference will prevent just a single human rights violation" (Wellman, 2012, p. 119). His main argument relies on a thought experiment. Imagine, says Wellman, that he is unjustly convicted to a long-term prison sentence in a legitimate state. Setting aside for the moment the risk of harming or wronging others in the process, would he be permitted to escape? Of course he is. And this is so despite the legitimacy of the state. What if someone else, knowing all the facts, could safely set him free? Wellman answers that surely this too would be permissible. And this would be no different if his rescuer lived in a foreign country or were a government official. This thought experiment shows, says Wellman, that if foreign regimes can undo human rights violations without thereby causing any wrongful harm, even legitimate states have no right against it.

In a similar manner, Fernando Tesón constructs what he calls the Green Button Test (Lomasky and Tesón, forthcoming; see also Tesón, Chapter 26 in this volume). Suppose we could undo human rights violations in a foreign state by simply pressing a green button, without thereby violating any rights or causing other harms. Would pressing the button be morally wrong? Tesón says no. Does it matter if the state in question is legitimate? Again, no. But if this is right, then it seems that even legitimate states have no principled right against interference.

What are we to make of these arguments? Do they show that states have no rights against interference? The crucial issue here is whether we can infer from (a) the (true) claim that *individuals* are morally permitted to safely free unjustly imprisoned persons and press green buttons; that (b) *other states* would be permitted to do the same. The inference goes through only if there are no morally important reasons for treating states and individuals differently.

But of course there are many good reasons for treating them differently. To name just a few of the more obvious ones: individuals typically do not have the armies of states, individuals typically do not occupy the positions of great domestic power and privilege of states, individuals typically do not make their decisions like states do, and so on. These facts about states make it the case that, even if it would be morally permissible for individuals to press green buttons, it need not be permissible for state officials to do so. Given the nature of states, it is not at all clear that state officials conscientiously acting on such rules would bring about morally desirable states of affairs.

More on this below. For now, my point is that these facts about states ought to play a central role in our thinking about the morality of interstate conduct. They are features of interstate action that critically affect the ways in which the rules of international morality will lead even conscientious agents to behave. To put it bluntly: we do not live in a world of green buttons. In our world, when states intervene they almost always cause death, harm, destruction of property, and other forms of suffering. In our world, states are run by officials who are inevitably under mainly domestic political pressures. In our world, states are liable to retaliate against what they might reasonably see as uncalled for interventions. And so on. Asking what agents might permissibly do in a world of green buttons, then, is simply not helpful for understanding what they might permissibly do in a world like ours. Indeed, it might lead us to a set of rules that is so permissive as to become counterproductive, leading to avoidable human rights violations.

These crucial differences between states and individuals mean that the morality of actions by individuals (in their private capacities) may be different from the morality of actions by government officials (in their official capacities). When one acts as a government official, one occupies a role in a state institution, uses its coercive apparatus, and is subject to the pressures and biases that come with that. Our actions as state officials are importantly unlike those we undertake as private individuals. As state officials, we use different means and operate in a different moral context. Different rules are appropriate for evaluating such actions. And these rules must take into account the relevant facts about states, officials, and their decision-making.

The approach that I am proposing cuts across two related distinctions that are commonly made in discussions of intervention. The first is between moral rules as *guides* and *standards*. It is common to say that the facts to which I have been alluding should inform moral guides – how responsible agents should deliberate, or how good institutions and international law are to be designed – but are irrelevant to moral standards – which settle whether particular actions are prohibited, obligatory, or permissible. This is to miss the more fundamental point. For, as the analogy with property rights above shows, these facts also crucially inform the moral rules as such, the standards of morality. Morality itself is sensitive to structural facts that affect (individual or state) action in ways influencing the outcomes that conscientious agents will end up bringing about.

The second distinction is between *prudential* and *principled* reasons. States have prudential reasons not to intervene, for example, when intervention would be counterproductive or when there is significant risk of failure. The reasons to which I have alluded are often said to be merely prudential reasons, and not capable of supporting principled reasons against intervention, such as the rights of legitimate states against intervention. Again, however, this moves too quickly. The distinction between prudential and principled reasons must be drawn with great care. Given that international moral rules and rights are supposed to function as regulatory norms applying to state institutions, certain important features of states and international interaction as such should be reflected in the content of these rules.

None of this, of course, is to deny that moral rules can serve as guides, or that there is a class of truly prudential reasons. However, we need to distinguish between those reasons that are *merely incidental* to particular international decisions or acts states might undertake (Is a state's military apparatus capable of executing the task of ending a humanitarian crisis? Will the local population welcome or resist intervention?) and those that are *structural* features of the international scene. The former are genuine prudential reasons, and these can inform our moral guides while remaining outside our moral standards. But the latter, if sufficiently important, should be reflected in the content of our international moral rules themselves, and can therefore come to make up principled reasons against intervention.

Those rules that appropriately negotiate the direct effects of particular state actions and these structural features of international action, therefore, are the moral rules of international conduct. Some of these rules will quite straightforwardly prohibit acts that directly lead to human rights violations. But these rules do not, on this approach, exhaust international morality. Other rules prohibit acts because doing so is (indirectly) instrumental for protecting the human rights of all. All such rules must display the sensitivity to the important structural features of international (or better: interstate) action, including that: (a) states are coercive institutions, (b) interventions employ military means, and (c) state decision-making processes inevitably suffer from predictable biases.

Of course this is only the barest of sketches of the kind of approach I am proposing. In the next two sections I will try to explicate it further by applying the view to two questions concerning humanitarian intervention. Before turning to that, let me stress one important implication of this view: Moral rules and rights are necessarily relative to the nature of the context and agents to which they apply. The rights of states vis-à-vis other states may require different things (have different contours) than the rights of individuals vis-à-vis states, or the rights of states vis-à-vis individuals, NGOs, or international institutions. This is a result of the fact that different structural facts and features enter into the consideration of these questions. (This, of course, is merely the reverse of my point about the green button example: it does not follow from the claim that states should not push green buttons that NGOs or individuals should not either.)

Legitimacy and Non-intervention

The first question to address is whether legitimate states have rights against outside intervention. As we saw above, some now deny that even legitimate states enjoy rights-based protections against interventions. One might be attracted to this view because we said that states enjoy their rights of sovereignty, when they do, in virtue of protecting people's rights. But if that is the point of states, then why should they be protected against interventions aimed at undoing violations of these rights?

However, as it stands, this moves too quickly. We should keep separate what are the *grounds* of the rights of legitimate states with the *extent* of the protections they provide. And while it is true that the importance of human rights grounds the rights of legitimate states, this does not entail that any human rights violation licenses forceful action. We must address the question head-on, therefore. I will do so by considering, first, what it would mean if legitimate states had no rights against interventions aimed at ending wrongdoing whatsoever. And after showing that this proves unacceptable, I will argue that allowing interventions only to end human rights violations more specifically is no better.

Consider first, then, the possibility of legitimate states lacking all rights against interventions aimed at ending wrongdoing. The problem with this is that this makes the right of non-intervention of states objectionably trivial. It takes the right of non-intervention to protect only a state's actions that are morally permissible. This is trivial since, given that permissible actions are those that a state may, or is supposed to, undertake, these are actions with which outside parties will have very little reason to interfere. On this view, a state's "right of non-intervention" protects precisely nothing. When a legitimate state does what it is supposed to do, no right against interference is required. And when a right against intervention might be needed to block outside interference, no right exists. Surely the rights of legitimate states against outside interference are more robust than this, precluding certain acts that other states might otherwise undertake.

There are a number of reasons why such more robust rights against interference are important. Some of these presuppose the framework outlined above, some of these do not. Here is an example of the latter kind. Suppose that there are processes of collective self-determination that can take place within legitimate states, and that these processes are morally valuable. If so, then this value can support robust protections against outside interference. Self-determination is not valuable only when the outcomes it achieves are substantively just or correct. Its value can be present, and call for respect, even when wrongdoing is involved. Moreover, there can be additional value, of both instrumental and non-instrumental kinds, in a society (governed by a legitimate state) finding its own way towards a better state of affairs, without external interference (Tasioulas, 2010). The value of self-determination, therefore, can give outsiders principled reasons to refrain from intervening in legitimate states even when such interventions are aimed at ending injustice.

Other reasons of this kind can be given. However, I will set these aside and use the framework outlined above to formulate some additional reasons for a robust right of non-intervention. The first two of these are familiar, and for that reason I will not spend much time on them. The third is less familiar and will therefore be discussed more extensively. In line with the argument above, I will focus on structural features of state action and decision-making that tell in favor of a moral right of non-intervention for legitimate states against other states.

The first of these refers to the benefits that come from the existence of a generally accepted international norm of non-interference with legitimate states. There is evidence suggesting that a practice of non-interference between states that are recognized as legitimate fosters both peace among states and decent internal behavior (Clark, 2005). It is not hard to see why. Interventions involve the use of military force, inevitably imposing risks of loss of life, harmful social instability, potential retaliatory acts, damage to the legitimate state's ability to govern itself, and so on. Allowing interventions in legitimate states, then, not only opens the door to the many harmful effects of military action, it will likely harm reasonably well-functioning states.

Second, denying the right of legitimate states against intervention fails the test of what we might call incentive-compatibility. It does so at both ends of the intervention. On the one hand, the decision-making processes internal to intervening states are predictably subject to important biases. Political decisions about intervention reflect the interests of the rulers, citizens, and various pressure groups internal to the intervening state, much less (if at all) the interests of the people in the foreign country. Opening the door to more interventions, then, means opening the door to more actions where the interests of stable, militarily powerful states take precedence over the interests of everyone else.

On the other hand, domestic groups looking to overthrow or upset the legitimate government under which they live would come to face perverse incentives. Domestic groups looking to overthrow their government, for example, could be tempted to initiate or escalate violent conflict in the hope that this will trigger outside intervention on their behalf. This is to invite dangerous international instability and to imperil the prospects of success for weak but legitimate regimes.

Third, these problems are exacerbated if legitimate states have no right against interventions. For in these circumstances all decision-making about interventions is to be made on the basis of judgments about the *justice or injustice* of other states and their policies. But these judgments are prone to mistakes, especially if made from a distance, and deeply contested to boot. By contrast, judgments of state legitimacy, while by no means without problems, are less susceptible to this. Part of the difference here is due to the fact that in this context judgments of justice are scalar: we can easily view states as more or less just. If decisions about intervention are based on justice-judgments, then such decisions will come down to fairly fine-grained judgments about where to draw the line between acceptable government-action and injustice, or serious and not-so serious wrongs.

Judgments of state legitimacy are not similarly scalar. Or at least they are not if we understand them as referring to states having a moral right to rule that is consistent with (some) domestic wrongdoing. Such judgments of legitimacy are binary in nature (a state either has that right to rule, or it does not) and thus avoid some of the problems that plague justice-judgments. This character of legitimacy judgments follows, again, from their point. Legitimacy judgments help us overcome a particular kind of problem: how to coordinate action regarding states in light of widespread disagreement about justice (Buchanan and Keohane, 2006). Solving this problem requires two things. First, given that disagreement about justice is partly due to its scalar nature, legitimacy judgments must be different. Hence their binary nature. Second, to avoid the problems of justice-judgments, legitimacy must be error-tolerant. That is, states must be able to remain legitimate even when not perfectly just.

If legitimacy judgments are binary and error-tolerant, then conscientious persons can, despite disagreeing about the justice of a state, still coordinate their actions in support or resistance of it. Legitimacy judgments of this kind can function as benchmarks – signaling whether a state is good enough to support, or so bad it needs to be opposed – and thereby enable people with deeply different background moral, ethical, and religious convictions to coordinate their actions when they have reason to do so. And while people might of course also disagree about the legitimacy of states in any given case, the error-tolerance of legitimacy ensures that such disagreements will be less frequent.

Making the permissibility of intervention rely on judgments of legitimacy that are error-tolerant, therefore, has the potential of avoiding some of the worst problems mentioned above. Consequently, we have good reason to endorse moral rules according to which the permissibility of intervention depends on the legitimacy of states. In other words, we have good reason to endorse rules that hold legitimate states to have rights of non-intervention.

What, then, of restricting the proposal? Perhaps legitimate states have rights against all interventions except those aimed at ending violations of human rights? Would such a proposal avoid these problems? Unfortunately, the answer is no. None of the problems above turn on the severity or nature of the injustice that intervention is supposed to undo. Instead, they stem from the nature of state institutions (from which follow the general benefits of a norm of international non-intervention), the nature of political decision-making processes (from which follow problems of incentive-compatibility), and the problems of justice-judgments. The first two of these are not addressed by restricting intervention only to those that aim at undoing human rights abuses. Things are somewhat better concerning the latter: international human rights law provides a canonical statement of human rights that is widely regarded as authoritative, and this constrains disagreement. However, when we consider debates about so-called human rights inflation, the contentious distinction between basic and non-basic rights, arguments about Asian values, and many other issues, we can see that the idea of human rights is still deeply contested.

The correct conclusion, then, is the straightforward one: legitimate states have a right against outside interventions even in (some) cases of domestic wrongdoing, and even in (some) cases that involve human rights violations. Legitimate states can remain rights-protected against intervention even when they act in ways that are in an important sense wrong – indeed, beyond what their legitimate authority (or internal right to rule) permits or enables them to do.

Intervention in Illegitimate States

What, then, if we are dealing with an illegitimate state? When is intervention in such states permissible? Here the general permissibility conditions for the use of force apply. Reasonable non-violent measures achieving the same results must be unavailable, armed intervention must be proportional to the problem it aims to amend, and so on. However, many hold that there are also conditions particular to humanitarian intervention, such as that permissible intervention requires a severe humanitarian crisis (e.g. Wheeler, 2000).

In this section, I want to argue that this latter condition is too strict. In building my case, I will depart from parts of J.S. Mill's work on intervention (Mill, 1859). Mill's views are worth considering because they capture an important and often overlooked reason to be suspicious of interventionist policies. Employing the framework outlined above, I will argue that we can both honor Mill's central insight and relax the overly strict supreme humanitarian crisis-condition.

Mill opposed almost all interventionist policies, at least among what he considered "civilized" nations. Most controversial to modern readers is his prohibition of all interventions aimed at ending oppression by domestic powers. No matter how grave the oppression, Mill rejected intervention in these cases because he thought that freedom (the desired outcome) could not be achieved in this way. Freedom, he argued, cannot be brought about by outsiders, but must be earned by the people themselves:

[W]hen freedom has been achieved *for* them, they have little prospect indeed of escaping this fate [of despotism]. When a people has had the misfortune to be ruled by a government under which the feelings and virtues needful for maintaining freedom could not develop themselves, it is during an arduous struggle to become free by their own efforts that these feelings and virtues have the best chance to spring up. (Mill, 1859, pp. 122–123)

This argument raises some obvious questions. Can the virtues requisite for freedom really not be taught without bloodshed? Is freedom not good enough to be desirable to people who have known only oppression? Must outsiders really choose between taking over an entire country and doing nothing at all? Mill's view seems to lead to implausible answers to each of these. After all, we do know of successful nation-building, such as the postwar construction of Germany and Japan. And interventions can also support internal forces of resistance.

However, we can sidestep these problems by looking for another (and more charitable) reading of Mill's remarks. On this reading, Mill is making an important point: Free institutions require broad support in the population. This is suggested by Mill's explanation for his views, namely that "[n]o people ever was and remained free, but because it was determined to be so; because neither its rulers nor any other party in the nation could compel it to be otherwise" (1859, p. 122).

This point seems plausible enough. It is extremely difficult for free and legitimate institutions to be created and it is very difficult for these to be imposed from the top down. For such institutions to have a chance at working and surviving, they need to be supported by, and correctly align with, highly complex networks of formal and informal rules, norms, and shared expectations across civil society. These background norms, at least as much as the formal legal checks, are what actually prevent ruling elites from abusing their powers. These, at least as much as institutional design, are what make it possible for individuals to enjoy genuine freedom. After all, we are unfortunately all too familiar with unfree societies governed by what are formally free institutions. As Mill puts it, "unless the spirit of liberty is strong in a people, those who have the executive in their hands easily work *any* institutions to the purposes of despotism" (1859, p. 122).

The need for informal support for free institutions explains why examples of successful nation-building are, *pace* Germany and Japan, so hard to come by. Free institutions are typically created from the bottom-up; they are phenomena that emerge from the complex interplay of largely unpredictable social forces. Mill's point, then, need not be that these must be earned the hard way, but rather that top-down impositions of such institutions are too unlikely to succeed to warrant taking the risks of intervention. Unless there is domestic support for post-intervention free institutions, intervention will not be justified.

It is significant in this regard that the examples of Germany and Japan are highly exceptional. A critical element of the postwar success was the already highly developed political cultures of these countries. And, given the very real threats of not only the Soviet Union but also retaliation by neighboring countries, both the Japanese and Germans had compelling security interests to comply with requirements for democratization. Similar conditions are very unlikely to be present in cases of interventions. Most importantly, however, the postwar commitment of resources by the Allied powers, and in particular the United States, to the construction of these countries was far beyond what can be reasonably expected in any case of humanitarian intervention. Humanitarian interventions are most likely to be undertaken by democracies and democratic publics are unlikely to be willing to bear the costs of long-term occupation necessary to build legitimate institutions. The Allied campaigns in Germany and Japan were perceived not as primarily campaigns of humanitarian intervention, but of self-defense. And the decision-making biases mentioned above make it much easier for governments to commit significant resources to what is seen as self-defense than to what is seen as intervention on behalf of strangers.

Mill's point is subtle. Given that success in nation-building is very much the exception to a rule of many failures, the typical case of intervention to end domestic oppression will fail. And given the high cost of such failures – including loss of life, serious harm, and damage to property and infrastructure, but also diminishing support for future interventions – it may be better to discourage such interventions altogether. Allowing interventions in these cases may prove counterproductive, and for reasons that are not just incidental but *structural*. In light of the argument above, then, this strongly tells against the moral permissibility of such interventions. For there is a real sense in which states ought not to intervene in cases like this.

We can now see one important motivation for the view that intervention requires a supreme humanitarian crisis. These perils of intervention can be outweighed, it might be said, but only by something of momentous moral import. Whereas ending a massacre, genocide, and other acts of extreme or large-scale injustice are sufficiently serious, less significant injustices do not have the same potential. In those cases, the harm of letting the crisis continue is not likely to be greater than the harm of a failed intervention. And these, therefore, remain impermissible.

This argument deserves to be taken seriously. But it is a mistake to think that taking it seriously entails accepting the supreme humanitarian crisis-condition. We can introduce exceptions without violating Mill's insight. Such exceptions are necessary because of two important facts: (a) the great power differentials that exist, at least under modern conditions, between states equipped with extensive and sophisticated military capabilities, and potential movements of resistance, and (b) the ways in which the modern state system can render minorities within existing states vulnerable. In what follows I propose to relax the supreme humanitarian emergency-condition in two ways in order to accommodate these facts. Throughout, I will argue that these modifications are consistent with taking seriously Mill's insight about the structural features of intervention.

First, modern military capabilities create an enormous disparity between the power available to abusive regimes and those who might resist them. This disparity is so great that we often cannot expect people to rise up and earn freedom any more than we can expect them to commit suicide. When tyrants can use chemical weapons on their own people, when carefully planned genocidal campaigns happen, when air strikes can simply wipe out pockets of resistance, the option of replacing an oppressive regime with a legitimate one is often simply not available.

One problem with the supreme humanitarian emergency-condition is that it requires outsiders to sit by idly and either wait until the massacres begin or leave the oppressed without any possibility of escape. We might try to remedy this problem by introducing an exception. However, for this to be acceptable, it must acknowledge Mill's insight. One obvious way of doing this would be to require the presence of local support for legitimate government. This suggests the following condition: Humanitarian intervention in an illegitimate state is permissible only if there is either a supreme humanitarian crisis, or a politically viable movement with a sincere commitment to establishing and living under legitimate rule.

Let us call this *the weaker condition*. The weaker condition honors Mill's insight by combining the rationale behind the supreme humanitarian crisis-condition (that stopping such crises will likely do more good than harm), with a demand in all other cases for a viable movement committed to life under legitimate institutions (assuring that intervention will not result in failure). This condition has other advantages as well. It serves, for example, to remove a troubling element of Mill's view: that the price of free and legitimate institutions must somehow involve bloodshed. A politically viable movement committed to life under legitimate institutions need not be in the process of using, or even find acceptable, violent means to overthrow the government. In fact, peaceful movements can provide such support at least as well as violent ones, and perhaps even better.

That said, the weaker condition does not yet address the second concern mentioned above. The problem with the weaker condition is that it requires domestic support for sustaining legitimate institutions within the state *as it presently exists*. This leaves oppressed minorities unnecessarily vulnerable. Sometimes minorities that are committed to life under legitimate rule remain precluded from satisfying the weaker condition because of a mere lack of numbers within a larger state. Such minorities are willing, and possibly able, to sustain a free society if only they had a state of their own. By taking the current configuration of the state system as a given, then, the weaker condition leaves such minorities without the possibility to be freed from oppressive regimes.

This is a perverse implication. It condemns as impermissible intervention on behalf of oppressed minorities *even if* they are committed to life under free institutions. Such minorities could in principle satisfy Mill's insight, but are precluded from doing so by the current configuration of the state system. This raises a question: What reason do we have to take the current shape of the state system as a given? But the answer to this question is obvious: given that we are, *ex hypothesi*, dealing with illegitimate states, there can be no such reason. Only legitimate states are worthy of moral respect. Thus, if we can rescue people from oppression and help them build a safe political environment, the non-existent moral claims of illegitimate states cannot stand in our way.

This opens the possibility of adding another clause to the weaker condition. This clause aims to make intervention permissible in support of secessionist movements by oppressed minorities that are willing and capable of supporting their own free institutions in a newly found state. Fully specified, the condition I am proposing can be summarized as follows: humanitarian intervention in an illegitimate state is permissible only if at least one of the following conditions is satisfied: (a) there is a supreme humanitarian crisis, (b) there is a politically viable movement with a sincere commitment to establishing and living in the present state under legitimate government, or (c) there is a politically viable minority with a sincere commitment to seceding and establishing a new and legitimate state.

This condition, I claim, satisfactorily honors Mill's central insight. However, it also honors the need to address the two concerns mentioned above (uneven military capabilities, and the configuration of the state system). That is, it captures the truth that, contrary to what Mill thought, intervention can be permissible precisely *because* a group is unable to achieve freedom on its own.

414   **Bas van der Vossen**

Of course, this leaves many important questions unaddressed. For example, whether a viable movement that is sincerely committed to life under a legitimate regime exists in another country requires judgment on the part of would-be intervening states. Such judgment calls are extremely difficult. Groups can fail to be viable in a number of ways. They can be lack support of their fellow citizens. They can be committed to life under an illegitimate government of their own. They can be seen as the puppets of foreign oppressors. And so on. But groups can also be under pressure not to play up their commitment to legitimate rule, even if they would actually abide by it when in power. Reliance on judgment is inevitable here. However, fortunately, there may be clear cases as well. Intervention protecting groups undertaking a just secession may be a case in point.

That said, I believe that the view I have proposed here adequately balances two important elements that I have argued must be part of any acceptable approach to humanitarian intervention: the need to protect those whose human rights are being violated, on the one hand, and the incentive effects and standard risks of interventions, on the other.

Conclusion

International morality, I have argued, should reflect certain structural features of the international scene. Taking this seriously means we cannot think of states as directly analogous to individuals. The fact that states are large-scale political institutions, and potentially dangerous ones at that, has two important implications for the debate on humanitarian intervention. First, contrary to Tesón and Wellman, all legitimate states are rights-protected against outside intervention, even when their societies fall short of being fully just in the eyes of outsiders. But second, when dealing with illegitimate states, contrary to popular opinion, intervention can be justified even in the absence of a supreme humanitarian intervention – as long as certain other conditions are met.

In closing, I want to emphasize that these conclusions should be read as appropriate only under current conditions. I do not mean to deny that our world, and the institutions that govern it, might be and probably ought to be, significantly different than they are. The search for feasible proposals for changing institutions and their decision-making processes is tremendously important. With better institutions in place, different conclusions for state legitimacy, humanitarian intervention, and indeed military action more broadly, might follow. (For excellent discussion, see Buchanan, 2006.) However, until such changes come about, my conclusions reflect a truly tragic fact about the world: We are quite frequently in a position where we simply cannot successfully act to stop human rights violations.

References

Buchanan, A. (2006) Institutionalizing the just war. *Philosophy and Public Affairs* 34: 2–38.

Buchanan, A. and Keohane, R.O. (2006) The legitimacy of global governance institutions. *Ethics and International Affairs* 20: 405–437.

Clark, I. (2005) *Legitimacy in International Society*. Oxford: Oxford University Press.

Lomasky, L. and Tesón, F. (forthcoming) Justice at a Distance.

Mill, J.S. (1859) A few words on non-intervention. In *The Collected Works of John Stuart Mill, vol. XXI – Essays on Equality, Law, and Education*, ed. J.M. Robson (1984). London: Routledge & Kegan Paul.

Tasioulas, J. (2010) The legitimacy of international law. In *The Philosophy of International Law*, ed. S. Besson and J. Tasioulas, pp. 97–116. Oxford: Oxford University Press.

Wellman, C.H. (2012) Taking human rights seriously. *The Journal of Political Philosophy* 20: 119–130.

Wheeler, N. (2000) *Saving Strangers*. Oxford: Oxford University Press.

World hunger

CHAPTER TWENTY-EIGHT

Famine Relief: The Duties We Have to Others

Christopher Heath Wellman

> *In developing countries, 6 million children die each year, mostly from hunger-related causes.*
>
> Bread for the World Institute

> *Never doubt that a small group of thoughtful, committed citizens can change the world; indeed, it's the only thing that ever has.*
>
> Margaret Mead

Positive Duties

Any moral theory that requires one ceaselessly to sacrifice for the common good should be rejected as too demanding. In my view, we need not apologize for devoting the lion's share of our time and resources to our own self-regarding projects and the people we love. However, if another person is gravely imperiled and one can rescue her at no unreasonable cost to oneself, then one has a moral duty to do so.

Imagine, for instance, that you are lounging by the pool at the Hard Rock Hotel and Casino in Las Vegas. In one hand you have a frozen margarita, in the other you hold a copy of this book. Ordinarily, of course, the chapters in this volume would hold your undivided attention. On this occasion, however, you find yourself reading the same few sentences over and over again, as you repeatedly lift your head to check out the scantily clad, hard-bodied men and women frolicking in and around the pool. As you survey the "beautiful people," you notice that an unattended infant has just fallen into the water and will surely drown unless someone immediately saves her (Singer, 1972). Are you morally required to jump in and rescue the baby? Does it matter that she is not your child and that you have no special relationship with her?

I presume that virtually everyone reading this would agree that you ought to rescue the child, even if doing so would involve spilling your margarita and ruining the book.

Contemporary Debates in Applied Ethics, Second Edition. Edited by Andrew I. Cohen and Christopher Heath Wellman.
© 2014 John Wiley & Sons, Inc. Published 2014 by John Wiley & Sons, Inc.

Perhaps we would not be obligated to help if the baby were not imperiled (we need not come to the infant's aid if she merely needed another coat of sun screen or a long overdue diaper change, for instance) or if the assistance would be unreasonably costly (as it might be if one was holding the Mona Lisa, rather than a copy of this book). Because the baby is sufficiently imperiled and you could save her without sacrificing anything significant, however, it does not matter that you are in no way related to or especially responsible for the child.[1] Thus, it is no defense to callously protest: "It's not my baby," or "I never agreed to baby-sit that kid." These defenses might be relevant in some instances (if someone questioned why you had not changed the baby's diaper, for instance), but they are not germane in this case because all of us have positive moral duties to rescue even anonymous strangers when they are sufficiently imperiled and we can do so without significant cost to ourselves.[2]

I take the preceding analysis to be merely commonsensical, and thus I presume that most people reading this chapter will not seriously object to anything at this early stage. Notice, however, that surprising implications follow from granting that we have moral duties to rescue others when they are sufficiently imperiled and we can assist them at no unreasonable cost. This is because there are currently masses of children starving to death, and virtually everyone reading this book is wealthy enough to save some of them without sacrificing anything significant. Thus, for the very same reasons that you would be morally required to save the drowning infant at the Hard Rock pool, you are morally required to contribute a modest amount, say $100, to saving the lives of a few children who are currently starving to death.

At this point, one might object that there is a huge difference between saving a drowning child in your immediate presence and sending money to help anonymous foreign children who are starving in some unfamiliar place, thousands of miles away. I acknowledge that these two scenarios are likely to *feel* different to many of us, but I suggest that there is no morally relevant difference between them. In other words, whatever effect the difference in nationality, the physical distance, or the use of mediating devices might make in *motivating* us to rescue someone else, the moral relations between you and the starving distant foreigner are the same as those between you and the drowning infant (Singer, 1972).

To see that common nationality is not necessary to ground a duty to rescue, think again of the drowning infant at the pool. Suppose that you are American: does it matter whether or not the infant is also American? I presume not. Imagine, for instance, if an American who sat and watched the infant drown defended herself in the following fashion: "Ordinarily I would have leapt in to save the child, but I did not do so in this case because I knew she was Australian." Would this strike you as an adequate defense? I assume that most people reading this book would not accept this justification because the infant's nationality is irrelevant. As long as the infant is sufficiently imperiled and one can rescue her without sacrificing anything significant, it makes no difference what nationality the two parties are because Samaritan duties are owed to fellow *human beings*, not just to *compatriots*.[3] (Notice, for instance, that the biblical story from which Samaritan duties derive their name involves a gentleman from Samaria saving an imperiled stranger, not a fellow Samaritan.)

Moreover, it is worth adding that it is equally irrelevant whether the rescuer and the imperiled person are on the same country's soil. Imagine, for instance, that the pool in question is not in Las Vegas but is on a desert resort that straddles the US/Mexico border.

Suppose that in order to create a "Swim to Mexico" gimmick, the resort designed the small pool so that one side is in the United States and the other in Mexico. Would it make a difference whether the infant fell in the American or the Mexican portion of the pool? Presumably not. Combining these two points, a Canadian tourist lounging on the American side of the pool who saw an Australian infant fall in the Mexican portion of the pool would be just as morally obligated to perform the rescue as an American tourist on the American side of the pool who saw an American infant drowning in the American portion of the pool. In short, both the citizenship of the parties and the country in which the rescue must be performed are morally irrelevant. What is crucial is whether the rescuee is sufficiently imperiled and can be saved at no unreasonable cost to the rescuer; where both of these conditions obtain, neither nationality nor national location makes a difference.

At this point, one might object that while the national location of the two parties is irrelevant, their spatial location does make a difference because one can be bound only to assist those in one's close proximity. To appreciate the moral relevance of distance, this critic might ask us to imagine that one is lounging beside the ocean rather than a pool. Suppose that one sees (perhaps through binoculars) an infant fall off the back of a boat ten miles offshore. (And suppose that those on the boat did not notice the infant's fall and that there is no one else on the beach at the time.) Under these circumstances, when the imperiled person is no longer right under one's nose, so to speak, it is not so clear that one has a moral duty. And this is explained, the skeptic suggests, by the distance between oneself and the infant.

I acknowledge that there may be no Samaritan duty in this case, but I deny that this is due merely to the physical distance separating the two parties. In my view, the distance itself is not morally significant; if one has no duty to rescue a drowning infant ten miles offshore, it is either because one is unable to do so (since the infant would no doubt drown before one could swim out there) or because doing so would be unreasonably costly (since the rescuer might reasonably fear drowning or being attacked by sharks). To see that the distance itself is morally irrelevant, though, imagine that one has freakishly long arms that enable one to pull the baby out of the ocean without even getting out of one's chair on the beach (Kamm, 2000). (Or, if such long arms are too difficult to fathom, imagine that one has a super speedboat, a jetpack, or even a giant crane that would enable one safely to retrieve the infant in a matter of seconds.) Under these circumstances, I suspect that most would agree that one has a duty to save the drowning infant. Thus, once we strip this scenario of the features that undermine one's capacity to perform the rescue at no unreasonable cost, we see that the issue of distance is not in itself morally relevant.

Finally, notice that it makes no moral difference whether one's rescue is mediated by devices or other people. Imagine, for instance, that after spending a couple of hours by the Hard Rock pool, you decide that you had better return to your hotel room before you get sun-burned. Fortunately, the hotel has closed-circuit television coverage of the pool, so you can continue to check out the lively scene from the comfort of your air-conditioned room. While watching on your room's television, you notice the infant fall into the pool. Because you are staying on the 30th floor, there is no way that you could make it down to the pool in time to save her yourself. Without getting out of your chair, however, you could pick up your cell phone and call the bartender at the poolside bar, who – once alerted – could easily rescue the infant herself. It seems to me that you are just as obligated to make that call (even if there would be a substantial charge on your cell bill) as you would be to dive into the pool yourself. It makes no difference, in other words, whether one can personally rescue the drowning child all on one's own, or whether one can merely play a part in the rescue by calling others who, once informed, can complete the rescue.[4]

But notice: once one recognizes that neither nationality, distance, nor the use of mediating devices and people in any way diminishes one's duty to rescue imperiled strangers, it is clear that one's duty to rescue starving infants on another part of the planet is just as pressing as the initial poolside rescue with which we began. Indeed, the last scenario of using one's cell phone to initiate a rescue of someone whom one sees drowning on a television monitor is very much like a situation that many of us routinely experience. We are watching something entertaining on television when a commercial alerts us that starving children desperately need our help. If we have a duty to jump into the pool to save the infant, and we have a duty to make a relatively expensive cellular phone call to the poolside bar, then why do we not equally have a duty to use our cell phone to make a modest donation (say, $100) to the institution saving the starving children? If (1) the fact that the children are citizens of another country is irrelevant; if (2) the physical distance between you and them makes no difference; if (3), like the loss of the margarita and the damage to one's book, the loss of $100 is not an unreasonable sacrifice; and if (4) the use of mediating devices like cell phones, credit cards, and international relief agencies is not important, then it is hard not to conclude that one's moral duty to send money to famine relief is just as strong as one's duty to jump in the pool to save a drowning child.

At this point one might protest that there remains a big difference between saving a single drowning infant and sending money to help masses of starving children: the number of people imperiled. Numbers might be thought to matter because when there is only one imperiled person, her peril becomes salient in a way that explains why you as a potential rescuer have no discretion but to help her. When there are numerous imperiled people (so many, in fact, that you could not possibly rescue all of them), no single individual's peril is salient, and thus one retains the discretion as to whether or not to help.

I agree that numbers can sometimes matter, but I do not think they can make the type of difference that this objection supposes. More specifically, I acknowledge that one enjoys some discretion when there are more imperiled people than one could possibly save, but it is not the discretion of whether or not to perform the rescue; rather, it is merely the choice of whom to rescue.

Most who believe that we have a duty to assist others do not couch their arguments in the language of rights, but I would explain this discretion in terms of the correlative rights to assistance. Thus, to return to our initial example, I would say that the drowning infant in Hard Rock pool has a Samaritan right that you rescue her. If the situation were altered slightly so that there were two babies in the pool, and you could not possibly save both, would you say that you no longer have any duty to rescue at all? Presumably not. The more sensible conclusion, I think, is that you must still rescue one of the babies, and you may choose which to rescue. In terms of the infant's rights, obviously neither of the two drowning babies has a right that you save her in particular, but I would say that each has a right that you save one of them (Feinberg, 1984). Thus, just as a lounger by the Hard Rock pool could not justify rescuing neither of the infants with the lame excuse that "Once the second child fell in, I knew that I could not save both," the fact that we cannot save all of the world's people from starving to death provides no justification for not rescuing some.[5] In short, while the world's current situation is admittedly much more messy and heartbreaking than our imagined situation of a single drowning baby who is seen by a single sunbather, there is nothing about the complexity of the actual world's crises that makes our duty to rescue any less stringent.

Finally, let me comment on my suggestion that each of us has a duty to donate $100 to famine relief. I suspect that virtually everyone reading this book could easily give substantially more than $100 without sacrificing anything significant, but I chose this conservative sum because it is only a little bit more than what it would cost to replace the drink and book that I imagined might be ruined in the initial rescue situation. Let me quickly respond, however, to those who might object that $100 is too large an amount to expect people, especially students, to sacrifice.

There will invariably be exceptional cases, of course, of people who could not give up $100 without sacrificing something morally significant. Some students are working parents, for instance, who have too little money even to buy the assigned texts (they either check the books out of the library or routinely borrow them from patient classmates), and who could not part with $100 and still manage to pay for their children's health insurance. If that sounds something like your situation, then it seems only reasonable to conclude that you could not contribute to famine relief without sacrificing something morally significant. If we are being honest, however, the vast majority of us must admit that we could charge $100 to our credit card and still shop at A&F, buy our coffee at Starbucks, order our dinner from Domino's, watch MTV on cable television, and talk with friends on our cell phones. If so, then it is hard to say with a straight face that we have no duty to save the lives of starving children because doing so would require us to sacrifice something significant.

Before moving on, let me acknowledge that in the past there was a profound difference between our moral responsibilities to an infant drowning in our midst and a child starving to death in some distant land. This difference stemmed from our lack of information regarding, and capacity to save, the latter. Times have changed, however, and so has the scope of our moral responsibilities (Singer, 1972). We do not have freakishly long arms that enable us literally to reach out and feed people thousands of miles away, but we do have other instruments that are just as effective. We have international media that can inform us about distant tragedies, we have international relief agencies dedicated to performing acts of rescue, and we have phones and credit cards that enable us conveniently to transfer our funds to these agencies. Thus, if you are unwilling to contribute money to help save the lives of several starving children, it is hard to see why there is any difference, morally speaking, between you and a lounger by the Hard Rock pool who cannot be bothered to put down her drink and book to save the drowning infant.

Negative Duties

One of the most frequent objections to sending money to the masses of famine-stricken people around the world is that these famines are not strictly accidents; rather, they are brought on at least in part by inefficient or corrupt political and business institutions.[6] The twofold thought behind this observation is: "Why should I have to bail out these people when they played a part in creating their own misfortune and are likely to do so again?" It is common to argue in response either that the specific famine in question was in fact an unforeseeable accident or that, however much political and/or business leaders might be to blame for the severity of the problem, surely those actually starving to death are no more responsible for the unforgiving conditions that caused their peril than we are for the favorable conditions that (largely) explain our wealth. Here I will pursue neither of these routes. Instead, I shall concede that much of the world's poverty is at least exacerbated and prolonged (if not outright caused) by national and international institutions, but I will argue that this fact only strengthens the case for the duty to offer assistance because it illustrates that we have negative as well as positive duties to assist the world's most needy.

Before exploring the relationship between political institutions and world hunger, I would like to suggest that we have a negative duty to neither support nor profit from institutions that wrongly harm others (Pogge, 2002).[7] The basic idea behind this claim is merely that, just as we should not personally harm others, nor should we either support or profit from institutions that do so. Imagine, for instance, that your parents own slaves and therefore are able to provide a comfortable life for you. Among other things, they pay for your college tuition with the profits they garner from the slave labor. Should you accept this money from them? What would you think of a daughter of slave-owners who defended her privileged life by saying: "I agree that owning slaves is morally repulsive, but that provides no reason to criticize me because *I* don't own any slaves!"

I can understand why someone might contend either that children should not accept money from slave-owning parents or even that adult children should have nothing to do with their slave-owning parents, but I would argue for a more modest claim. Because children have limited influence over their parents, and because it would be an enormous sacrifice for most college-age children to have nothing to do with, or perhaps even to accept no financial support from, their parents, I suggest merely the following: if one is going to accept money from one's slave-owning parents, then one must at least make a conscientious effort to persuade one's parents that owning slaves is wrong. In other words, accepting the benefits of an unjust institution like slavery requires one, at the very least, to work to eliminate the unjust institution.

As I indicated above, the rationale for this conclusion is the commonsensical position that one should not be an accessory to injustice. As a historical example of someone who took this moral directive to heart, consider Henry David Thoreau. Both because of its support for the practice of slavery and because of its engagement in the Mexican War, Thoreau was convinced that the United States government was a powerful instrument for injustice. Not wanting to support such an institution with his actions or money, Thoreau retreated to Walden Pond, where he lived in relative isolation, refusing to pay any taxes to the US government. In my view, Thoreau is to be applauded for his concerns about supporting an unjust institution, but he went above and beyond the call of duty by completely divorcing himself from political society. According to the modest view I am advocating here, one could not have objected to Thoreau's enjoying the benefits of political life as long as he worked to reform US policy.

For a more recent example of how one might try to influence an unjust institution, consider the student activism during apartheid South Africa. When I was an undergraduate, South Africa had an oppressive system of apartheid, wherein the whites oppressed the blacks. Despite being a numerical minority, the whites were able effectively to exploit the blacks because they controlled the political and financial institutions. What is more, the international community effectively buttressed the whites' privileged position by investing in their businesses and recognizing their government as legitimate. At the University of North Carolina, where I was in school, there was a relatively small group of well-informed students who were disturbed by the injustices being perpetrated in South Africa. (I regret to say that I was not among their number.) Distraught that their university was contributing to the injustice by investing in some of the South African companies that played a part in this oppressive system, these students lobbied the relevant authorities to divest the university of all South African holdings. As you might imagine, however, a few students did not wield a great deal of influence over the university's investment portfolio. Rather than give up, however, these students built a "shanty town" in a prominent place on campus (on the main quad, right below the Chancellor's office, actually). The students lived in these makeshift huts for months to call attention to the plight of blacks in South Africa who were forcibly relegated to ghettos where they lived in similar conditions. Over time, these huts attracted more and more embarrassing attention until the university finally decided to divest itself of all South African companies.

In my view, this story provides a prime example of how one might work to make one's institutions more just. Had these students been more like Thoreau, they might have simply withdrawn from school, so as not to play a supporting role in the perpetuation of injustice. Leaving school is a huge sacrifice, though, especially when one considers that virtually all schools were invested in South African companies, and thus there was nowhere else that these conscientious students could have enrolled. Under these conditions, it is enough for these students to make a concerted effort to reform their university. (Indeed, I should think that living in makeshift huts goes well beyond what could reasonably be asked of an average student, and thus they could have stayed in school in good conscience even if they had done considerably less – such as merely sponsoring petitions and organizing rallies.) Notice also that it is too much to require that students continue their efforts until they prevail. Students typically exert very little influence over university policy, and thus all one can ask is that they make a concerted effort to get their school to stop supporting major injustices. Finally, I would suggest that remaining within an institution and working for its reform is in many ways preferable to completely withdrawing from the institution because the former involves being an agent for positive change. Therefore, while it is sometimes thought to be better to keep one's hands entirely clean of injustice, working from the inside to improve an unjust institution can often be the best way to fight the good fight. (Indeed, if no one worked from within to reform corrupt institutions, these institutions would be left under the exclusive control of those who were either ignorant of or indifferent to injustice. Thus, it is perhaps best if some fight from without and others fight from within.) With this in mind, let us now return to the objection that we cannot be expected to save the victims of famines that were at least partly caused by institutional mismanagement.

Recent research confirms that there is indeed a correlation between the quality of one's government and the degree to which one is protected from famine (Dreze and Sen, 1989). In particular, evidence indicates that effective democratic governance virtually ensures that a country will not be ravaged by a widespread famine with which it cannot internally cope.[8] This might seem counter-intuitive to those of us who think of famines as natural disasters but, on reflection, this claim makes perfect sense. Most of us have various qualms with our governments, but those of us fortunate enough to live in liberal democratic states take it for granted that governments are designed to be mutually beneficial institutions that more or less serve their constituents. In far too many instances, however, political power is not democratically distributed, and the government is a powerful institution designed to serve the few elite who happen to wield the political power. Just as apartheid South Africa was designed maximally to benefit the politically empowered whites, for instance, some governments are ruled so as to work to the greatest advantage of the dictator and her closest friends and family. It does not take much imagination to see why a government designed to benefit just a small fraction of the population would be uninterested and/or unable effectively to prevent famines, but it does require some explanation as to how such a government can stay in power. Think of it this way: if people more than 200 years ago in France and the American colonies were able to overthrow oppressive governments, why are there currently so many people in the world who are either uninterested in or unable to establish effective democratic governments?

The answer to this last question is simply "brute force." Dictators are often able to maintain their oppressive regimes simply because they control the military, and they ruthlessly use this power to suppress anyone who seeks democratic reform. Of course, staying in power requires a vicious circle because the dictators are typically able to retain the military's loyalty only as long as they have the money to pay them, and they can acquire the necessary funds only if they continue to exploit their political power. What I want to call attention to now, though, is more specifically how these dictators are able to use their political power to generate wealth. Part of the answer, of course, simply comes from taxes that (in so far as the funds are used to benefit the ruler rather than the people themselves) essentially enslave the political subjects. Another important part of the equation, however, is that dictators frequently amass huge sums of money by selling the country's natural resources to foreign companies and governments.[9] Thus, if a dictator's country has extensive oil reserves, for instance, then the dictator can sell this oil and use the money to secure her military stranglehold over her subjects.[10]

Here, two points clearly emerge. First and most obviously, the mere fact that a dictator effectively controls the country's natural resources does not make her *morally entitled* to those resources any more than a slave-owner's effective control over her slaves implies that she is morally entitled to the fruits of these slaves' labor. Second, and more importantly for our purposes here, foreign companies are an integral part of the problem because, in seeking to acquire natural resources as cheaply as possible, they are giving the undemocratic leaders the money necessary to continue their unjust domination over their political subjects. In a very real sense, it is as if these companies were buying cheap cotton from slave-owners who were using this money to buy more guns and slaves.

If all of this is right, where does it leave you and me? Where does it leave those of us who enjoy our clothes from A&F, our coffee from Starbucks, our dinners from Domino's, our cell phones from Sprint, and our MTV on cable television?[11] Certainly, part of the reason we are able to enjoy these luxuries is because we work extremely hard in order to be able to buy these things for ourselves and those we love. But equally certainly, another part of the reason we enjoy these luxuries is because we benefit from an economic system that utilizes natural resources bought very cheaply from political leaders who have control over these resources only because they happen to have the military power to suppress their compatriots. Thus, you and I profit from an overall economic system that plays a prominent role in propping up military dictators who in turn create the political conditions that play a causal role in the world's worst famines. In the end, then, the role that political and business institutions play in contributing to famine does not undermine our duty to send money to famine relief; on the contrary, it explains why we have not only positive duties to help those who are currently starving to death, but also negative duties to work to change the system so that future famines do not occur. In other words, just as Thoreau felt the need to divorce himself from an unjust political institution and my fellow students felt compelled to reform an unjust university, you and I should recognize our obligation to either withdraw from, or seek to reform, the current political and economic environment.

Now, just as it was extremely costly for Thoreau to withdraw from political society and it would have been a huge sacrifice for my fellow students to withdraw from school, virtually none of us is willing entirely to divorce herself from the existing international economic system. But if we are going to continue helping ourselves to the spoils of an unjust political and economic environment, then we have a responsibility to work conscientiously to make this system a more just one. If we continue to participate in the system without working diligently for its reform, on the other hand, then we are morally no different from the daughter of slave-owners who defends her willingness to accept gifts made possible only via the exploitation of slaves by saying: "Don't blame me; I don't own any slaves." Just as it would clearly not be too much to ask this daughter to try to persuade her parents of the injustice of slavery, it is not too much to ask you and me to work to make the international economic and political order more just.

At this point, it is tempting to protest that there is nothing one can do. Calling an international relief agency such as Oxfam or Unicef and giving $100 on one's credit card is a relatively simple act that will make a real difference for people who would otherwise starve to death, but how in the world is one supposed to change the international economic and political order?

This worry is understandable, but it is important to remember that we are not morally required to change the system; we are merely obligated conscientiously to work to reform it. Even so, one might object, it is not even clear how to begin![12] I concede that it is hard not to feel impotent in the face of such enormous institutions, but notice that the world has already experienced wave after wave of moral reform, and each of these changes had to start somewhere. Think, for instance, of Henry David Thoreau. It is unrealistic to suppose that Thoreau thought he could single-handedly get the United States to abolish slavery, but there is no question that the integrity with which he lived his life had a profound influence on others who, over time, were able successfully to abolish slavery. Similarly, my fellow college students who built the shanty town on campus were among those who raised awareness of the horrors of apartheid South Africa until the international community gradually ceased supporting and ultimately began placing reformist pressures on the relevant political and economic institutions. More recently still, think about what a profound change has occurred regarding recycling in the United States. Not very long ago, one could not help but think that there was nothing substantial one could do. Over a remarkably brief period of time, however, environmental and political activists were able to change the system so that municipalities now routinely provide services that make it easy (if not mandatory) for each of us to contribute to a large-scale recycling effort.

If these and countless other monumental reform movements can succeed, then there is no reason to suppose that each of us cannot do our part in a movement to change international business and politics so that military dictators are no longer able to oppress their constituents in ways that, among other things, contribute to the frequency and severity of famines. I am not the most imaginative person, but it strikes me that anyone reading this chapter for a class could begin by trying to raise awareness on her own campus. Perhaps with the help of the professor who teaches the class, one might begin by organizing a student forum to publicize the issue and form a group on campus that can subsequently come up with additional ideas to spread the word and inspire constructive action. I cannot promise that you will change the world, but I do know that the incentives to perpetuate the current system are strong, so the world will not change without people like you dedicating their time and energy to making it a more just place.

Conclusion

Virtually everyone agrees that we have negative and positive duties toward one another. Negative duties prohibit us from harming others, and positive duties require us to assist others when they are gravely imperiled and we can rescue them at no unreasonable cost to ourselves. In this chapter I have sought to show that each of these types of duty explains why we are morally bound to help those famine victims who are starving to death. The positive duty to provide easy rescues obligates us at the very least to send money to those international relief agencies which have assigned themselves the task of ministering to those who are starving to death, and the negative duty not to benefit from an institution that wrongly harms others requires us to work to reform the current practice of international politics and business. It would be very easy to find out how to do so: search the web for any of the leading international relief organizations such as CARE, Oxfam America, or UNICEF (Unger, 1996, p. 175). It's your call . . .

Notes

This chapter is inspired by, and draws heavily upon, the previous work of a number of authors, especially Peter Singer and Thomas Pogge. I am grateful to Andrew Altman and Hugh LaFollette for helpful comments on an earlier version of this chapter.

1 For the purposes of this chapter, I treat "insignificant costs" and "not unreasonable costs" as interchangeable. Readers familiar with Peter Singer's landmark article, "Famine, affluence, and morality," will recognize this language from Singer's second, less demanding principle that we should contribute to famine relief until we sacrifice something "morally significant." (I do not mean to defend Singer's more demanding principle that we ought to contribute until we are sacrificing something "morally comparable.")

2 "Positive" duties require us to assist others; they are to be contrasted with "negative" duties, which require merely that we not harm or interfere with others.

3 I do not deny that one might have more robust responsibilities to one's compatriots; I insist only that being a fellow citizen is not necessary for one to have a minimal Samaritan duty to another.

4 One reason that you may be less motivated to make the call than to personally save the drowning child is because the former act would be less public. Thus, whereas you would be publicly applauded for diving in to save the drowning child (and perhaps condemned for failing to do so), your relatively private decision to call the bartender need not have these same social consequences. But, while these types of considerations can no doubt affect one's motivations, they are clearly irrelevant to what morality requires. To see this, notice that we might have much less motivation to refrain from murdering an enemy when we can do so in private without any social repercussions, but clearly this does not mean that our moral duty against clandestine murder is any less weighty.

5 Indeed, not only does each imperiled person have no right that you save her in particular, it is not clear that the most gravely imperiled have a right that you help someone who is at least as imperiled. If (as some argue) we can sometimes make a greater marginal difference by contributing to those who are less imperiled, then it would not seem objectionable to do so.

6 A similar objection is that we should not all give our money away to save foreigners because this would ruin our national economy and, as a consequence, render us unable to help other foreigners (or perhaps even our compatriots) in the future. This objection need not be taken seriously. It is true that our economy depends upon a certain amount of spending, but this would counsel us against *saving* too much, not against spending our money *on others*. More importantly, the dire economic consequences invoked in this objection could only come to fruition if the great majority of us gave considerably more than the $100 I am advocating here. In short, there are many things about which it is legitimate to worry, but excessive altruism to foreigners is not among them.

7 I should stress that this is separate from the Samaritan duty. Samaritan duties are positive (as is the general duty to make the world a more just place), but the duty to refrain from either supporting or benefiting from injustice is a negative one.

8 There is also considerable evidence that extreme poverty and various problems tied to population growth are directly related to the standing of women. Societies that give women control over their bodies as well as access to education, economic opportunities, and reproductive technologies tend to have reduced birth rates and higher standards of living.

9 Of course, buying natural resources is only one of the more obvious ways in which the international community can help a dictator strengthen her domination over a population. As Thomas Pogge explains: "Local elites can afford to be oppressive and corrupt, because, with foreign loans and military aid, they can stay in power even without popular support. And they are often so oppressive and corrupt, because it is, in light of the prevailing extreme international inequalities, far more lucrative for them to cater to the interests of foreign governments and firms than to those of their impoverished compatriots. Examples abound. There are, in the poor countries, plenty of governments that came to power and/or stay in power only thanks to foreign support. And there are plenty of politicians and bureaucrats who, induced or even bribed by foreigners, work against the interests of their people: *for* the development of a tourist-friendly sex industry (whose forced exploitation of children and women they tolerate and profit from), *for* the importation of unneeded, obsolete, or overpriced products at public expense, *for* the permission to import hazardous products, wastes, or productive facilities, *against* laws protecting employees or the environment, etc." (2002, p. 244).

10 One might protest that, while an illegitimate ruler undeniably has no right to her country's natural resources, neither do her compatriots. According to this objection, the world's

natural resources are owned jointly by all of the world's population. I will not contest this claim here. Rather, I suggest that if everyone is equally entitled to the world's natural resources, then this constitutes an argument in favor of something like a "global resources dividend." This dividend, recommended by Pogge, would be paid for by those of us who use the world's natural resources and would be owed to the world's poor who are involuntarily not using their share of these natural resources (Pogge, 2002, pp. 196–215).

11   Let me be clear: I am *not* alleging that A&F, Starbucks, Domino's, Sprint, and MTV are particularly corrupt companies; each may do absolutely nothing immoral on its own. My point is that companies like these are part of an international system that benefits from the inexpensive natural resources purchased from undemocratic, illegitimate rulers.

12   Notice how awkward it is to protest that those of us who are privileged cannot be obligated to change the system because we are impotent in the face of its enormity, while simultaneously suggesting that those who are starving to death are entitled to no assistance because *they* are responsible for the political and economic institutions which led to their ruin.

References

Bread for the World Institute. http://www.bread.org (last accessed 6/17/13).

Dreze, J. and Sen, A. (1989) *Hunger and Public Action*. Oxford: Oxford University Press.

Feinberg, J. (1984) *Harm to Others*. New York: Oxford University Press.

Kamm, F. (2000) Does distance matter morally to the duty to rescue? *Law and Philosophy* 19: 655–681.

Pogge, T. (2002) *World Poverty and Human Rights*. Cambridge: Polity.

Singer, P. (1972) Famine, affluence, and morality. *Philosophy and Public Affairs* 1: 229–243.

Unger, P. (1996) *Living High and Letting Die: Our Illusion of Innocence*. New York: Oxford University Press.

Further Reading

Hardin, G. (1974) Lifeboat ethics: the case against helping the poor. *Psychology Today Magazine*.

LaFollette, H. (2003) World hunger. In *A Companion to Applied Ethics*, ed. R. Frey and C. Wellman, pp. 238–253. Oxford: Blackwell.

Schmidtz, D. and Goodin, R. (1998) *Social Welfare and Individual Responsibility*. Cambridge: Cambridge University Press.

Shue, H. (1996) *Basic Rights: Subsistence, Affluence, and US Foreign Policy*. Princeton, NJ: Princeton University Press.

Singer, P. (2002) *One World: The Ethics of Globalization*. New Haven, CT: Yale University Press.

CHAPTER TWENTY-NINE

Famine Relief and Human Virtue

Andrew I. Cohen

Much of the philosophical literature on world hunger draws analogies to life-threatening emergencies. Writers ask us to imagine babies drowning in various bodies of water. It seems we should rescue them – especially when it is easy to do so.

Such fanciful examples have a compelling appeal. Virtuous persons automatically help others in immediate and profound need when they are in a position to do so. We do whatever we can to help without dwelling on, for instance, the nuances of the value of saving babies versus saving the perfect martini, a good hair-do, or a fine work of art. The morally mature person lifts drowning babies out of the water.

And so, this well-intentioned argument continues, moral decency similarly has us alleviating world hunger when we can. There is suffering and death – we know about it, and we can do something about it at little cost to ourselves. More than that: we *ought* to contribute to famine relief. A failure to do so is blameworthy; we may even *owe* such relief to distant suffering peoples.

I believe, however, that there are important moral differences between famine relief and tending to easily fixed nearby suffering. While it might be true that we ought to provide easy rescue, it is not clear that we have any similar moral responsibilities to distant hungry persons. But even if we had some duties to aid distant hungry persons, such duties must not be *enforceable*.

In what follows, I argue that the drowning baby analogy tells us very little about duties of famine relief. I explore the place for charity in a good life, arguing that enforceable duties of charity are incompatible with the key moral concern that every person should have the best chance to define and live a life of her own. I discuss how a virtuous commitment to alleviate suffering should have us focusing more on local problems. I close with some general remarks about economic and political considerations, noting how breaking down barriers to free markets would the best way to promote everyone's prosperity – especially for the world's poorest peoples.

Contemporary Debates in Applied Ethics, Second Edition. Edited by Andrew I. Cohen and Christopher Heath Wellman.
© 2014 John Wiley & Sons, Inc. Published 2014 by John Wiley & Sons, Inc.

Drowning Babies

Let us return to the type of example that launched this and many other discussions. Some writers, such as Peter Singer (1972), Peter Unger (1996), Christopher Heath Wellman (Chapter 28 in this volume, 2014, pp. 419–430), and others, argue that just as we ought to rescue nearby drowning babies when we can do so at little cost to ourselves, so too we have a duty to alleviate distant suffering when we can do so without incurring unreasonable costs. The moral reasons to alleviate suffering are the same in each case. You ought to help – especially when you might easily redirect resources from some more frivolous pursuits.

The appeal to drowning babies gives little guidance for our responsibilities to distant suffering peoples. There are significant moral differences between the two cases. The cases warrant different moral reasoning and different responses.

At stake here is whether there are any "positive duties" to provide aid (as opposed to mere "negative duties" to abstain from performing certain actions). Arguments about rescue take different forms depending on the moral requirements they impose on potential benefactors. To simplify matters, we can speak of three forms of argument:

1. *Weak* versions of such arguments say that rescue, though morally commendable, is at your *discretion*. Rescue is above and beyond the call of duty, so should you choose not to rescue, you are not blameworthy.

2. *Moderate* versions remove any moral discretion for rescue: rescue is morally *required*; a failure to rescue is blameworthy. Other persons may at most blame you should you choose not to rescue. They may not use physical force to compel you to rescue or to punish your failure to do so. The positive duties implied by such arguments may be called *moderate duties*.

3. *Strong* versions of the argument, like moderate versions, say that rescue is morally required. And like moderate versions, blame is fitting should you choose not to rescue. But unlike moderate versions, your responsibility to rescue is *enforceable* in a powerful way. Potential beneficiaries of your aid, or those acting on their behalf, may use physical force to compel your assistance or otherwise punish you for your failure to act. The positive duties implied by such arguments may be called *strong duties*.

The question is what sort of responsibilities, if any, we have regarding distant hungry people. Following many proponents of either moderate or strong duties of famine relief, let us then start by considering babies drowning at our feet. I think we might best understand appropriate responses here by considering what good persons do in such cases. Leaving off fanciful counter-examples, virtuous people rescue drowning babies when they can do so at little cost or risk. Notice too that they do so in a certain way. They *automatically* and unhesitatingly take steps to rescue. They take steps to rescue as an expression of a certain commendable character. Their character is marked by tendencies or dispositions to do the right thing in the right way at the right time and for the right reasons. For such persons, doing the right thing is second nature (Aristotle, 1984, II.4). And so, such persons automatically lend a hand in dire emergencies at their feet.[1] This is why there is (at least) a *moderate* requirement that a person provide easy rescue. Those who fail to do so are rightly regarded as despicable. We would understandably take their failure to rescue as a moral failing. Though I doubt there is a *strong* requirement of easy rescue – that physical force is appropriate to compel easy rescue or to punish the failure to do so – I will not argue the point.[2] This much nevertheless seems clear: such persons lack important virtues. To put it another way, their character is not sufficiently defined by dispositions to do the right thing in the right way at the right time and for the right reasons. We do not want to be such persons, nor, other things equal, do we want them as neighbors, colleagues, or friends.[3]

So far I have claimed that a failure to provide easy rescue shows that a person lacks important moral virtues. Notice, though, that saying we should provide easy rescue is actually shorthand for saying that we have excellent moral reasons for being the sort of persons who would unhesitatingly take steps to rescue. Why, then, can we not say the same things about helping distant suffering peoples?

To start, notice that we do not believe requirements of aid are the same in each case. Typically, we think that there are moderate requirements to rescue babies drowning at our feet, but there are only weak requirements to alleviate world hunger. If our thinking is correct about this, then there would be neither moderate nor strong duties to provide aid to distant hungry people. Relieving distant hunger might then be commendable, but a failure to do so would not make us fitting objects of scorn – and it would certainly not make us candidates for being coerced by potential beneficiaries of our care or by people acting on their behalf.

Critics may respond in the spirit of Singer (1972) and say that this line of argument at best *reports* moral beliefs; it does not defend them as legitimate. This is true. Some critics may then defend a sort of moral revisionism: we should, they might say, revise our moral beliefs or otherwise be more consistent in a way that favors our treating world hunger just as we treat drowning babies. I believe, however, that we rightly treat world hunger differently.

Consider a key morally relevant difference between babies drowning at our feet and distant peoples suffering from hunger. One is an *emergency* calling for immediate action; the other is a *chronic* condition calling for reflection on complex moral, political, and economic considerations. As Paul Gomberg argues, "Hunger raises issues of causation and remedy that are not present in our duty to rescue" (Gomberg, 2002, p. 30). But an easily rescued baby drowning in a shallow puddle is quite different. Such a case is so exceptional and presents such immediate need that it would be vastly inappropriate to consider the relative costs and benefits of rescue (Gomberg, 2002, p. 37). It is inappropriate to *pause* to determine the cause of the drowning baby and all circumstances surrounding the drowning. Typically, none of these questions is appropriate beforehand, or even at all: Was she left here deliberately? Where are her parents? How wet will I get by rescuing her? If I rescue her, will she fall in again next week? Will she grow up to have children of her own who might happen to fall into puddles along my path? Will she grow up to become a mass murderer, or simply a mean person? What are the pH and temperature of this water? And, how can I rescue her best to promote my career?

World hunger, however, is more complicated and calls for us to consider its causes and circumstances (Gomberg, 2002, p. 37). Given that world hunger is chronic, it is also a good idea to consider how best to alleviate it (Schmidtz, 1998). We might even understandably fault someone who indiscriminately attempted to alleviate world hunger, by, for instance, giving $100 to some self-described representative of a relief organization without doing a little reflection and background work first (Kekes, 1987, p. 27). Donors must first consider important questions: is this person a genuine representative of the organization and not some charlatan? Does the organization have low administrative costs, or is it just a make-work scheme under the pretense of charity? Are there better uses of my money? *Does this organization do more harm than good to the people it claims to help?*

In short, the "moral logic" of the two cases differs. Typically, a case of a drowning baby calls for us to act immediately; but typically, a case of distant starving people calls for us to pause to consider causes and consequences. Since the moral logic differs, we ought to reason about the situations differently. Since the situations call for different reasoning, the one cannot be a moral analogue for the other without much more argument. It is thus not enough to justify a duty to aid distant suffering people by pointing to our intuitions about drowning babies needing easy rescue.

Of course, this has only shown that cases of nearby drowning babies are not necessarily analogous to cases of distant suffering peoples. It is still possible that there is a moderate (or even a strong) requirement to alleviate the suffering of distant hungry peoples. To assess whether there is such a requirement, we need to consider the function, the place, and the proper target for charity in a good life.

Charity, Personal Autonomy, and the Right to do Wrong

Charity as a virtue

I regard charity, in the sense relevant here, as a disposition to sympathize appropriately with, and to aid, persons in need. This is not the place for a full discussion of the nature and grounding of charity. Here we need only consider how charity is a virtue and what conditions are required for it to be a part of a good life. My arguments will address the possibility of *strong* duties of charity, that is, duties that are physically enforceable. Later in the chapter I raise some worries about moderate duties.

A charitable person is someone disposed to feel and act toward needy people in the right way, at the right time, and for the right reasons. Charity is a virtue mainly because it is a desirable character trait. Speaking quite generally, a person's life tends to go better if one is the sort of person who feels sympathy for others' suffering and is disposed to mitigate their neediness when possible. How much sympathy one feels, how one manifests concern, and how much aid one provides will all vary from one person to another for many reasons, such as different temperaments, different financial circumstances, different abilities to have insight into others' lives, and variable understandings of the conditions in which people live. Speaking again generally, we can still say that charitable people will be sympathetic and helpful toward the right other persons in the right way at the right time in the right amount and for the right reasons.

A virtuous person determines how and when to be charitable after reflecting on particular circumstances and the personalities involved. She must consider her own situation and the situations of needy others. She must also reflect on alternative uses of her property and emotional energy in light of other moral demands. For this sort of particularized reflection to be effective, though, prospective benefactors need the space to explore and deliberate about how they will be charitable. They then have the best opportunity to feel the spontaneous and correctly targeted charitable impulses that are central affective components of the virtue of charity. Without such opportunities for reflection, deliberation, and affective response, they lose a key motivational basis for cultivating the virtue of charity. More sharply, if they lack the opportunity *not* to be charitable, they are deprived of the fullest chance to define themselves as charitable.

Suppose a morally mature person, Allie, lives in reasonable comfort, while Bryce does not. Suppose further that Allie does not enjoy any protected opportunity to withhold her property or her time when providing them might benefit some very needy persons. There might be various institutions or norms in place to ensure that Allie provides aid to persons such as Bryce. Perhaps the state taxes Allie and sends the money to the "Bryce fund." Or maybe Allie could help out by preparing a sandwich and delivering it to Bryce, so imagine that she is required to do so. She is not free not to do so; if she withholds her money or her time, she can be physically forced to provide them or punished for her failure to do so.

When Allie complies with the requirement that she assist Bryce or others like him, there is very little charity involved. Allie lacks the fullest opportunity to feel spontaneous sympathy for Bryce. There is little point to her gathering information about the merits of Bryce's case because, after all, she has to give anyway. Reflection on how much to give (at least regarding what she *must* give) is irrelevant; she has little choice about the matter. But if she does not enjoy any protected opportunity to study Bryce's case, reflect on its merits, and decide how much *if anything* to give, then her acts of giving are morally cheapened or entirely emptied of virtue. We cannot coerce the virtue of charity.

Charity and the right to do wrong

Individual rights are special moral norms that define and protect certain opportunities to reflect, choose, and act. If Allie may not withhold her resources when providing them might help Bryce, then she does not enjoy a right not to give. The right to make a choice in this situation is crucial for Allie to have the chance to define and cultivate a charitable character. Self-definition and personal integrity demand self-directed practice with the possibility of failure. (Cohen, 1997, p. 48)[4]

We might even suppose that a virtuous person in Allie's situation *would have* given to Bryce after learning about his plight and reflecting on the merits of his case in light of her circumstances. Suppose also that Allie would have freely given exactly what she had no choice but to give to Bryce anyway. Her failure to give in such a situation would then have been wrong: it would have manifested the vice of stinginess. But without a right to be uncharitable, Allie has little reason to discover this. She would lack the fullest opportunity to decide. Absent what we might then call a *right to do wrong*, she is not in the correct moral position to study and reflect on Bryce's situation. Such research and reflection are key for Allie to experience appropriate sympathetic feelings and to determine whether giving is appropriate (and how much) in light of other possible uses and moral demands for her time and property.

This is not just an issue of facilitating virtuous self-development, but, one of making it possible for individuals to live their own lives. If Allie is not entitled to her property or her time when there are others who might need such resources, then her life is not hers to define and live. Consider that it is not just Bryce's needs that are at issue. There is also Callandra. And Doris. And Eunice. And countless others. All such persons may be worse off than Allie, so they may have a stake in Allie's property or time. There is no point, however, to Allie's taking any steps to live her own life when her productivity is mortgaged to the bottomless needs of others (Schmidtz, 2000, p. 693). For Allie to have her own life to lead, she must enjoy the right to make choices – including some wrong ones.[5]

Critical rejoinders

Critics may raise at least three possible objections at this point.

Rejoinder 1: There is no serious danger to liberty from enforceable positive duties A critic may say that the foregoing arguments exaggerate the threat to a potential benefactor's liberty from enforceable positive duties. For instance, Wellman (2014, p. 420) only defends duties to give "a modest amount, say $100, to saving the lives of a few children who are currently starving to death." But it is not clear why such a duty can only demand so little. Philosophers such as Singer believe a person may be obligated to reduce herself to penury as long as others are worse off (Singer, 1972, pp. 231, 241). It is then difficult to see why a mere $100 absolves us of an obligation to aid distant starving people. There are always people starving, and at least until we are dead, there is always something more we could give (Schmidtz, 2000, p. 693).

No matter what the amount, though, there are at least four sorts of moral costs involved in obligating a person to provide any portion of her time or wealth.[6]

1   Such compulsion may clash with other important moral values, such as respecting each person's freedom to live her own life. A person who is obligated to give to the needy is deprived of the fullest opportunity to decide whether to use that money for famine relief, for an AIDS research fund, for cancer research, for some books for her child, for a gift for her lover, or even to save it for a rainy day. The point is that if she is to have a protected opportunity to define and live a life of her own, these must be her decisions to make.
2   The compulsion may not be the best way to satisfy the relevant moral demands. Allie may, for instance, do a better job at being charitable if she is not forced to give money or time.
3   It may be hard for anyone to know how to do the right thing or to know what exactly is the right thing to do, so enforcing duties to give may be misguided.
4   Such compulsion hinders the development of the virtues that are important for personal moral development. When we obligate a person to give, she has less of a chance to be fully and authentically charitable with that money.[7] And for reasons I discuss later, I believe facilitating such authentic charity is the best way to minimize the need for it.

Rejoinder 2: Enforcing positive duties makes a person better This brings us to a second possible objection. Critics may agree that Allie should have some discretion to make choices, but they may draw the line at certain obviously wrong choices, such as, say, Allie's decision to use $100 not for famine relief efforts but on a new outfit from Abercrombie & Fitch (which duplicates three others she already has, but would be in a different color). Not only would it be better overall for Allie's money instead to go to charity, the argument may run, but it would be better *for Allie* if that money were redirected. Here Allie's rights would be constrained by what might make her a better person (or by what might best help her to do the right thing).

In response, perhaps Allie is mistaken to devote her money to seemingly frivolous purchases instead of other uses that might better enhance the condition of others. But the question is whether this is something for Allie to decide. There is an important difficulty in saying that a person's life goes better when others impose a certain plan on her. Unless that person may make these and other key choices, she does not enjoy the fullest chance to learn about and understand just what moral reasons bind her. A person has the best chance for a good life only if she leads it according to her own values (Kymlicka, 1989, p. 12).

Critics may insist that forcing Allie to hand over her money for a better cause is something she *would* endorse if only she were to think long and hard about it. Perhaps taking the money from her would even give her the chance to reflect about how frivolous the A&F purchase would have been in light of how the money may have helped distant starving people. Taking her money without or despite her consent could then be morally edifying.

This might all be true. But the problem with this approach is that it is better suited to children than to mature adults. Mature adults may make their own choices – even if their choices might go wrong – precisely because we grant that they should define and live lives of their own. This does not mean we should always ignore someone's offensive choices. It is always open to us to persuade that person to change. We can also protest or repudiate someone's actions. But, absent wrongful injuries of others, it is inappropriate to use physical force to compel someone to do something we think is right – even if it seems to be for her own good (Mill, 1978, p. 9). Doubtless we want people to make better choices – especially when significant moral values are at stake. But the only way human beings can be in the right relationship with prospective moral truths is if they are free to explore and discover them on their own (Locke, 1993, pp. 394–395; Hampton, 2003, p. 224).

There are three further reasons to reject strong duties of charity. The first has to do with reasonable differences about what counts as good. The second has to do with finding effective ways to reduce the need for charity. The third has to do with the danger of giving anyone the power to make such decisions for us. I discuss each in turn.

First, people often disagree about morality. Such disagreement is not necessarily a sign of some vice or poor reasoning; reasonable and conscientious persons often differ on moral matters. This disagreement is a function of different life experiences, different perspectives, and different knowledge about the world. More often than not, such differences are permanent and track fundamentally different world views – not just about what is good, but even about what should be the *standard* of good (Rawls, 1993, pp. 54–58). If we are committed to letting each person live her own life, a healthy humility about moral knowledge along with a constructive openness to reasonable differences provide strong reasons for guaranteeing each person a morally protected space in which to decide whether, how much, and how often to be charitable. Otherwise someone arrogates to herself an inappropriate moral authority, and the rest of us lose the chance to live our own lives.

Second,[8] reasonable people disagree about how best to solve the problem of chronic hunger. But coercing people to give a certain amount, in a certain way, at a certain time, just about guarantees that people will discover no better way to respond to hunger. Experience has shown that people are best able to come up with innovative solutions to problems when they have the freedom to experiment with and discuss alternatives. This freedom to experiment – which requires a freedom not to give according to some single formula – will produce institutions and norms that differ depending upon the context of need and the situations of prospective benefactors.

Third, defenses of strong duties of charity suppose there are trustworthy and reliable moral experts whose dictates, if imposed, would help us better to do the right thing with regard to hunger. But I doubt there are such persons. Given political and psychological realities, no one should be trusted with such power. Even if someone may seem to deserve such trust, this should be something each person gets to decide for herself. Moreover, no one person has such extensive knowledge about your circumstances and the circumstances of others that she reliably knows *better than you* how it would be morally best for you to allocate your resources with regard to hunger. Perhaps there are, then, moral authorities whose advice we would do well to heed on such matters, but whom to put in such a role is something we should each be free to decide for ourselves. Our lives do not go better if they are foisted on us.

Rejoinder 3: Welfare is more important than liberty Now we come up against an important third objection. So far the arguments have shown that a right to do wrong – which may include a right not to give to needy people – is an important component of protecting each person's opportunity to define and live and life of her own. A critic may say that all this talk of self-definition is overblown. What should really count in a moral theory, the critic may claim, is well-being – and not just the individual's well-being, but the well-being of everyone overall. So a prospective benefactor's liberty must sometimes (or always?) give way to the greater benefits that would come from redistributing her resources to other persons who are seen as needier (where judgments about who is needier are made and enforced by some authority with political power).

In response, note that this view all but rejects the importance of individual self-definition and choice. On this view, Allie can do as she pleases provided her conduct complies with the calculations of overall well-being by someone in power. But this cuts against Allie's having a chance to live her own life.

Admittedly, this will not persuade the critic who takes self-definition lightly. But then the critic has to explain what, if anything, limits the goal of advancing overall well-being. May innocent persons be killed to quell a bloodthirsty mob bent on lynching someone for a crime? May babies be drowned in mud puddles in order to use their tissues for lifesaving medical procedures? If the answers to such similar questions are no (as I hope), then we need to hear why.

Presumably some principles or policies are necessary to guide and limit just how we seek to promote overall well-being for everyone. Typically individual rights serve this purpose: they "trump" the pursuit of net welfare. Though respecting rights may sometimes seem to close off gains to well-being overall, we can do better in the long run by letting rights define protected liberties (Schmidtz, 2000).

438   **Andrew I. Cohen**

At least two possible sorts of critics may speak up here. One sort says that rights *are* important, but that what rights should protect is not just a domain of choice but human welfare. This sort of critic might then defend a positive *right* to famine relief. Such a right typically correlates with some enforceable positive duties. The other sort of critic may dismiss talk of rights and simply argue for an enforceable positive *duty* of famine relief. In either case, the critics defend strong duties of famine relief.

But such critics face an important challenge: they must show that the relevant positive duties do more good than harm. There is a danger that such duties may create what policy theorists sometimes call a "moral hazard." While intended to alleviate suffering, guaranteeing aid to people who suffer may in the long run create more suffering people (Schmidtz, 1998; Shapiro, 2002, p. 23; Wenar, 2010).

Requiring famine relief undermines benefactors' chances to decide how they shall live their own lives, and this cuts against overall well-being. Forcing people to hand over money to famine relief also threatens to "crowd out" better directed (and better motivated) giving (Shapiro, 2002). At its worst, such compulsion threatens to create an oppressive state and an institutionalized network of busybodies – and these would also cut against overall well-being. Even more, allowing for involuntary transfers of money or time for famine relief may further entrench the corrupt persons and institutions that often directly or indirectly contribute to widespread famine in distant countries. But most important here is that the beneficiaries of such relief may lose incentives to live their lives as best as they can (Schmidtz, 2000, pp. 684–688). The beneficiaries can fall into a culture of dependency that undermines the families, communities, and sense of personal responsibility that are crucial for human beings to live good lives.

Of course, none of this shows that duties of famine relief – especially moderate duties – do not on balance promote overall well-being. Nor is this a decisive argument against any right to famine relief. Here we merely see how much is required to establish that there are such positive rights or positive duties. Philosophers must await the data from scholars and researchers in fields as diverse as economics, public policy, social psychology, political science, agricultural technology, and many others. But there is still much room for philosophy before the data come in. We can argue (as I did above) that personal liberty is of sufficient moral importance that it warrants protection from fallible human beings acting on limited knowledge who believe they know better than we do how best to dispose of our money and time. They rarely do. Given reasonable disagreement about what properly counts as a standard for the "best" use of money and time, it is far from clear that anyone can ever properly be in a position to make these decisions for us.

Many readers may still worry that without enforceable positive duties of famine relief, distant starving peoples will be consigned to certain death. But this worry itself reflects a widespread concern for the suffering of others. Since wealthier people do care, they can be persuaded to create and/or support institutions to alleviate the suffering of distant hungry peoples.[9] Though we cannot rob Peter to feed Paul, each of us is free to take steps to teach Paul how to feed himself.

As many as 11 million children may die each year before their fifth birthday, mostly from diseases and conditions traceable to poverty and malnutrition (Cowley, 2003, p. 78). Note though that private, voluntarily funded relief efforts – especially when organized locally – have often been quite effective at helping people to get back on their feet (Beito, 2000; Shapiro, 2002, pp. 21–31). So once again, we confront important and complex empirical questions about how best to alleviate hunger in the long run. We need to find out about the political, economic, and moral causes of chronic hunger, and we need consider whether certain sorts of relief efforts do more harm than good. This is a vastly complicated issue, but given the empirical uncertainties, reasonable disagreements, and importance of personal liberty and moral virtue, it is something that each person should be left to decide for herself after reflection and research.

In this section I have discussed how enforceable duties of famine relief are morally inappropriate and require daunting empirical support. Readers may think there is yet much room for moderate duties of famine relief. Even if we cannot be forced to reduce ourselves to penury on behalf of others, perhaps we still *ought* to devote some or all of our available resources to distant famine relief. In the next section, I will argue that distant famine relief, while sometimes commendable, must often take second place to addressing more local needs.

Local Versus Distant Needs

As recently as 2010–12, nearly 870 million people throughout the world were chronically undernourished (FAO, WFP and IFAD, 2012). These are staggering numbers. But nearly 16 million people in Western countries were similarly malnourished (FAO, WFP and IFAD, 2012). Even if they are not our neighbors, we sometimes pass such persons on the street. The problems we face extend beyond hunger. There are ghastly statistics about battered wives, illiteracy, healthcare for the poor, and innocent children who endure horrific illnesses. The numbers in these and other categories refer to far too many persons in our communities who are victims, who suffer, and who could benefit from a helping hand.

I argue that our charitable energies are more constructively focused on local needs. This is not necessarily because our neighbors are any more deserving or needier than distant starving persons. There need not be anything morally significant, in and of itself, about the fact that some person is *your neighbor*. Because our knowledge of local conditions is typically deeper than that of distant contexts, and because the actual costs of administering aid locally is typically lower, our charitable impulses are usually (though certainly not always) more constructively directed toward local contexts.

Some needy persons live among us. We are more likely to know about their plight and to have better insight into how we might effectively help them while neither insulting their dignity nor fostering any "moral hazards." So it seems that typically (though again, certainly not always), our moral reasons to contribute to distant famine relief would be outweighed by moral reasons to contribute in some way to a *local* rape crisis center, or a *local* children's cancer ward, or a *local* soup kitchen, or a *local* literacy campaign. Again, this is not because being local is in itself morally significant. But in so far as we are concerned with alleviating need, we are best in a position to do that for familiar people in familiar situations. Typically (though certainly not always), these are people who are near or dear to us.[10]

Of course, sometimes people do have excellent knowledge of distant conditions – perhaps even more so than of local conditions. In today's age of global communication and easy travel, we are sometimes better positioned to address some distant needs than to address those that may be closer to us. The point here is that because there is so much local need, it is difficult to see why we can always fault someone for failing to contribute to distant famine relief.

Note, though, that addressing charitable need is not a person's sole moral function. There are many moral demands on us – demands that come from various sources. As intimacy increases in relationships, for instance, there are greater legitimate expectations for care, attention, and devotion. What forms these take will vary from one relationship to another and from one moment to the next. But our resources are finite. We have to decide how best to satisfy all the moral reasons that bind us. Being required to direct our resources to the suffering of distant people may then jeopardize our ability to do fully what we ought to do in nearer and dearer contexts about which we have better knowledge (Kekes, 1987). There is a danger of falling into the pattern of the Dickens character Mrs Jellyby, who focused her caring energies on the natives of Borrioboola-Gha at the expense of her own children (Dickens, 2002).

Each of us is involved in many relationships of different levels of intimacy. To various persons you might be a sibling, a parent, a child, a spouse, a dear friend, a neighbor, a colleague, a teammate, or a fellow citizen. Each of these relationships may impose demands on us. Meanwhile, each of us has commitments to various other personal projects that shape a life. You might be a painter, a runner, a dancer, or a musician. You might enjoy poetry, travel, science fiction, or basketball. Pursuing and cultivating such interests are also part of what gives richness and meaning to a good life. If we devote ourselves to relieving distant situations (about which we know little) at the expense of our own interests (about which we each have a privileged understanding), we do violence to ourselves and undermine our chances to live a good life. There is more to life than alleviating distant need.

Does this mean that contributing to famine relief is *wrong*? Certainly not. Sometimes we can do much good by helping a well-organized relief effort – especially one that has good insight into local conditions, has very low administrative costs, and has taken great pains not to unwittingly prop up corrupt governments or create a culture of dependency. But given how much each of us differs in our understandings of distant conditions, and given our reasonably different conceptions of how each of us might best fashion a good life, contributing to famine relief, assuming it is done conscientiously, is commendable at most but not morally required – even in the moderate sense.

Local Reform and Distant Suffering

Writers on world hunger sometimes defend duties of famine relief as a way to compensate for having benefited from supposedly unjust institutions. Certainly decent persons must not blithely enjoy the fruits of oppression. But we need to consider just what is oppressive.

It would be vastly inappropriate to support or benefit from an industry whose products were manufactured exclusively in Nazi concentration camps. The labor force would consist of brutally oppressed prisoners; the products might be drawn from the property or body parts of slaughtered captives. If there are contemporary analogues to such Nazi concentration camp industries, then we do indeed have a responsibility to withdraw our support from them. We may even have a compelling reason to take active steps toward reform. But it is unclear just what the contemporary analogues are.

Consider just one example of a disturbing Western practice that props up the rich at the expense of poor people in distant countries. It is hardly as ghastly as a concentration camp, but it still unjustly robs the poor of a livelihood. I am speaking of protective tariffs and domestic agricultural and industrial subsidies. US cotton subsidies are a fine example. In the first decade of the twenty-first century, over $24 billion of subsidies went to propping up the US cotton industry by providing low-cost water and grants to large cotton conglomerates. The upshot is that domestic cotton farmers have greater incentives to plant a crop that would be less expensively grown abroad in developing countries. The subsidies give an unfair advantage to wealthy American cotton growers. They also drive down global prices for cotton. This deprives farmers in developing countries of much-needed income (Jowit, 2010).

This is not an isolated phenomenon. Any time the government subsidizes or protects an industry, it bypasses the market mechanisms that would otherwise direct resources toward their most efficient use. More often than not, the people who lose most are those with the least to lose. So if we are concerned about helping distant persons rise out of poverty, one step is to disassociate ourselves from these American industries benefiting from unfair advantages. Since it is nearly impossible to sort out how to do this, perhaps we might simply work to eliminate such protective measures.

Well-meaning people sometimes unfavorably compare the working conditions in developing economies with those in the West. True, workers in much of the world earn a fraction of what is earned by those in the West. They often work longer hours. Sometimes they start working at a young age. Critics of such conditions sometimes call for boycotts or the closure of "sweatshops" as a way of ending what they take to be oppression.

This is certainly a complex and controversial topic in social and economic theory. But we might note that many workers in developing economies eagerly embrace work in "sweatshops" as a chance to improve their lives and the lives of their families. Workers often complain that a patronizing, misguided elitism motivates Westerners who believe the developing world would be better without such industries (Langewiesche, 2000, see, e.g., p. 46; see also Zwolinski, 2007). Here we speak not of inmates in Soviet gulags, but people for whom working at a factory manufacturing Nike running shoes would quadruple their family's income, increase their caloric intake, and give the children a chance to be literate. Low-wage industries overseas are often a key step in improving the lives of terribly poor peoples (Myerson, 1997). If anything, often one of the kindest things we can do for the distant poor is to spend some of our money on a new outfit from A&F – especially if it or its components (the fabric, dyes, or fasteners) were manufactured in low-wage factories overseas. Doing so supports distant economies and gives workers there the opportunities to build better lives for themselves and their families. We show a lot more respect for a person by trading with her and treating her as a productive equal than by merely sending her grain and treating her as a helpless open mouth or outstretched hand.

Economists have repeatedly discovered that the easiest way to improve the condition of the world's poor is to eliminate barriers to free markets and establish a rule of law that respects property rights (Bray, 1996; Simon 1996; Simon and Moore, 2000; Lomborg, 2001, part II; Gollin *et al.*, 2002). If we in the West do have any moderate duties to relieve distant suffering, then maybe we are responsible for *opposing* subsidies and other government price supports and *supporting* foreign aid policies reasonably calculated to foster markets and the rule of law in impoverished states.[11] What this duty means for any given person – especially with such a matter of public policy – will of course vary considerably depending upon circumstances. One possibility is that we ought to deepen and apply our understanding of the social, moral, political, and economic institutions that allow people to live successful lives here and abroad. In the meantime, the best thing we could do for others might be to have a productive career and a successful life.

To paraphrase Aristotle (1984, II.6), it is not easy being good. There are so many ways to go wrong, and only one or a few ways of doing the right thing. Living well requires a lot of practice, and it is something we must do for ourselves. Each of us must reflect on all the competing moral considerations that vie for our attention, and each of us must decide how best to allocate our energies and how to forge a life for ourselves. Whether and how charity fits in that life is a deeply personal decision that we must be free to decide for ourselves.[12]

Notes

1   I stress that to simplify matters, I pass over possible exceptions here. We can suppose that there are no mitigating circumstances for the prospective benefactor such as: she is closely chased by a homicidal maniac, or, is rushing her own dying child to the hospital, or, is not able-bodied, or, can only attempt rescue at grave risk to her life and limb . . . and so forth. Proponents of a duty to rescue distant starving peoples must also grant this simplifying assumption in order to show the duty in its clearest light. Otherwise the drowning baby analogy never gets off the ground.

2   Good Samaritan laws, which punish those who fail to provide easy rescues, must be based on such strong requirements. (Thanks to Mark LeBar for pointing this out.)

3   Consider the disturbing 1997 case of the teenager David Cash, Jr, who did nothing to stop his friend Jeremy Strohmeyer from raping and murdering 7-year-old Sherrice Iverson in the bathroom of a Nevada casino. Cash apparently saw his friend assaulting Iverson and muffling her screams in a bathroom stall, but he took no effective steps to stop the assault. Strohmeyer now serves life without parole in a Nevada prison; Cash went on to study nuclear physics at UC-Berkeley. Cash's inaction and remarks to the press have repeatedly illustrated that he is, to put it mildly, morally underdeveloped. Neither his classmates at UC-Berkeley nor the law were able to use physical force to punish him. But he rightly met with the deep scorn of his classmates at Berkeley. For further details on the story, see, for instance McDermott (1998).

4   A person who does not enjoy the right not to be charitable might still express charity in the acts she is forced to perform. She might simply authentically identify with them anyway. She may also give (in the right way and for the right reasons) over and above what she is required to do. My point is simply that rights to choose must include a right to withhold, and such rights are important moral norms that facilitate self-definition by protecting opportunities to choose freely. See Cohen (1997, p. 48).

5   Certain wrong choices could never be protected by right. For example, no one can enjoy a right to be *unjust* (Cohen, 1997, pp. 44–45). Here I talk only about non-rights-violating wrongs, and I also wish to argue that no one can, or should, have a right to another person's charity.

6   My thanks to Mark LeBar and George Rainbolt for a discussion of the issues in this paragraph.

7   Much depends on whether the duty to provide aid is moderate or strong. Here I only speak of *strong* positive duties (i.e., physically enforceable obligations). Some proponents of duties of famine relief, however, are not committed one way or the other on this issue. Wellman (2014), for instance, merely speaks of a positive duty to provide minimal aid (as if to suggest that people who fail to provide such aid deserve our scorn but no more). But at other times he speaks of prospective beneficiaries' "samaritan right" to aid (2014, p. 422). I think saying that people have a right to such aid suggests a strong duty on others to provide it. Later I raise doubts about whether there are even grounds for a *moderate* duty of famine relief.

8   My thanks to Mark LeBar for suggesting a discussion of the issues in this paragraph.

9   Not only can they be persuaded, but they often are. The data on this is extensive and complex, but here's one snapshot: individuals in the United States give far more than corporations (and often several hours each week of their own time). Charitable giving has remained constant at around 1.9 percent of personal income since the 1970s (Lang, 1998). Private donors from the United States give about $35.1 billion in oversees aid, which is 3½ times what the US government provides in Official Development Assistance. And the United States (privately and publicly) provides the most direct foreign investment and foreign aid and generates the bulk of the world's research and development (Adelman, 2003).

10 Interestingly, Wellman (2000, pp. 545–547, here at 545), who defends Samaritan duties of famine relief, elsewhere appeals to similar considerations when discussing "redistributive policies that favor compatriots." Such policies, Wellman argues, can help us better to comply with other significant moral reasons that bear on our cases.

11 Thanks to Andy Altman for suggesting this point.

12 I am grateful to Andrew Altman, Harry Dolan, Eric Karch, Mark LeBar, George Rainbolt, and Kit Wellman, each of whom provided many helpful comments on an earlier version of this chapter.

References

Adelman, C.C. (2003) The privatization of foreign aid: reassessing national largesse. *Foreign Affairs* 82 (November/December): 9–14.

Aristotle (1984) Nichomachean ethics. In *The Complete Works of Aristotle*, vol. II, ed. J. Barnes, pp. 1729–1867. Princeton, NJ: Princeton University Press.

Beito, D. (2000) *From Mutual Aid to the Welfare State: Fraternal Societies and Social Services, 1890–1967*. Chapel Hill, NC: University of North Carolina Press.

Bray, A.J. (1996) Hunger's real cure? Freedom. *Investor's Business Daily*, Nov. 22, 1996, 1.

Cohen, A.I. (1997) Virtues, opportunities, and the right to do wrong. *Journal of Social Philosophy* 28: 43–55.

Cowley, G. (2003) Where living is lethal. *Newsweek*, Sept. 22, 2003, 78–80.

Dickens, C. (2002) *Bleak House*. New York: Modern Library (Original work published in monthly parts Mar. 1852–Sep. 1853).

FAO, WFP and IFAD (2012) The state of food insecurity in the world 2012. Economic growth is necessary but not sufficient to accelerate reduction of hunger and malnutrition. Rome, FAO. http://www.fao.org/docrep/016/i3027e/i3027e.pdf (last accessed 6/17/13).

Gollin, D., Parente, S., and Rogerson, R. (2002) The role of agriculture in development. *The American Economic Review* 92: 160–164.

Gomberg, P. (2002) The fallacy of philanthropy. *Canadian Journal of Philosophy* 32: 29–66.

Hampton, J. (2003) The liberals strike back. In *Justice: Alternative Political Perspectives*, 4th edn, ed. J.P. Sterba, pp. 218–225. Belmont, CA: Wadsworth.

Jowit, J. (2010). Cotton subsidies costing west African farmers £155m a year, report reveals. *The Guardian*, November 14, 2010. http://www.guardian.co.uk/environment/2010/nov/15/cotton-subsidies-west-africa (last accessed 6/17/13).

Kekes, J. (1987) Benevolence: a minor virtue. *Social Philosophy & Policy* 4: 21–36.

Kymlicka, W. (1989) *Liberalism, Community and Culture*. Oxford: Clarendon Press.

Lang, J. (1998). In U.S., giving is national pastime. *Washington Times*, September 24, 1998.

Langewiesche, W. (2000) The shipbreakers. *The Atlantic Monthly* 286: Aug–2000.

Locke, J. (1993) A letter concerning toleration. In *Political Writings of John Locke*, ed. D. Wootton, pp. 390–436. New York: Mentor (Original work published 1689).

Lomborg, B. (2001) *The Skeptical Environmentalist*. Cambridge: University Press.

McDermott, A. (1998). A silent friend, and a debate over good Samaritan laws. Posted Sep. 4, 1998 on cnn.com at http://www.cnn.com/SPECIALS/views/y/1998/09/mcdermott.casino/ (last accessed 6/17/13).

Mill, J.S. (1978). *On Liberty*, ed. E. Rapaport. Indianapolis: Hackett (Original work published 1859).

Myerson, A.R. (1997). In principle, a case for more 'sweatshops'. *The New York Times*, June 22, 1997, section 4, 5.

Rawls, J. (1993) *Political Liberalism*. New York: Columbia University Press.

Schmidtz, D. (1998) Taking responsibility. In *Social Welfare and Individual Responsibility*, ed. D. Schmidtz and R. Goodin, pp. 3–96. Cambridge: University Press.

Schmidtz, D. (2000) Islands in a sea of obligation. *Law and Philosophy* 6: 683–705.

Shapiro, D. (2002) Egalitarianism and welfare-state redistribution. *Social Philosophy & Policy* 19: 1–35.

Simon, J. (1996) *The Ultimate Resource 2*. Princeton, NJ: Princeton University Press.

Simon, J. and Moore, S. (2000) *It's Getting Better All the Time: 100 Greatest Trends of the 20th Century*. Washington, DC: Cato Institute.

Singer, P. (1972) Famine, affluence, and morality. *Philosophy and Public Affairs* 1: 229–243.

Unger, P. (1996) *Living High and Letting Die Our Illusion of Innocence*. New York: Oxford University Press.

Wellman, C.H. (2000) Relational facts in liberal political theory: is there magic in the pronoun "my"? *Ethics* 110: 537–562.

Wellman, C.H. (2014) Famine relief: the duties we have to others. In *Contemporary Debates in Applied Ethics*, 2nd edn, ed. A.I. Cohen and C.H. Wellman, pp. 419–430. Malden, MA: Wiley-Blackwell.

Wenar, L. (2010) Poverty is no pond: challenges for the affluent. In *Giving Well*, ed. P. Illingworth, T. Pogge, and L. Wenar. New York: Oxford University Press.

Zwolinski, M. (2007) Sweatshops, choice, and exploitation. *Business Ethics Quarterly* 17(4): 689–727.

Further Reading

Beito, D., Gordon, P., and Tabarrok, A., eds (2002) *Voluntary City: Choice, Community, and Civil Society*. Ann Arbor: University of Michigan Press.

Den Uyl, D.J. (1995) The right to welfare and the virtue of charity. In *Liberty for the 21st Century*, ed. T.R. Machan and D.B. Rasmussen, pp. 305–334. Lanham, Maryland: Rowman & Littlefield.

Den Uyl, D.J. and Rasmussen, D.B. (1995) Rights' as metanormative principles. In *Liberty for the 21st Century*, ed. T.R. Machan and D.B. Rasmussen, pp. 59–75. Lanham, Maryland: Rowman & Littlefield.

Gilder, G. (1981) *Wealth and Poverty*. New York: Basic Books.

Hasnas, J. (1995) From cannibalism to caesareans: two conceptions of fundamental rights. *Northwestern University Law Review* 89: 900–941.

Hayek, F.A. (1945) The use of knowledge in society. *The American Economic Review* 35: 519–530.

Lomasky, L. (1987) *Persons, Rights, and the Moral Community*. Oxford: University Press.

Murray, C.A. (1994a) *Losing Ground: American Social Policy, 1950–1980*. New York: Basic Books.

Murray, C.A. (1994b) *In Pursuit of Happiness and Good Government*. San Francisco: ICS Press.

Nozick, R. (1974) *Anarchy, State, and Utopia*. New York: Basic Books.

Semple, K. (2003). Tiniest of loans bring big payoff, aid group says. *The New York Times*, November 3, 2003, A6.

Smith, T. (1995) *Moral Rights and Political Freedom*. Lanham, MD: Rowman & Littlefield.

Index

Contemporary Debates in Applied Ethics, Second Edition. Edited by Andrew I. Cohen and Christopher Heath Wellman.
© 2014 John Wiley & Sons, Inc. Published 2014 by John Wiley & Sons, Inc.

Printed and bound by CPI Group (UK) Ltd, Croydon, CR0 4YY